Series Editors
*Trevor Johnson & Tony Clough*

**Student Book**

# Edexcel GCSE Mathematics (Modular)

## Higher Tier

# Contents

## MODULE 2

## MODULE 3

# Introduction

Welcome to *Edexcel GCSE Mathematics Modular Higher Student Book and ActiveBook*. Written by Edexcel as an exact match to the new Edexcel GCSE Mathematics Higher Tier specification these materials give you more chances to succeed in your examinations

## The Student Book

Each chapter has a number of units to work through, with full explanations of each topic, numerous worked examples and plenty of exercises, followed by a chapter summary and chapter review questions.

There are some Module 3 topics that may also be assessed in Modules 2 or 4. These are identified in the contents list with the symbol: +M4

These topics are also highlighted within the chapters themselves,

**also assessed in Module 4**

using this flag by the relevant unit headings:

The text and worked examples in each unit have been written to explain clearly the ideas and techniques you need to work through the subsequent exercises. The questions in these exercises have all been written to progress from easy to more difficult.

At the end of each chapter, there is a Chapter Summary which will help you remember all the key points and concepts you need to know from the chapter and tell you what you should be able to do for the exam.

Following the Chapter Summary is a Chapter Review which comprises further questions. These are either past exam questions, or newly written exam-style questions – written by examiners for the new specifications. Like the questions in the exercise sections, these progress from easy to hard.

In the exercise sections and Chapter Reviews

 by a question shows that you may use a calculator for this question or those that follow.

 by a question shows that you may NOT use a calculator for this question or those that follow.

## The ActiveBook

The ActiveBook CD-ROM is found in the back of this book. It is a digital version of this Student Book, with links to additional resources and extra support. Using the ActiveBook you can:

- Find out what you need to know before you can tackle the unit
- See what vocabulary you will learn in the unit
- See what the learning objectives are for the unit
- Easily access and display answers to the questions in the exercise sections (these do not appear in the printed Student Book)
- Click on glossary words to see and hear their definitions
- Access a complete glossary for the whole book
- Practice exam questions and improve your exam technique with *Exam Tutor* model questions and answers. Each question that has an *Exam Tutor* icon beside it links to a worked solution with audio and visual annotation to guide you through it

### Recommendation specification

Pentium 3 500 Mhz processor
128MB RAM
8× speed CD-ROM
1GB free hard disc space
800 × 600 (or 1024 × 768) resolution screen at 16 bit colour
sound card, speakers or headphones

Windows 2000 or XP. This product has been designed for Windows 98, but will be unsupported in line with Microsoft's Product Life-Cycle policy.

### Installation

Insert the CD. If you have autorun enabled the program should start within a few seconds. Follow on-screen instructions. Should you experience difficulty, please locate and review the readme file on the CD.

### Technical support

If after reviewing the readme you are unable to resolve your problem, contact customer support:

- telephone 0870 6073777 (between 8.00 and 4.00)
- email schools.cd-romhelpdesk@pearson.com
- web http://centraal.uk.knowledgebox.com/kbase/

# Scatter graphs

## 1.1 Scatter graphs and relationships

**Scatter graphs** can be used to investigate whether there is a relationship between two quantities.
For example, a scatter graph could show the marks of twelve students who took a test in science and in maths. One of the students, Cath, scored 27 in science and 16 in maths.
On the grid, the cross shows her pair of marks.

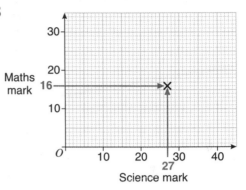

The table shows the test marks of the other eleven students.

| Science mark | 5 | 19 | 35 | 7 | 33 | 29 | 23 | 9 | 36 | 17 | 32 |
|---|---|---|---|---|---|---|---|---|---|---|---|
| Maths mark | 6 | 15 | 21 | 10 | 22 | 21 | 18 | 7 | 25 | 13 | 19 |

Here is the completed scatter graph for all twelve students.

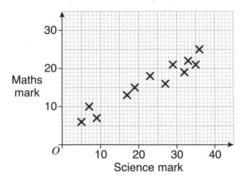

The pattern of the crosses on the scatter graph suggests that there is a relationship between the science marks and the maths marks.

One way of describing the relationship is

*As the science mark increases, the maths mark increases.*

Scatter graphs can suggest other sorts of relationship between two quantities or even that there is no relationship at all.

### Example 1

The table shows, for each of ten men, his age and the number of his own teeth he has.

| Age (years) | 22 | 28 | 33 | 37 | 41 | 49 | 52 | 56 | 64 | 68 |
|---|---|---|---|---|---|---|---|---|---|---|
| Number of own teeth | 29 | 29 | 26 | 28 | 26 | 21 | 23 | 21 | 17 | 15 |

a   Draw a scatter graph to show this information.
b   Describe the relationship between the men's ages and the number of their own teeth they have.

**Solution 1**

**a**

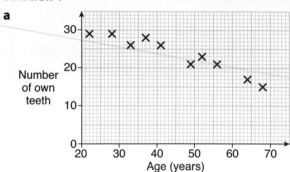

**b** The older a man is, the fewer of his own teeth he has.

(This relationship is true for these ten men but, in the whole country, there are many men who do not fit this pattern.)

---

## Example 2

The table shows the ages, in years, and the heights, in cm, of twelve women.

| Age (years) | 21 | 25 | 30 | 34 | 39 | 42 | 46 | 54 | 57 | 63 | 66 | 68 |
|---|---|---|---|---|---|---|---|---|---|---|---|---|
| Height (cm) | 175 | 162 | 173 | 154 | 177 | 148 | 157 | 152 | 168 | 164 | 143 | 171 |

**a** Draw a scatter graph to show this information.
**b** Describe the relationship, if any, between the women's ages and their heights.

**Solution 2**

**a**

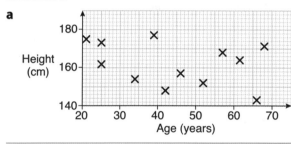

**b** There is no pattern to the points and so there is no obvious relationship between the women's ages and their heights.

---

## Exercise 1A

**1** The table shows the height, in cm, and the weight, in kg, of each of ten men.

| Height (cm) | 173 | 153 | 187 | 183 | 179 | 166 | 176 | 165 | 181 | 158 |
|---|---|---|---|---|---|---|---|---|---|---|
| Weight (kg) | 68 | 57 | 92 | 97 | 78 | 76 | 81 | 67 | 85 | 59 |

**a** On the resource sheet, complete the scatter graph to show the information in the table. The first two points in the table have been plotted for you.

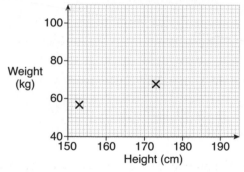

**b** Describe the relationship between the men's heights and their weights.

**2** The table shows the birth weights, in kg, of ten babies and the average number of cigarettes per day their mothers smoked during pregnancy.

| Number of cigarettes | 2 | 19 | 11 | 29 | 0 | 6 | 14 | 23 | 5 | 8 |
|---|---|---|---|---|---|---|---|---|---|---|
| Birth weight (kg) | 3.53 | 3.37 | 3.48 | 3.24 | 3.62 | 3.51 | 3.38 | 3.31 | 3.56 | 3.46 |

**a** On the resource sheet, complete the scatter graph to show the information in the table. The first two points in the table have been plotted for you.

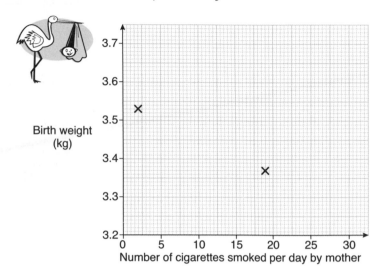

**b** Describe the relationship between the number of cigarettes smoked per day by the mothers during pregnancy and their babies' birth weights.

**3** The table shows the number of goals scored and the number of goals conceded by each of twelve Premier League football teams in a recent season.

| Number of goals scored | 45 | 52 | 49 | 53 | 47 | 47 | 45 | 42 | 40 | 52 | 47 | 32 |
|---|---|---|---|---|---|---|---|---|---|---|---|---|
| Number of goals conceded | 46 | 41 | 44 | 46 | 39 | 41 | 52 | 58 | 46 | 60 | 57 | 43 |

**a** On the resource sheet, complete the scatter graph to show the information in the table. The first point in the table has been plotted for you.

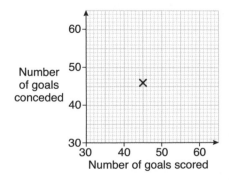 

**b** Describe the relationship, if any, between the number of goals scored by the teams and the number of goals they conceded.

**4** Nick sells ice creams. The table shows the noon temperature and the number of ice creams he sells each day for a fortnight.

| Noon temperature (°C) | 20 | 28 | 18 | 24 | 30 | 22 | 21 | 16 | 29 | 19 | 27 | 26 | 23 | 27 |
|---|---|---|---|---|---|---|---|---|---|---|---|---|---|---|
| Number of ice creams | 70 | 86 | 58 | 76 | 97 | 78 | 65 | 58 | 91 | 63 | 93 | 91 | 79 | 82 |

**a** On the resource sheet, complete the scatter graph to show the information in the table. The first two points in the table have been plotted for you.

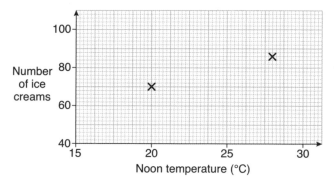

**b** Describe the relationship between the noon temperatures and the number of ice creams Nick sells.

**5** The table shows the number of hours of sunshine and the rainfall, in mm, each month in England in 2004.

| Month | Jan | Feb | Mar | Apr | May | Jun | Jul | Aug | Sep | Oct | Nov | Dec |
|---|---|---|---|---|---|---|---|---|---|---|---|---|
| No of hours of sunshine | 51 | 87 | 107 | 136 | 204 | 200 | 169 | 176 | 160 | 97 | 50 | 53 |
| Rainfall (mm) | 152 | 49 | 48 | 78 | 43 | 56 | 67 | 148 | 53 | 131 | 47 | 59 |

**a** On the resource sheet, complete the scatter graph to show the information in the table. The first two points in the table have been plotted for you.

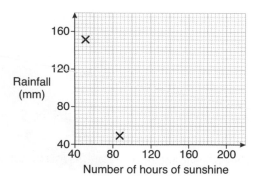

**b** Describe the relationship, if any, between the number of hours of sunshine and the rainfall.

**6** The table shows some annual mileages of a car and the running costs per mile, in pence, of the car for those mileages.

| Annual mileage | 4000 | 6500 | 8000 | 10 000 | 12 000 | 14 500 | 18 500 | 22 000 | 27 500 | 29 500 |
|---|---|---|---|---|---|---|---|---|---|---|
| Running costs per mile (pence) | 53 | 48 | 42 | 34 | 32 | 31 | 25 | 24 | 21 | 23 |

  **a** On the resource sheet, complete the scatter graph to show the information in the table. The first two points in the table have been plotted for you.

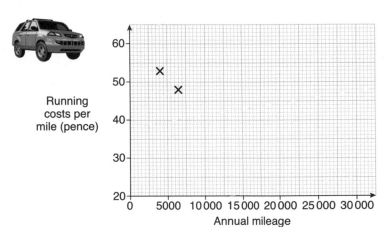

  **b** Describe the relationship between the car's annual mileage and its running costs per mile.

## 1.2 Lines of best fit and correlation

On the scatter graph at the beginning of this chapter, it is possible to draw a straight line which passes near all the points. This line is called a **line of best fit**.

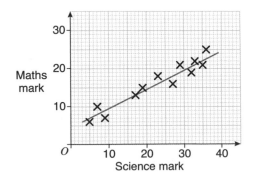

Use a ruler to draw a line of best fit.

A line of best fit does not have to pass through any of the points, although it can, and there should be roughly equal numbers of points on each side of the line.

If a line of best fit can be drawn, then there may be a relationship, called a **correlation**, between the quantities.

Maths marks increase as science marks increase. When one quantity increases as the other increases, the correlation is called **positive correlation**.

When one quantity decreases as the other increases, the line of best fit slopes in the opposite direction to the one above and the correlation is called **negative correlation**. The scatter graph in Example 1 shows negative correlation.

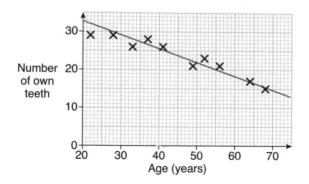

It is not possible to draw a line of best fit on the scatter graph in Example 2, as there is no pattern to the points. So there is **no correlation** or **zero correlation** between a woman's age and her height.

If the points are all close to the line of best fit, the correlation is called **high** or strong.

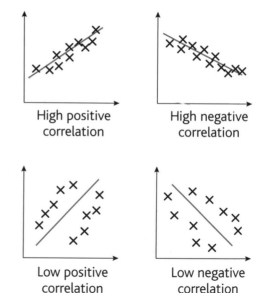

High positive
correlation

High negative
correlation

If the points are not all close to the line of best fit, the correlation is called **low** or weak.

Low positive
correlation

Low negative
correlation

## 1.3 Using lines of best fit

If the value of only one of the two quantities is known, lines of best fit can be used to estimate the value of the other quantity.

For example, to estimate the maths mark of a student whose science mark is 13, draw a vertical line up from 13 to the line of best fit. Then draw a horizontal line across and read off the maths mark, 11.

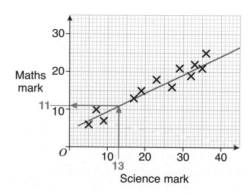

Using lines of best fit outside the range of the plotted points can give estimates which are unreliable or even ridiculous. Extending the line of best fit on the scatter graph in Example 1 and using it to estimate the number of teeth for a 20-year-old man gives an answer of 33 but people with *all* their own teeth normally have only 32!

### Exercise 1B

1 The table shows the marks scored by ten students in an English exam and a history exam.

| English | 20 | 87 | 56 | 42 | 96 | 92 | 41 | 32 | 86 | 52 |
|---------|----|----|----|----|----|----|----|----|----|----|
| History | 8  | 49 | 32 | 22 | 55 | 52 | 14 | 10 | 52 | 23 |

a On the resource sheet, draw a scatter graph to show the information in the table.

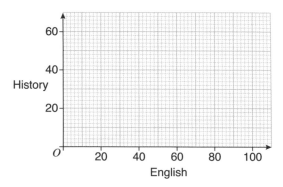

b Describe the correlation between the marks scored in the two exams.

c Draw a line of best fit on your scatter graph.

d Use your line of best fit to estimate
 i the history mark of a student whose English mark is 70
 ii the English mark of a student whose history mark is 44

> 'Describe the correlation' means state whether it is positive or negative. It is not necessary to say whether it is high or low. A description of the *relationship* is not acceptable.

2 The table shows the outdoors temperature, in °C, at noon on ten days and the number of units of electricity used in heating a house on each of those days.

| Noon temperature (°C)   | 9  | 2  | 0  | 4  | 11 | 10 | 12 | 5  | 3  | 1  |
|-------------------------|----|----|----|----|----|----|----|----|----|----|
| Units of electricity used | 25 | 39 | 44 | 34 | 23 | 24 | 21 | 32 | 36 | 42 |

a On the resource sheet, draw a scatter graph to show the information in the table.

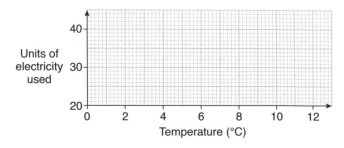

b Which of these three terms best describes the relationship between the temperature and the number of units of electricity used?

positive correlation     negative correlation     no correlation

c Draw a line of best fit on your scatter graph.

d Use your line of best fit to estimate
 i the number of units of electricity used when the outdoors temperature was 8°C,
 ii the outdoors temperature when 30 units of electricity were used.

3 The table shows the inflation rate and the unemployment rate in the UK every two years from 1982 to 2000

| Year | 1982 | 1984 | 1986 | 1988 | 1990 | 1992 | 1994 | 1996 | 1998 | 2000 |
|---|---|---|---|---|---|---|---|---|---|---|
| Inflation rate (%) | 8.6 | 5.0 | 3.4 | 4.9 | 9.5 | 3.7 | 2.5 | 2.5 | 3.4 | 3.0 |
| Unemployment rate (%) | 9.0 | 10.1 | 10.5 | 7.6 | 5.5 | 9.2 | 8.8 | 7.0 | 4.5 | 3.6 |

a On the resource sheet, draw a scatter graph to show the information in the table.

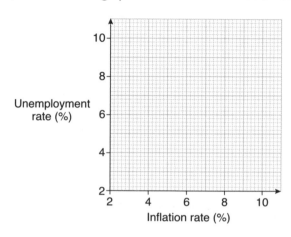

b Which of these three terms best describes the relationship between the inflation rate and the unemployment rate?

positive correlation     negative correlation     zero correlation

4 Complete the table on the resource sheet to show whether there is a positive correlation, a negative correlation or no correlation between the quantities. The first one has been done for you.

| | Positive correlation | Negative correlation | No correlation |
|---|---|---|---|
| ages of cars and their value | | ✓ | |
| heights of women and their weekly pay | | | |
| the number of students and the number of teachers in a school | | | |
| the distance a motorist drives and the amount of petrol used | | | |
| the hat sizes of students and their GCSE maths marks | | | |
| the amount of rain one day at a seaside town and the number of people on the beach | | | |

**5** The table shows, for ten cities, the highest temperature, in °C, and the number of hours of sunshine one day in September.

| Highest temperature (°C) | 20 | 33 | 14 | 21 | 29 | 25 | 16 | 26 | 19 | 24 |
|---|---|---|---|---|---|---|---|---|---|---|
| Number of hours of sunshine | 6 | 10 | 5 | 7 | 9 | 8 | 5 | 8 | 6 | 7 |

**a** On the resource sheet, draw a scatter graph to show the information in the table.

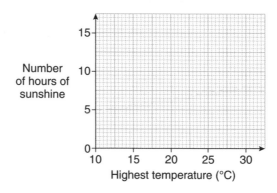

**b** On the same day, in St. Petersburg, the highest temperature was 16°C and there were 13 hours of sunshine. Plot this information on the scatter graph.

**c** For the eleven cities, comment on the relationship between the highest temperature and the number of hours of sunshine.

**6** The scatter graph shows the age and the price of 15 *Vector* cars. A line of best fit has been drawn.

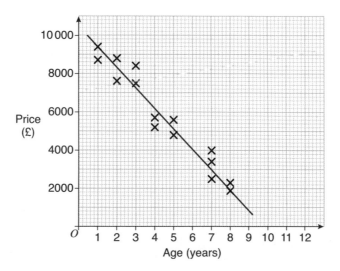

**a** Describe the correlation between the age of the *Vector* and its price.

**b** Use the line of best fit to find
   **i** the price of a 6-year-old *Vector*,
   **ii** the age of a *Vector* that costs £6200

**c** Why would it not be sensible to use the line of best fit to estimate the price of an 11-year-old *Vector*?

## Chapter summary

**You should now know:**

★ how to draw a **scatter graph**

★ how to describe the relationship, if any, suggested by a scatter graph

★ how to draw a **line of best fit**

★ how to use a line of best fit

★ the meaning of **correlation** and how a line of best fit can be used to show it

★ how to recognise **positive** correlation, **negative** correlation and **no** correlation or **zero** correlation

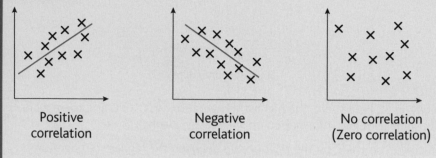

Positive correlation    Negative correlation    No correlation (Zero correlation)

## Chapter 1 review questions

**1** The scatter graph shows information about eight countries.

For each country, it shows the birth rate and the life expectancy, in years.

The table shows the birth rate and the life expectancy for six more countries.

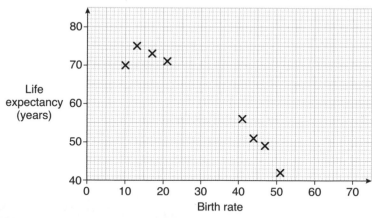

| Birth rate | 25 | 28 | 30 | 31 | 34 | 38 |
|---|---|---|---|---|---|---|
| Life expectancy (years) | 68 | 65 | 62 | 61 | 65 | 61 |

**a** On the scatter graph on the resource sheet, plot the information from the table.

**b** Describe the relationship between the birth rate and the life expectancy.

**c** Draw a line of best fit on the scatter graph.

The birth rate in a country is 42

**d** Use your line of best fit to estimate the life expectancy in that country.

The life expectancy in a different country is 66 years.

**e** Use your line of best fit to estimate the birth rate in that country.

(1385 June 2000)

**2** Ten men took part in a long jump competition.
The table shows the heights of the ten men and the best jumps they made.

| Best jump (m) | 5.33 | 6.00 | 5.00 | 5.95 | 4.80 | 5.72 | 4.60 | 5.80 | 4.40 | 5.04 |
|---|---|---|---|---|---|---|---|---|---|---|
| Height of men (m) | 1.70 | 1.80 | 1.65 | 1.75 | 1.65 | 1.74 | 1.60 | 1.75 | 1.60 | 1.67 |

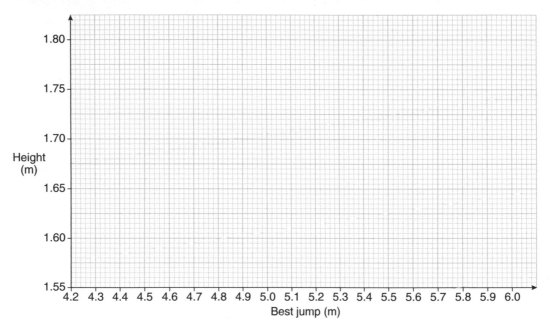

**a** On the grid on the resource sheet, plot the points as a scatter diagram.

**b** Describe the relationship between the height and the best jump.

**c** Draw in a line of best fit.

**d** Use your line of best fit to estimate
   **i** the height of a man who could make a best jump of 5.2 m,
   **ii** the best jump of a man of height 1.73 m.                    (1385 November 1998)

**3** The scatter graph shows some information
about six new-born baby apes.

For each baby ape, it shows the mother's
leg length and the baby ape's birth
weight.

The table shows the mother's leg
length and the birth weight for two
more baby apes.

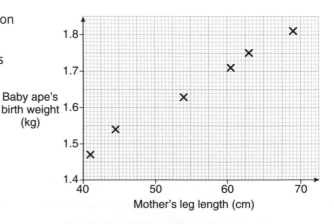

**a** On the scatter graph on the resource
sheet, plot the information from the
table.

**b** Describe the correlation between
a mother's leg length and the
baby ape's birth weight.

**c** Draw a line of best fit on the diagram.

| Mother's leg length (cm) | 50 | 65 |
|---|---|---|
| Baby ape's birth weight (kg) | 1.6 | 1.75 |

A mother's leg length is 55 cm.

**d** Use your line of best fit to estimate the birth weight of the baby ape.     (1387 June 2005)

**4** A park has an outdoor swimming pool.
The scatter graph shows the maximum
temperature and the number of people who
used the pool on ten Saturdays in summer.

**a** Describe the correlation between the
maximum temperature and the number
of people who used the pool.

**b** Draw a line of best fit on the scatter
graph on the resource sheet.

The weather forecast for next Saturday gives
a maximum temperature of 27 °C.

**c** Use your line of best fit to estimate the
number of people who will use the pool.

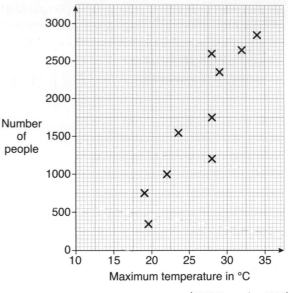

(1385 November 1999)

**5** The scatter graph shows some information about seven children.
It shows the age of each child and the number of hours sleep each child had last night.

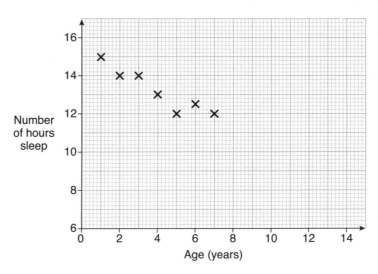

The table shows the ages of four more children and the number of hours sleep each of them had
last night.

| Age (years) | 10 | 11 | 12 | 13 |
|---|---|---|---|---|
| Number of hours sleep | 11 | 10 | 10.5 | 9.6 |

**a** On the scatter graph on the resource sheet, plot the information from the table.

**b** Describe the correlation between the age, in years, of the children and the number of hours
sleep they had last night.

**c** Draw a line of best fit on the diagram.

**d** Use your line of best fit to estimate the number of hours sleep for an 8 year old child.

(1385 June 2002)

**6 a** Here is a scatter graph. One axis is labelled 'weight'.

    **i** For this graph state the type of correlation.

    **ii** From this list choose an appropriate axis for the other axis.

        shoe size, length of hair, height, hat size, length of arm.

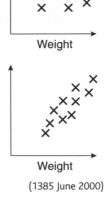

**b** Here is another scatter graph with one axis labelled 'weight'.

    **i** For this graph state the type of correlation.

    **ii** From this list, choose an appropriate label for the other axis.

        shoe size, distance around neck, waist measurement, GCSE Maths mark

(1385 June 2000)

**7** The table shows information about the percentage share of daily viewing hours of ITV1 and cable/satellite channels from 1992 until 2002.

| Year | ITV1 | Cable/satellite channels |
|------|------|--------------------------|
| 1992 | 41 | 5 |
| 1993 | 40 | 6 |
| 1994 | 39 | 7 |
| 1995 | 37 | 9 |
| 1996 | 35 | 10 |
| 1997 | 33 | 12 |
| 1998 | 32 | 13 |
| 1999 | 31 | 14 |
| 2000 | 29 | 17 |
| 2001 | 27 | 20 |
| 2002 | 24 | 22 |

THE LIBRARY
NORTH WEST KENT COLLEGE
DERING WAY, GRAVESEND

Here is a scatter graph for the information from 1992 to 1998

**a** On the scatter graph on the resource sheet, plot the information for 1999, 2000, 2001 and 2002

**b** Describe the relationship between ITV1's percentage share of the daily viewing hours and the percentage share of cable/satellite channels.

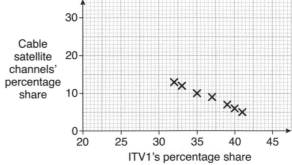

**c** Draw a line of best fit on the scatter graph.

In 2003, cable/satellite channels had a 24% share of daily viewing hours.

**d** Use your line of best fit to estimate ITV1's percentage share in 2003

# Collecting and recording data

**2**

## 2.1 Introduction to statistics

Statistics is the area of mathematics in which information is collected, recorded, displayed and then used.

The table shows some information about second hand cars.

| Make | Model | Year | Engine | Doors | Miles | Colour | Price |
|------|-------|------|--------|-------|-------|--------|-------|
| Ford | Fiesta | 2004 | 1.2 litres | 3 | 7050 | Silver | £5890 |
| Vauxhall | Astra | 2002 | 1.6 litres | 5 | 10 000 | Silver | £5990 |
| VW | Golf | 2005 | 1.4 litres | 5 | 6040 | Black | £10 990 |
| Peugeot | 206 | 2002 | 1.4 litres | 3 | 27 500 | Blue | £5590 |
| Audi | A4 | 2000 | 2.0 litres | 4 | 41 000 | Red | £7590 |

In statistics information like this is called **data**.
In the table, some of the data is given in words, for example the colour of each car.
This is called **qualitative data**. Some of the data is given as numerical values, for example the number of doors or the number of miles travelled. This is called **quantitative data**.
Numerical data can either be **discrete data** or **continuous data**.
Discrete data has a definite value, for example shoe size, number of doors, scores on a dice.
Continuous data can be measured; it can take <u>any</u> value.
Height, weight and time are all examples of continuous data.

## 2.2 Data by observation and by experiment

Which type of fruit do most people buy?
To answer this question, data needs to be collected. Observing the type of fruit people buy and recording this information is called *collecting data by* **observation**.
A **data collection sheet** is a way of recording information.

### Example 1

Twenty people bought the following types of fruit.

| | | | | |
|---|---|---|---|---|
| Apples | Bananas | Plums | Apples | Bananas |
| Oranges | Oranges | Apples | Pears | Pears |
| Bananas | Bananas | Oranges | Bananas | Apples |
| Bananas | Bananas | Pears | Apples | Oranges |

Draw a completed data collection sheet to show this information.

## Solution 1

| Type of fruit | Tally | Frequency |
|---|---|---|
| Apples | IIII | 5 |
| Bananas | IIII II | 7 |
| Oranges | IIII | 4 |
| Pears | III | 3 |
| Plum | I | 1 |

**Frequency** shows the total number of people buying each type of fruit. 4 people bought oranges.

These are **tally marks**. There is one tally for each type of fruit chosen.

IIII is another way of writing 5 tally marks.

The total of the frequency column = 20 This is the total number of people who bought fruit.

Since tally marks are used, this is also called a **tally chart**.
It is also sometimes called a **frequency table** because it shows the number of times each type of fruit was bought.

Another way of collecting data is **by experiment**.
An experiment can be carried out to find out if this spinner is biased or fair.

The spinner is fair if it lands on each of the four numbers about the same number of times. Otherwise, the spinner is said to be biased.

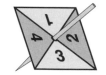

The spinner is spun 80 times.
For each spin the number that the spinner lands on is recorded in a tally chart and the frequency of each number is found.

| Number | Tally | Frequency |
|---|---|---|
| 1 | IIII IIII IIII IIII | 19 |
| 2 | IIII IIII IIII IIII II | 22 |
| 3 | IIII IIII IIII IIII | 20 |
| 4 | IIII IIII IIII IIII | 19 |

Since the four frequencies are about the same, the spinner is probably fair.

Data can also be recorded in a **two-way table**.

This two-way table shows some information about the number of girls and the number of boys in Year 10 and in Year 11 in a school.

|  | Year 10 | Year 11 | Total |
|---|---|---|---|
| Girls | 80 | 90 | 170 |
| Boys | 75 | 105 | 180 |
| Total | 155 | 195 | 350 |

One way, the table shows information about the number of girls and boys.
The other way shows information about the number of students in Year 10 and the number of students in Year 11
It shows, for example, that the number of **boys** in **Year 11** is **105**

## Example 2

Here is some information about the ways that the boys and girls in class 10C travel to school.

| Boys | | | | | Girls | | | | |
|---|---|---|---|---|---|---|---|---|---|
| Walk | Bus | Car | Bus | | Bus | Bus | Car | Walk | Bus | Bus |
| Car | Bus | Bus | Bus | | Bus | Car | Walk | Bus | Car | Car |
| Walk | Walk | Bus | Walk | | Bus | Bus | Car | Walk | Bus | Car |

Show this information in a suitable two-way table.

*Solution 2*

|        | Walk | Bus | Car | Total |
|--------|------|-----|-----|-------|
| Boys   | 4    | 6   | 2   | 12    |
| Girls  | 3    | 9   | 6   | 18    |
| Total  | 7    | 15  | 8   | 30    |

## 2.3 Grouping data

When there is a large amount of data, the data is often grouped.
A **grouped frequency table** contains data that has been put into different groups.

The table shows information about the heights of 30 people.
The heights have been grouped.

$160 \leqslant h < 165$ means 160 cm or more but less than 165 cm.

Each group is called a **class interval** and in this table each
class interval is of equal width.

| Height ($h$ cm) | Frequency |
|-----------------|-----------|
| $160 \leqslant h < 165$ | 9 |
| $165 \leqslant h < 170$ | 15 |
| $170 \leqslant h < 175$ | 6 |

There are 9 people with heights in the class interval $160 \leqslant h < 165$

The **modal class interval** is the class interval with the highest frequency.
So the modal class interval for the heights is $165 \leqslant h < 170$

See Chapter 15 Inequalities
< means less than
≤ means less than or equal to.

### Example 3

20 students took a maths test. Their test marks are shown below

| 17 | 9  | 6  | 12 | 19 |
| 4  | 13 | 12 | 15 | 16 |
| 20 | 14 | 5  | 10 | 11 |
| 17 | 8  | 14 | 12 | 18 |

Complete the grouped frequency table.

| Mark  | Tally | Frequency |
|-------|-------|-----------|
| 1–5   |       |           |
| 6–10  |       |           |
| 11–15 |       |           |
| 16–20 |       |           |

For discrete data, for example test marks,
the class interval for marks 1, 2, 3, 4 and 5
is written as 1–5

*Solution 3*

| Mark  | Tally | Frequency |
|-------|-------|-----------|
| 1–5   | ‖     | 2 |
| 6–10  | ‖‖    | 4 |
| 11–15 | ⅏ ‖‖  | 8 |
| 16–20 | ⅏ ‖   | 6 |

## Example 4

The grouped frequency table shows information about the heights of 42 students.

**a** Write down the modal class interval.

**b** Jenny's height is 177.2 cm. In which class interval is her height recorded?

**c** Charlie's height is exactly 180 cm. In which class interval is his height recorded?

| Height ($h$ cm) | Frequency |
|---|---|
| $160 \leqslant h < 165$ | 10 |
| $165 \leqslant h < 170$ | 14 |
| $170 \leqslant h < 175$ | 8 |
| $175 \leqslant h < 180$ | 5 |
| $180 \leqslant h < 185$ | 3 |
| $185 \leqslant h < 190$ | 2 |

### Solution 4

**a** The modal class interval is $165 \leqslant h < 170$

> This class interval has the highest frequency, 14

**b** Jenny's height is in the class interval $175 \leqslant h < 180$

> 177.2 is greater than 175 but less than 180

**c** Charlie's height is in the class interval $180 \leqslant h < 185$

> 180 is shown at the end of one class interval and at the beginning of another. The sign for 'less than or equal to' ($\leqslant$) shows that 180 should go in the class interval $180 \leqslant h < 185$

### Exercise 2A

**1** Write down whether each of the following data is qualitative data or quantitative data.
  **a** The numbers of wheels on some vehicles.
  **b** The colours of toy bricks in a bag.
  **c** The heights of buildings in a town.
  **d** The areas of kitchen floors.
  **e** The countries in Europe.
  **f** The ages of members of your family.

**2** Write down whether each of the following data is discrete data or continuous data.
  **a** The numbers of gears on some bicycles.
  **b** The numbers of flowers in a garden.
  **c** The lengths of branches on a tree.
  **d** The heights of students in a school.
  **e** The IQs of students in a school.

**3** Cindy is going to carry out a survey of favourite Pop stars of each of her friends.
   Draw a suitable data collection sheet that Cindy could use.

**4** Tyler asked 25 people where they spent their last holiday in England. Here are their answers.

| | | | | |
|---|---|---|---|---|
| Blackpool | Newquay | Brighton | Blackpool | Newquay |
| Whitby | Skegness | Blackpool | Brighton | Skegness |
| Whitby | Brighton | Newquay | Newquay | Blackpool |
| Skegness | Blackpool | Skegness | Blackpool | Skegness |
| Blackpool | Brighton | Newquay | Newquay | Skegness |

   Record this data in a frequency table.

**5** Nadia throws a dice 40 times.
   Here are the scores on the dice for each of the 40 throws.
  **a** Draw a tally chart to show this information.
  **b** Is the dice fair or biased?
     Explain your answer.

| | | | | | | | | | |
|---|---|---|---|---|---|---|---|---|---|
| 3 | 4 | 1 | 3 | 5 | 6 | 3 | 1 | 2 | 1 |
| 5 | 5 | 1 | 3 | 4 | 4 | 2 | 4 | 1 | 3 |
| 3 | 2 | 6 | 4 | 3 | 1 | 2 | 3 | 5 | 4 |
| 1 | 2 | 1 | 4 | 3 | 6 | 3 | 2 | 3 | 1 |

**6** A group of 17 boys and 15 girls were asked at what time they went to lunch yesterday.
12 boys went to lunch before 1230
13 students, of which 8 were girls, went to lunch from 1230 to 1300
3 girls went to lunch after 1300

Copy and complete the two-way table.

|  | Before 1230 | From 1230 to 1300 | After 1300 | Total |
|---|---|---|---|---|
| **Boys** |  |  |  |  |
| **Girls** |  |  |  |  |
| **Total** |  |  |  | 32 |

**7** Jason recorded how many copies of a daily newspaper his shop sold each day in September.
Here are his results.

| 56 | 43 | 51 | 54 | 49 | 42 |
|---|---|---|---|---|---|
| 57 | 60 | 67 | 52 | 52 | 47 |
| 42 | 61 | 65 | 58 | 50 | 54 |
| 65 | 54 | 63 | 62 | 61 | 53 |
| 55 | 59 | 61 | 45 | 68 | 61 |

| Number of newspapers | Tally | Frequency |
|---|---|---|
| 41–45 |  |  |
| 46–50 |  |  |
| 51–55 |  |  |
| 56–60 |  |  |
| 61–65 |  |  |
| 66–70 |  |  |

**a** Copy and complete the grouped frequency table.

**b** Write down the modal class interval.

**8** Some boys run a 100 m race.
This grouped frequency table shows information about the times of all the boys.

| Time ($t$ seconds) | Frequency |
|---|---|
| $10 \leqslant t < 12$ | 6 |
| $12 \leqslant t < 14$ | 5 |
| $14 \leqslant t < 16$ | 5 |
| $16 \leqslant t < 18$ | 3 |
| $18 \leqslant t < 20$ | 1 |

**a** Write down the modal class.

**b** Work out the total number of boys.

**c** Find the number of boys whose time was less than 16 seconds.

## 2.4 Questionnaires

What types of fruit do people buy most in a large town?
This can be answered by carrying out a survey using a questionnaire.
A **questionnaire** is a list of questions which help to collect data in a survey.
These are examples of good and bad questions that could appear on the questionnaire.

**1** 'Which types of fruit do you eat?'
(You may choose more than one.)

> This question is short and clear.
> It will help show which types of fruit are eaten.

Apples   Bananas   Pears   Oranges   Plums   Others (please state)   Do not eat fruit
☐   ☐   ☐   ☐   ☐   ☐   ☐

> Every question on a questionnaire must have **response boxes**.
> A response box is where a person can answer the question, usually by putting a tick (✓) in the box.
> This question has seven response boxes.

> These two response boxes make sure that everyone can complete at least 1 box in answer to the question.

**2** 'How often do you eat fruit?'

Often      Sometimes      Now and then

☐      ☐      ☐

> In this question the response boxes could mean different things to different people. Using, for example, 'less than once a week', 'once or twice a week' and 'more than twice a week' would be better.

**3** 'Have you ever stolen fruit?'

Yes      No

☐      ☐

> It is unlikely that anyone would admit to a question like this. The question will not help find out which types of fruit are bought.

**4** 'How much do you usually spend on fruit each week?'

£1 up to £3      £3 up to £5      £5 up to £10

☐      ☐      ☐

> '£1 up to £3' means 'an amount of £1 and over but less than £3' but there is no response box to tick for anyone spending less than £1 or £10 and over each week. This could be improved by adding two more response boxes 'less than £1' and '£10 and over'.
>
> It is important that the intervals cover ALL amounts and that each person has just ONE box to tick.

**5** 'Do you agree that to be healthy you should eat more fruit?'

Yes      No

☐      ☐

> By asking 'Do you agree ...', this question suggests that the correct answer is 'yes'. This is called a **biased** question.

## Example 5

The manager of a restaurant has made some changes. She uses these questions as part of a questionnaire to find out what the public think of these changes.

Write down what is wrong with each of these questions.

**a** 'What do you think of the improvements to the restaurant?'

Excellent      Very good      Good

☐      ☐      ☐

**b** 'How much money do you normally pay for a main course?'

A lot      Average

☐      ☐

**c** 'How often do you come to the restaurant?'

Not very often      Often

☐      ☐

### Solution 5

**a** Possible answers could be:
- the question is biased since it only allows positive responses to be recorded
- the question is too vague since excellent, very good and good may mean different things to different people
- there is no response box for her customers who do not like the changes.

**b** ● a lot and average may mean different things to different people.

**c** Possible answers could be:
- this question does not make it clear if she is asking how often a customer comes to the restaurant each month or each year
- the response boxes are not very helpful as again they are too vague.

## Example 6

Angela carries out a survey of the amount of television watched by the public.
She uses this question on a questionnaire.

'How much time do you spend watching television?'

A lot                           Average                          A bit

☐                               ☐                                ☐

**a**  Design a better question that Angela could use.

**b**  State if the data used here is discrete data or continuous data.

*Solution 6*

**a**  How much time do you spend watching television each week?

Under 5 hours        5 up to              10 up to             20 hours
                     10 hours             20 hours            and over

☐                    ☐                    ☐                    ☐

**b**  Since time can be measured and can take any value the data is continuous data.

# 2.5  Sampling

It is difficult to ask every person in a large town what fruit they eat. It is easier to ask a small number of people. This is called a **sample**. It is important that this sample is not **biased** towards any group of people – for example adults only or men only. In questionnaires personal information is often requested. Recording information about the person answering the questions will tell you whether your sample is biased or unbiased.

## Methods of sampling

A **random sample** is one in which each person has an equal chance of being chosen.

For example from a list of 30 students in a class, 5 are required in a sample.

Each student is allocated a number and the random selection facility on a calculator is used

or

> Check that you can do this on your own calculator.

the names of the students may be placed in a hat and a name selected. In this case it is important to replace each name after it has been chosen. The process is repeated until 5 different names are chosen.

To make sure a sample is random and unbiased choose the best place and time to ask questions. A good place to stand to ask about buying fruit is in a busy part of the town centre during the daytime. This is because most people shop in the town centre during the day.

## Example 7

The manager of a leisure centre wants to find out what sporting activities people prefer. He asks the first 100 people who enter the leisure centre one Saturday morning. Explain what is wrong about this way of sampling.

*Solution 7*

This way of sampling is <u>biased</u> because:

- choosing the first 100 people is no guarantee that males and females of all ages are represented; they may all be children
- just sampling on a Saturday morning may not include people who use the leisure centre at other times
- choosing only people who use this leisure centre may exclude people who prefer sporting activities that are not available at this leisure centre.

A **systematic sample** (sometimes called selective sampling) is a sample chosen from a population which has been arranged in some defined order. For example from a list of 180 students, 18 (=10%) are required in a sample. The names are arranged in alphabetic order and every 10th person is chosen systematically. This could be the 1st, 11th, 21st, ... names or the 5th, 15th, 25th, ... names.

A **stratified sample** is found by dividing the population to be sampled into groups based upon agreed criteria. These groups are called **strata** and a random sample from each strata is taken to give the complete stratified sample. For example, a sample of 100 students is required from a college population of 5000 students. The students could be grouped by Year and gender.

## Example 8

The two-way table shows the Year and gender of 5000 students in a college. A stratified sample of 100 students stratified by Year and gender is to be surveyed on the decoration of the student common room.

| | Year 1 | Year 2 | Year 3 | Total |
|---|---|---|---|---|
| **Male** | 800 | 900 | 850 | 2550 |
| **Female** | 750 | 842 | 858 | 2450 |
| **Total** | 1550 | 1742 | 1708 | 5000 |

a  Work out the number of male students in the stratified sample.

b  Work out the number of Year 1 female students in the stratified sample.

c  Work out the number of Year 2 female students in the stratified sample.

*Solution 8*

a  $\dfrac{2550}{5000}$

| | Year 1 | Year 2 | Year 3 | Total |
|---|---|---|---|---|
| **Male** | 800 | 900 | 850 | **2550** |
| **Female** | 750 | 842 | 858 | 2450 |
| **Total** | 1550 | 1742 | 1708 | 5000 |

5000 is the population to be sampled.
Write the number of male students as a fraction of the population.

$\dfrac{2550}{5000} \times 100 = 51$

Multiply this fraction by 100 since 100 is the size of the stratified sample.

The stratified sample has 51 male students.

b  $\dfrac{750}{5000} \times 100 = 15$

| | Year 1 | Year 2 | Year 3 | Total |
|---|---|---|---|---|
| **Male** | 800 | 900 | 850 | 2550 |
| **Female** | **750** | 842 | 858 | 2450 |
| **Total** | 1550 | 1742 | 1708 | 5000 |

The stratified sample has 15 Year 1 female students.

c  $\dfrac{842}{5000} \times 100 = 16.84$

| | Year 1 | Year 2 | Year 3 | Total |
|---|---|---|---|---|
| **Male** | 800 | 900 | 850 | 2550 |
| **Female** | 750 | **842** | 858 | 2450 |
| **Total** | 1550 | 1742 | 1708 | 5000 |

The stratified sample has 17 Year 2 female students.

Write 16.84 correct to the nearest whole number since only a whole number of students are possible.

### Exercise 2B

**1** Write down one thing that is wrong with each of the following questions.

   **a** 'How much time do you spend doing homework?'

   **b** 'How much pocket money do you get each week?'    A lot ☐    Not much ☐

   **c** 'Do you agree that exercise is good for you?'

       Yes       No
       ☐       ☐

   **d** 'How much do you usually spend on your lunch each day?'

       Under £1    £1–£2    £3–£4    Over £4
       ☐      ☐     ☐    ☐

   **e** 'Have you ever been in trouble with the police?'    Yes ☐    No ☐

   **f** 'What is your favourite colour?'    Red ☐    Blue ☐    Yellow ☐    Green ☐

**2** The manager of a cinema asks these two questions in a questionnaire.

   'Do you go to the cinema?'       Often    Sometimes    Never
                              ☐       ☐      ☐

   'How old are you?'    0 to 10 years    10 to 20 years    20 to 40 years    Over 40 years
                       ☐         ☐         ☐        ☐

   There is something wrong with each question.
   For each one design a better question for the manager to use.

**3** A company is going to open a new adventure playground. They design a questionnaire to find out what type of activity would be most popular.

   **a**  **i** Design a suitable question that could be asked to find out which activity most people would use in an adventure playground.

      **ii** Draw a suitable data collection sheet to record the answers.

   **b** A worker for the company stands in the town centre one Monday morning and asks the first 100 people she meets to complete the questionnaire. Explain fully what is wrong with this method of sampling.

**4** Melissa wants to find out how students in her year travel to school.

   **a** Design a suitable question to find out how students in her year travel to school.

   There are 240 students in her year. Melissa decides to use a sample of 40 students from her year. She gives a questionnaire to each of the 30 members of her class and also a questionnaire to each of 10 friends from different classes.

   **b**  **i** What is wrong with this sample?

      **ii** How could Melissa get a better sample?

**5** Jim carries out a survey of the types of holiday people prefer. He needs an unbiased sample of 100 people. He uses a questionnaire. Write down five questions that Jim might include in his questionnaire. Describe how this information might be collected and recorded.

**6** The two-way table shows the group and gender of 120 students in Year 9
A stratified sample of 30 students stratified by group and gender is to be surveyed on the venue for the summer activity day.

| | Group A | Group B | Group C | Group D | Total |
|---|---|---|---|---|---|
| **Male** | 13 | 17 | 14 | 13 | 57 |
| **Female** | 18 | 10 | 15 | 20 | 63 |
| **Total** | 31 | 27 | 29 | 33 | 120 |

**a** Work out the number of Group D female students in the stratified sample.

**b** Work out the number of Group A male students in the stratified sample.

**c** Work out the number of Group C female students in the stratified sample.

**7** A sports centre has 120 employees. Describe two ways in which you could choose a random sample of 20 of these employees.

## 2.6 Databases

A **database** is a collection of information. It is organised so that any required piece of data can be found quickly. This information is usually stored on a computer but it can also be stored on paper.

So far, all the data that has been considered has been collected and recorded by the person carrying out a particular survey or experiment. This is called **primary data**. The data in a database has usually been collected and recorded by someone else. Data which has been collected by someone else is called **secondary data**.

Primary data is often collected by **data logging**.
Data logging may be defined as the collection and storage of data for later use.

Data is usually collected electronically at equal time intervals. For example, in hospital, a patient's temperature, pulse rate and blood pressure can be monitored continuously throughout the day. Graphs of these results are electronically produced.

### Example 9

This database contains information about some students.

| Name | Gender | Age | Key stage 3 English level | Key stage 3 maths level | Key stage 3 science level |
|---|---|---|---|---|---|
| David | Male | 16 | 4 | 3 | 4 |
| Samantha | Female | 14 | 5 | 5 | 5 |
| Daniel | Male | 16 | 4 | 4 | 4 |
| Yousef | Male | 15 | 6 | 6 | 5 |
| Amir | Male | 15 | 7 | 6 | 6 |
| Kirsty | Female | 15 | 3 | 4 | 4 |
| Nadia | Female | 16 | 4 | 6 | 4 |
| Clive | Male | 14 | 5 | 6 | 6 |
| Natasha | Female | 16 | 6 | 8 | 7 |
| Joshua | Male | 16 | 3 | 3 | 4 |

Using the database,

**a** write down Kirsty's age.
**b** write down Clive's key stage 3 maths level.
**c** write down Nadia's key stage 3 English level.
**d** write down who got the only level 8
**e** which students got the same level in English, maths and science?
**f** how many students are under 16 years of age?
**g** how many boys got level 5 in key stage 3 English?
**h** list the girls in order of key stage 3 maths level, highest level first.
**i** write down the age of the boy whose maths level was higher than his English level.

*Solution 9*

Using the database,

**a** 15

| Name | Gender | Age |
|------|--------|-----|
| Kirsty | | 15 |

**b** 6

| Name | Gender | Age | Key stage 3 English level | Key stage 3 maths level |
|------|--------|-----|---------------------------|-------------------------|
| Clive | | | | 6 |

**c** 4

**d** Natasha

| Name | Gender | Age | Key stage 3 English level | Key stage 3 maths level |
|------|--------|-----|---------------------------|-------------------------|
| Natasha | | | | 8 |

**e** Samantha (5, 5, 5) and Daniel (4, 4, 4)

| Name | Gender | Age | Key stage 3 English level | Key stage 3 maths level | Key stage 3 science level |
|------|--------|-----|---------------------------|-------------------------|---------------------------|
| Samantha | | | 5 | 5 | 5 |
| Daniel | | | 4 | 4 | 4 |

**f** 5 (Samantha, Yousef, Amir, Kirsty and Clive)

**g** 1 (Clive)

**h** Natasha (8), Nadia (6), Samantha (5), Kirsty (4)

**i** 16

| Name | Gender | Age | Key stage 3 English level | Key stage 3 maths level |
|------|--------|-----|---------------------------|-------------------------|
| Clive | Male | 16 | 5 | **6** |

## Example 10

'Teenage girls watch more TV soaps than any other group.' Bill wants to test if this statement is true. Here is part of a database that Bill finds on the internet.

| Name | Age | Month of birth | Gender | Colour of hair | Favourite colour | Favourite music | Favourite sport | Favourite TV programme | TV hours watched per week | Number of pets |
|------|-----|----------------|--------|----------------|------------------|-----------------|-----------------|------------------------|---------------------------|----------------|
| Asif | 20 | April | Male | Black | Red | R & B | Rugby | *The Simpsons* | 42 | 0 |
| Ben | 15 | June | Male | Black | Orange | Pop | Rounders | *Coronation Street* | 14 | 2 |
| Cath | 16 | March | Female | Red | Black | Rock | Tennis | *The News* | 27 | 10 |
| Deb | 12 | May | Female | Blonde | Blue | Rock | Baseball | *Coronation Street* | 22 | 3 |
| Eric | 35 | June | Male | Brown | Pink | Rock | Judo | *Crimewatch* | 22 | 5 |

Bill does not need to use all of this database. Which parts of the database should Bill use?

**Solution 10**

| Age | Gender | Favourite TV programme | TV hours watched per week |
|-----|--------|------------------------|---------------------------|
| 20 | Male | *The Simpsons* | 42 |
| 15 | Male | *Coronation Street* | 14 |
| 16 | Female | *The News* | 27 |
| 12 | Female | *Coronation Street* | 22 |
| 35 | Male | *Crimewatch* | 22 |

## Exercise 2C

**1**  The table gives some information about the matches played by five football teams.

| Team | Played | Won | Drawn | Lost | Points |
|------|--------|-----|-------|------|--------|
| Chad Rovers | 16 | 11 | 3 | 2 | 36 |
| Turton Lane | 16 | 9 | 5 | 2 | 32 |
| Pine United | 16 | 9 | 2 | 5 | 29 |
| Grafton Town | 16 | 5 | 5 | 6 | 20 |
| Workston | 16 | 1 | 3 | 12 | 6 |

**a**  Write down the number of matches that Pine United won.

**b**  Write down the number of matches that Grafton Town drew.

**c**  Write down the number of points that Turton Lane have.

**d**  Which team lost 6 matches?

**e**  Which two teams drew more matches than they lost?

**2**  The database contains some information about second hand cars.

| Make | Model | Year | Engine | Doors | Miles | Colour | Price |
|------|-------|------|--------|-------|-------|--------|-------|
| Ford | Fiesta | 2004 | 1.2 litres | 3 | 7050 | Silver | £5890 |
| Vauxhall | Astra | 2002 | 1.6 litres | 5 | 10 000 | Silver | £5990 |
| VW | Golf | 2005 | 1.4 litres | 5 | 6040 | Black | £10 990 |
| Peugeot | 206 | 2002 | 1.4 litres | 3 | 27 500 | Blue | £5590 |
| Audi | A4 | 2000 | 2.0 litres | 4 | 41 000 | Red | £7590 |

Using this database

**a**  write down the colour of the Peugeot 206

**b**  write down the price of the Audi A4

**c**  write down the make of car which is silver and has 5 doors

**d**  write down the makes of the cars which have travelled more than 10 000 miles

**e**  write down the make of the car which has 5 doors and costs less than £6000

**3** The database contains some information about houses for sale.

|   | Type of house | Bedrooms | Bathrooms | Garage | Garden area in square metres | Price |
|---|---|---|---|---|---|---|
| A | Detached | 5 | 2 | Yes | 120 | £244 000 |
| B | Semi-detached | 3 | 1 | No | 40 | £109 000 |
| C | Town house | 2 | 1 | No | 20 | £98 000 |
| D | Semi-detached | 3 | 1 | Yes | 50 | £128 000 |
| E | Detached | 4 | 2 | Yes | 200 | £285 000 |
| F | Detached | 3 | 1 | No | 90 | £165 000 |
| G | Terraced | 2 | 1 | No | none | £95 000 |
| H | Semi-detached | 4 | 1 | Yes | 60 | £205 000 |
| I | Semi-detached | 5 | 3 | No | 240 | £324 000 |
| J | Flat | 1 | 1 | No | none | £130 000 |

Using this database

**a** write down the letter of the house
   **i** with a price of £95 000       **ii** which has a garden area greater than 200 square metres
**b** write down the letters of the houses that have
   **i** no garden and no garage     **ii** more than one bathroom
**c** write down the number of semi-detached houses that have no garage
**d** list the houses in order of price, starting with the lowest price
**e** list the houses which have a garden in order of garden area, starting with the largest area.

**4** The table shows information about 20 students.

| Name | Gender | Age | Colour of hair | Colour of eyes |
|---|---|---|---|---|
| Angus | Male | 16 | Blonde | Brown |
| Caroline | Female | 14 | Blonde | Blue |
| Stuart | Male | 16 | Ginger | Green |
| Mark | Male | 15 | Blonde | Blue |
| Tahir | Male | 15 | Black | Brown |
| Fatima | Female | 15 | Black | Brown |
| Brenda | Female | 16 | Brown | Hazel |
| Damian | Male | 14 | Brown | Blue |
| Shirley | Female | 16 | Ginger | Blue |
| Kyle | Male | 16 | Brown | Hazel |
| Mumtaz | Male | 15 | Black | Brown |
| Nosheen | Female | 16 | Brown | Hazel |
| Helen | Female | 14 | Blonde | Blue |
| Parvinder | Male | 16 | Black | Brown |
| Viv | Female | 15 | Brown | Hazel |
| Arshad | Male | 14 | Black | Brown |
| Alex | Male | 14 | Black | Green |
| Asia | Female | 16 | Blonde | Green |
| Summer | Female | 15 | Black | Blue |
| Dale | Male | 16 | Brown | Brown |

a  Write down the name of the 14 year old male who has brown eyes.

b  Write down the name of the 16 year old female who has blonde hair.

c  How many of the students are male?

d  How many of the female students are 15 years of age?

e  Write down the names of all the male students with blue eyes.

f  Write down the names of all the female students who are over 15 years of age and have brown hair.

g  Draw and complete a tally chart to show the different colours of eyes.

**5**  'Teenage boys who like Rugby prefer Rock music'.
Viv wants to test if this statement is true. Here is part of a database she finds on the internet.

| Name | Age | Month of birth | Gender | Colour of hair | Favourite colour | Favourite music | Favourite sport | Favourite TV programme | TV hours watched per week | Number of pets |
|------|-----|----------------|--------|----------------|------------------|-----------------|-----------------|------------------------|---------------------------|----------------|
| Asif | 20 | April | Male | Black | Red | R & B | Rugby | *The Simpsons* | 42 | 0 |
| Ben | 15 | June | Male | Black | Orange | Pop | Rounders | *Coronation Street* | 14 | 2 |
| Cath | 16 | March | Female | Red | Black | Rock | Tennis | *The News* | 27 | 10 |
| Deb | 12 | May | Female | Blonde | Blue | Rock | Baseball | *Coronation Street* | 22 | 3 |
| Eric | 35 | June | Male | Brown | Pink | Rock | Judo | *Crimewatch* | 22 | 5 |

Viv does not need to use all of this database.
Which parts of the database should she use?

## Chapter summary

**You should now know:**

★  **data** is information which can be recorded in words **(qualitative data)** and in numbers **(quantitative data)**

★  quantitative data is either **discrete data** (for example key stage maths level, number of goals scored, and shoe size) or **continuous data** (for example height, weight, and time)

★  data can be collected by **observation**, by **experiment** and by using **questionnaires**

★  how to record data using **data collection sheets**, **frequency tables**, **two-way tables** and **grouped frequency tables**

★  in a frequency table:
   • **tally marks** are used to record each piece of information
   • **frequency** means the same as total

★  in a grouped frequency table:
   • each group is called a **class interval**
   • the **modal class interval** is the class interval with the highest frequency

★  how to write questions for a questionnaire avoiding **bias** and including **response boxes** which cover all possible answers

★  the **sample** of people chosen to take part in a survey must be carefully chosen to avoid bias.

★  a **random sample** is one where each person has an equal chance of being chosen

★  a **systematic sample** (sometimes called selective sampling) is a sample chosen from a population which has been arranged in some defined order

★ a **stratified sample** is found by dividing the population to be sampled into groups based upon previously agreed criteria. These groups are called **strata** and a random sample from each strata is taken to give the complete stratified sample

★ **primary data** is data collected by yourself and **secondary data** is data collected by someone else

★ a **database** is an organised collection of information which can be stored in a computer or on paper

★ how to recognise the parts of a database that are needed to solve a problem and be able to answer questions using a database

★ **data logging** may be defined as the collection and storage of data for later use.

## Chapter 2 review questions

**1** Angela asked 20 people in which country they spent their last holiday. Here are their answers.

| France | Spain | Italy | England |
|--------|-------|-------|---------|
| Spain | England | France | Spain |
| Italy | France | England | Spain |
| Spain | Italy | Spain | France |
| England | Spain | France | Italy |

Design *and* complete a suitable data collection sheet that Angela could have used to show this information.

(1388 March 2004)

**2** A student wanted to find out how people used the cinema in his town. He used this question on a questionnaire.

'How many times have you visited the cinema in town?'

[ ]          [ ]

A lot          A bit

**a** Explain two things that are wrong with this question.

**b** Write an improved question. You should include some response boxes.

**3** Janie wants to collect information about the amount of sleep the students in her class get. Design a suitable question she could use.

(1387 June 2005)

**4** Hamid wants to find out what people in Melworth think about the sports facilities in the town. Hamid plans to stand outside the Melworth sports centre one Monday morning. He plans to ask people going into the sports centre to complete a questionnaire. Carol tells Hamid that his survey will be biased.

**i** Give *one* reason why the survey will be biased.

**ii** Describe *one* change Hamid could make to the way in which he is going to carry out his survey so that it will be less biased.

(1388 January 2004)

**5** Here are some examples of different types of data.

A The number of people in a café.     B The time it takes to go to work.     C The colour of a dress.

**a**   **i** Which one of these is continuous data?       **ii** Which one of these is qualitative data?

In a survey it is decided to use secondary data.

**b** Write down *one* advantage and *one* disadvantage of using secondary data.

**6** A researcher wants to choose a random sample of 300 people from her town.
She considers choosing the people that live near her house.

  **a** Write down one advantage and one disadvantage in this method of choosing her sample.

There are 30 000 people on the electoral register in her town.

  **b** Describe how she should choose a systematic sample of 300 people from the electoral register.

**7** The table shows the number of students in each year group at a school.

| Year group | 7 | 8 | 9 | 10 | 11 |
|---|---|---|---|---|---|
| **Number of students** | 190 | 145 | 145 | 140 | 130 |

Jenny is carrying out a survey for her GCSE mathematics project.
She uses a stratified sample of 60 students according to year group.
Calculate the number of Year 11 students that should be in her sample.     (1387 November 2005)

**8** There are 800 pupils at Hightier School.
The table shows information about the pupils.

| Year group | Number of boys | Number of girls |
|---|---|---|
| 7 | 110 | 87 |
| 8 | 98 | 85 |
| 9 | 76 | 74 |
| 10 | 73 | 77 |
| 11 | 65 | 55 |

An inspector is carrying out a survey into pupils' views about the school.
She takes a sample, stratified both by Year group and by gender, of 50 of the 800 pupils.

  **a** Calculate the number of Year 9 boys to be sampled.

Toni stated 'There will be twice as many Year 7 boys as Year 11 girls to be sampled.'

  **b** Is Toni's statement correct?
    You must show how you reached your decision.

**9** The table shows the number of students in each of the four Year 11 maths classes in a school.

| Maths class | Number of pupils |
|---|---|
| Class 1 | 35 |
| Class 2 | 30 |
| Class 3 | 20 |
| Class 4 | 10 |

A sample of size 30 is to be taken from Year 11
Omar suggests that three of the classes are chosen at random and 10 students selected at random from each class.

  **a** Would this method give a random sample?
    *Explain* your answer.

Nesta suggests a stratified sample of size 30 from the whole of Year 11 according to each maths class.

  **b** How many students from Class 1 should be in the sample?     (1385 June 2002)

**10** The table shows the numbers of boys and the numbers of girls in Year 10 and Year 11 of a school.

| | Year 10 Group | Year 11 Group |
|---|---|---|
| **Boys** | 100 | 50 |
| **Girls** | 90 | 60 |

The headteacher wants to find out what pupils think about a new Year 11 common room.
A stratified sample of size 50 is to be taken from Year 10 and Year 11.

Calculate the number of pupils to be sampled from Year 10.                    (1385 June 1999)

**11** The table shows the number of students in each year group at Mathstown High School.

Ben is carrying out a survey about the students' favourite television programmes. He uses a stratified sample of 50 students according to year group.

Calculate the number of Year 12 students which should be in his sample.

| Year group | Number of students |
|---|---|
| 9 | 300 |
| 10 | 290 |
| 11 | 340 |
| 12 | 210 |
| 13 | 180 |

(1385 November 1998)

# Averages and spread

## 3.1 Mean, mode and median

The heights, in metres, of the 11 football players are

    1.78   1.68   1.88   1.82   1.80   2.05
    1.83   1.97   1.81   1.88   1.74

What is the average height of these 11 football players?

In statistics there are three different measures of
average, the **mean**, the **mode** and the **median**.

To find the **mean** height, add the eleven heights together and divide by 11

The mean height $= \dfrac{1.78 + 1.68 + 1.88 + 1.82 + 1.80 + 2.05 + 1.83 + 1.97 + 1.81 + 1.88 + 1.74}{11}$

                    $= 20.24 \div 11 = 1.84$ metres

In general, to find the mean of a set of numbers

$$\text{mean} = \frac{\text{sum of all the numbers}}{\text{how many numbers there are}}$$

This can be written as $\bar{x} = \dfrac{\sum x}{n}$

where $\bar{x}$ is the mean, $\sum x$ is the sum of all the numbers and $n$ is the number of numbers.

The **mode** of a set of numbers is the number that occurs most often.
The mode of the heights (modal height) of the football team is 1.88 metres.

To find the median of a set of numbers, list the numbers in order; the **median** is then the middle number.

If there are two middle numbers the median is halfway between the two numbers.

In increasing order the heights are

| 1.68 | 1.74 | 1.78 | 1.80 | 1.81 | 1.82 | 1.83 | 1.88 | 1.88 | 1.97 | 2.05 |

The median height is 1.82 metres.
In general the median is the $\frac{1}{2}(n + 1)$th number.
Here, $n = 11$, so $\frac{1}{2}(11 + 1) = $ 6th number. The 6th height is 1.82 metres.

### Example 1

In one week a salesman makes 9 journeys from his office.
Here is the length, in kilometres, of each of the journeys.

    120   135   117   10   140   150   50   130   28

**a** Find the median length of journey.
**b** Find the mean length of journey.
   Give your answer to an appropriate degree of accuracy.

*Solution 1*

**a** 10   28   50   117   **120**   130   135   140   150   | List the numbers in order and select the middle number.

Median length of journey = 120 km

**b** $\dfrac{120 + 135 + 117 + 10 + 140 + 150 + 50 + 130 + 28}{9}$   | Mean = $\dfrac{\Sigma x}{n}$

$= \dfrac{880}{9} = 97.77777\ ...$   | 97.77777 … km cannot be measured so is not an appropriate answer.

Mean length of journey = 98 km (to the nearest km)   | All the given lengths are whole number of kilometres so an appropriate degree of accuracy is to the nearest km.

Some calculators can be used to calculate the mean directly. Check that you know how to calculate the mean using your own calculator.

## Example 2

Seven people work in an office. The table shows their earnings last year.

**a** Work out the mean earnings last year.
**b** Find the median earnings last year.
**c** Leaving out Julian's earnings, work out, for the remaining six office workers,
   **i** their mean earnings
   **ii** their median earnings.

| Name | Earnings last year |
|---|---|
| Arthur | £12 000 |
| Bob | £ 8 500 |
| Bradley | £30 000 |
| Jim | £11 000 |
| Julian | £73 000 |
| Pamela | £29 500 |
| Tracey | £11 000 |

*Solution 2*

**a** Mean earnings $= \dfrac{12\,000 + 8500 + 30\,000 + 11\,000 + 73\,000 + 29\,500 + 11\,000}{7}$

$= \dfrac{175\,000}{7} = £25\,000$

**b** List the seven earnings in order

| £8500   £11 000   £11 000 | £12 000 | £29 500   £30 000   £73 000 |

Median earnings = £12 000

**c** **i** Mean earnings $= \dfrac{12\,000 + 8500 + 30\,000 + 11\,000 + 29\,500 + 11\,000}{6}$

$= \dfrac{102\,000}{6} = £17\,000$

**ii** List the six earnings in order

median

| £8500   £11 000 | £11 000 ● £12 000 | £29 500   £30 000 |

Find the number which is halfway between 11 000 and 12 000

$\dfrac{11\,000 + 12\,000}{2} = \dfrac{23\,000}{2} = 11\,500$   | 11 500 is the mean of the middle two numbers.

So the median earnings = £11 500

Look at the means for the two sets of earnings in part **a** and part **c i**.
One extreme value (£73 000) has a big effect on the mean but little effect on the median.
The median can be used for average earnings to avoid this effect.

### Example 3

The mean height of a group of eight girls is 1.56 m.
When another girl joins the group the mean height is 1.55 m.
Work out the height of this girl.

*Solution 3*

$$\text{Mean height} = \frac{\sum x}{n} \text{ so}$$

Mean height $\times n = \sum x =$ sum of all the heights

$1.56 \times 8 = 12.48$    Find the sum of the eight heights.

$1.55 \times 9 = 13.95$    Find the sum of the nine heights.

$13.95 - 12.48 = 1.47$    Find the difference between these totals.

Height of the ninth girl = 1.47 m.

### Exercise 3A

**1** Here is a list of five numbers.

    2   6   3   7   2

  **a** Write down the mode.    **b** Find the median.    **c** Work out the mean.

**2** Here is a list of six numbers.

    2   5   3   7   2   11

  **a** Find the median.          **b** Work out the mean.

**3** The list shows the number of cars sold at a garage in the last ten days.

    3   2   7   8   4   9   7   5   7   3

  **a** Write down the mode.    **b** Find the median.

  **c** Work out the mean number of cars sold per day.

**4** A rugby team plays 12 games. Here are the number of points they scored.

    24   10   23   16   12   8   19   23   16   37   16   27

  **a** Write down the mode.    **b** Work out the mean number of points per game.

**5** Here are the lengths, in centimetres, of five used matchsticks.

    2.7   2.8   3.0   3.2   2.8

  **a** Work out the mean length of these matchsticks.

  **b** The mean length of ten other matchsticks is 3.0 cm.
    Find the total of the lengths of these ten matchsticks.

**6** Five people work in a canteen.
The table shows their earnings last year.

  **a** Work out the mean earnings last year.

  **b** Find the median earnings last year.

  **c** Pamela is the canteen manager.
    How much more were Pamela's earnings than the
    mean earnings of the remaining four workers?

| Name | Earnings last year |
|------|--------------------|
| Leanne | £2500 |
| Mike | £3200 |
| Nazia | £5800 |
| Owen | £4100 |
| Pamela | £22 400 |

**7** The mean weight of a group of five boys is 56 kg.

  **a** Work out the total weight of these five boys.

  When a sixth boy joins the group the mean weight is 58 kg.

  **b** Work out the weight of the sixth boy.

**8** The mean number of runs scored by a cricketer in his last 10 innings is 47.3
Work out the number of runs that the cricketer must score in the next innings for the mean to be exactly 50.

**9** Ted had 20 DVDs. The mean playing time for these 20 DVDs was 145 minutes.
Ted gave away 4 of his DVDs. The mean playing time of the 16 DVDs left was 152 minutes.
Work out the mean playing time of the 4 DVDs that Ted gave away.

**10** 8 men and 5 women work in an office. The mean weekly wage of the men is £338
The mean weekly wage of the women is £289
Work out the mean weekly wage of all 13 workers.

## 3.2 Using frequency tables to find averages

### Example 4

A dice is rolled ten times. The frequency table shows the scores.

**a** Write down the modal score.

**b** Find the median for the ten scores.

**c** Find the mean for the ten scores.

| Score | Frequency |
|---|---|
| 1 | 1 |
| 2 | 3 |
| 3 | 2 |
| 4 | 2 |
| 5 | 1 |
| 6 | 1 |

*Solution 4*

**a** Modal score = 2

> Modal score is the score with the greatest frequency.

**b** Median = $\frac{1}{2}(10 + 1) = 5.5$

The median will be halfway between the 5th and 6th scores.

> Median = $\frac{1}{2}(n + 1)$th number.

5th score is 3, 6th score is 3

> From the frequency table the scores in order are 1 2 2 2 3 3 4 4 5 6

Median = 3

**c**

| Score ($x$) | Frequency ($f$) | $f \times x$ |
|---|---|---|
| 1 | 1 | 1 |
| 2 | 3 | 6 |
| 3 | 2 | 6 |
| 4 | 2 | 8 |
| 5 | 1 | 5 |
| 6 | 1 | 6 |
| | $\sum f = 10$ | $\sum fx = 32$ |

> Mean = $\dfrac{\text{sum of the scores}}{\text{number of scores}}$
>
> $x$ is the score; $f$ is the frequency.
>
> The total of each score is the score multiplied by its frequency, $(= f \times x)$ for example, a score of **2** occurs **3** times so the total of **3** lots of **2** is $3 \times 2 = 6$
>
> Find $\sum fx$, the sum of the ten scores, and $\sum f$, the number of scores.

Mean = $\frac{32}{10}$ = 3.2

> Mean = $\dfrac{\text{sum of the ten scores}}{10}$

In general, for data given in a frequency table, mean = $\dfrac{\sum fx}{\sum f}$

### Example 5

The frequency table shows information about the number of certificates awarded to each student in a class last month.

**a** How many students were in the class?

**b** Work out the total number of certificates awarded.

**c** Work out the mean number of certificates awarded.

**d** Work out the median number of certificates awarded.

| Number of certificates | Frequency |
|---|---|
| 0 | 3 |
| 1 | 7 |
| 2 | 3 |
| 3 | 9 |
| 4 | 8 |

### Solution 5

**a** $3 + 7 + 3 + 9 + 8 = 30$ students in the class.

| Find $\sum f$ |
|---|

**b** $(3 \times 0) + (7 \times 1) + (3 \times 2) + (9 \times 3) + (8 \times 4)$
$= 0 + 7 + 6 + 27 + 32 = 72$

| Find $\sum fx$ |
|---|

Total number of certificates awarded $= 72$

**c** $72 \div 30 = 2.4$

Mean number of certificates awarded $= 2.4$

| Find $\dfrac{\sum fx}{\sum f}$ |
|---|

**d** Median $= \frac{1}{2}(30 + 1) = 15.5$

| Median $= \frac{1}{2}(n + 1)$th number. |
|---|

The median will be halfway between the 15th and 16th numbers.

| Number of certificates | Frequency | |
|---|---|---|
| 0 | 3 | 1st to 3rd |
| 1 | 7 | 4th to 10th |
| 2 | 3 | 11th to 13th |
| 3 | 9 | 14th to 22nd |
| 4 | 8 | 23rd to 30th |

The 15th student received **3** certificates.

The 16th student received **3** certificates.

The median number of certificates $= 3$

### Exercise 3B

**1** The table shows the results of rolling a dice 10 times.

    **a** Find the median score.

    **b** Work out the mean score.

| Score | Frequency |
|---|---|
| 1 | 3 |
| 2 | 3 |
| 3 | 1 |
| 4 | 2 |
| 5 | 1 |
| 6 | 0 |

**2** The table shows the numbers of cakes sold in a shop to the first 30 customers.

    **a** Write down the modal number of cakes sold.

    **b** Work out the total number of cakes sold to these 30 customers.

    **c** Work out the mean number of cakes sold.

    **d** Find the median number of cakes sold.

| Number of cakes | Number of customers |
|---|---|
| 0 | 2 |
| 1 | 9 |
| 2 | 6 |
| 3 | 6 |
| 4 | 5 |
| 5 | 2 |

**3** The table shows the numbers of goals scored by a hockey team in each of 25 matches.

    **a** Write down the mode of the number of goals scored.

    **b** Work out the mean number of goals scored per match.

| Number of goals | Frequency |
|---|---|
| 0 | 9 |
| 1 | 5 |
| 2 | 5 |
| 3 | 4 |
| 4 | 2 |

**4** The table shows the numbers of planes landing at a small airport during each hourly period yesterday.

    **a** Work out the total number of planes that landed at the airport yesterday.

    **b** Work out the mean number of planes landing per hour.

| Number of planes | Frequency |
|---|---|
| 0 | 5 |
| 1 | 0 |
| 2 | 1 |
| 3 | 2 |
| 4 | 4 |
| 5 | 12 |

**5** Mrs Fox did a survey of the number of books each pupil in her class borrowed from the library last month. The frequency table shows her results.

    **a** Work out the number of pupils that Mrs Fox asked.

    **b** Ben thinks that the average number of books borrowed in this survey is 6
       Explain why Ben cannot be correct.

    **c** Find the median number of books borrowed.

    **d** Work out the mean number of books borrowed.

| Number of books borrowed | Frequency |
|---|---|
| 0 | 4 |
| 1 | 6 |
| 2 | 6 |
| 3 | 8 |

**6** A wedding photographer recorded the number of weddings he attended each week last year. The table shows his results.

    **a** Find the median number of weddings he attended per week.

    **b** Find the mean number of weddings he attended per week.

    **c** The photographer is paid £250 for each wedding he attends. Work out the total amount the photographer was paid last year.

| Number of weddings | Frequency |
|---|---|
| 0 | 10 |
| 1 | 14 |
| 2 | 4 |
| 3 | 2 |
| 4 | 13 |
| 5 | 5 |
| 6 | 3 |
| 7 | 1 |

## 3.3 Range and interquartile range

In statistics, **range** is a measure of how spread out numerical data is.

To find the range of a set of numbers, work out the difference between the highest number and the lowest number.

The heights, in metres, of 11 football players listed in order are

      1.68   1.74   1.78   1.80   1.81   1.82   1.83   1.88   1.88   1.97   2.05

The range of these heights is $2.05 - 1.68 = 0.37$ metres.

The median height is the 6th ($= \frac{1}{2}(11 + 1)$th) height, which is 1.82 metres.

The median is the value that is halfway through the data.
The **lower quartile** is the value that is a quarter of the way through the data.
The **upper quartile** is the value that is three-quarters of the way through the data.

For $n$ numbers listed in order:
- the median is the $\frac{1}{2}(n+1)$th number
- the lower quartile is the $\frac{1}{4}(n+1)$th number
- the upper quartile is the $\frac{3}{4}(n+1)$th number.

It is sometimes useful to know how **spread** out values are over the middle 50% of data.
This is called the **interquartile range**.

**interquartile range = upper quartile − lower quartile**

For the 11 heights in order,

1.68  1.74  **1.78**  1.80  1.81  **1.82**  1.83  1.88  **1.88**  1.97  2.05

The lower quartile is the $\frac{1}{4}(11+1)=$ 3rd number = **1.78**
The upper quartile is the $\frac{3}{4}(11+1)=$ 9th number = **1.88**
Interquartile range = 1.88 − 1.78 = 0.1 metres.

## Example 6

The table shows information about the ages, in years, of junior members of a tennis club.

**a** Find the range of their ages.

**b** Find the interquartile range of their ages.

| Age in years | Frequency |
|---|---|
| 9 | 30 |
| 10 | 40 |
| 11 | 19 |
| 12 | 38 |
| 13 | 11 |
| 14 | 18 |
| 15 | 13 |

*Solution 6*

**a** 15 − 9

Range = 6 years

Range = highest number − lowest number.

**b**

Lower quartile is the $\frac{1}{4}(169+1)=42\frac{1}{2}$th number.
The 42nd age is 10, the 43rd age is 10

Frequency table gives ages in order.

Lower quartile is $\frac{1}{4}(n+1)$th number
$n=\Sigma f=169$

Lower quartile = 10

Upper quartile is the $\frac{3}{4}(169+1)=127\frac{1}{2}$th number.
The 127th age is 12, the 128th age is 13

Upper quartile is $\frac{3}{4}(n+1)$th number.

Upper quartile = 12.5

Halfway between 12 and 13

12.5 − 10 = 2.5

Interquartile range = upper quartile − lower quartile.

Interquartile range = 2.5 years

## 3.4 Stem and leaf diagrams

A **stem and leaf diagram** is a diagram that shows data in a systematic way.
The mode, median and range of a set of data can be found easily from a stem and leaf diagram.

The ages, in years, of 11 people are

     12  9  20  24  15  17  31  4  15  17  28

Here is a stem and leaf diagram showing these ages. The key is part of the stem and leaf diagram.

| Stem | Leaves |
|------|--------|
| 0 | 4　9 |
| 1 | 2　5　5　7　7 |
| 2 | 0　4　8 |
| 3 | 1 |

Key
1 | **2** means age 12

In this case, the key makes it clear that the **stem** shows the **tens** and the **leaves** show the **units**.

The data is written so that every number has a tens and units value:

    12  09  20  24  15  17  31  04  15  17  28

The data is ordered from 0 | **4** (= age 4) to 3 | **1** (= age 31)

### Example 7

Here is a stem and leaf diagram showing the ages, in years, of 15 office workers.
Find

**a** the range of the 15 ages

**b** the median age.

| 1 | 6　7　9 |
|---|--------|
| 2 | 1　2　5　5　9　9 |
| 3 | 0　4　5　8 |
| 4 | 1　7 |

Key
1 | 6 means age 16

*Solution 7*

**a** Range = 47 − 16 = 31 years

The data is ordered from 1 | **6** (= age 16) to 4 | **7** (= age 47)

**b** The median age is 29 years.

The median is the eighth age which is 2 | 9
(The middle of 15 numbers is the eighth number.)

### Example 8

Fifteen boys are timed over a 10 metre sprint. Here are their times to the nearest tenth of a second.

    2.6  3.8  3.0  4.7  4.1  2.1  3.1  3.9
    4.0  2.4  3.7  2.7  5.1  2.8  4.7

**a** Draw a stem and leaf diagram to show these results.

**b** Use your stem and leaf diagram to find
  **i** the range of the times    **ii** the median time    **iii** the interquartile range.

*Solution 8*

**a**

| 2 | 1　4　6　7　8 |
|---|--------|
| 3 | 0　1　7　8　9 |
| 4 | 0　1　7　7 |
| 5 | 1 |

Key
3 | 9 means 3.9 seconds

**b  i** $5.1 - 2.1$

> Range = highest number − lowest number.

Range = 3.0 seconds

**ii**

> Stem and leaf diagram gives times in order.

Median = $\frac{1}{2}(15 + 1)$th number

= **8th number**

> Median = $\frac{1}{2}(n + 1)$th number.
> n = total number of boys = 15

Median = 3.7 seconds

> $3 \mid 7 = 3.7$ seconds.

**iii** Lower quartile is $\frac{1}{4}(15 + 1)$th = **4th number**

> Lower quartile is $\frac{1}{4}(n + 1)$th number.

Lower quartile = 2.7

> $2 \mid 7 = 2.7$ seconds.

Upper quartile is $\frac{3}{4}(15 + 1)$ = **12th number**

> Upper quartile is $\frac{3}{4}(n + 1)$th number.

Upper quartile = 4.1

> $4 \mid 1 = 4.1$ seconds.

$4.1 - 2.7 = 1.4$

> Interquartile range = upper quartile − lower quartile.

Interquartile range = 1.4 seconds

---

Stem and leaf diagrams can be used to compare two sets of data. In these cases two diagrams are combined to make a **back-to-back stem and leaf diagram**.

## Example 9

David chooses a sample of 30 girls and 30 boys in Year 10 to compare their weights.
This back-to-back stem and leaf diagram shows his results.

| Girls | | Boys |
|---|:---:|---|
| 9 9 8 7 7 0 | **4** | |
| 9 9 9 8 7 7 6 4 4 3 2 1 | **5** | 6 7 7 8 9 |
| 8 7 6 6 6 5 4 4 3 2 | **6** | 0 1 1 2 4 5 6 6 6 **6 8** 9 |
| 2 1 | **7** | 0 0 1 2 2 3 4 4 5 6 7 |
| | **8** | 0 1 |

Key  **5** $\mid$ 1 means 51 kg

Compare the weights of the girls and boys by finding the range and the median of each distribution.

### Solution 9

For the girls, the data is ordered from 0 $\mid$ **4** (= 40 kg) to 2 $\mid$ **7** (= 72 kg) The range = 72 − 40 = 32 kg.
For the boys, the data is ordered from **5** $\mid$ 6 (= 56 kg) to  **8** $\mid$ 1 (= 81 kg) The range = 81 − 56 = 25 kg.
This shows that the weights of the girls are more spread out.
The median is in the middle of the 30 weights for each distribution.
This is halfway between the 15th weight and 16th weight.

For the girls,
   the 15th weight is **8** $\mid$ **5** which is 58 kg
   the 16th weight is 9 $\mid$ **5** which is 59 kg
   The median for the girls = 58.5 kg

For the boys,
   the 15th weight is **6** $\mid$ **6** which is 66 kg
   the 16th weight is **6** $\mid$ **8** which is 68 kg
   The median for the boys = 67 kg

This shows that in general the boys are heavier than the girls. Look back at the stem and leaf diagram. The distribution of the weights of the boys is towards the higher values.

**Exercise 3C**

**1** The weights, in grams, of nine potatoes are

262   234   208   248   239   210   206   227   254

    **a** Find the range of the weights of these potatoes.

    **b** Find the interquartile range of the weights of these potatoes.

**2** In an experiment some people were asked to estimate the length, in centimetres, of a piece of string. The frequency table shows their estimates.

    **a** Find the range of their estimates.

    **b** Find the interquartile range of their estimates.

| Length (cm) | Frequency |
|:-----------:|:---------:|
| 11 | 1 |
| 12 | 2 |
| 13 | 4 |
| 14 | 18 |
| 15 | 16 |
| 16 | 14 |
| 17 | 20 |
| 18 | 18 |
| 19 | 0 |
| 20 | 2 |

**3** The stem and leaf diagram shows the number of minutes taken by each student in a class to complete a puzzle.

    **a** Find the number of students in the class.

    **b** Write down how many students took 24 minutes to complete the puzzle.

    **c** Write down how many students took 10 minutes longer than the quickest student to complete the puzzle.

    **d** Find the range of the times.

    **e** Find the median time.

    **f** Find the interquartile range.

```
0 | 8  9  9
1 | 0  1  3  3  8  8  9
2 | 1  4  4  4  6  6  9
3 | 0  2  3  4  7  7  8  8
4 | 1  1  2  2  4
```

Key
4 | 1 means for 41 minutes

**4** Here are the number of minutes a sample of 15 patients had to wait before seeing a hospital doctor.

    **a** Draw a stem and leaf diagram to show this information.

    **b** Use your stem and leaf diagram to find
      **i** the range of the times    **ii** the median time.

    **c** Find the interquartile range.

|    |    |    |
|----|----|----|
| 49 | 23 | 34 |
| 10 | 28 | 28 |
| 25 | 45 | 20 |
| 39 | 35 | 15 |
| 14 | 48 | 10 |

**5** Tony records the number of emails he receives each day. Here are his results for the last 20 days.

| | | | | |
|-----|-----|-----|-----|-----|
| 178 | 189 | 147 | 147 | 166 |
| 171 | 153 | 171 | 164 | 158 |
| 189 | 166 | 165 | 155 | 152 |
| 147 | 158 | 148 | 152 | 172 |

    **a** Draw a stem and leaf diagram to show this information.

    **b** Use your stem and leaf diagram to find the median number of emails.

    **c** Find the interquartile range.

**6** The table gives the number of days each of 25 people had been on holiday in Spain.

|    |    |    |    |    |    |    |    |    |
|----|----|----|----|----|----|----|----|----|
| 0  | 7  | 14 | 38 | 26 | 16 | 13 | 21 | 35 |
| 30 | 9  | 10 | 21 | 25 | 22 | 7  | 15 | 7  |
| 21 | 14 | 25 | 31 | 30 | 7  | 7  |    |    |

   **a** Draw a stem and leaf diagram to show this information.

   **b** Use your stem and leaf diagram to find the median number of days.

   **c** Find the interquartile range.

**7** Nicki weighs 20 parcels. Here are her results in kilograms.

|     |     |     |     |     |     |     |     |     |     |
|-----|-----|-----|-----|-----|-----|-----|-----|-----|-----|
| 0.7 | 1.6 | 2.3 | 3.4 | 2.8 | 1.7 | 1.5 | 1.1 | 1.4 | 0.8 |
| 3.3 | 2.6 | 1.6 | 1.1 | 2.7 | 2.7 | 1.8 | 2.0 | 0.9 | 3.0 |

   **a** Draw a stem and leaf diagram to show this information.

   **b** Use your stem and leaf diagram to find the range of the weights.

   **c** Use your stem and leaf diagram to find the median weight.

**8** The back-to-back stem and leaf diagram shows the percentage marks of a group of boys and girls in a science test.

| Girls | | Boys |
|------:|:-:|:-----|
| 9 7 6 1 | **3** | |
| 9 8 7 5 5 0 | **4** | 0 2 8 |
| 9 9 8 8 7 7 3 3 2 1 1 | **5** | 2 3 3 8 |
| 6 5 5 4 3 2 0 | **6** | 0 1 1 2 3 5 5 6 6 7 8 9 |
| 2 1 | **7** | 0 0 1 2 2 3 4 4 |
| | **8** | 0 1 6 |

Key **5** | 2 means 52%

   **a** Find the range of percentage marks for   **i** the boys   **ii** the girls.

   **b** Find the median of percentage marks for   **i** the boys   **ii** the girls.

   **c** Compare and comment on the marks for the boys and girls.

## 3.5 Estimating the mean of grouped data

This table appeared in the *Burwich Guardian*.

It shows some information about the number of road accidents in Burwich each day last September.

There were 7 days on which there were fewer than 5 accidents. It is impossible to tell from the table the exact number of accidents on each day so an exact value for the average number of accidents per day cannot be found.

| Number of road accidents | Number of days |
|:------------------------:|:--------------:|
| 0 to 4  | 7  |
| 5 to 9  | 14 |
| 10 to 14 | 8 |
| 15 to 19 | 1 |

It is possible to find an estimate for the mean number of accidents per day. We use the middle of each class interval. For example, we assume that there were $\dfrac{0+4}{2} = 2$ accidents on each of the 7 days. This gives a total of $7 \times 2 = 14$ accidents.

## Example 10

The table shows some information about the number of road accidents in Burmage last April.

| Number of road accidents | Number of days | | |
|---|---|---|---|
| 0 to 4 | 7 | | |
| 5 to 9 | 14 | | |
| 10 to 14 | 8 | | |
| 15 to 19 | 1 | | |

Work out an estimate for the mean number of accidents per day in Burmage last April.

### Solution 10

| Number of road accidents | Number of days ($f$) | Middle of class interval ($x$) | Totals of the numbers of accidents ($fx$) |
|---|---|---|---|
| 0 to 4 | 7 | $\dfrac{0+4}{2}=2$ | $7 \times 2 = 14$ |
| 5 to 9 | 9 | $\dfrac{5+9}{2}=7$ | $9 \times 7 = 63$ |
| 10 to 14 | 12 | $\dfrac{10+14}{2}=12$ | $12 \times 12 = 144$ |
| 15 to 19 | 2 | $\dfrac{15+19}{2}=17$ | $2 \times 17 = 34$ |

The 'number of days' is the frequency, $f$.
Let the middle of class interval be $x$.
The total number of accidents in April is $\sum fx = 14 + 63 + 144 + 34 = 255$
The total number of days in April is $\sum f = 7 + 9 + 12 + 2 = 30$
Estimated mean $= \dfrac{\sum fx}{\sum f} = \dfrac{255}{30} = 8.5$ accidents per day.

## Example 11

The table shows some information about the annual earnings of 140 employees of a company.

| Annual earnings (£$P$) | Frequency |
|---|---|
| $0 < P \leqslant 10\,000$ | 30 |
| $10\,000 < P \leqslant 20\,000$ | 42 |
| $20\,000 < P \leqslant 30\,000$ | 28 |
| $30\,000 < P \leqslant 40\,000$ | 20 |
| $40\,000 < P \leqslant 50\,000$ | 18 |
| $50\,000 < P \leqslant 60\,000$ | 2 |

$10\,000 < P \leqslant 20\,000$ means earnings above £10 000 up to and including £20 000 (See Section 27.1)

a Work out an estimate for the mean earnings.
b Find the class interval that contains the median earnings.

**Solution 11**

a

| Annual earnings (£P) | Frequency (f) | Middle of class interval (x) | Totals of earnings (fx) |
|---|---|---|---|
| $0 < P \leqslant 10\,000$ | 30 | 5 000 | 5 000 × 30 = **150 000** |
| $10\,000 < P \leqslant 20\,000$ | 42 | 15 000 | 15 000 × 42 = **630 000** |
| $20\,000 < P \leqslant 30\,000$ | 28 | 25 000 | 25 000 × 28 = **700 000** |
| $30\,000 < P \leqslant 40\,000$ | 20 | 35 000 | 35 000 × 20 = **700 000** |
| $40\,000 < P \leqslant 50\,000$ | 18 | 45 000 | 45 000 × 18 = **810 000** |
| $50\,000 < P \leqslant 60\,000$ | 2 | 55 000 | 55 000 × 2 = **110 000** |
| | $\sum f = 140$ | | $\sum fx = $ **3 100 000** |

$$\text{Estimated mean} = \frac{\sum fx}{\sum f} = \frac{3\,100\,000}{140} = £22\,143 \text{ (to the nearest £)}.$$

**b** $\frac{1}{2}(140 + 1) = 70.5$   The median is in the middle of the 70th and the 71st earnings.

| Annual earnings (£P) | Frequency | |
|---|---|---|
| $0 < P \leqslant 10\,000$ | 30 | 1st to the 30th earnings |
| $10\,000 < P \leqslant 20\,000$ | 42 | 31st to 72nd earnings |

Both the 70th and the 71st earnings lie in the class interval $10\,000 < P \leqslant 20\,000$

So the median lies in the class interval $10\,000 < P \leqslant 20\,000$

---

## Exercise 3D

**1** Twenty people took part in a competition. The points scored are grouped in the frequency table.

  **a** Find the class interval which contains the median.

  **b** Work out an estimate for the mean number of points scored.

| Points scored | Number of people |
|---|---|
| 1 to 5 | 1 |
| 6 to 10 | 2 |
| 11 to 15 | 2 |
| 16 to 20 | 6 |
| 21 to 25 | 7 |
| 26 to 30 | 2 |

**2** The table shows information about the number of minutes that 125 students spent doing homework yesterday.

  **a** Find the class interval which contains the median.

  **b** Work out an estimate for the mean number of minutes that the students spent doing homework yesterday.

| Number of minutes (t) | Frequency (f) |
|---|---|
| $0 < t \leqslant 20$ | 10 |
| $20 < t \leqslant 40$ | 20 |
| $40 < t \leqslant 60$ | 30 |
| $60 < t \leqslant 80$ | 35 |
| $80 < t \leqslant 100$ | 25 |
| $100 < t \leqslant 120$ | 5 |

**3** The table shows some information about the lifetimes, in hours, of 50 light bulbs.

   **a** Find the class interval which contains the median.

   **b** Work out an estimate for the mean number of hours.

| Number of hours ($t$) | Number of light bulbs |
|---|---|
| $0 < t \leqslant 50$ | 2 |
| $50 < t \leqslant 100$ | 3 |
| $100 < t \leqslant 150$ | 6 |
| $150 < t \leqslant 200$ | 9 |
| $200 < t \leqslant 250$ | 19 |
| $250 < t \leqslant 300$ | 11 |

**4** The table shows some information about the number of text messages Simon received each day in December.

   **a** Find the class interval which contains the median.

   **b** Work out an estimate for the mean number of messages received.
      Give your answer correct to one decimal place.

| Number of text messages | Frequency ($f$) |
|---|---|
| 0 to 4 | 3 |
| 5 to 9 | 11 |
| 10 to 14 | 12 |
| 15 to 19 | 3 |
| 20 to 24 | 0 |
| 25 to 29 | 2 |

**5** Jack grows onions. The table shows some information about the weights ($w$) of some onions.

   **a** Work out an estimate for the mean weight of these onions.
      Give your answer correct to the nearest gram.

   **b** Find the class interval that contains the median.

| Weight ($w$ grams) | Frequency ($f$) |
|---|---|
| $0 \leqslant w < 40$ | 10 |
| $40 \leqslant w < 60$ | 16 |
| $60 \leqslant w < 80$ | 25 |
| $80 \leqslant w < 100$ | 28 |
| $100 \leqslant w < 120$ | 17 |
| $120 \leqslant w < 150$ | 8 |

## 3.6 Moving averages

The table shows the number of units of gas used in a village, in each season, over a three-year period.

| Season | Year 1 | Year 2 | Year 3 |
|---|---|---|---|
| Winter (W) | 5900 | 5500 | 5100 |
| Spring (Sp) | 2400 | 2000 | 1600 |
| Summer (Su) | 1600 | 1200 | 1200 |
| Autumn (A) | 5700 | 5300 | 4900 |

This information can be plotted on a graph.

The graph shows the changes in the number of units of gas used over the three-year period. It is called a **time series** graph. A time series is a set of readings taken over a period of time.

In this time series graph there are high points, showing the number of units of gas used in Winter, and low points, showing the number of units of gas used in Summer. These are seasonal variations in the gas used.

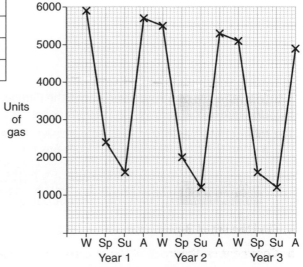

Has the amount of gas used in the village increased or decreased over this three-year period? This cannot be answered easily from the graph. The question can be answered using the following calculations:

The mean of the first four values (W, Sp, Su and A of Year 1)

$$\mathbf{5900}\quad \mathbf{2400}\quad \mathbf{1600}\quad \mathbf{5700}\quad 5500\quad 2000\quad 1200\quad 5300\quad 5100\quad 1600\quad 1200\quad 4900$$

$$= \frac{5900 + 2400 + 1600 + 5700}{4} = \frac{15\,600}{4} = 3900 \text{ units of gas.}$$

The mean of the four values (Sp, Su and A of Year 1 and W of Year 2)

$$5900\quad \mathbf{2400}\quad \mathbf{1600}\quad \mathbf{5700}\quad \mathbf{5500}\quad 2000\quad 1200\quad 5300\quad 5100\quad 1600\quad 1200\quad 4900$$

$$= \frac{2400 + 1600 + 5700 + 5500}{4} = \frac{15\,200}{4} = 3800 \text{ units of gas.}$$

The mean of the 4 values (Su and A of Year 1 and W and Sp of Year 2)

$$5900\quad 2400\quad \mathbf{1600}\quad \mathbf{5700}\quad \mathbf{5500}\quad \mathbf{2000}\quad 1200\quad 5300\quad 5100\quad 1600\quad 1200\quad 4900$$

is then found, and so on.

These are called **moving averages**. Since 4 consecutive values are used in each case, they are called 4-point moving averages. The moving averages are then plotted on the time series graph, and joined. For example the point representing the 1st moving average is plotted halfway between Spring and Summer of Year 1 at a height of 3900.

It is clear from this graph that the number of units used in the village over the three-year period decreases. This is called the **trend** of the graph.

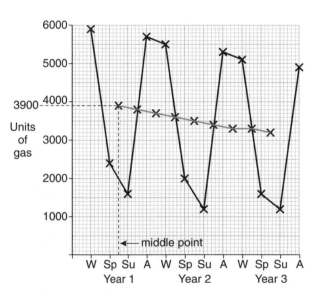

---

## Example 12

The table shows the attendances at a cinema during each day of one week.

|                   | Mon | Tues | Wed | Thurs | Fri | Sat  | Sun  |
| ----------------- | --- | ---- | --- | ----- | --- | ---- | ---- |
| **Number of people** | 170 | 380  | 530 | 560   | 500 | 1250 | 1160 |

**a** Find the set of 3-point moving averages for this information.

**b** Comment on the trend of the number of attendances throughout the week.

## Solution 12

a  $\dfrac{170 + 380 + 530}{2} = \dfrac{1080}{3} = 360$    Find the mean of Mon, Tues and Wed attendances.

$\dfrac{380 + 530 + 560}{3} = \dfrac{1470}{3} = 490$    Find the mean of Tues, Wed and Thurs attendances.

$\dfrac{530 + 560 + 500}{3} = \dfrac{1590}{3} = 530$    Find the mean of Wed, Thurs and Fri attendances.

$\dfrac{560 + 500 + 1250}{3} = \dfrac{2310}{3} = 770$    Find the mean of Thurs, Fri and Sat attendances.

$\dfrac{500 + 1250 + 1160}{3} = \dfrac{2910}{3} = 970$    Find the mean of Fri, Sat and Sun attendances.

b  Attendances increase throughout the week.    Compare 360, 490, 530, 770, 970

---

## Exercise 3E

**1** The table shows the number of tyres fitted each month by a local garage.

| Month | May | June | July | Aug | Sept | Oct | Nov | Dec |
|---|---|---|---|---|---|---|---|---|
| Number of tyres fitted | 32 | 27 | 34 | 19 | 24 | 16 | 12 | 23 |

Work out the first two 4-month moving averages for this data.

**2** The table shows the number of books sold in a shop each month from January to June.

| Jan | Feb | Mar | Apr | May | Jun |
|---|---|---|---|---|---|
| 149 | 156 | 154 | 161 | 159 | 169 |

Work out the 3-month moving averages for this information.

**3** A shop sells televisions.
The table shows the number of televisions sold in every quarter in a two-year period.

**a** Calculate the set of 4-point moving averages for this data.

**b** What do your moving averages in part **a** tell you about the trend in the sale of televisions?

| | Months | Number of televisions sold |
|---|---|---|
| Year 1 | Jan–Mar | 46 |
| | Apr–Jun | 23 |
| | Jul–Sep | 36 |
| | Oct–Dec | 71 |
| Year 2 | Jan–Mar | 54 |
| | Apr–Jun | 31 |
| | Jul–Sep | 44 |
| | Oct–Dec | 71 |

**4** The table shows the total charges, in £, for gas used by Mr Smith, in each season, over a three-year period.

**a** Plot this information on a time series graph.

**b** Work out the 4-point moving averages for this period.

| Season | Year 1 | Year 2 | Year 3 |
|---|---|---|---|
| Winter (W) | 218 | 228 | 242 |
| Spring (Sp) | 163 | 166 | 138 |
| Summer (Su) | 68 | 43 | 38 |
| Autumn (A) | 179 | 139 | 122 |

**c** On your time series graph plot the moving averages.

**d** Comment on the seasonal gas charges during this period.

# Chapter summary

**You should now know that:**

★ for a list of numbers,

- the **mean** = $\dfrac{\text{sum of all the numbers}}{\text{how many numbers there are}}$

- the **mode** is the number which occurs most often

- the **median** is the middle number when the numbers are written in order

- the **range** is the difference between the highest and lowest numbers

★ when data is arranged in order,

- the median is the value that is halfway through the data,

- the **lower quartile** is the value that is a quarter of the way through the data

- the **upper quartile** is the value that is three-quarters of the way through the data

★ **interquartile range** = upper quartile − lower quartile and represents the middle 50% of the data.

**You should also be able to:**

★ find the mean, mode and median of a list of numbers

★ find the mean, mode and median of data given in a frequency table

★ find the range and interquartile range of data given in a list or in a frequency table

★ draw single and back-to-back **stem and leaf diagrams** and use them to find the median, range and interquartile range

★ use average and range to compare distributions

★ estimate the mean of grouped data by using the middle value of each class interval

★ draw **time series** graphs from given data

★ find **moving averages** and use these to describe the **trend** for the data over a given time period.

# Chapter 3 review questions

**1 a** Kuldip recorded the numbers of people getting off his tram at 10 stops.
Here are his results for Monday.

3   5   4   7   4
7   4   9   8   12

For these 10 numbers, work out

  **i** the range

 **ii** the median

**iii** the mean.

**b** On Tuesday, the mean number of people getting off his tram at the same ten stops is 8.6
How many more people got off the tram on Tuesday at these ten stops?

**2** Andy did a survey of the number of cups of coffee some pupils in his school had drunk yesterday.
The frequency table shows his results.

  **a** Work out the number of pupils that Andy asked.

Andy thinks that the average number of drinks pupils in his survey had drunk is 7.

  **b** Explain why Andy cannot be correct.

| Number of cups of coffee | Frequency |
|---|---|
| 2 | 1 |
| 3 | 3 |
| 4 | 5 |
| 5 | 8 |
| 6 | 5 |

(1387 June 2003)

**3** Rosie had 10 boxes of drawing pins.
She counted the number of drawing pins in each box.
The table gives information about her results.

  **a** Write down the modal number of drawing pins in a box.

  **b** Work out the range of the number of drawing pins in a box.

  **c** Work out the mean number of drawing pins in a box.

| Number of drawing pins | Frequency | |
|---|---|---|
| 29 | 2 | |
| 30 | 5 | |
| 31 | 2 | |
| 32 | 1 | |

(1387 June 2003)

**4** Amy had 30 CDs.
The mean playing time of these 30 CDs was 42 minutes.
Amy sold 5 of her CDs.
The mean playing time of the 25 CDs left was 42.8 minutes.
Calculate the mean playing time of the 5 CDs that Amy sold.

(1387 June 2004)

**5** Mary recorded the heights, in centimetres, of the girls in her class.
She put the heights in order.

    132    144    150    152    160    162    162    167
    167    170    172    177    181    182    182

Find

  **a** the lower quartile

  **b** the upper quartile.

**6** Mrs Chowdery gives her class a maths test. Here are the test marks for the girls.

    7   5   8   5   2   8   7   4   7   10   3   7   4   3   6

  **a** Work out the mode.

  **b** Work out the median.

The median mark for the boys was 7 and the range of the marks for the boys was 4.
The range of the girls' marks was 8.

  **c** By comparing the results explain whether the boys or the girls did better in the test.

(1385 June 1998)

**7** Here are the weights, in kilograms, of 15 parcels.

    1.1    1.7    2.0    1.0    1.1    0.5    3.3    2.0
    1.5    2.6    3.5    2.1    0.7    1.2    0.6

Draw a stem and leaf diagram to show this information.

(1388 March 2004)

**8 a** Ahmed recorded the number of lorries entering a motorway service station every three minutes between 9 am and 10 am yesterday. The stem and leaf diagram shows this information.

Number of lorries

| 0 | 1 3 3 5 7 |
|---|---|
| 1 | 0 0 2 3 4 6 7 9 |
| 2 | 0 1 1 1 3 5 6 |

Key

1 | 3 means 13 lorries

**i** Find the median number of lorries.　　**ii** Work out the range of the number of lorries.

Uzma recorded the number of cars entering the motorway service station in each of the same three-minute periods. The list below shows this information.

| 13 | 21 | 17 | 25 | 32 | 19 | 17 | 30 | 31 | 24 |
|---|---|---|---|---|---|---|---|---|---|
| 24 | 14 | 32 | 18 | 19 | 27 | 20 | 30 | 37 | 21 |

**b** Draw a stem and leaf diagram to show this information.

**9** Mark recorded the number of e-mails he received each day for 21 days. The stem and leaf diagram shows this information.

Number of e-mails

| 0 | 4 5 5 6 7 7 8 9 |
|---|---|
| 1 | 0 1 2 3 3 4 6 7 8 |
| 2 | 0 1 3 6 |

Key

2 | 6 means 26 e-mails

**a** Find the median number of e-mails that Mark received in the 21 days.
**b** Work out the range of the number of e-mails Mark received in the 21 days.
**c** Find the interquartile range of the number of e-mails Mark received in the 21 days.

(1388 March 2005)

**10** The table shows information about the number of hours that 120 children used a computer last week.

Work out an estimate for the mean number of hours that the children used a computer.
Give your answer correct to two decimal places.

| Number of hours ($h$) | Frequency |
|---|---|
| $0 < h \leqslant 2$ | 10 |
| $2 < h \leqslant 4$ | 15 |
| $4 < h \leqslant 6$ | 30 |
| $6 < h \leqslant 8$ | 35 |
| $8 < h \leqslant 10$ | 25 |
| $10 < h \leqslant 12$ | 5 |

(1388 June 2005)

**11** Charles found out the length of reign of each of 41 kings.
He used the information to complete the frequency table.

| Length of reign ($L$ years) | Number of kings | | |
|---|---|---|---|
| $0 < L \leqslant 10$ | 14 | | |
| $10 < L \leqslant 20$ | 13 | | |
| $20 < L \leqslant 30$ | 8 | | |
| $30 < L \leqslant 40$ | 4 | | |
| $40 < L \leqslant 50$ | 2 | | |

Use the copy of the table on the resource sheet.

**a** Write down the class interval that contains the median.

**b** Calculate an estimate for the mean length of reign.

(1387 November 2003)

**12** The table shows the number of orders received each month by a small company.

| Month | Jan | Feb | Mar | Apr | May | Jun | Jul | Aug |
|---|---|---|---|---|---|---|---|---|
| Number of orders received | 23 | 31 | 15 | 11 | 19 | 16 | 20 | 13 |

Work out the first two 4-month moving averages for this data.

(1388 June 2003)

**13** Paul and Carol open a new shop in the High Street.
The table shows the monthly takings in each of the first four months.

| Month | Jan | Feb | March | April |
|---|---|---|---|---|
| Monthly takings (£) | 9375 | 8907 | 9255 | 9420 |

Work out the 3-point moving averages for this information

(1388 November 2005)

**14** The table shows information about the quarterly gas bill, in £s, for Samira's house, over a period of two years.

| | Quarter | | | |
|---|---|---|---|---|
| Year | 1 | 2 | 3 | 4 |
| 1 | £200 | £162 | £80 | £130 |
| 2 | £216 | £166 | £96 | £142 |

The data has been plotted as a time series.

**a** The first three 4-point moving averages are £143, £147 and £148.
  **i** Work out the last two 4-point moving averages.
  **ii** Plot all five of the moving averages on the copy of the graph on the resource sheet.

**b** What do the moving averages show about the trend of the quarterly gas bills?

(1389 June 2005)

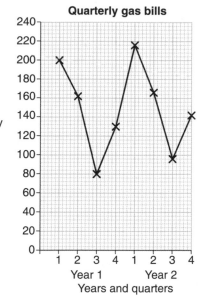

Quarterly gas bill (£)

Quarterly gas bills

Years and quarters

Year 1        Year 2

**15** A youth club has 60 members.
40 of the members are boys.
20 of the members are girls.
The mean number of videos watched last week by all 60 members was 2.8
The mean number of videos watched last week by the 40 boys was 3.3

**a** Calculate the mean number of videos watched last week by the 20 girls.

Ibrahim has two lists of numbers.
The mean of the numbers in the first list is $p$.
The mean of the numbers in the second list is $q$.
Ibrahim combines the two lists into one new list of numbers.

Ibrahim says 'The mean of the new list of numbers is equal to $\dfrac{p+q}{2}$.'

One of two conditions must be satisfied for Ibrahim to be correct.

**b** Write down each of these two conditions.

(1387 November 2004)

# Processing, representing and interpreting data

## 4.1 Frequency polygons

A **frequency diagram** can be drawn from grouped discrete data. A frequency diagram looks the same as a bar chart except that the label underneath each bar represents a group.

The information in this frequency table can be represented as a frequency diagram.

| Mark | Frequency |
|------|-----------|
| 1–5 | 2 |
| 6–10 | 4 |
| 11–15 | 8 |
| 16–20 | 6 |

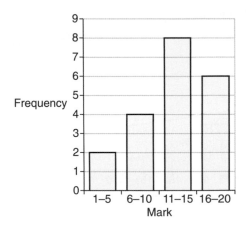

A **histogram** can be drawn from grouped continuous data. A **histogram** is similar to a bar chart but represents continuous data so there is no gap between the bars. There is a scale on the horizontal axis rather than a label under each bar.

---

### Example 1

The grouped frequency table shows information about the heights of 42 students.

**a** Write down the modal class interval.

**b** Jenny's height is 177.2 cm. In which class interval is her height recorded?

**c** Charlie's height is exactly 180 cm. In which class interval is his height recorded?

**d** Draw a histogram to represent this information.

| Height ($h$ cm) | Frequency |
|-----------------|-----------|
| $160 \leqslant h < 165$ | 10 |
| $165 \leqslant h < 170$ | 14 |
| $170 \leqslant h < 175$ | 8 |
| $175 \leqslant h < 180$ | 5 |
| $180 \leqslant h < 185$ | 3 |
| $185 \leqslant h < 190$ | 2 |

**Solution 1**

**a** The modal class is $165 \leqslant h < 170$

> This class interval has the highest frequency, 14

**b** Jenny's height is in the class interval $175 \leqslant h < 180$

> 177.2 cm is greater than 175 cm but less than 180 cm.

**c** Charlie's height is in the class interval $180 \leqslant h < 185$

> 180 is shown at the end of one class interval and at the beginning of another. The sign for 'less than or equal to' ($\leqslant$) shows that 180 should go in the class interval $180 \leqslant h < 185$

**d** The histogram has a scale on the horizontal axis and no gaps between the bars.

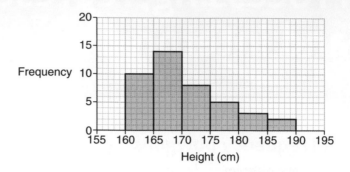

A **frequency polygon** is another graph which shows data.

To draw a frequency polygon for the data in Example 1, mark the midpoint of the top of each bar and join these points with straight lines.

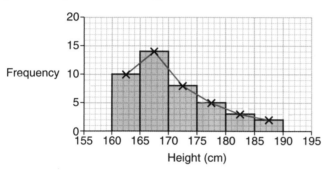

This is the frequency polygon.

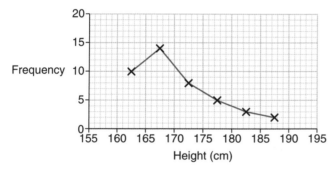

## Example 2

The frequency table gives information about the weights, in grams, of 30 apples.

**a** Write down the modal class.

**b** Use the information to draw a histogram.

**c** Draw a frequency polygon to represent the information.

| Weight ($w$ grams) | Frequency |
|---|---|
| $90 \leqslant w < 95$ | 6 |
| $95 \leqslant w < 100$ | 6 |
| $100 \leqslant w < 105$ | 7 |
| $105 \leqslant w < 110$ | 4 |
| $110 \leqslant w < 115$ | 5 |
| $115 \leqslant w < 120$ | 2 |

## Solution 2

**a** The modal class is $100 \leqslant w < 105$

**b** and **c**

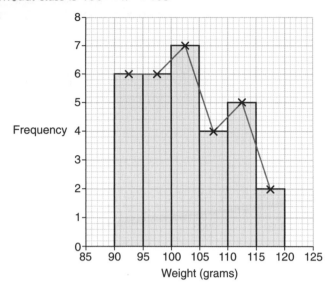

> As the question has asked for both a histogram and a frequency polygon to be drawn, draw the frequency polygon on the histogram.

---

More than one frequency polygon can be drawn on the same grid to compare data.

## Example 3

The two frequency polygons show the heights of a group of girls and the heights of a group of boys.

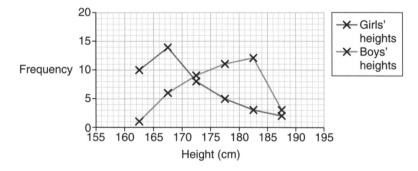

Compare the heights of the two groups. Give a reason for your answers.

## Solution 3

The boys are generally taller than the girls.

> The line showing the boys' heights is above the line for the girls' heights towards the right of the graph.

There are more tall boys than tall girls.

> There are two girls and three boys in the 185–190 cm class interval.

There are more short girls than short boys.

> There are ten girls and one boy in the 160–165 cm class interval.

### Example 4

This graph shows the percentage of Mathstown buses that arrived on time each year from 2000 to 2003
Explain why this graph is misleading.

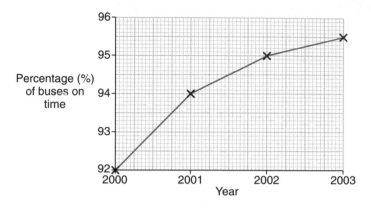

### Solution 4

The steep line between 2000 and 2001 suggests that a much higher percentage of buses were on time in 2001 than in 2000 but it is only 2% more.

If the vertical scale starts from 0 the lines are all less steep, showing that the increases are small. The disadvantage of this is that a large area of the graph is empty.

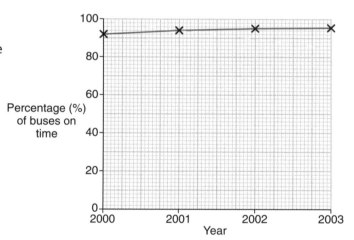

A zig-zag in the vertical axis can be used to show that the scale does not start at 0

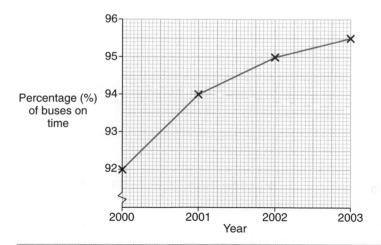

**Exercise 4A**

**1** The grouped frequency table shows the test marks of a class of 32 students.

 **a** Write down the modal class.

 **b** Draw a frequency diagram for this information.

| Test marks | Frequency |
|---|---|
| 1–4 | 2 |
| 5–8 | 5 |
| 9–12 | 6 |
| 13–16 | 11 |
| 17–20 | 8 |

**2** The histogram shows information about the times taken by some girls to run 100 m.

 **a** Write down the modal class.

 **b** Work out the total number of girls.

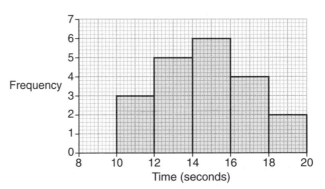

Some boys also ran 100 m. The grouped frequency table shows information about their times.

 **c** Draw a histogram to show this information.

 **d** Draw a frequency polygon to show this information.

| Time (*t* seconds) | Frequency |
|---|---|
| $10 \leqslant t < 12$ | 6 |
| $12 \leqslant t < 14$ | 5 |
| $14 \leqslant t < 16$ | 5 |
| $16 \leqslant t < 18$ | 3 |
| $18 \leqslant t < 20$ | 1 |

**3** In one month Julie went to the post office 20 times and to the bank 20 times.

The frequency polygons show information about the amount of time Julie spent waiting in a queue at the post office and at the bank.

 **a** At the bank how many times did Julie wait for between 15 and 20 minutes?

 **b** At the post office how many times did Julie wait for between 5 and 10 minutes?

 **c** For what fraction of the times Julie went to the post office did she wait for less than 10 minutes? Give your fraction in its simplest form.

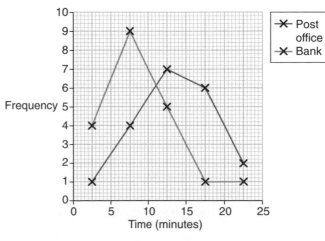

 **d** Where did Julie generally have to wait the longest time, the bank or the post office? You must give a reason for your answer.

**4** The graph shows the number of cars sold in 3 years. Give two reasons why the graph is misleading.

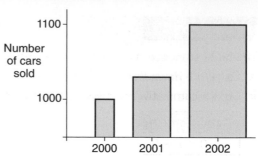

## 4.2 Cumulative frequency

The grouped frequency table shows information about the amount of time 160 students spent doing homework one evening.

| Time ($x$ minutes) | Frequency |
|---|---|
| $0 < x \leqslant 10$ | 4 |
| $10 < x \leqslant 20$ | 12 |
| $20 < x \leqslant 30$ | 46 |
| $30 < x \leqslant 40$ | 68 |
| $40 < x \leqslant 50$ | 20 |
| $50 < x \leqslant 60$ | 10 |

The information in the table can be used to find, for example, the total number of students who spent up to and including 20 minutes which is 16 (= 4 + 12).
16 is the **cumulative frequency** for the interval $0 < x \leqslant 20$
(In other words cumulative frequency is the 'running total'.)
Similarly the cumulative frequency for the interval $0 < x \leqslant 30$ is 62 (= 4 + 12 + 46)

Here is the complete **cumulative frequency table**.

| Time ($x$ minutes) | Cumulative frequency |
|---|---|
| $0 < x \leqslant 10$ | 4 |
| $0 < x \leqslant 20$ | (4 + **12** =)  **16** |
| $0 < x \leqslant 30$ | (16 + 46 =)  **62** |
| $0 < x \leqslant 40$ | (**62** + 68 =) 130 |
| $0 < x \leqslant 50$ | (130 + 20 =) 150 |
| $0 < x \leqslant 60$ | (150 + 10 = ) 160 |

The last number in the cumulative frequency column is 160, the total number of students.
A cumulative frequency table can be used to draw a **cumulative frequency graph**.
The cumulative frequency 4 for the interval $0 < x \leqslant 10$ is plotted at (10, 4), that is, at the *top end* of the interval to ensure that all 4 students have been included.
Similarly the cumulative frequency 16 for the interval $0 < x \leqslant 20$ is plotted at (20, 16). The remaining points are plotted at (30, 62), (40, 130), (50, 150) and (60, 160).

The points can be joined with a smooth curve or with line segments to give a **cumulative frequency graph**.

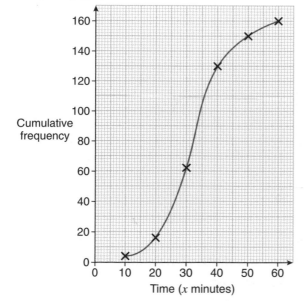

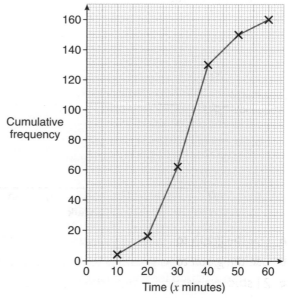

## Example 5

The table gives information about the times in minutes 30 people in a doctor's surgery waited.

a  Complete the cumulative frequency table.

b  Draw a cumulative frequency graph for your table.

| Time ($t$ minutes) | Frequency |
|---|---|
| $0 < t \leqslant 5$ | 1 |
| $5 < t \leqslant 10$ | 5 |
| $10 < t \leqslant 15$ | 12 |
| $15 < t \leqslant 20$ | 10 |
| $20 < t \leqslant 25$ | 2 |

| Time ($t$ minutes) | Cumulative frequency |
|---|---|
| $0 < t \leqslant 5$ | |
| $0 < t \leqslant 10$ | |
| $0 < t \leqslant 15$ | |
| $0 < t \leqslant 20$ | |
| $0 < t \leqslant 25$ | |

*Solution 5*

a

| Time ($t$ minutes) | Cumulative frequency |
|---|---|
| $0 < t \leqslant 5$ | 1 |
| $0 < t \leqslant 10$ | 6 |
| $0 < t \leqslant 15$ | 18 |
| $0 < t \leqslant 20$ | 28 |
| $0 < t \leqslant 25$ | 30 |

b

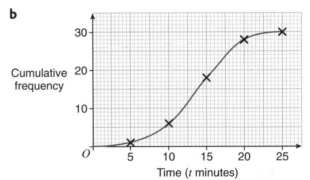

The values at the end of each class interval are 5, 10, 15, 20 and 25  So the coordinates used for plotting this cumulative frequency graph are (5, 1), (10, 6), (15, 18), (20, 28) and (25, 30).

Cumulative frequency graphs can be used to find estimates for the number of items up to a certain value.

## Example 6

40 students took a test.
The cumulative frequency graph gives information about their marks.

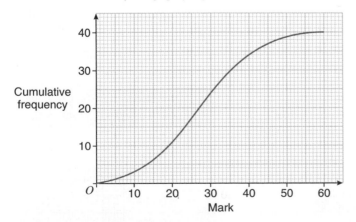

a  Use the graph to find an estimate for the number of students whose mark is *less* than 30

b  Use the graph to work out an estimate for the number of students whose mark is *more* than 45

c  21 students passed the test. Work out the pass mark for the test.

*Solution 6*

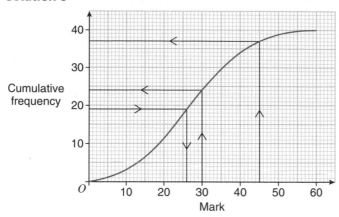

**a** 24 students have a mark less than 30

> Find a mark of 30 on the $x$-axis. Move vertically up to the curve. Read the value off the cumulative frequency axis.

**b** 37 students have a mark less than 45

> Find a mark of 45 on the $x$-axis. Move vertically up to the curve. Read the value off the cumulative frequency axis.

$40 - 37 = 3$
3 students have a mark more than 45

> The number of students whose mark was *more than* 45 is required so subtract 37 from 40

**c** 21 students passed the test
so $40 - 21 = 19$ students did not pass the test.
The pass mark for the test is 26

> Find 19 on the cumulative frequency axis. Move horizontally across to the curve.
> Read the value off the Mark axis.

## Example 7

The grouped frequency table gives information about the number of people in a bus on 30 mornings.

| Number of passengers | Frequency |
|---|---|
| 0–4 | 4 |
| 5–9 | 7 |
| 10–14 | 11 |
| 15–19 | 5 |
| 20–24 | 3 |

**a** Complete the cumulative frequency table.
**b** Draw a cumulative frequency diagram for your table.

| Number of passengers | Cumulative frequency |
|---|---|
| 0–4 | |
| 0–9 | |
| 0–14 | |
| 0–19 | |
| 0–24 | |

**Solution 7**

a

| Number of passengers | Cumulative frequency |
|---|---|
| 0–4 | 4 |
| 0–9 | 11 |
| 0–14 | 22 |
| 0–19 | 27 |
| 0–24 | 30 |

b

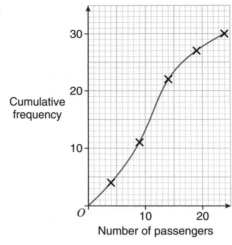

Cumulative frequency

Number of passengers

The data in the table is discrete. The coordinates used for plotting this cumulative frequency graph are (4, 4), (9, 11), (14, 22), (19, 27) and (24, 30).

## Exercise 4B

1 The grouped frequency table gives information about the number of minutes 60 music students practised last week.

| Minutes ($m$) | Frequency |
|---|---|
| $0 < m \leqslant 15$ | 3 |
| $15 < m \leqslant 30$ | 8 |
| $30 < m \leqslant 45$ | 12 |
| $45 < m \leqslant 60$ | 18 |
| $60 < m \leqslant 75$ | 8 |
| $75 < m \leqslant 90$ | 5 |
| $90 < m \leqslant 105$ | 4 |
| $105 < m \leqslant 120$ | 2 |

a Copy and complete the cumulative frequency table.
b Draw a cumulative frequency graph for your table.
c Use your graph to find an estimate for the number of music students who practised for
   i *less* than 40 minutes    ii *more* than 40 minutes.

| Minutes ($m$) | Cumulative frequency |
|---|---|
| $0 < m \leqslant 15$ | |
| $0 < m \leqslant 30$ | |
| $0 < m \leqslant 45$ | |
| $0 < m \leqslant 60$ | |
| $0 < m \leqslant 75$ | |
| $0 < m \leqslant 90$ | |
| $0 < m \leqslant 105$ | |
| $0 < m \leqslant 120$ | |

**2** The grouped frequency table gives information about the time taken for 32 people to solve a mathematics problem.

| Time ($t$ seconds) | Frequency |
|---|---|
| $0 < t \leqslant 10$ | 1 |
| $10 < t \leqslant 20$ | 4 |
| $20 < t \leqslant 30$ | 11 |
| $30 < t \leqslant 40$ | 14 |
| $40 < t \leqslant 50$ | 2 |

  **a** Copy and complete the cumulative frequency table.

  **b** Use your table to draw a cumulative frequency graph.

| Time ($t$ seconds) | Cumulative frequency |
|---|---|
| $0 < t \leqslant 10$ | |
| $0 < t \leqslant 20$ | |
| $0 < t \leqslant 30$ | |
| $0 < t \leqslant 40$ | |
| $0 < t \leqslant 50$ | |

**3** The grouped frequency table gives information about the number of people waiting at a bus stop on 20 mornings.

| Number of people | Frequency |
|---|---|
| 0–5 | 2 |
| 6–11 | 5 |
| 12–17 | 9 |
| 18–23 | 3 |
| 24–29 | 1 |

  **a** Copy and complete the cumulative frequency table.

  **b** Use your table to draw a cumulative frequency graph.

| Number of people | Cumulative frequency |
|---|---|
| 0–5 | |
| 0–11 | |
| 0–17 | |
| 0–23 | |
| 0–29 | |

**4** The graph shows information about the weights in grams of 80 plums.

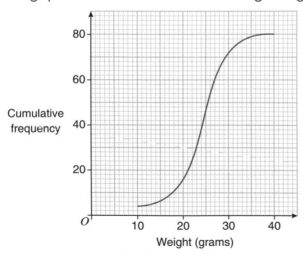

  **a** Use the graph to find an estimate for the number of plums that weigh *less* than 25 grams.

  **b** Use the graph to work out an estimate for the number of plums that weigh *more* than 30 grams.

**5** The cumulative frequency diagram below gives information about the prices of some houses.

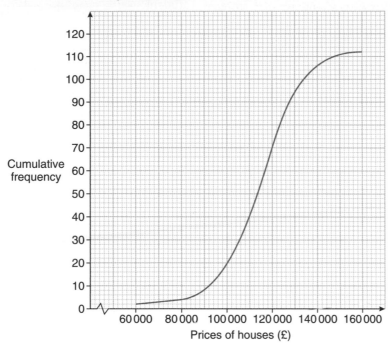

Use the graph to find

**a** the total number of houses

**b** an estimate for the number of houses that cost *less* than £130 000

**c** an estimate for the number of houses that cost *more* than £100 000 but *less* than £150 000

**6** The cumulative frequency graph shows information about the English exam marks of 100 students.

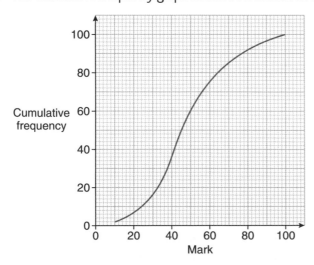

**a** Use the graph to find an estimate for the number of students who scored *less* than 40 marks.

**b** 32 students passed the English exam. Use the graph to work out an estimate for the pass mark.

## Median

The grouped frequency table from the start of Section 4.2 shows information about the amount of time 160 students spent doing homework one evening.

To find the class interval containing the **median** time, the position in the table of the $\left(\frac{n+1}{2}\right)^{th}$ student is needed (see Section 3.3).

The median time is therefore the time of the $80\frac{1}{2}^{th}$ $\left(=\dfrac{160+1}{2}\right)$ student.

To find the class interval containing the $80\frac{1}{2}^{th}$ student use the frequency table along with the **cumulative frequency**.

| Time ($x$ minutes) | Frequency | |
|---|---|---|
| $0 < x \leqslant 10$ | 4 | 4 |
| $10 < x \leqslant 20$ | 12 | 16 |
| $20 < x \leqslant 30$ | 46 | 62 |
| $30 < x \leqslant 40$ | 68 | 130 |
| $40 < x \leqslant 50$ | 20 | 150 |
| $50 < x \leqslant 60$ | 10 | 160 |

62 students had a time of 30 minutes or less.

130 students had a time of 40 minutes or less.

So the $80\frac{1}{2}^{th}$ student will be in the $30 < x \leqslant 40$ class interval.

The class interval containing the median is $30 < x \leqslant 40$

The cumulative frequency graph can be used to find an estimate for the median.

As this graph has been drawn from grouped data the individual time of each student is not known. So it is only possible to find an estimate for the median rather than the actual median.

The estimate for the median will be found from the cumulative frequency graph. The scale makes it difficult to tell the difference between 80 and $80\frac{1}{2}$ on the cumulative frequency axis.

When finding an estimate for the median from a cumulative frequency graph it is acceptable to use the $\left(\dfrac{n}{2}\right)^{th}$ value rather than the $\left(\dfrac{n+1}{2}\right)^{th}$ value since $n$ is large .

To find an estimate for the median in this example use the graph to find the time of the $80^{th}$ $\left(=\dfrac{160}{2}\right)$ student.

## Quartiles and interquartile range

Quartiles and interquartile range were introduced in Section 3.3
Estimates for the **lower quartile** and **upper quartile** can be read off a cumulative frequency graph.

The values for the upper and lower quartiles are estimates so the $(\frac{1}{4} \times 160 =)$ $40^{th}$ value can be used to give an estimate for the lower quartile and the $(\frac{3}{4} \times 160 =)$ $120^{th}$ value can be used to give an estimate for the upper quartile.

When finding estimates from a cumulative frequency graph for the median, lower quartile and upper quartile, for a large data set containing $n$ values read off the cumulative frequency axis at

- $\frac{1}{2}n$ for the median
- $\frac{1}{4}n$ for the lower quartile
- $\frac{3}{4}n$ for the upper quartile.

The estimates for the lower quartile and the upper quartile from a cumulative frequency graph can be used to find an estimate for the **interquartile range**.

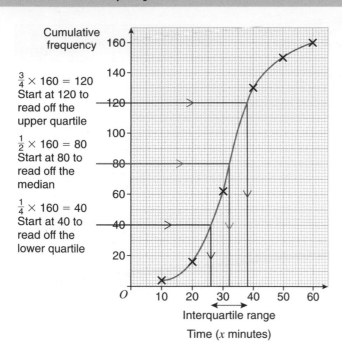

Cumulative frequency

$\frac{3}{4} \times 160 = 120$
Start at 120 to read off the upper quartile

$\frac{1}{2} \times 160 = 80$
Start at 80 to read off the median

$\frac{1}{4} \times 160 = 40$
Start at 40 to read off the lower quartile

Interquartile range

Time ($x$ minutes)

For this example an estimate for the

- **lower quartile** is **26**
- **median** is **32**
- **upper quartile** is **38**

Interquartile range = **38** − **26** = 12

## Exercise 4C

**1** Use the graph in Exercise 4B question **4** to find estimates for the
  **i** median　　　　　　　　**ii** lower quartile　　　　　　**iii** upper quartile.

**2** Use the graph in Exercise 4B question **5** to find estimates for the
  **i** median　　　　　　　　**ii** lower quartile　　　　　　**iii** upper quartile.

**3** Use the graph in Exercise 4B question **6** to find estimates for the
  **i** median　　　　　　　　**ii** interquartile range.

**4** 80 workers were asked what distance they travelled to get to work. The grouped frequency table shows this information.

  **a** Copy and complete the cumulative frequency table.

  **b** Find the class interval in which the median lies.

  **c** Use your table to draw a cumulative frequency graph.

  **d** Use your cumulative frequency graph to find an estimate for the median.

| Distance ($d$ kilometres) | Frequency |
|---|---|
| $0 < d \leqslant 5$ | 4 |
| $5 < d \leqslant 10$ | 25 |
| $10 < d \leqslant 15$ | 29 |
| $15 < d \leqslant 20$ | 13 |
| $20 < d \leqslant 25$ | 7 |
| $25 < d \leqslant 30$ | 2 |

| Distance ($d$ kilometres) | Cumulative Frequency |
|---|---|
| $0 < d \leqslant 5$ | |
| $0 < d \leqslant 10$ | |
| $0 < d \leqslant 15$ | |
| $0 < d \leqslant 20$ | |
| $0 < d \leqslant 25$ | |
| $0 < d \leqslant 30$ | |

  **e** Use your cumulative frequency graph to find an estimate for the interquartile range.

**5** The table gives information about the amount of time in minutes 100 adults spent preparing last night's meal.

    **a** Use the information in the table to draw a cumulative frequency graph.

    **b** Use your cumulative frequency graph to find estimates for the

      **i** lower quartile

      **ii** upper quartile

      **iii** interquartile range.

| Time ($t$ minutes) | Cumulative frequency |
|---|---|
| $0 < t \leqslant 10$ | 3 |
| $0 < t \leqslant 20$ | 19 |
| $0 < t \leqslant 30$ | 66 |
| $0 < t \leqslant 40$ | 90 |
| $0 < t \leqslant 50$ | 97 |
| $0 < t \leqslant 60$ | 100 |

**6** The table gives information about the luggage weight of 200 passengers.

    **a** Use the information in the table to draw a cumulative frequency graph.

    **b** Use your cumulative frequency graph to find estimates for the

      **i** number of passengers whose luggage weighs *more* than 15 kg

      **ii** median

      **iii** interquartile range.

| Luggage weight ($w$ kg) | Cumulative frequency |
|---|---|
| $0 < w \leqslant 4$ | 6 |
| $0 < w \leqslant 8$ | 22 |
| $0 < w \leqslant 12$ | 83 |
| $0 < w \leqslant 16$ | 155 |
| $0 < w \leqslant 20$ | 196 |
| $0 < w \leqslant 24$ | 200 |

## 4.3 Box plots

**Box plots** (sometimes called box and whisker diagrams) are diagrams that show the spread of a set of data. The median, lower and upper quartiles along with the minimum and maximum value are used to draw a box plot.

For example, for some data about times (in minutes) the

- minimum value = 8
- maximum value = 57
- lower quartile = 26
- median = 32
- upper quartile = 38

Here is the box plot for this data.

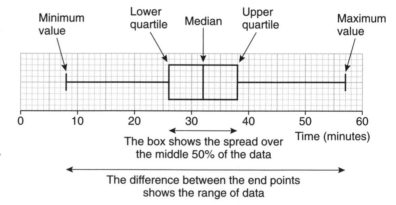

The box shows the spread over the middle 50% of the data

The difference between the end points shows the range of data

---

*Example 8*

The times in seconds taken by 15 students to solve a problem are listed in order.

    5  6  9  11  17  21  22  23  25  26  27  27  29  30  31

Draw a box plot for this data.

*Solution 8*

Median is the $\dfrac{(15 + 1)}{2} = 8^{\text{th}}$ number

> The numbers are in order. $n$ is not large so the median is the $\frac{1}{2}(n + 1)^{\text{th}}$ value.

    5    6    9   (11)   17   21   22   (23)   25   26   27   (27)   29   30   31

Median = 23

Lower quartile is the $\frac{1}{4} \times (15 + 1) = 4^{th}$ number

**Lower quartile = 11**

> Lower quartile is the $\frac{1}{4}(n + 1)^{th}$ value.

Upper quartile is the $\frac{3}{4} \times (15 + 1) = 12^{th}$ number

**Upper quartile = 27**

> Upper quartile is the $\frac{3}{4}(n + 1)^{th}$ value.

Minimum value = 5

Maximum value = 31

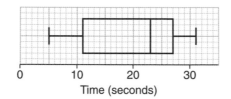

Time (seconds)

## 4.4 Comparing distributions

Box plots are useful for comparing the distribution of data sets.

Example 9

80 seedlings were divided into 2 groups.

Group A were grown in a greenhouse.
Group B were grown outside.

|  | Group A | Group B |
|---|---|---|
| Shortest seedling (cm) | 1.6 | 0.3 |
| Tallest seedling (cm) | 4.4 | 3.8 |

After a period of time the heights of the seedlings were measured.
The heights were used to draw two cumulative frequency graphs.

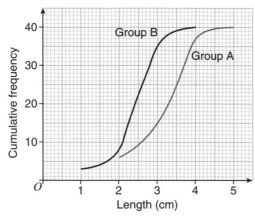

Length (cm)

**a** Use the information provided in the table and the cumulative frequency graphs to draw a box plot of the heights of seedlings in group A and a box plot of the heights of the seedlings in group B.

**b** Compare the heights of the seedlings in the two groups.

*Solution 9*

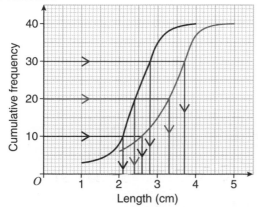

Length (cm)

|  | Group A | Group B |
|---|---|---|
| Shortest seedling (cm) | 1.6 | 0.3 |
| Tallest seedling (cm) | 4.4 | 3.8 |
| **Lower quartile** | **2.6** | **2.1** |
| Median | 3.3 | 2.4 |
| **Upper quartile** | **3.7** | **2.8** |

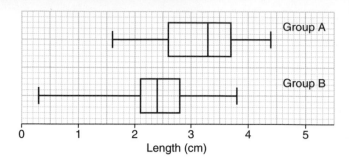

Length (cm)

The heights of the seedlings in group B are more spread out than the heights of the seedlings in group A.

| Range for A $4.4 - 1.6 = 2.8$ cm. |
| Range for B $3.8 - 0.3 = 3.5$ cm. |

The seedlings in group A are generally taller than the seedlings in group B.

| Median for A 3.3 cm. |
| Median for B 2.4 cm. |

The middle 50% of the seedlings in group A have a wider spread than the middle 50% of the seedlings in group B.

| Interquartile range for A $3.7 - 2.6 = 1.1$ |
| Interquartile range for B $2.8 - 2.1 = 0.7$ |

## Exercise 4D

**1** Sarah measured the lengths in centimetres of the hands of some of her friends.
The table shows some information about the lengths.
Draw a box plot to show this information.

| Length of shortest hand | 15.2 cm |
|---|---|
| Lower quartile | 16.7 cm |
| Median | 17.5 cm |
| Upper quartile | 18.1 cm |
| Length of longest hand | 19.8 cm |

**2** Some students took a test.
The table shows information about their marks.
Use this information to draw a box plot.

| Minimum mark | 13 |
|---|---|
| Lower quartile | 29 |
| Interquartile range | 32 |
| Median mark | 41 |
| Range | 53 |

**3** The number of letters delivered to an office for each of 11 days are listed in order.

9    12    13    18    21    21    24    26    28    30    31

**a** Find **i** the median **ii** the interquartile range.
**b** Draw a box plot for this data.

**4** Ahmed recorded the heights of some of his friends. He used some of this information to draw a box plot.

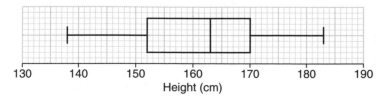

Height (cm)

Copy and complete the table.

| Minimum height | |
|---|---|
| Lower quartile | |
| Median height | |
| Upper quartile | |
| Maximum height | |

**5** The box plot shows some information about the number of patients seen in a day by some dentists.

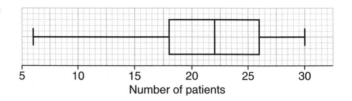

Number of patients

  **a** Write down the lower quartile.

  **b** Work out the interquartile range.

**6** Students in class 8P and class 8Q took the same test. Their results were used to draw the following box plots.

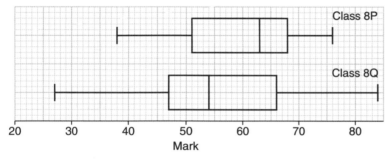

Mark

  **a** In which class was the student who scored the highest mark?

  **b** In which class did the students perform better in the test? You must give a reason for your answer.

**7** 60 boys and 60 girls each answered a number of questions. The information about the times in seconds that they took was used to draw two cumulative frequency graphs.

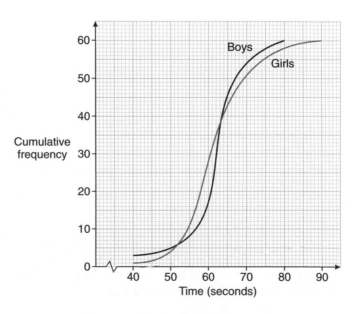

Time (seconds)

The table shows the minimum and maximum time taken by the boys and the girls to complete the questions.

|  | Girls | Boys |
|---|---|---|
| Minimum time to complete questions (seconds) | 35 | 37 |
| Maximum time to complete questions (seconds) | 85 | 76 |

**a** Draw a box plot to represent the girls' times and a box plot to represent the boys' times.

**b** Make two comparisons between the girls' times and the boys' times.

## 4.5 Frequency density and histograms

Section 4.1 introduced histograms. All the histograms drawn in Section 4.1 had class intervals of equal width and so the bars were of equal width.

Histograms can be drawn with unequal class intervals. The vertical axis is labelled **frequency density** where frequency density $= \dfrac{\text{frequency}}{\text{class width}}$

For example, the table gives some information about the ages of the audience at a concert.

| Age ($x$) in years | Frequency |
|---|---|
| $0 < x \leqslant 15$ | 12 |
| $15 < x \leqslant 25$ | 66 |
| $25 < x \leqslant 35$ | 90 |
| $35 < x \leqslant 40$ | 45 |
| $40 < x \leqslant 70$ | 60 |

To draw a histogram to represent this information

- work out the width of each class interval (the class width)
- divide the frequency by the class width to find the frequency density which gives the height of each bar.

| Age ($x$) in years | Frequency | Class width | Frequency density $= \dfrac{\text{frequency}}{\text{class width}}$ |
|---|---|---|---|
| $0 < x \leqslant 15$ | 12 | $15 - 0 = 15$ | $\frac{12}{15} = 0.8$ |
| $15 < x \leqslant 25$ | 66 | $25 - 15 = 10$ | $\frac{66}{10} = 6.6$ |
| $25 < x \leqslant 35$ | 90 | $35 - 25 = 10$ | $\frac{90}{10} = 9$ |
| $35 < x \leqslant 40$ | 45 | $40 - 35 = 5$ | $\frac{45}{5} = 9$ |
| $40 < x \leqslant 70$ | 60 | $70 - 40 = 30$ | $\frac{60}{30} = 2$ |

On a grid, label the horizontal axis 'Age (years)' and the vertical axis 'Frequency density'.

Scale the horizontal axis from 0 to 70 and the vertical axis from 0 to 10

Draw the bars with no gaps between them. The first bar goes from 0 to 15 and has a height of 0.8 The second bar goes from 15 to 25 and has a height of 6.5 and so on.

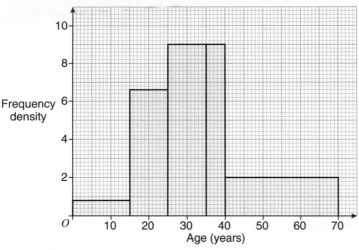

Rearranging

$$\text{frequency density} = \frac{\text{frequency}}{\text{class width}} \text{ gives}$$

$$\text{frequency} = \text{class width} \times \text{frequency density}$$

For each bar the 'width' is the class width and the 'height' is the frequency density.
So the area of each bar gives the frequency.

## Example 10

Use the information in the frequency table and the histogram to complete the frequency table.

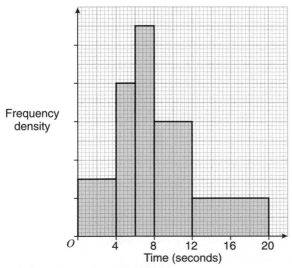

| Time ($t$ seconds) | Frequency |
|---|---|
| $0 < t \leqslant 4$ | |
| $4 < t \leqslant 6$ | 16 |
| $6 < t \leqslant 8$ | |
| $8 < t \leqslant 12$ | |
| $12 < t \leqslant 20$ | |

*Solution 10*
**Method 1**
Frequency density $= \frac{16}{2} = 8$

The frequency for the class interval $4 < t \leqslant 6$ is given as 16

The width of this class interval is 2 $(= 6 - 4)$

Frequency density $= \dfrac{\text{frequency}}{\text{class width}}$

That is, the height of this bar $= 8$

The vertical scale can now be inserted so that the frequency density for each bar can be read off.
For each class interval, work out class width $\times$ frequency density to give the frequency.

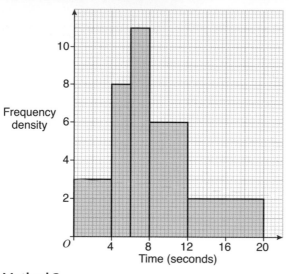

| Time (t seconds) | Class width | Frequency density | Frequency |
|---|---|---|---|
| 0 < t ≤ 4 | 4 | 3 | (4 × 3 =) 12 |
| 4 < t ≤ 6 | 2 | 8 | 16 |
| 6 < t ≤ 8 | 2 | 11 | (2 × 11 =) 22 |
| 8 < t ≤ 12 | 4 | 6 | (4 × 6 =) 24 |
| 12 < t ≤ 20 | 8 | 2 | (8 × 2 =) 16 |

represents 1 cm²

## Method 2

The frequency for the class interval $4 < t \leq 6$ is given as 16

On the grid when drawn to full size, the area of this bar is 8 cm²

So an area of 8 cm² represents a frequency of 16

An area of 1 cm² represents a frequency of $\frac{16}{8} = 2$

For the class interval $0 < t \leq 4$ the area of the bar is 6 cm² so the frequency for this class interval is $6 \times 2 = 12$

For the class interval $6 < t \leq 8$ the area of the bar is 11 cm² so the frequency for this class interval is $11 \times 2 = 22$

For the class interval $8 < t \leq 12$ the area of the bar is 12 cm² so the frequency for this class interval is $12 \times 2 = 24$

For the class interval $12 < t \leq 20$ the area of the bar is 8 cm² so the frequency for this class interval is $8 \times 2 = 16$

A key is sometimes used with a histogram to show how many data items an area on the histogram represents.

## Example 11

This histogram shows information about the distances in metres that a number of people threw a ball. Complete the table.

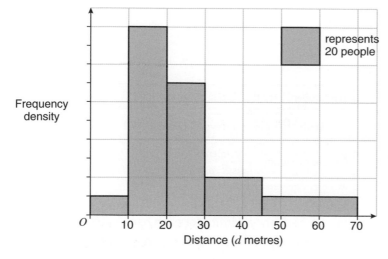

represents 20 people

| Distance (d metres) | Frequency |
|---|---|
| 0 < d ≤ 10 | |
| 10 < d ≤ 20 | |
| 20 < d ≤ 30 | |
| 30 < d ≤ 45 | |
| 45 < d ≤ 70 | |

*Solution 11*
Each square represents 20 people.
To work out the frequency for each class interval

- find the number of squares in each bar
- multiply the number of squares by 20

| Distance ($d$ metres) | Number of squares | Frequency |
|---|---|---|
| $0 < d \leqslant 10$ | $\frac{1}{2}$ | $\frac{1}{2} \times 20 = 10$ |
| $10 < d \leqslant 20$ | 5 | $5 \times 20 = 100$ |
| $20 < d \leqslant 30$ | $3\frac{1}{2}$ | $3\frac{1}{2} \times 20 = 70$ |
| $30 < d \leqslant 45$ | $1\frac{1}{2}$ | $1\frac{1}{2} \times 20 = 30$ |
| $45 < d \leqslant 70$ | $1\frac{1}{4}$ | $1\frac{1}{4} \times 20 = 25$ |

## Exercise 4E

**1** The histogram shows information about the number of hours of television some students watched one evening.

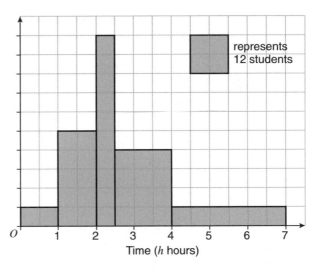

Copy and complete the frequency table.

| Time ($h$ hours) | Frequency |
|---|---|
| $0 < h \leqslant 1$ | |
| $1 < h \leqslant 2$ | |
| $2 < h \leqslant 2\frac{1}{2}$ | |
| $2\frac{1}{2} < h \leqslant 4$ | |
| $4 < h \leqslant 7$ | |

**2** The histogram shows information about the ages of the customers in a restaurant one evening. Copy and complete the frequency table.

| Age ($x$ years) | Frequency |
|---|---|
| $20 < x \leqslant 25$ | |
| $25 < x \leqslant 30$ | 25 |
| $30 < x \leqslant 40$ | |
| $40 < x \leqslant 55$ | |
| $55 < x \leqslant 65$ | |

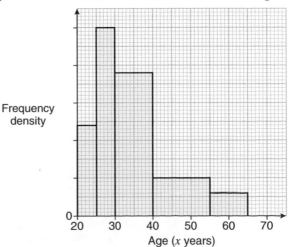

**3** The table gives information about the amount of time, in minutes, taken by some passengers to check in at the airport.
Draw a histogram to illustrate this information.

| Time (*m* minutes) | Frequency |
|---|---|
| $0 < m \leqslant 20$ | 16 |
| $20 < m \leqslant 40$ | 44 |
| $40 < m \leqslant 50$ | 55 |
| $50 < m \leqslant 60$ | 62 |
| $60 < m \leqslant 90$ | 18 |

**4** The table gives information about the lengths, in centimetres, of 45 leaves.
Draw a histogram to illustrate this information.

| Length (*x* centimetres) | Frequency |
|---|---|
| $5 < x \leqslant 6$ | 3 |
| $6 < x \leqslant 7$ | 9 |
| $7 < x \leqslant 7.5$ | 15 |
| $7.5 < x \leqslant 8$ | 14 |
| $8 < x \leqslant 10$ | 4 |

**5** The table gives information about the times, in minutes, that 60 runners took to complete a cross country race.
Draw a histogram to illustrate this information.

| Time (*x* minutes) | Frequency |
|---|---|
| $10 < x \leqslant 11$ | 4 |
| $11 < x \leqslant 12$ | 5 |
| $12 < x \leqslant 14$ | 21 |
| $14 < x \leqslant 16$ | 22 |
| $16 < x \leqslant 20$ | 8 |

**6** The histogram gives information about the weights of some apples.
The shaded bar represents 40 apples.
   **a** Work out how many of the apples weigh 100 grams or less.
   **b** Work out how many of the apples weigh more than 110 grams.

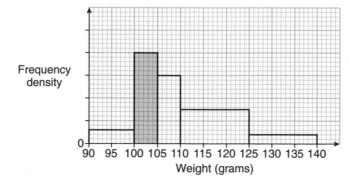

# Chapter summary

**You should now know:**

★ that **frequency diagrams** can be used to represent grouped discrete data

★ how to draw and interpret frequency diagrams

★ that **histograms** can be used to represent continuous data

★ how to draw and interpret histograms

★ that **frequency polygons** can be used to compare two or more sets of data

★ how to draw frequency polygons

★ that the **cumulative frequency** is the running total of the frequency

★ how to draw a **cumulative frequency graph** from a cumulative frequency table

★ how to use a **cumulative frequency graph** to find estimates for the lower quartile, median and upper quartile of a data set

★ that a **box plot** is a diagram showing the minimum, maximum, lower quartile, median and upper quartile of a set of data

★ how to use box plots to compare two or more sets of data

★ that in a **histogram** the area of a bar gives the frequency of the class interval

★ how to work out the **frequency densities** for a histogram, using

$$\text{frequency density} = \frac{\text{frequency}}{\text{class width}}$$

# Chapter 4 review questions

1 The table gives information about the number of computer games sold by a shop each day for a month. Draw a frequency diagram to illustrate this information.

| Number of computer games | Frequency |
|---|---|
| 0–4 | 5 |
| 5–9 | 8 |
| 10–14 | 9 |
| 15–20 | 6 |
| 21–24 | 2 |

2 The table shows the frequency distribution of student absences for a year.

On the resource sheet draw a frequency polygon for this frequency distribution.

| Absences (d days) | Frequency |
|---|---|
| $0 \leqslant d < 5$ | 4 |
| $5 \leqslant d < 10$ | 6 |
| $10 \leqslant d < 15$ | 8 |
| $15 \leqslant d < 20$ | 5 |
| $20 \leqslant d < 25$ | 4 |
| $25 \leqslant d < 30$ | 3 |

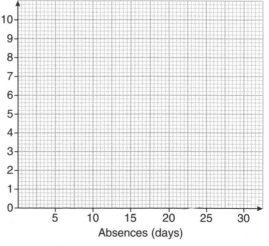

(1385 June 2001)

**3** A railway company wanted to show the improvements in its train service over 3 years.
This graph was drawn.

Explain why this graph may be misleading.

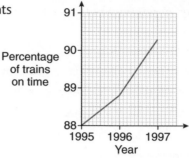

(1385 November 2001)

**4** 72 students took an examination.
The grouped frequency table shows information about their results.

| Mark ($x$) | Cumulative frequency |
|---|---|
| $0 < x \leqslant 10$ | |
| $0 < x \leqslant 20$ | |
| $0 < x \leqslant 30$ | |
| $0 < x \leqslant 40$ | |
| $0 < x \leqslant 50$ | |
| $0 < x \leqslant 60$ | |

| Mark ($x$) | Frequency |
|---|---|
| $0 < x \leqslant 10$ | 2 |
| $10 < x \leqslant 20$ | 9 |
| $20 < x \leqslant 30$ | 19 |
| $30 < x \leqslant 40$ | 28 |
| $40 < x \leqslant 50$ | 11 |
| $50 < x \leqslant 60$ | 3 |

**a** Complete the cumulative frequency table.

**b** Draw a cumulative frequency graph for your table.

**c** Use your graph to find an estimate for the median mark.

The pass mark for the examination was 22

**d** Use your graph to find an estimate for the number of students who passed the examination.

**5** 60 office workers recorded the number of words per minute they could type.
The grouped frequency table gives information about the number of words per minute they could type.

| Number of words ($w$) per minute | Frequency |
|---|---|
| $0 \leqslant w < 20$ | 6 |
| $20 \leqslant w < 40$ | 18 |
| $40 \leqslant w < 60$ | 16 |
| $60 \leqslant w < 80$ | 15 |
| $80 \leqslant w < 100$ | 3 |
| $100 \leqslant w < 120$ | 2 |

**a** Find the class interval in which the median lies.

The cumulative frequency graph for this information has been drawn on the grid.

**b** Use this graph to work out an estimate for the interquartile range of the number of words per minute.

**c** Use this graph to work out an estimate for the number of workers who could type *more* than 70 words per minute.

(1388 January 2005)

**6** Mary recorded the heights in centimetres of the girls in her class.
She put the heights in order.

    132   144   150   152   160   162   162   167
    167   170   172   177   181   182   182

**a** Find **i** the lower quartile   **ii** the upper quartile.

**b** On the grid on the resource sheet draw a box plot for this data.

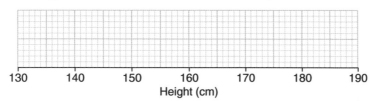

Height (cm)

(1387 June 2003)

**7** 30 students took part in a National Science quiz. The quiz was in two parts.
The cumulative frequency graph on the grid below gives information about the marks scored in Part One.
The lowest mark was 5 and the highest mark was 47

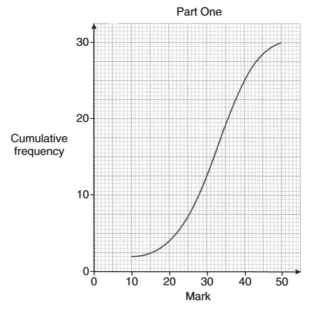

**a** Use the copy of the grid on the resource sheet to draw a box plot using the cumulative frequency graph for the results of Part One.

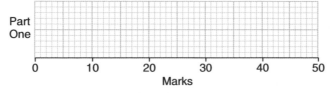

Below there is also a box plot for the results of Part Two.
Use the box plots to compare the two distributions.

**b** Give **two** differences between them.

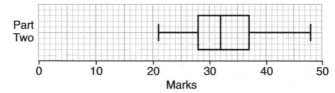

(1387 November 2005)

**8** The table gives information about the heights, in centimetres, of some 16 year old students.

Use the table to draw a histogram.

| Height ($h$ cm) | Frequency |
|---|---|
| $145 < h \leqslant 155$ | 15 |
| $155 < h \leqslant 175$ | 60 |
| $175 < h \leqslant 190$ | 18 |
| $190 < h \leqslant 195$ | 2 |

**9** A teacher asked some students how much time they spent using a mobile phone one week. The histogram was drawn from this information.

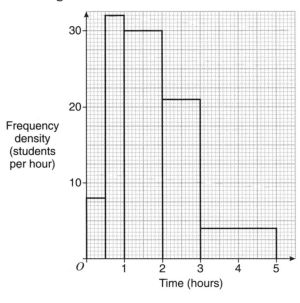

Use the histogram to complete the table.

| Time ($t$) hours | Frequency |
|---|---|
| $0 \leqslant t < \frac{1}{2}$ | |
| $\frac{1}{2} \leqslant t < 1$ | |
| $1 \leqslant t < 2$ | 30 |
| $2 \leqslant t < 3$ | |
| $3 \leqslant t < 5$ | |

(1388 March 2005)

# Probability

## Favourites to seize the Olympic flame

As the day for decision approaches it seems **unlikely** that London will win the battle to host the 2012 Olympic Games. The **probability** that Paris will win this race has always been high. It is felt that Madrid, Moscow and New York have little **chance** of success as the final presentations are made.

## London defy all the odds

London won with their bid to host the 2012 Summer Olympic Games. Yesterday's vote saw **likely** winners Paris stumble at the final hurdle. A spokesperson said 'Everyone thought that Paris was **certain** to win the vote but I always felt that we had a greater than **even chance** of success.'

## 5.1 Writing probabilities as numbers

The diagram shows a three-sided spinner.

The spinner can land on red or blue or yellow. If it is equally likely to land on each of the three colours the spinner is said to be **fair**.

This spinner, which is fair, is spun once. This is called a single **event**.

The colour it lands on is called the **outcome**.

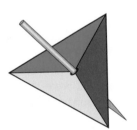

The outcome can be red or blue or yellow. There are three possible outcomes and each possible outcome is equally likely.

The **probability** of an outcome to an event is a measure of how likely it is that the outcome will happen.

The probability that the spinner will land on blue $= \dfrac{1 \text{ successful outcome}}{3 \text{ possible outcomes}} = \dfrac{1}{3}$

Similarly the probability that the spinner will land on red $= \dfrac{1}{3}$

and the probability that the spinner will land on yellow $= \dfrac{1}{3}$

When all the possible outcomes are equally likely to happen

$$\text{probability} = \frac{\text{number of successful outcomes}}{\text{total number of possible outcomes}}$$

Probability can be written as a fraction or a decimal.

For an event:
- the probability of an outcome which is **certain** to happen is 1   For example the probability that the spinner will land on red or blue or yellow is 1 since the spinner is certain to land on one of these three colours
- the probability of an outcome which is **impossible** is 0   For example the probability that the spinner will land on green is 0 since green is not a colour on the spinner
- all other probabilities lie between 0 and 1

## Example 1

A fair five-sided spinner is numbered 1 to 5
Jane spins the spinner once.

**a** Find the probability that the spinner will land on the number 4

**b** Find the probability that the spinner will land on an even number.

### Solution 1

**a** The possible outcomes are the numbers 1, 2, 3, 4 and 5  So the total
number of possible outcomes is **5** and they are all equally likely.
The **1** successful outcome is the number 4

$$\text{Probability} = \frac{\text{number of successful outcomes}}{\text{total number of possible outcomes}}$$

Probability that the spinner will land on the number $4 = \frac{1}{5}$ or 0.2

**b** 2 and 4 are the even numbers on the spinner.
The number of successful outcomes is **2**
The total number of possible outcomes is **5**
So the probability that the spinner will land on an even number $= \frac{2}{5}$ or 0.4

In the following example the term **at random** is used. This means that each possible outcome is
equally likely.

## Example 2

Six coloured counters are in a bag.
3 counters are red, 2 counters are green
and 1 counter is blue.

One counter is taken at random from the bag.

**a** Write down the colour of the counter which is
  **i** most likely to be taken          **ii** least likely to be taken

**b** Find the probability that the counter taken will be
  **i** red          **ii** green          **iii** blue

### Solution 2

**a**  **i** Red is the most likely colour to be taken since the number of red counters is greater than the
number of counters of any other colour.
  **ii** Blue is least likely to be taken since the number of blue counters is less than the number of
counters of any other colour.

**b** There are 6 counters so there are 6 possible outcomes.

Outcome    1      2      3      4      5      6

  **i** Number of successful outcomes = 3 (3 red counters)
Probability that the counter taken will be red $= \frac{3}{6} = \frac{1}{2}$
  **ii** Number of successful outcomes = 2 (2 green counters)
Probability that the counter taken will be green $= \frac{2}{6} = \frac{1}{3}$
  **iii** Number of successful outcomes = 1 (1 blue counter)
Probability that the counter taken will be blue $= \frac{1}{6}$

> Final answers for probabilities
> written as fractions should be
> given in their simplest form.

## Exercise 5A

**1** Nicky spins the spinner. The spinner is fair. Write down the probability that the spinner will land on a side coloured

   **a** blue            **b** red            **c** green

**2** John spins the fair spinner. Write down the probability that the spinner will land on

   **a** 2                   **b** a number greater than 5

   **c** an even number       **d** a number greater than 10

**3** Samantha Smith has 8 cards which spell 'Samantha'.

She puts the cards in a bag and chooses one of the cards at random.
Find the probability that she will choose a card showing a

   **a** letter S          **b** letter A          **c** letter which is also in her surname 'SMITH'

**4** Ben has 15 ties in a drawer. 7 of the ties are plain, 3 of the ties are striped and the rest are patterned. Ben chooses a tie at random from the drawer. What is the probability that he chooses a tie which is

   **a** plain          **b** striped          **c** patterned?

**5** Peter has a bag of 8 coins. In the bag he has one 10p coin, five 20p coins and the rest are 50p coins. Peter chooses one coin at random. What is the probability that Peter will choose a

   **a** 10p coin       **b** 20p coin       **c** 50p coin

   **d** £1 coin         **e** coin worth more than 5p?

**6** Rob has a drawer of 20 socks. 4 of the socks are blue, 6 of the socks are brown and the rest of the socks are black. Rob chooses a sock at random from the drawer. Find the probability that he chooses

   **a** a blue sock     **b** a brown sock     **c** a black sock     **d** a white sock

**7** Verity has a box of pens. Half of the pens are blue, 11 of the pens are green, 10 of the pens are red and the remaining 4 pens are black. Verity chooses a pen at random from the box. Find the probability that she chooses

   **a** a blue pen     **b** a green pen     **c** a red pen     **d** a black pen

## 5.2 Sample space diagrams

A **sample space** is all the possible outcomes of one or more events.

A **sample space diagram** is a diagram which shows the sample space.

For the three-sided spinner, the sample space when the spinner is spun once is
1   2   3

## Example 3

The three-sided spinner is spun and a coin is tossed at the same time.
Write down the sample space of all possible outcomes.

### Solution 3
There are 6 possible outcomes. For example the spinner landing on 1 and the coin showing heads is written as (1, head). The sample space is

| (1, head) | (1, tail) | (2, head) | (2, tail) | (3, head) | (3, tail) |

## Example 4

Two fair dice are thrown.

**a** Write down the sample space showing all the possible outcomes.
**b** Find the probability that the numbers on the two dice will be
   **i** both the same       **ii** both even numbers       **iii** both less than 3

### Solution 4
**a** (1, 1)  (2, 1)  (3, 1)  (4, 1)  (5, 1)  (6, 1)
   (1, 2)  (2, 2)  (3, 2)  (4, 2)  (5, 2)  (6, 2)
   (1, 3)  (2, 3)  (3, 3)  (4, 3)  (5, 3)  (6, 3)
   (1, 4)  (2, 4)  (3, 4)  (4, 4)  (5, 4)  (6, 4)
   (1, 5)  (2, 5)  (3, 5)  (4, 5)  (5, 5)  (6, 5)
   (1, 6)  (2, 6)  (3, 6)  (4, 6)  (5, 6)  (6, 6)

   The total number of possible outcomes is 36

**b** **i** (1, 1) (2, 2) (3, 3) (4, 4) (5, 5) and (6, 6) are the successful outcomes with both numbers the same.
     Probability that the numbers on the two dice will be both the same $= \frac{6}{36} = \frac{1}{6}$

  **ii** (2, 2) (2, 4) (2, 6) (4, 2) (4, 4) (4, 6) (6, 2) (6, 4) (6, 6) are the successful outcomes with both numbers even.
     Probability that the numbers on the two dice will both be even numbers $= \frac{9}{36} = \frac{1}{4}$

 **iii** (1, 1) (2, 1) (1, 2) (2, 2) are the successful outcomes with both numbers less than 3
     Probability that the numbers on the two dice will both be less than 3 is $\frac{4}{36} = \frac{1}{9}$

### Exercise 5B

In each of the questions in this exercise give all probabilities as fractions in their simplest forms.

**1** Two coins are spun at the same time.
  **a** Write down a sample space to show all possible outcomes.
  **b** Find the probability that both coins will come down heads.
  **c** Find the probability that one coin will come down heads and the other coin will come down tails.

**2** A bag contains 1 blue brick, 1 yellow brick, 1 green brick and 1 red brick.
  A brick is taken at random from the bag and its colour noted.
  The brick is then replaced in the bag.
  A brick is again taken at random from the bag and its colour noted.
  **a** Write down a sample space to show all the possible outcomes.
  **b** Find the probability that
    **i** the two bricks will be the same colour
   **ii** one brick will be red and the other brick will be green

**3** Two fair dice are thrown. The sample space is shown in Example 4
The numbers on the two dice are added together.

    **a** Find the probability that the sum of the numbers on the two dice will be

       **i** greater than 10        **ii** less than 6        **iii** a square number.

    **b**  **i** Which sum of the numbers on the two dice is most likely to occur?

       **ii** Find the probability of this sum.

**4** Daniel has four cards, the ace of hearts, the ace of diamonds, the ace of spades and the ace of clubs.
Daniel also has a fair dice.
He rolls the dice and takes a card at random.

    **a** Write down the sample space showing all possible outcomes.
One possible outcome, ace of Diamonds and 4 has been done for you, (D, 4).

    **b** Find the probability that a red ace will be taken.

    **c** Find the probability that he will take the ace of spades and roll an even number on the dice.

**5** Three fair coins are spun.

    **a** Draw a sample space showing all eight possible outcomes.

    **b** Find the probability that the three coins will show the same.

    **c** Find the probability that the coins will show two heads and a tail.

    **d** Write down the total number of possible outcomes when

       **i** four coins are spun        **ii** five coins are spun

## 5.3 Mutually exclusive outcomes and the probability that the outcome of an event will not happen

Nine coloured counters are in a bag.

3 counters are red, 2 counters are green and 4 counters are yellow.
One counter is chosen at random from the bag.
The probability that the counter will be red, $P(\text{red}) = \frac{3}{9}$

> Notation: 'P(red)' means the probability of red.

$P(\text{green}) = \frac{2}{9}$    $P(\text{yellow}) = \frac{4}{9}$

**Mutually exclusive outcomes** are outcomes which cannot happen at the same time.

For example when one counter is chosen at random from the bag the outcome 'red' cannot happen at the same time as the outcome 'green' or the outcome 'yellow'. So the three outcomes are mutually exclusive.

$P(\text{red}) + P(\text{green}) + P(\text{yellow}) = \frac{3}{9} + \frac{2}{9} + \frac{4}{9} = \frac{9}{9} = 1$

> **The sum of the probabilities of all the possible mutually exclusive outcomes of an event is 1**

There are 9 possible outcomes, 2 of which are green.
The probability that the counter will be green is $\frac{2}{9}$
Out of the 9 possible outcomes $9 - 2 = 7$ outcomes are NOT green.
The probability that the counter will NOT be green is $1 - \frac{2}{9} = \frac{7}{9}$

> **If the probability of an outcome of an event happening is $p$ then the probability of it NOT happening is $1 - p$**

If the counter is not green it must be either red or yellow.

So, P(not green) = P(either red or yellow)

The probability that the counter will be either red or yellow is $\frac{7}{9}$

$\frac{7}{9} = \frac{3}{9} + \frac{4}{9} = $ P(red) + P(yellow).

So P(either red or yellow) = P(red) + P(yellow)

In general when two outcomes A and B, of an event are mutually exclusive

> **P(A or B) = P(A) + P(B)**

This can be used as a quicker way of solving some problems.

## Example 5

David buys one newspaper each day. He buys the *Times* or the *Telegraph* or the *Independent*. The probability that he will buy the *Times* is 0.6   The probability that he will buy the *Telegraph* is 0.25

**a**  Work out the probability that David will buy the *Independent*.

**b**  Work out the probability that David will buy either the *Times* or the *Telegraph*.

*Solution 5*

**a**  P(*Times*) = 0.6
     P(*Telegraph*) = 0.25
     P(*Independent*) = ?

> P(*Times*) means the probability that David will buy the *Times*.

As David buys only one newspaper each day, the three outcomes are mutually exclusive.

P(*Independent*) + 0.6 + 0.25 = 1
P(*Independent*) + 0.85 = 1
P(*Independent*) = 1 − 0.85

The probability that David will buy the *Independent* = 0.15

**b**  P(*Times* or *Telegraph*) = P(*Times*) + P(*Telegraph*)
                          =      0.6   +      0.25          = 0.85

The probability that David will buy either the *Times* or the *Telegraph* = 0.85

## Example 6

The probability that Julie will pass her driving test next week is 0.6
Work out the probability that Julie will *not* pass her driving test next week.

*Solution 6*
The probability that Julie will *not* pass = 1 − 0.6 = 0.4

### Exercise 5C

**1** Nosheen travels from home to school. She travels by bus or by car or by tram. The probability that she travels by bus is 0.4   The probability that she travels by car is 0.5

  **a**  Work out the probability that she travels by tram.

  **b**  Work out the probability that she travels by car or by bus.

**2** Roger's train can be on time or late or early. The probability that his train will be on time is 0.15 The probability that his train will be early is 0.6

  **a**  Work out the probability that Roger's train will be late.

  **b**  Work out the probability that Roger's train will be either on time or early.

3  The probability that Lisa will pass her Maths exam is 0.8
   Work out the probability that Lisa will *not* pass her Maths exam.

4  A company makes batteries. A battery is chosen at random. The probability that the battery will not be faulty is 0.97
   Work out the probability that the battery will be faulty.

5  Four athletes Aaron, Ben, Carl and Des take part in a race.
   The table shows the probabilities that Aaron or Ben or Carl will win the race.

| Aaron | Ben | Carl | Des |
|-------|-----|------|-----|
| 0.2 | 0.14 | 0.3 | |

   a  Work out the probability that Aaron will not win the race.
   b  Work out the probability that Ben will not win the race.
   c  Work out the probability that Des will win the race.
   d  Work out the probability that either Aaron or Ben will win the race.
   e  Work out the probability that either Aaron or Carl or Des will win the race.

6  The table shows the probabilities of a dice landing on each of the numbers 1 to 6 when thrown.
   The dice is thrown once.

   a  Work out the probability that the dice will land on either 1 or 3

   b  Work out the probability that the dice will land on either 2 or 4

   c  Work out the probability that the dice will land on
      i   an even number
      ii  an odd number

| Number | Probability |
|--------|-------------|
| 1 | 0.2 |
| 2 | 0.15 |
| 3 | 0.25 |
| 4 | 0.18 |
| 5 | 0.05 |
| 6 | 0.17 |

7  A roundabout has four roads leading from it. Michael is driving round the roundabout. The roads lead to Liverpool or Trafford Park or Eccles or Bolton. The table shows the probabilities that Michael will take the road to Liverpool or Trafford Park or Bolton.

| Liverpool | Trafford Park | Eccles | Bolton |
|-----------|---------------|--------|--------|
| 0.49 | 0.18 | $x$ | 0.23 |

   a  Work out the probability that Michael will not take the road to Liverpool.
   b  Work out the value of $x$.
   c  Work out the probability that Michael will take either the road to Trafford Park or the road to Bolton.

8  Sam has red, white, yellow and green coloured T-shirts only. She chooses a T-shirt at random.
   The probabilities that Sam will choose a red T-shirt or a white T-shirt are given in the table.
   Sam is twice as likely to choose a green T-shirt as she is to choose a yellow T-shirt.
   Work out the value of $x$.

| Red | White | Yellow | Green |
|-----|-------|--------|-------|
| 0.5 | 0.14 | $x$ | $2x$ |

## 5.4 Estimating probability from relative frequency

The diagram shows two three-sided spinners.
One spinner is fair and one is biased.
A spinner is **biased** if it is not equally likely to land on each of the
numbers. This can be tested by experiment.
If a spinner is spun 300 times it is fair if it lands on each of the
numbers approximately 100 times.

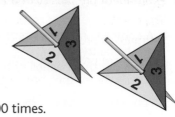

John spins one spinner 300 times and Mary spins the other spinner 300 times.

| John's spinner | | | Mary's spinner | | |
|---|---|---|---|---|---|
| 300 spins | | | 300 spins | | |
| 1 | 2 | 3 | 1 | 2 | 3 |
| 97 | 104 | 99 | 147 | 96 | 57 |
| **FAIR** spinner | | | **BIASED** spinner | | |

John's spinner is fair because it lands on each of the three numbers approximately the same number
of times.

Mary's spinner is biased because it is more likely to land on the number **1**
It is least likely to land on the number **3**

To estimate the probability that Mary's spinner will land on each number, the **relative frequency** of
each number is found using

$$\text{relative frequency} = \frac{\text{number of times the spinner lands on the number}}{\text{total number of spins}}$$

Relative frequency that Mary's spinner will land on the number **1** $= \frac{147}{300} = 0.49$

Relative frequency that Mary's spinner will land on the number **2** $= \frac{96}{300} = 0.32$

Relative frequency that Mary's spinner will land on the number **3** $= \frac{57}{300} = 0.19$

An estimate of the probability that the spinner will land on the number **1** is 0.49
An estimate of the probability that the spinner will land on the number **2** is 0.32
An estimate of the probability that the spinner will land on the number **3** is 0.19
If Mary spins the spinner a further 500 times, an estimate for the number of times the spinner lands
on the number 2 is 0.32 × 500 = 160
In general

**if the probability that an experiment will be successful is $p$ and the experiment is carried
out $N$ times, then an estimate for the number of successful experiments is $p \times N$.**

### Example 7

In a statistical experiment Brendan
throws a dice 600 times.
The table shows the results.
Brendan throws the dice again.

| Number on dice | 1 | 2 | 3 | 4 | 5 | 6 |
|---|---|---|---|---|---|---|
| Frequency | 48 | 120 | 180 | 96 | 54 | 102 |

**a** Find an estimate of the probability that he will throw a 2
**b** Find an estimate of the probability that he will throw an even number.
Zoe now throws the same dice 200 times.
**c** Find an estimate of the number of times she will throw a 6

**Solution 7**

**a** Estimate of probability of a 2 is $\frac{120}{600} = 0.2$

**b** The number of times an even number is thrown = 120 + 96 + 102 = 318

Estimate of probability of an even number = $\frac{318}{600} = 0.53$

**c** Estimate of probability of a 6 is $\frac{102}{600} = 0.17$

An estimate for the number of times Zoe will throw a 6 in 200 throws = 0.17 × 200 = 34

---

### Exercise 5D

**1** A coin is biased. The coin is tossed 200 times.
It lands on heads 140 times and it lands on tails 60 times.

  **a** Write down the relative frequency of the coin landing on tails.

  **b** The coin is to be tossed again. Estimate the probability that the coin will land on
    **i** tails                    **ii** heads.

**2** A bag contains a red counter, a blue counter, a
white counter and a green counter. Asif takes a
counter at random. He does this 400 times.

| Red | Blue | White | Green |
|-----|------|-------|-------|
| 81  | 110  | 136   | 73    |

The table shows the number of times each of the coloured counters is taken.

  **a** Write down the relative frequency of Asif taking the *red* counter.

  **b** Write down the relative frequency of Asif taking the *white* counter. Asif takes a counter one
    more time. Estimate the probability that this counter will be

  **c** **i** blue                    **ii** green.

**3** Tyler carries out a survey about the words in a newspaper. He chooses an article at random.
He counts the number of letters in each of the first 150 words of the article.
The table shows Tyler's results.

| Number of letters in a word | 1 | 2 | 3 | 4 | 5 | 6 | 7 | 8 | 9 | 10 |
|------------------------------|---|----|----|----|----|----|----|---|---|----|
| Frequency | 7 | 14 | 42 | 31 | 21 | 13 | 10 | 6 | 4 | 2 |

A word is chosen at random from the 150 words.

  **a** Write down the most likely number of letters in the word.

  **b** Estimate the probability that the word will have
    **i** 1 letter                **ii** 7 letters                **iii** more than 5 letters.

  **c** The whole article has 1000 words. Estimate the total number of 3-letter words in this article.

**4** A bag contains 10 coloured bricks. Each brick is white or
red or blue. Alan chooses a brick at random from the 10
bricks in the bag and then replaces it in the bag.

| White | Red | Blue |
|-------|-----|------|
| 290   | 50  | 160  |

He does this 500 times. The table shows the numbers of each coloured brick chosen.

  **a** Estimate the number of red bricks in the bag.

  **b** Estimate the number of white bricks in the bag.

**5** The probability that someone will pass their driving test at the first attempt is 0.45
On a particular day, 1000 people will take the test for the first time.
Work out an estimate for the number of these 1000 people who will pass.

**6** Gwen has a biased coin. When she spins the coin the probability that it will come down tails is $\frac{3}{5}$
Work out an estimate for the number of tails she gets when she spins her coin 400 times.

**7** The probability that a biased dice will land on a 1 is 0.09   Andy is going to roll the dice 300 times.
Work out an estimate for the number of times the dice will land on a 1

## 5.5 Independent events

In Example 3 a fair three-sided spinner is spun and a fair coin is tossed at the same time. The outcomes from spinning the spinner do not affect the outcomes from tossing the coin. The outcomes from tossing the coin do not affect the outcomes from spinning the spinner.

These are **independent events** since an outcome of one event does not affect the outcome of the other event.

What is the probability that the spinner will land on 3 and the coin will land on tails? This is written as P(3, tail).

$$P(3) = \tfrac{1}{3} \quad P(\text{tail}) = \tfrac{1}{2}$$

To work out P(3, tail) the sample space could be used. The sample space is

  (1, head)     (1, tail)     (2, head)     (2, tail)     (3, head)     (3, tail)

$P(3, \text{tail}) = \tfrac{1}{6}$ since this is 1 out of 6 possible outcomes.

But  $P(3, \text{tail}) = \tfrac{1}{3} \times \tfrac{1}{2} = \tfrac{1}{6}$

so  $P(3, \text{tail}) = P(3) \times P(\text{tail})$

In general when the outcomes, A and B, of two events are independent

**P(A and B) = P(A) × P(B)**

## Example 8

A bag contains 4 green counters and 5 red counters. A counter is chosen at random and then replaced in the bag. A second counter is then chosen at random. Work out the probability that for the counters chosen

**a** both will be green     **b** both will be red     **c** one will be green and one will be red

*Solution 8*

**a**     $P(G) = \tfrac{4}{9}$

| Find the probability that a green counter will be chosen. |

$P(G \text{ and } G) = \tfrac{4}{9} \times \tfrac{4}{9} = \tfrac{16}{81}$

| The choosing of the two counters are two independent events so use P(A and B) = P(A) × P(B) |

The probability that both counters chosen will be green $= \tfrac{16}{81}$

**b**     $P(R) = \tfrac{5}{9}$

| Find the probability of choosing a red counter. |

$P(R \text{ and } R) = \tfrac{5}{9} \times \tfrac{5}{9} = \tfrac{25}{81}$

| Use P(A and B) = P(A) × P(B) |

The probability that both counters chosen will be red $= \tfrac{25}{81}$

**c** P(one G and one R) = P(first G and second R or first R and second G)

| Use P(A or B) = P(A) + P(B) |

= P(first G and second R) + P(first R and second G)

| *Hint:* A *or* B → *Add* probabilities   A *and* B → *Multiply* probabilities |

$= \tfrac{4}{9} \times \tfrac{5}{9} + \tfrac{5}{9} \times \tfrac{4}{9}$

$= \tfrac{20}{81} + \tfrac{20}{81}$

| Use P(A and B) = P(A) × P(B) |

P(one G and one R) $= \tfrac{40}{81}$

| *Note:* (G, G) (R, R) (G, R) (R, G) is the full sample space so answers to parts **a**, **b** and **c** must add up to 1 |

The probability that one of the counters chosen will be green and one will be red $= \tfrac{40}{81}$

**Exercise 5E**

**1** A biased coin and a biased dice are thrown. The probability that the coin will land on heads is 0.6
   The probability that the dice will land on an even number is 0.7

   **a** Write down the probability that the coin will not land on heads.

   **b** Find the probability that the coin will land on heads and that the dice will land on an even number.

   **c** Find the probability that the coin will *not* land on heads and that the dice will *not* land on an even number.

**2** A basket of fruit contains 3 apples and 4 oranges. A piece of fruit is picked at random and then returned to the basket. A second piece of fruit is then picked at random.
   Work out the probability that for the fruit picked

   **a** both will be apples

   **b** both will be oranges

   **c** one will be an apple and one will be an orange.

**3** Eric and Frank each try to hit the bulls-eye.
   They each have one attempt.
   The events are independent.

   The probability that Eric will hit the bulls-eye is $\frac{2}{3}$

   The probability that Frank will hit the bulls-eye is $\frac{3}{4}$

   **a** Find the probability that both Eric and Frank will hit the bulls-eye.

   **b** Find the probability that just one of them will hit the bulls-eye.

   **c** Find the probability that neither of them will hit the bulls-eye.

**4** When Edna rings the health centre the probability that the phone is engaged is 0.35
   Edna needs to ring the health centre at 9 am on both Monday and Tuesday.
   Find the probability that at 9 am the phone will

   **a** be engaged on both Monday and Tuesday

   **b** not be engaged on Monday but will be engaged on Tuesday

   **c** be engaged on at least one day

**5** Mrs Rashid buys a car.

| Fault | Engine | Brakes |
|-------|--------|--------|
| Probability | 0.05 | 0.1 |

The table shows the probability of different mechanical faults.
Find the probability that the car will have

   **a** a faulty engine and faulty brakes

   **b** no faults

   **c** **exactly** one fault

## 5.6 Probability tree diagrams

It is often helpful to use **probability tree diagrams** to solve probability problems.
A probability tree diagram shows all of the possible outcomes of more than one event by following
all of the possible paths along the branches of the tree.

### Example 9

Mumtaz and Barry are going for an interview.
The probability that Mumtaz will arrive early is 0.7
The probability that Barry will arrive early is 0.4
The two events are independent.

a Complete the probability tree diagram.
b Work out the probability that Mumtaz and Barry will both arrive early.
c Work out the probability that just one person will arrive early.

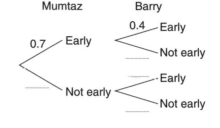

### Solution 9

a Mumtaz not early: $1 - 0.7 = 0.3$    | Must be either early or not early. |

Barry not early: $1 - 0.4 = 0.6$    | Must be either early or not early. |

Barry early: $0.4$    | The two events are independent. |

P(Early, Early) — This probability is found in part **b**

P(Early, Not early)
P(Not early, Early) — These probabilities are added in part **c**

P(Not early, Not early)

b P(Mumtaz early and Barry early)
= P(Early, Early) = $0.7 \times 0.4 = 0.28$

Use $P(A \text{ and } B) = P(A) \times P(B)$ When moving along a path multiply the probabilities on each of the branches.

c P(just one person early)
= P(Mumtaz early and Barry not early OR Mumtaz not early and Barry early)

Possible ways for just one person to be early. Use $P(A \text{ or } B) = P(A) + P(B)$

= P(Early, Not early) + P(Not early, Early)
= $(0.7 \times 0.6) + (0.3 \times 0.4)$

Use $P(A \text{ and } B) = P(A) \times P(B)$

= $0.42 + 0.12 = 0.54$
P(just one person early) = 0.54

### Exercise 5F

1 Amy and Beth are going to take a driving test tomorrow.
The probability Amy will pass the test is 0.7
The probability Beth will pass the test is 0.8
The probability tree diagram shows this information.
Use the probability tree diagram to work out the probability that

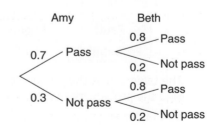

a both women will pass the test
b only Amy will pass the test
c neither woman will pass the test.

**2** A bag contains 10 coloured counters,
4 of which are yellow.
A box also contains 10 coloured counters,
7 of which are yellow.
One counter is chosen at random from the bag
and one counter is chosen at random from the box.

   **a** Copy and complete the probability tree diagram.

   **b** Find the probability that

      **i** both counters will be yellow

      **ii** the counter from the bag will be yellow and the counter from the box will not be yellow

     **iii** at least one counter will be yellow.

**Bag**      **Box**

$\frac{4}{10}$ Yellow

$\frac{7}{10}$ Yellow

Not yellow

Not yellow

Yellow

Not yellow

**3** The probability that a biased coin will show heads when thrown is 0.4
Tina throws the coin twice and records her results.

   **a** Draw a probability tree diagram.

   **b** Use your diagram to work out the probability that the coin will show

      **i** heads on both throws      **ii** heads on *exactly* one throw.

**4** The probability that Jason will receive one DVD for his birthday is $\frac{4}{5}$
The probability that he will receive one DVD for Christmas is $\frac{3}{8}$
These two events are independent.
Find the probability that Jason will receive at least one DVD.

**5** Stuart and Chris each try to score a goal in a penalty shoot-out.
They each have one attempt.
The probability that Stuart will score a goal is 0.75
The probability that Chris will score a goal is 0.64

   **a** Work out the probability that both Stuart and Chris will score a goal.

   **b** Work out the probability that exactly one of them will fail to score a goal.

## 5.7 Conditional probability

The probability of an outcome of an event that is dependent on the outcome of a previous event is called **conditional probability**. For example when choosing two pieces of fruit without replacing the first one, the choice of the second piece of fruit is dependent on the choice of the first.

### Example 10

A bowl of fruit contains 3 apples and 4 bananas. A piece of fruit is chosen at random and eaten. A second piece of fruit is then chosen at random. Work out the probability that for the two pieces of fruit chosen

**a** both will be apples

**b** the first will be an apple and the second will be a banana

**c** at least one apple will be chosen.

*Solution 10*

**a** 1st choice:            $P(A) = \frac{3}{7}$

> Find the probability that the *first* piece of fruit will be an apple. There is a total of 7 pieces, 3 of which are apples.

    2nd choice:         $P(A) = \frac{2}{6}$

> 1st choice was apple so there are now only 6 pieces of fruit, 2 of which are apples. Find the probability that the *second* piece of fruit will be an apple.

$$P(A \text{ and } A) = \frac{3}{7} \times \frac{2}{6} = \frac{1}{7}$$

Probability both will be apples $= \frac{1}{7}$

> Multiply the probabilities.

**b** 1st choice:                                   $P(A) = \frac{3}{7}$

| Find the probability that the *first* piece of fruit will be an apple. There is a total of 7 pieces, 3 of which are apples. |

2nd choice:                                   $P(B) = \frac{4}{6}$

| 1st choice was apple so there are now only 6 pieces of fruit, 4 of which are bananas. Find the probability that the *second* piece of fruit will be a banana. |

$P(A \text{ and } B) = \frac{3}{7} \times \frac{4}{6} = \frac{2}{7}$

Probability the first will be an apple and the second will be a banana $= \frac{2}{7}$

| Multiply the probabilities. |

**c** $P(\text{At least one apple}) = 1 - P(B, B)$

| (A, A) (A, B) (B, A) (B, B) are all the possible outcomes so P(At *least* one apple) + P(B, B) = 1 |

1st choice:          $P(B) = \frac{4}{7}$

| Find the probability that the *first* piece of fruit will be a banana. There is a total of 7 pieces, 4 of which are bananas. |

2nd choice:          $P(B) = \frac{3}{6}$

| 1st choice was banana so there are now only 6 pieces of fruit, 3 of which are bananas. Find the probability that the *second* piece of fruit will be a banana. |

$P(B, B) = \frac{4}{7} \times \frac{3}{6} = \frac{2}{7}$

$P(\text{At least one apple}) = 1 - \frac{2}{7} = \frac{5}{7}$

| Multiply the probabilities. |

## Example 11

There are 4 red crayons, 3 blue crayons and 1 green crayon in a box.
A crayon is taken at random and *not* replaced. A second crayon is then taken at random.
**a** Draw and complete a probability tree diagram.
**b** Find the probability that both crayons taken will be     **i** blue     **ii** the same colour.
**c** Find the probability that *exactly* one of the crayons will be red.

### Solution 11

**a** First crayon: total of 8 crayons out of which 4 are red, 3 are blue, 1 is green.
Second crayon: total of 7 crayons (since 1st not replaced)
When 1st crayon is red, 3 red, 3 blue, 1 green remain. When 1st crayon is blue, 4 red, 2 blue, 1 green remain. When 1st crayon is green, 4 red, 3 blue, 0 green remain

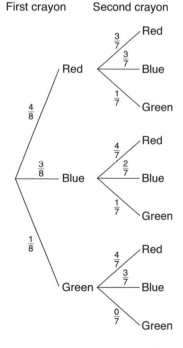

| These results are used later in the question. |

**b** **i** $P(B \text{ and } B) = \frac{3}{8} \times \frac{2}{7} = \frac{3}{28}$

Probability that both colours will be blue $= \frac{3}{28}$

> Follow the branches 'blue to blue' and multiply the probabilities.

**ii** Probability that colours will be the same

$= P(R \text{ and } R \text{ or } B \text{ and } B \text{ or } G \text{ and } G)$

$= P(R, R) + P(B, B) + P(G, G)$

$= \frac{4}{8} \times \frac{3}{7} + \frac{3}{8} \times \frac{2}{7} + \frac{1}{8} \times \frac{0}{7}$

$= \frac{12}{56} + \frac{6}{56} + \frac{0}{56} = \frac{18}{56}$

Probability that colours will be the same $= \frac{9}{28}$

> Colours are either both red or both blue or both green.

**c** Probability of *exactly* one red

$= P(R, B) + P(R, G) + P(B, R) + P(G, R)$

$= \frac{4}{8} \times \frac{3}{7} + \frac{4}{8} \times \frac{1}{7} + \frac{3}{8} \times \frac{4}{7} + \frac{1}{8} \times \frac{4}{7}$

$= \frac{12}{56} + \frac{4}{56} + \frac{12}{56} + \frac{4}{56} = \frac{32}{56}$

Probability of *exactly* one red $= \frac{4}{7}$

> Follow the pairs of branches which have just one red.

---

## Example 12

When driving to the shops Rose passes through
two sets of traffic lights.
If she stops at the first set of lights the probability that she stops at the
second set of lights is 0.25
If she does not stop at the first set of lights the probability that she
stops at the second set is 0.35
The probability that Rose stops at the first set of lights is 0.4

**a** Draw and complete a probability tree diagram.

**b** Find the probability that when Rose next drives to the shops she will not
stop at the second set of traffic lights.

*Solution 12*

**a**

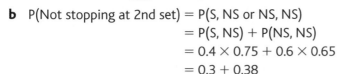

1st set    2nd set

0.4 — Stop → 0.25 — Stop
         → 0.75 — Not stop

0.6 — Not stop → 0.35 — Stop
         → 0.65 — Not stop

**b** P(Not stopping at 2nd set) = P(S, NS or NS, NS)

$= P(S, NS) + P(NS, NS)$

$= 0.4 \times 0.75 + 0.6 \times 0.65$

$= 0.3 + 0.38$

Probability that Rose will not stop at the 2nd set = 0.68

---

## Exercise 5G

**1** A box of chocolates contains 10 milk chocolates and 12 plain chocolates. Two chocolates are
taken at random without replacement.
Work out the probability that

   **a** both chocolates will be milk chocolates

   **b** at least one chocolate will be a milk chocolate

**2** Anil has 13 coins in his pocket, 6 pound coins, 3 twenty-pence coins and 4 two-pence coins. He picks two coins at random from his pocket. Work out the probability that the two coins each have the same value.

**3** Mandy has these five cards
Each card has a number on it.
She chooses two cards at random
without replacement and records
the number on each card.

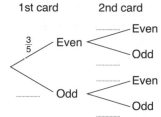

**a** Copy and complete the probability tree diagram.

**b** Find the probability that both numbers are even.

**c** Find the probability that the sum of the two numbers is an even number.

**4** Michael returns to school tomorrow.
If it is raining the probability that Michael walks to school is 0.3
If it is not raining the probability that Michael walks to school is 0.8
The probability that it will rain tomorrow is 0.1

**a** Draw a probability tree diagram.

**b** Find the probability that Michael will walk to school tomorrow.

**5** A box contains 3 tins of soup. 2 of the tins are chicken soup and 1 is tomato soup.
Betty wants tomato soup. She picks a tin at random from the box. If it is not tomato she gives the tin to her son and then picks another tin at random from the box.

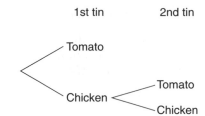

**a** Copy and complete the probability tree diagram.

**b** Find the probability that Betty does not pick the tin of tomato soup.

**6** The probability that a biased dice when thrown will land on 6 is $\frac{1}{4}$  In a game Patrick throws the biased dice until it lands on 6  Patrick wins the game if he takes no more than three throws.

**a** Find the probability that Patrick throws a 6 with his second throw of the dice.

**b** Find the probability that Patrick wins the game.

## Chapter summary

**You should now know:**

★ that **probability** is a measure of how likely the outcome of an event is to happen

★ that probabilities are written as fractions or decimals between 0 and 1

★ that an outcome which is **impossible** has a probability of 0

★ that an outcome which is **certain** to happen has a probability of 1

★ for an event, outcomes which are equally likely have equal probabilities

★ that when calculating probabilities you can use

$$\text{probability} = \frac{\text{number of successful outcomes}}{\text{total number of possible outcomes}}$$

when all outcomes of an event are equally likely to happen
For example the probability of throwing a six on a normal fair dice is $\frac{1}{6}$

★ that a **sample space** is all the possible outcomes of one or more events and a **sample space diagram** is a diagram which shows the sample space

★ how to list all outcomes in an ordered way using sample space diagrams. For example when two coins are tossed the outcomes are

(H, H) (T, H)
(H, T) (T, T)

★ that for an event **mutually exclusive outcomes** are outcomes which cannot happen at the same time

★ that the sum of the probabilities of all the possible mutually exclusive outcomes is 1

★ that if the probability of something happening is $p$, then the probability of it *NOT* happening is $1 - p$

★ that when two outcomes, A and B, of an event are mutually exclusive

P(A or B) = P(A) + P(B)

★ that from a statistical experiment for each outcome

$$\text{relative frequency} = \frac{\text{number of times the outcome happens}}{\text{total number of trials of the event}}$$

★ that relative frequencies give good estimates to probabilities when the number of trials is large

★ that if the probability that an experiment will be successful is $p$ and the experiment is carried out a number of times, then an estimate for the number of successful experiments is $p \times$ number of experiments

For example if the probability that a biased coin will come down heads is 0.7 and the coin is spun 200 times, then an estimate for the number of times it will come down heads is $0.7 \times 200 = 140$

★ that for **independent events** an outcome from one event does not affect the outcome of the other event

★ that when the outcomes, A and B, of two events are independent

P(A and B) = P(A) × P(B)

★ that a **probability tree diagram** shows all of the possible outcomes of more than one event by following all of the possible paths along the branches of the tree. When moving along a path multiply the probabilities on each of the branches

★ that **conditional probability** is the probability of an outcome of an event that is dependent on the outcome of a previous event. For example choosing two pieces of fruit without replacing the first one where the choice of the second piece of fruit is dependent on the choice of the first.

## Chapter 5 review questions

1  Shreena has a bag of 20 sweets.
10 of the sweets are red.
3 of the sweets are black.
The rest of the sweets are white.
Shreena chooses one sweet at random.
What is the probability that Shreena will choose a

**a** red sweet            **b** white sweet?        (1385 June 1999)

**2** 80 students each study one of three languages.
The two-way table shows some information about these students.

|        | French | German | Spanish | Total |
|--------|--------|--------|---------|-------|
| **Female** | 15     |        |         | 39    |
| **Male**   |        | 17     |         | 41    |
| **Total**  | 31     | 28     |         | 80    |

**a** Copy and complete the two-way table.

One of these students is to be picked at random.

**b** Write down the probability that the student picked studies French.          (1387 June 2005)

**3** Here are two sets of cards.
Each card has a number on it as shown.
A card is selected at random from set A and a card is
selected at random from set B.
The difference between the number on the card
selected from set A and the number on the card
selected from set B is worked out.

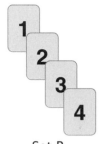

Set A          Set B

**a** Copy and complete the table started below to
show all the possible differences.

Set A

|       |   | 1 | 2 | 3 | 4 |
|-------|---|---|---|---|---|
|       | **1** | 0 |   | 2 |   |
| Set B | **2** |   | 0 |   |   |
|       | **3** |   | 1 |   |   |
|       | **4** |   |   |   |   |

**b** Find the probability that the difference will be zero.
**c** Find the probability that the difference will *not* be 2

**4** There are 20 coins in a bag.
7 of the coins are pound coins.
Gordon is going to take a coin at random from the bag.

**a** Write down the probability that he will take a pound coin.

**b** Find the probability that he will take a coin which is *NOT* a pound coin.

**5** Mr Brown chooses one book from the library each week.
He chooses a crime novel or a horror story or a non-fiction book.
The probability that he chooses a horror story is 0.4
The probability that he chooses a non-fiction book is 0.15
Work out the probability that Mr Brown chooses a crime novel.          (1387 June 2005)

**6** Here is a 4-sided spinner. The sides of the spinner are labelled 1, 2, 3 and 4
The spinner is biased.
The probability that the spinner will land on each of
the numbers 2 and 3 is given in the table.
The probability that the spinner will land on 1 is
equal to the probability that it will land on 4

**a** Work out the value of $x$.

| Number | 1 | 2 | 3 | 4 |
|--------|---|---|---|---|
| Probability | $x$ | 0.3 | 0.2 | $x$ |

Sarah is going to spin the spinner 200 times.

**b** Work out an estimate for the number of times it will land on 2          (1387 June 2005)

**7** Meg has a biased coin.
When she spins the coin the probability that it will come down heads is 0.4
Meg is going to spin the coin 350 times.
Work out an estimate for the number of times it will come down heads.

**8** A dice has one red face and the other faces coloured white.
The dice is biased.
Sophie rolled the dice 200 times.
The dice landed on the red face 46 times.
The dice landed on a white face the other times.
Sophie rolls the dice again.

  **a** Estimate the probability that the dice will land on a white face.

Each face of a different dice is either rectangular or hexagonal.
When this dice is rolled the probability that it will land on a rectangular face is 0.85
Billy rolls this dice 1000 times.

  **b** Estimate the number of times it will land on a rectangular face.

**9** Julie does a statistical experiment.
She throws a dice 600 times.
She scores six 200 times.

  **a** Is the dice fair? Explain your answer.

Julie then throws a fair red dice once and a fair blue dice once.

  **b** Copy and complete the probability tree diagram to show
     the outcomes.
     Label clearly the branches of the probability tree diagram.
     The probability tree diagram has been started.

  **c**  **i** Julie throws a fair red dice once and a fair blue dice once.
         Calculate the probability that Julie gets a six on both the
         red dice and the blue dice.
      **ii** Calculate the probability that Julie gets at least one six.

Red dice    Blue dice

$\frac{1}{6}$  Six

Not six

(1387 June 2003)

**10** Lauren and Yasmina each try to score a goal.
They each have one attempt.
The probability that Lauren will score a goal is 0.85
The probability that Yasmina will score a goal is 0.6

  **a** Work out the probability that *both* Lauren *and* Yasmina will score a goal.

  **b** Work out the probability that Lauren *will* score a goal *and* Yasmina *will not* score a goal.

(1385 June 1998)

**11** Amy has 10 CDs in a CD holder.
Amy's favourite group is Edex.
She has 6 Edex CDs in the CD holder.
Amy takes one of these 10 CDs at random.
She writes down whether or not it is an Edex CD.
She puts the CD back in the holder.
Amy again takes one of these 10 CDs at random.

  **a** Copy and complete the probability tree diagram.

Amy had 30 CDs.
The mean playing time of these 30 CDs
was 42 minutes.
Amy sold 5 of her CDs.
The mean playing time of the 25 CDs left
was 42.8 minutes.

  **b** Calculate the mean playing time of the 5 CDs that Amy sold.

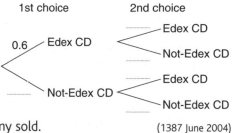

1st choice       2nd choice

0.6   Edex CD

Edex CD

Not-Edex CD

Not-Edex CD

Edex CD

Not-Edex CD

(1387 June 2004)

**12** A bag contains 3 black beads, 5 red beads and 2 green beads.
Gianna takes a bead at random from the bag, records its colour and replaces it.
She does this two more times.
Work out the probability that of the three beads Gianna takes, exactly two are the same colour.

(1387 June 2003)

**13** Amy is going to play one game of snooker and one game of billiards.
The probability that she will win the game of snooker is $\frac{1}{3}$
The probability that she will win the game of billiards is $\frac{3}{4}$
The probability tree diagram shows this information.

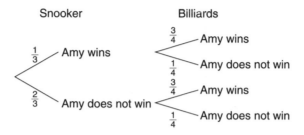

Amy played one game of snooker and one game of billiards on a number of Fridays.
She won at *both* snooker and billiards on 21 Fridays.
Work out an estimate for the number of Fridays on which Amy did not win either game.

(1388 June 2005)

**14** A bag contains 10 coloured discs.
4 of the discs are red and 6 of the discs are black.
Asif is going to take two discs at random from the bag,
*without* replacement.

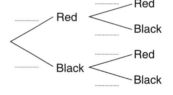

  **a** Copy and complete the tree diagram.
  **b** Work out the probability that Asif will take two black discs.
  **c** Work out the probability that Asif takes two discs of the same colour.

(1385 May 2002)

**15** Ali has twenty socks in a sock drawer.
10 of them are grey, 6 of them are black and 4 of them are red.
Ali takes two socks at random without replacement from the drawer.
Calculate the probability that he takes two socks that have the same colour.

(1385 November 2001)

**16** There are $n$ beads in a bag.
6 of the beads are black and all the rest are white.
Heather picks one bead at random from the bag and does not replace it.
She picks a second bead at random from the bag.
The probability that she will pick 2 white beads is $\frac{1}{2}$
Show that $n^2 - 25n + 84 = 0$

(1385 June 2002)

# Number

## CHAPTER 6

## 6.1 Properties of whole numbers

A **factor** of a number, $x$, is a number which divides into $x$ an exact number of times.
So 3 is a factor of 12 because $3 \times 4 = 12$
3 and 4 are called a **factor pair** of 12
3 is not a factor of 10 because 10 leaves a remainder when divided by 3

A **multiple** of a number, $y$, is a number which divides exactly by $y$.
The first three multiples of 5 are 5, 10 and 15
21 is not a multiple of 5 because 21 leaves a remainder when divided by 5

### Example 1

Find all the factors of 20   Write them in factor pairs.

*Solution 1*
1, 20
2, 10      In each case the product
4, 5       of the two numbers is 20

A number which has *exactly* two factors is called a **prime number**.
So, 2, 3, 5 and 23 are prime numbers but 1, 4 and 15 are not.
1 is not a prime number because it only has one factor, 1

### Example 2

2 is the only even prime number. Explain why.

*Solution 2*
2 is a prime number because it has exactly two factors, 1 and 2

All other even numbers have at least 3 factors, 1, 2 and the number itself.

A **common factor** of two numbers, $x$ and $y$, is a number which is both a factor of $x$ and is also a factor of $y$.

So 4 is a common factor of 12 and 20
3 is not a common factor of 12 and 20 because 3 is not a factor of 20

### Example 3

Find all the common factors of 12 and 20

*Solution 3*
1, 2, 3, 4, 6, 12                    These are the factors of 12

1, 2, 4, 5, 10, 20                   These are the factors of 20

1, 2, 4 are the common factors of 12 and 20   These three factors appear in both lists.

**Exercise 6A**

**1** Which of the following numbers are factors of 18?

    **a** 12         **b** 6         **c** 9         **d** 3         **e** 36

**2** Which of the following numbers are factors of 30?

    **a** 1         **b** 20         **c** 15         **d** 3         **e** 6

**3** Find all the factors of 50   Write them in factor pairs.

**4** List all of the factors of the following numbers.

    **a** 8         **b** 10         **c** 16         **d** 24         **e** 28

    **f** 32         **g** 36         **h** 40         **i** 60         **j** 100

**5** List all the common factors of

    **a** 6 and 8         **b** 6 and 9         **c** 6 and 10         **d** 8 and 12

    **e** 12 and 15         **f** 10 and 20         **g** 15 and 20         **h** 18 and 24

**6 a** Write down the first three multiples of 4

    **b** Write down the first three multiples of 10

    **c** Write down the first four multiples of 8

    **d** Write down the first four multiples of 7

    **e** Write down the first three multiples of 23

**7** State whether the following statements are true or false.

    **a** 12 is a multiple of 2         **b** 14 is a factor of 7

    **c** 24 is a multiple of 3         **d** 72 is a multiple of 9

    **e** 12 is both a multiple of 6 and a factor of 36     **f** 9 is a factor of 27

    **g** 6 is a multiple of 1         **h** 4 is a multiple of 12

**8** Show that 33 is a factor of 3003

**9** Find the first multiple of 29 which is greater than 2000

**10** Find two prime numbers between 110 and 120

**11** Bertrand's theorem states that 'Between any two numbers $n$ and $2n$, there always lies at least one prime number, providing $n$ is bigger than 1.
Show that Bertrand's theorem is true **i** for $n = 10$     **ii** for $n = 34$

**12** Find a number which has exactly

    **a** 4 factors         **b** 3 factors         **c** 7 factors         **d** 10 factors

## 6.2 Multiplication and division of directed numbers

*also assessed in Module 4*

A directed number is a number with a + or a − sign. +4 and −3 are examples of directed numbers. Often a directed number is written in brackets, for example (+4) and (−3).

### *Multiplication*

(+3) is the same as 3     (+2) is the same as 2

so (+3) × (+2) is the same as 3 × 2 = 6

3 × 2 also means 2 + 2 + 2     3 × (−2) means (−2) + (−2) + (−2)

so 3 × (−2) or (+3) × (−2) = (−6)

Look at the patterns in the multiplications on the right.

The blue numbers are decreasing by 1

The orange numbers are increasing by 2

The pattern continues like this.

The rules are

$(+) \times (+) = (+)$   positive $\times$ positive $=$ positive
$(+) \times (-) = (-)$   positive $\times$ negative $=$ negative
$(-) \times (+) = (-)$   negative $\times$ positive $=$ negative
$(-) \times (-) = (+)$   negative $\times$ negative $=$ positive

$$(+3) \times (-2) = (-6)$$
$$(+2) \times (-2) = (-4)$$
$$(+1) \times (-2) = (-2)$$
$$0 \times (-2) = 0$$
$$(-1) \times (-2) = (+2)$$
$$(-2) \times (-2) = (+4)$$
$$(-3) \times (-2) = (+6)$$

## Division

$3 \times 2 = 6$    $6 \div 3 = 2$    and    $6 \div 2 = 3$

$(+3) \times (-2) = (-6)$,    so    $(-6) \div (+3) = (-2)$    and    $(-6) \div (-2) = (+3)$

The rules are

$(+) \div (+) = (+)$   positive $\div$ positive $=$ positive
$(+) \div (-) = (-)$   positive $\div$ negative $=$ negative
$(-) \div (+) = (-)$   negative $\div$ positive $=$ negative
$(-) \div (-) = (+)$   negative $\div$ negative $=$ positive

When multiplying or dividing two directed numbers you can remember the rules by the following
if the signs are the **same**, the answer is **positive**
if the signs are **different**, the answer is **negative**.

### Example 4

**a** Work out $(+5) \times (-3)$

**b** Work out $(-16) \div (-2)$

### Solution 4

**a** $(+5) \times (-3) = (-15)$
$(-15)$ is also written as $-15$

> The signs are different so the answer is negative and $5 \times 3 = 15$

**b** $(-16) \div (-2) = (+8)$
$(+8)$ is also written as 8

> The signs are the same so the answer is positive and $16 \div 2 = 8$

### Exercise 6B

**1** Work out

**a** $(+2) \times (-4)$    **b** $(-3) \times (-5)$    **c** $(-4) \times (-6)$
**d** $(+3) \times (+5)$    **e** $(-2) \times (+5)$    **f** $(-4) \times (+5)$
**g** $(-3) \times (+8)$    **h** $(-1) \times (+9)$    **i** $(-4) \times (-4)$

**2** Work out

**a** $(+6) \div (+3)$    **b** $(-8) \div (+4)$    **c** $(+10) \div (-5)$    **d** $(-12) \div (-3)$
**e** $(-8) \div (+4)$    **f** $(-12) \div (-12)$    **g** $(-14) \div (+2)$    **h** $(+12) \div (+4)$

**3** Find the missing directed number.

**a** $(+10) \div (\ ) = (-2)$    **b** $(-8) \div (\ ) = (+2)$    **c** $(-3) \times (\ ) = (+12)$
**d** $(-5) \times (\ ) = (+20)$    **e** $(+5) \times (\ ) = (-25)$    **f** $(\ ) \times (-4) = (+20)$
**g** $(\ ) \div (+3) = (+4)$    **h** $(\ ) \div (-4) = (-5)$    **i** $(+16) \div (\ ) = (-2)$

**4** Work out the product of

a $(-6)$ and $(+3)$   b $(-5)$ and $(+4)$   c $(+3)$ and $(-5)$   d $(-6)$ and $(-6)$

e $(+3)$ and $(+4)$   f $(-4)$ and $(+9)$   g $(-5)$ and $(-4)$   h $(+3)$ and $(-2)$

## 6.3 Squares and cubes

The **square** of a number, $x$, is the number which is the product $x \times x$.
The square of the number $x$ is written $x^2$.
So the square of 10 is written as $10^2 = 10 \times 10$
The square of 10 is 100

The **cube** of a number, $y$, is the number which is the product $y \times y \times y$.
The cube of the number $y$ is written $y^3$.
So the cube of 4 is written $4^3 = 4 \times 4 \times 4$
The cube of 4 is 64

The **square root** of a number $n$ is the number which when squared gives $n$.
The square root of the number $n$ is written $\sqrt{n}$.
So the square root of 16, written $\sqrt{16}$, is 4 since $4^2 = 16$

Since $(-4)^2 = 16$, the **negative square root** of 16 is $-4$
It *is not* possible to find the square root of a negative number.

The **cube root** of a number $m$ is the number which when cubed gives $m$.
The cube root of a number $m$ is written $\sqrt[3]{m}$.
So the cube root of 1000, written $\sqrt[3]{1000}$, is 10 since $10^3 = 1000$
It *is* possible to find the cube root of a negative number.

### Example 5

Work out $2^3 + \sqrt[3]{-27}$

*Solution 5*

$2^3 = 8$          $\boxed{2 \times 2 \times 2 = 8}$

$\sqrt[3]{-27} = -3$     $\boxed{\text{because } (-3) \times (-3) \times (-3) = -27}$

$2^3 + \sqrt[3]{-27} = 5$     $\boxed{8 + (-3) = 5}$

### Exercise 6C

**1** Work out

a $3^2$      b $5^2$      c $11^2$      d $13^2$      e $15^2$      f $100^2$

**2** Work out

a $(-2)^2$      b $(-4)^2$      c $(-10)^2$      d $(-12)^2$

**3** Work out

a $3^3$      b $1^3$      c $5^3$      d $(-10)^3$      e $(-4)^3$

**4** Work out

a $\sqrt{4} \times 3^2$      b $5^2 \times \sqrt{100}$      c $2^3 + 3^2$      d $\sqrt[3]{-8} + 4^2$

e $\sqrt[3]{1000} \div \sqrt{100}$      f $4^3 \div 2^3$      g $(-1)^3 + 2^3 - (-3)^3$

h $4^2 + (-3)^3$      i $\sqrt[3]{125} \times \sqrt{81}$      j $\dfrac{6^2}{2^2}$

**5** Here is a number pattern
$$1^2 = 1^3$$
$$(1 + 2)^2 = 1^3 + 2^3$$
$$(1 + 2 + 3)^2 = 1^3 + 2^3 + 3^3$$

Show that the next line of the number pattern is also true.

## 6.4 Index laws

As well as squares and cubes it is possible to represent a number multiplied by itself any number of times. For example,

- $2^4$ (2 raised to the **power** 4) means $2 \times 2 \times 2 \times 2$
- $3^6$ (3 raised to the power 6) means $3 \times 3 \times 3 \times 3 \times 3 \times 3$

Another name for power is **index**.

---

**Example 6**

Work out **a** $3^4$ **b** $2^6$

*Solution 6*

**a** $3^4 = 3 \times 3 \times 3 \times 3 = 81$

**b** $2^6 = 2 \times 2 \times 2 \times 2 \times 2 \times 2 = 64$

---

To work out one number raised to a power multiplied by the *same* number raised to a second power you **add** the powers.

For example $2^3 \times 2^4 = 2^7$ because $2^3 = 2 \times 2 \times 2$ and $2^4 = 2 \times 2 \times 2 \times 2$
and $(2 \times 2 \times 2) \times (2 \times 2 \times 2 \times 2) = 2^{3+4} = 2^7$

To divide one number raised to a power by the **same** number raised to a second power you **subtract** the powers.

For example, $3^6 \div 3^4 = \dfrac{3 \times 3 \times 3 \times 3 \times 3 \times 3}{3 \times 3 \times 3 \times 3} = \dfrac{\cancel{3} \times \cancel{3} \times \cancel{3} \times \cancel{3} \times 3 \times 3}{\cancel{3} \times \cancel{3} \times \cancel{3} \times \cancel{3}}$ cancelling all the 3s

on the bottom with four of the 3s on the top.

So $3^6 \div 3^4 = 3^{6-4} = 3^2$

---

**Example 7**

**a** Work out $2^4 \times 2^5$. Give your answer as a power of 2
**b** Work out $5^8 \div 5^5$. Give your answer as a power of 5
**c** Work out $(3^2)^4$. Give your answer as a power of 3
**d** Work out $4 \times 4^7$. Give your answer as a power of 4

*Solution 7*

**a** $2^4 \times 2^5 = 2^{4+5} = 2^9$

**b** $5^8 \div 5^5 = 5^{8-5} = 5^3$

**c** $(3^2)^4 = 3^2 \times 3^2 \times 3^2 \times 3^2 = 3^{2+2+2+2} = 3^8$

**d** $4 \times 4^7 = 4^{1+7} = 4^8$

---

## Example 8

Work out $\dfrac{7^4 \times 7^6}{7^8}$

### Solution 8

$$\dfrac{7^4 \times 7^6}{7^8} = \dfrac{7^{10}}{7^8} = 7^2 = 49$$

> 'Work out' means 'evaluate' the expression rather than leaving the answer as a power of 7

### Exercise 6D

**1** Write as a power of 2

   **a** $2^4 \times 2^5$     **b** $2^3 \times 2^4$     **c** $2^2 \times 2^6$     **d** $2^4 \times 2^3$     **e** $2^4 \times 2^6$

**2** Write as a power of 3

   **a** $3^4 \div 3^2$     **b** $3^5 \div 3^2$     **c** $3^4 \div 3$     **d** $3^6 \div 3^2$     **e** $3^{10} \div 3^4$

**3** Write as a power of a single number

   **a** $4^4 \div 4^2$     **b** $5^7 \div 5^2$     **c** $3^4 \times 3^2$     **d** $6^4 \times 6^3$     **e** $10^4 \div 10^2$

**4** Find the value of $n$

   **a** $3^n \div 3^2 = 3^3$     **b** $8^5 \div 8^n = 8^2$     **c** $2^5 \times 2^n = 2^{10}$     **d** $3^n \times 3^5 = 3^9$     **e** $2^6 \times 2^3 = 2^n$

**5** Work out

   **a** $3^4 \div 3^2$     **b** $4^5 \div 4^3$     **c** $2^5 \div 2^2$     **d** $10^4 \times 10^2$     **e** $6^5 \div 6^5$

**6** Write as a power of 3

   **a** $\dfrac{3^3 \times 3^5}{3^4}$     **b** $(3^3)^2$     **c** $\dfrac{3 \times 3^7}{3^4}$     **d** $\dfrac{3^9}{3^4 \times 3^3}$     **e** $\dfrac{3^2 \times 3^{10}}{3^2 \times 3^5}$

**7** Write as a power of a single number

   **a** $\dfrac{2^3 \times 2^4}{2^5}$     **b** $\dfrac{3^4 \times 3^3}{3^4}$     **c** $\dfrac{5^3 \times 5^5}{5^6}$     **d** $\dfrac{10^8 \times 10^3}{10^7}$     **e** $\dfrac{4^5 \times 4}{4^2}$

**8** Work out

   **a** $\dfrac{5^5}{5^2 \times 5^2}$     **b** $\dfrac{3^4}{3^2 \times 3^2}$     **c** $\dfrac{4^7}{4^2 \times 4^3}$     **d** $\dfrac{2^3 \times 2^4}{2^4 \times 2^2}$     **e** $\dfrac{3 \times 3^7}{3^4 \times 3^2}$

**9** Work out the value of $n$ in the following.

   **a** $40 = 5 \times 2^n$     **b** $32 = 2^n$     **c** $50 = 5^n \times 2$     **d** $48 = 3 \times 2^n$     **e** $54 = 2 \times 3^n$

## 6.5 Order of operations

Some expressions include powers and other operations.
**BIDMAS** gives the order in which operations should be carried out.

Remember that BIDMAS stands for

     **B**rackets
     **I**ndices
     **D**ivision
     **M**ultiplication
     **A**ddition
     **S**ubtraction

- If there are brackets, work out the value of the expression in the brackets first.
- Square roots are carried out at the same stage as indices.
- If there are no brackets, do multiplication and division before addition and subtraction no matter where they come in an expression.
- If an expression has only addition and subtraction then work it out from left to right.

## Example 9

Work out

**a** $2 \times 6^2$  **b** $10^2 + 10^3$  **c** $16 \div 2^4$

### Solution 9

**a** $6^2 = 6 \times 6 = 36$  
$2 \times 6^2 = 72$

> Squaring is carried out before multiplication.

**b** $10^2 + 10^3 = 100 + 1000 = 1100$

> Squaring and cubing are carried out before addition.

**c** $16 \div 2^4 = 16 \div 16 = 1$

> Working out indices is carried out before division.

## Example 10

Work out

**a** $(2 \times 4)^2$  **b** $(12 - 2 \times 5)^3$  **c** $\sqrt{9} \times 4$

### Solution 10

**a** $2 \times 4 = 8$, then $8^2 = 64$

> Expressions in brackets are worked out first, then indices.

**b** $(12 - 2 \times 5)^3 = (12 - 10)^3 = 2^3 = 8$

> Multiplication is carried out before subtraction.

**c** $3 \times 4 = 12$

> Square root is found before multiplication is carried out.

## Exercise 6E

**1** Work out

   **a** $2 \times 5^2$  **b** $2 \times 4^2$  **c** $3 \times 2^2$  **d** $5 \times 10^2$

   **e** $4 \times 3^2$  **f** $6 \times 2^2$  **g** $8 \times 1^2$  **h** $5 \times 2^2$

**2** Work out

   **a** $(2 + 5)^2$  **b** $(7 - 3)^2$  **c** $(5 + 5)^2$  **d** $(12 - 5)^2$

   **e** $(4 + 4)^2$  **f** $(8 - 5)^2$  **g** $(20 - 10)^2$  **h** $(18 - 9)^2$

**3** Work out

   **a** $20 - 3^2$  **b** $17 + 4^2$  **c** $17 - 4^2$  **d** $27 - 4^2$

   **e** $36 + 4^2$  **f** $25 - 5^2$  **g** $14 - 4^2$  **h** $22 - 5^2$

**4** Work out

   **a** $24 \div 2^2$  **b** $44 \div 2^2$  **c** $50 \div 5^2$  **d** $100 \div 5^2$  **e** $36 \div 3^2$

   **f** $(10 \div 2)^2$  **g** $(12 \div 2)^2$  **h** $(15 \div 5)^2$  **i** $(21 \div 3)^2$

**5** Work out

   **a** $4^2 - 3^2$  **b** $4^2 + 3^2$  **c** $5^2 - 4^2$  **d** $6^2 - 3^2$  **e** $10^2 - 3^2$

   **f** $6^2 + 1^2$  **g** $6^2 - 4^2$  **h** $7^2 - 3^2$  **i** $6^2 + 8^2$

**6** Work out

a $2 \times \sqrt{9}$  b $4 \times \sqrt{4}$  c $5 \times \sqrt{100}$  d $\sqrt{4} \times 2$

e $\sqrt{9} \times 5$  f $\sqrt{25} \times 6$  g $\sqrt{36} \times 5$  h $\sqrt{100} \times 7$

## 6.6 Using a calculator

Arithmetical expressions can be worked out using a scientific calculator.

To work out $6.4^2$ key in $\boxed{6}$ $\boxed{.}$ $\boxed{4}$ $\boxed{x^2}$ $\boxed{=}$ which gives 40.96

For working out cubes, some scientific calculators have a cube key $\boxed{x^3}$.

To work out $1.1^3$ key in $\boxed{1}$ $\boxed{.}$ $\boxed{1}$ $\boxed{x^3}$ $\boxed{=}$ which gives 1.331

All scientific calculators have a power or index key. This comes in two forms.

- The first form is a $\boxed{y^x}$ key.

  To work out $1.1^3$ key in $\boxed{1}$ $\boxed{.}$ $\boxed{1}$ $\boxed{y^x}$ $\boxed{3}$ $\boxed{=}$ which gives 1.331

- The second form is an upwards arrow key $\boxed{\wedge}$.

  To work out $1.1^3$ key in $\boxed{1}$ $\boxed{.}$ $\boxed{1}$ $\boxed{\wedge}$ $\boxed{3}$ $\boxed{=}$ which gives 1.331

Not all calculators are the same so make sure you know how to calculate powers on your own calculator.

### Example 11

Use a calculator to work out

a $4.5^2$    b $2.3^3 + 3.1^4$

*Solution 11*

a $4.5^2 = 20.25$    Key in $\boxed{4}$ $\boxed{.}$ $\boxed{5}$ $\boxed{x^2}$ $\boxed{=}$

    Key in $\boxed{2}$ $\boxed{.}$ $\boxed{3}$ $\boxed{y^x}$ $\boxed{3}$ $\boxed{=}$ or key in $\boxed{2}$ $\boxed{.}$ $\boxed{3}$ $\boxed{\wedge}$ $\boxed{3}$ $\boxed{=}$

b $2.3^3 = 12.167$

    Write down the answer 12.167

    Key in $\boxed{3}$ $\boxed{.}$ $\boxed{1}$ $\boxed{y^x}$ $\boxed{4}$ $\boxed{=}$ or key in $\boxed{3}$ $\boxed{.}$ $\boxed{1}$ $\boxed{\wedge}$ $\boxed{4}$ $\boxed{=}$

$3.1^4 = 92.3521$

    Write down the answer 92.3521

$2.3^3 + 3.1^4 = 104.5191$    Finally add the two results.

A more efficient method would be to key in

$\boxed{2}$ $\boxed{.}$ $\boxed{3}$ $\boxed{y^x}$ $\boxed{3}$ $\boxed{+}$ $\boxed{3}$ $\boxed{.}$ $\boxed{1}$ $\boxed{y^x}$ $\boxed{4}$ $\boxed{=}$

which does not involve writing down the separate results for $2.3^3$ and $3.1^4$.

Sometimes an answer has to be given correct to one decimal place. The answer to **b** correct to one decimal place is 104.5

The usual method of finding a square root is to use the square root key on the calculator.

All scientific calculators have a square root key $\boxed{\sqrt{}}$.

### Example 12

Work out $\dfrac{2.4^3 + \sqrt{43.56}}{3.5^2 - 2.5^2}$

### Solution 12

The numerator is 20.424

Key in [2] [.] [4] [$y^x$] [3] [+] [$\sqrt{\ }$] [4] [3] [.] [5] [6] [=]

The denominator is 6

Key in [3] [.] [5] [$x^2$] [−] [2] [.] [5] [$x^2$] [=]

The answer is 3.404

Finally divide 20.424 by 6

The expression can be worked out more efficiently using the key sequence

[(] [2] [.] [4] [$y^x$] [3] [+] [$\sqrt{\ }$] [4] [3] [.] [5] [6] [)] [÷]

[(] [3] [.] [5] [$x^2$] [−] [2] [.] [5] [$x^2$] [)] [=]

The **reciprocal** of a number $n$ is the number $\dfrac{1}{n}$ or $1 \div n$.

The reciprocal of 2 is $\frac{1}{2}$ or 0.5

The reciprocal of 1.25 is 0.8

When a number is multiplied by its reciprocal the answer is always 1
All numbers, except 0, have a reciprocal.

The reciprocal button on a calculator is usually shown as $\boxed{\frac{1}{x}}$ or $\boxed{x^{-1}}$.

Dividing an expression by a number is the same as multiplying the expression by the reciprocal of that number.

### Exercise 6F

**1** Work out
    **a** $8.4^2$
    **b** $9.2^2$
    **c** $(-3.6)^2$
    **d** $24^2$
    **e** $15.4^2$

**2** Work out
    **a** $134 + 21^2$
    **b** $231 + 31^2$
    **c** $37 + 23^2$
    **d** $502 + 35^2$

**3** Work out
    **a** $23^2 + 31^2$
    **b** $25^2 + 23^2$
    **c** $19^2 + 22^2$
    **d** $35^2 + 45^2$

**4** Work out
    **a** $200 - 14^2$
    **b** $20 - 1.4^2$
    **c** $356 - 17^2$
    **d** $366 - 16^2$

**5** Work out
    **a** $\sqrt{576}$
    **b** $\sqrt{1024}$
    **c** $\sqrt{625}$
    **d** $\sqrt{1296}$

**6** Work out, giving your answers correct to one decimal place
    **a** $\sqrt{200}$
    **b** $\sqrt{300}$
    **c** $\sqrt{80}$
    **d** $\sqrt{128}$
    **e** $\sqrt{125}$

**7** Work out, giving your answers correct to one decimal place
    **a** $2.4^3$
    **b** $3.7^4$
    **c** $0.95^3$
    **d** $(-1.7)^6$
    **e** $1.05^{10}$

**8** Work out, giving your answers correct to one decimal place
    **a** $3.3^3 \times 2.5$
    **b** $2.3^3 + 5.6^2$
    **c** $3.4^3 \div \sqrt{12}$
    **d** $6.4^2 \div 2.5^3$

**9** Work out, giving your answers correct to one decimal place
    **a** $6.37 - 2.4^2$
    **b** $(12.5 - 9.8)^3$
    **c** $3.4 \times (-2.7)^2$
    **d** $20 \div 1.6^2$

**10** Work out, giving your answers correct to one decimal place

**a** $\dfrac{6.3^2 + 3.3^2}{7.5}$
**b** $\dfrac{80.6}{2.5^3 + 10}$
**c** $\dfrac{3.5^3 + 8.5}{2.6}$
**d** $\dfrac{8.7 + \sqrt{30}}{6.5}$

**e** $\dfrac{17.4 - 2.4^2}{\sqrt{4.5}}$
**f** $\dfrac{5.5^2 - 1.5^2}{2.2^2 + 5}$
**g** $\dfrac{4.5^3 - 18}{3.4^2 - 10}$
**h** $14.6 + \dfrac{3.9^3}{2.6}$

**i** $\dfrac{6.4^2 - \sqrt{20}}{\sqrt{20} + \sqrt{30}}$
**j** $\dfrac{3.6^3 + 4 \times \sqrt{20}}{3.6^3 - 4 \times \sqrt{20}}$

**11** Find the reciprocal of each of the following numbers.
**a** 4      **b** 8      **c** 40      **d** 0.625      **e** 3.2

# 6.7 Prime factors, HCF and LCM

A **prime factor** of the number $n$, is a prime number which is a factor of $n$.

The factors of 30 are 1, 2, 3, 5, 6, 10, 15 and 30
The prime numbers in this list are 2, 3 and 5
So the prime factors of 30 are 2, 3 and 5

Prime numbers can be thought of as the basis of all whole numbers because all whole numbers are either prime or can be written as a product of prime numbers.

For example, 15 is not prime, but can be written as the product $3 \times 5$

12 is not prime, but can be written as the product $2 \times 2 \times 3$

For small numbers it is easy to see what prime numbers to use.
For larger numbers use the following method.

## Example 13

Write 72 as

**a** the product of its prime factors      **b** the product of powers of its prime factors.

*Solution 13*

**a** The prime factors of 72 are 2 and 3

| 2 | 72 | Divide 72 by 2 |
|---|----|----|
| 2 | 36 | Divide 36 by 2 |
| 2 | 18 | Divide 18 by 2 |
| 3 | 9  | Divide 9 by 3 |
| 3 | 3  | Divide 3 by 3 |
|   | 1  | |

$72 = 2 \times 2 \times 2 \times 3 \times 3$

**b** $72 = 2^3 \times 3^2$

The **highest common factor (HCF)** of two numbers is the largest number which is a factor of both of the numbers.

For example, the highest common factor (HCF) of 8 and 12 is 4 because it is the largest number that is a factor of both 8 and 12

For larger numbers, it is useful to list the factors of each number and then pick out the largest number that appears in all the lists.

## Example 14

Find the highest common factor (HCF) of 24 and 36

### Solution 14
The factors of 24 are **1, 2, 3, 4, 6**, 8, **12**, 24
The factors of 36 are **1, 2, 3, 4, 6**, 9, **12**, 18, 36

The numbers which appear in both lists, that is the common factors, are 1, 2, 3, 4, 6 and 12
So 12 is the highest common factor of 24 and 36

---

The **lowest common multiple (LCM)** of two numbers is the smallest number which is a multiple of both numbers.

For example, the lowest common multiple of 8 and 12 is 24 because it is the smallest number which is a multiple of both 8 and 12

For larger numbers, it is useful to list the multiples of each number and then pick out the smallest number that appears in both lists.

## Example 15

Find the lowest common multiple (LCM) of 15 and 20

### Solution 15
The first few multiples of 15 are    15, 30, 45, **60**, 75, ...
The first few multiples of 20 are    20, 40, **60**, 80, 100, ...

The smallest number which appears in both lists is 60
So the lowest common multiple of 15 and 20 is 60

---

The HCF can be worked out for large numbers if each of the numbers is written as a product of its prime factors.

For example, for the numbers 120 and 144 the products are

$$120 = 2 \times 2 \times 2 \times 3 \times 5$$
$$144 = 2 \times 2 \times 2 \times 2 \times 3 \times 3$$

So $2 \times 2 \times 2 \times 3 = 24$ is the highest common factor of 120 and 144

In terms of products of powers of their prime factors,

$$120 = 2^3 \times 3 \times 5$$
$$144 = 2^4 \times 3^2$$

Their highest common factor 24 ($= 2^3 \times 3$) is the product of the *lowest* power of each of their common prime factors.

## Example 16

Find the highest common factor (HCF) of 750 and 225

### Solution 16
$750 = 2 \times 3 \times 5^3$
$225 = 3^2 \times 5^2$

Write 750 and 225 as the product of powers of their prime factors.

$HCF = 3 \times 5^2 = 75$

The common prime factors are 3 and 5
The lowest power of 3 is 1 (as $3 = 3^1$) and the lowest power of 5 is 2

To find the LCM of 36 and 120, list the multiples of 36 and 120 until the same multiple appears in both lists.

The multiples of 36 are 36, 72, 108, 144, 180, 216, 252, 288, 324, **360**, ...
The multiples of 120 are 120, 240, **360**, ...
The LCM of 36 and 120 is 360
As a product of its prime factors, $360 = 2 \times 2 \times 2 \times 3 \times 3 \times 5$

As a product of their prime factors the numbers 36 and 120 are

$36 = 2 \times 2 \times 3 \times 3$
$120 = 2 \times 2 \times 2 \times 3 \times 5$

The LCM of 36 and 120 (360) is the product of the common prime factors *and* all other prime factors, that is $2 \times 2 \times 3 \times 3 \times 2 \times 5$

In terms of products of powers of their prime factors,

$36 = 2^2 \times 3^2$
$120 = 2^3 \times 3 \times 5$

Their lowest common multiple, 360 ($= 2^3 \times 3^2 \times 5$), is the product of the *highest* power of all their prime factors.

## Example 17

Find the lowest common multiple (LCM) of 750 and 225

*Solution 17*
$750 = 2 \times 3 \times 5^3$

$225 = 3^2 \times 5^2$

| Write 750 and 225 as the product of powers of their prime factors. |

$LCM = 2 \times 3^2 \times 5^3 = 2250$

| The highest power of 2 is 1 (as $2 = 2^1$)<br>The highest power of 3 is 2<br>The highest power of 5 is 3 |

### Exercise 6G

**1** Find the two prime numbers that are between 30 and 40

**2** Find two prime numbers which have a sum of 7

**3** Find two prime numbers which have a product of 14

**4** Find two prime numbers which are factors of 20

**5** Find two prime numbers which are factors of 24

**6** Find two prime numbers which are factors of 33

**7** Write the following numbers as a product of two prime factors.
   **a** 10      **b** 15      **c** 21      **d** 22      **e** 33      **f** 39

**8** Which of the following show a number written correctly as a product of prime factors?
   **a** $12 = 2 \times 2 \times 3$      **b** $18 = 2 \times 9$      **c** $20 = 2 + 2 + 5$
   **d** $16 = 2 \times 2 \times 2 \times 2$      **e** $56 = 2, 2, 2, 7$      **f** $10 = 2 \times 5 \times 1$

**9** Write each of these numbers as a product of its prime factors.
   **a** 30      **b** 42      **c** 48      **d** 36      **e** 60      **f** 63
   **g** 54      **h** 80      **i** 76      **j** 88      **k** 68      **l** 66

**10** Find the highest common factor (HCF) of the following pairs of numbers.

   **a** 12 and 14     **b** 6 and 9     **c** 6 and 8     **d** 8 and 10     **e** 6 and 10

**11** Find the highest common factor (HCF) of the following pairs of numbers.

   **a** 12 and 18     **b** 10 and 15     **c** 16 and 20     **d** 18 and 24     **e** 24 and 30

**12** Find the lowest common multiple (LCM) of the following pairs of numbers.

   **a** 6 and 8     **b** 6 and 9     **c** 6 and 10     **d** 9 and 12     **e** 10 and 15

**13** Find the lowest common multiple (LCM) of the following pairs of numbers.

   **a** 12 and 15     **b** 12 and 24     **c** 12 and 18     **d** 18 and 24     **e** 20 and 24

**14 a** Find the number of multiples of 3 that are less than 100

   **b** Find the number of multiples of 5 that are less than 100

**15** Frank has two flashing lamps. The first lamp flashes every 4 seconds.
The second lamp flashes every 6 seconds. Both lamps start flashing together.

   **a** After how many seconds will they again flash together?

   **b** How many times in a minute will they flash together?

**16** As a product of its prime factors, $360 = 2 \times 2 \times 2 \times 3 \times 3 \times 5$
Write 720 as a product of its prime factors.

**17** The number 48 can be written in the form $2^n \times 3$    Find the value of $n$.

**18** The number 189 can be written in the form $3^n \times p$ where $n$ and $p$ are prime numbers.
Find the value of $n$ and the value of $p$.

**19** The number 120 can be written in the form $2^n \times m \times p$ where $n$, $m$ and $p$ are prime numbers.
Find the value of each of $n$, $m$ and $p$.

**20** $x = 2 \times 3^2 \times 5, y = 2^3 \times 3 \times 7$

   **a** Find the highest common factor (HCF) of $x$ and $y$.

   **b** Find the lowest common multiple (LCM) of $x$ and $y$.

**21** 2    3    5    7    21    22    24

   **a** Which of the numbers in the list are factors of 288?

   **b** Which of the numbers in the list are factors of 550?

**22** Write each of these numbers as a product of its prime factors.

   **a** 105             **b** 539             **c** 231

   **d** 847             **e** 1001

**23** Find the lowest common multiple of these pairs of numbers.

   **a** 24 and 30        **b** 27 and 36        **c** 28 and 35

   **d** 36 and 42        **e** 54 and 72

**24** Find all the integer values of $n$ less than or equal to 10 for which $2^n - 1$ is a prime number.

**25 a** Any square number which is even is always a multiple of 4. Explain why.

   **b** Investigate what the corresponding answer is for square numbers which are odd.

   **c** Explain why the number $10^{50} + 3$ cannot be a square number.

# Chapter summary

**You should now know that:**

★ the **power** or **index** of a number shows how many of the number are multiplied together, for example, in $2^5$ the 5 is the power or index and five 2s are multiplied together, that is $2 \times 2 \times 2 \times 2 \times 2$

★ the **square root** of a number is that number which when squared gives the original number

★ the **negative square root** of a number is that negative number which when squared gives the original number

★ the correct order of working out an expression is obtained by using **BIDMAS**

★ the **reciprocal** of a whole number $n$ is the fraction $\dfrac{1}{n}$

★ the **highest common factor (HCF)** of two numbers is the largest number which is a factor of both of the numbers

★ the **lowest common multiple (LCM)** of two numbers is the smallest number which is a multiple of both numbers.

**You should also be able to:**

★ use the rule for adding powers when two of the same number raised to a power are multiplied together

★ use the rule for subtracting powers when one number raised to a power is divided by the same number raised to a power

★ use a calculator to work out the square root of a number

★ use a calculator to work out powers of a number

★ write any whole number as a product of its prime factors

★ find the highest common factor of two or more numbers

★ find the lowest common multiple of two or more numbers.

# Chapter 6 review questions

**1** Work out

   **a** $40^2$       **b** $7000^2$       **c** $\sqrt{100}$       **d** $\sqrt{4900}$

**2** Work out

   **a** $2^3$       **b** $10^3$       **c** $5^3$       **d** $1^3$       **e** $30^3$

**3** Work out

   **a** $4^2 + \sqrt{25}$       **b** $2^3 + 2^3$       **c** $3^3 + 2^3$       **d** $5^3 + 10^3$

**4** Work out

   **a** $3 \times \sqrt{25}$       **b** $4 \times \sqrt{36}$       **c** $10 \times \sqrt{9}$       **d** $8 \times \sqrt{49}$

**5** Work out

   **a** $4^2 \div 8$       **b** $4^3 \div 8^2$       **c** $10^3 \div 10^2$       **d** $3^3 \div 3$

**6** Work out

 **a** $\sqrt{5^2}$          **b** $\sqrt{10^2}$          **c** $(\sqrt{9})^2$          **d** $(5-1)^2$          **e** $(2+3)^2$

**7** Write each of the following as a single power of 2

 **a** $2^3 \times 2^4$          **b** $2^6 \div 2^4$          **c** $(2^3)^2$          **d** $2 \times 2^8$          **e** $2^6 \div 2$

**8** Write each of the following as a single power of 3

 **a** $\dfrac{3 \times 3^4}{3^2}$          **b** $\dfrac{3^2 \times 3^3}{3^4}$          **c** $\dfrac{3^5 \times 3^4}{3^2 \times 3^3}$

**9** Each of the following represents a number written as a product of powers of its prime factors. Find the number.

 **a** $2^3 \times 3^2$          **b** $2 \times 3^3 \times 5$          **c** $2^3 \times 3 \times 7$

**10 a** Express 108 as a product of powers of its prime factors.
 **b** Find the highest common factor (HCF) of 108 and 24          (1387 June 2004)

**11 a** Express 120 as a product of its prime factors.
 **b** Find the lowest common multiple (LCM) of 120 and 150          (1387 November 2003)

**12** Find the reciprocal of 3.5   Give your answer as simply as possible.

**13** The number 40 can be written as $2^m \times n$, where $m$ and $n$ are prime numbers. Find the value of $m$ and the value of $n$.          (1387 June 2005)

**14 a** Write as a power of 5
   **i** $5^4 \times 5^2$     **ii** $5^9 \div 5^6$
 **b** $2^x \times 2^y = 2^{10}$ and $2^x \div 2^y = 2^4$
   Work out the value of $x$ and the value of $y$.          (1387 June 2005)

**15** Work out

 **a** $1.3^2$          **b** $\sqrt{13.69}$          **c** $25^3$          **d** $\sqrt{14}$

**16 a** The length of the side of a square is 4.8 cm. Work out the area of the square.
 **b** The area of a second square is 576 cm². Work out the length of one side of the square.

**17 a** Use your calculator to work out

   $$\dfrac{(6.2 - 3.9)^2}{1.25}$$

   Write down all the figures on your calculator display.
 **b** Put brackets in the expression so that the statement is true.
   $14.5 - 2.6 \times 4.5 - 3.6 = 49.95$          (Mock 2003)

**18** Work out the value of each of the following. Give each answer correct to one decimal place.

 **a** $\dfrac{14.7 \times 21.2}{2.5}$          **b** $\dfrac{18.7 \times 21.8}{2.5 + 3.7}$          **c** $\dfrac{\sqrt{200} + 8.6^2}{12.5}$

**19** Work out the value of $3.8^2 - \sqrt{75}$
 Write down all the figures on your calculator display.          (1387 June 2005)

**20** $y^2 = \dfrac{x^2 + 4x}{2t^2 - 6t}$

$x = 6.4, t = 4.6, y$ is a positive number.

**a** Work out the value of $y$. Write down all the figures on your calculator display.

**b** Round off your answer to an appropriate degree of accuracy.

**21** $p$ is a prime number not equal to 7

**a** Write down the highest common factor (HCF) of $49p$ and $7p^2$

$x$ and $y$ are different prime numbers.

**b**   **i** Write down the highest common factor (HCF) of the two expressions
$x^2y$      $xy^2$

**ii** Write down the lowest common multiple (LCM) of the two expressions
$x^2y$      $xy^2$

(1388 January 2005)

# Angles (1)

## 7.1 Triangles

| | | |
|---|---|---|
| Draw a triangle on paper and label its angles $a$, $b$ and $c$. 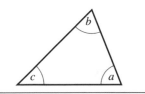 | Tear off its corners. 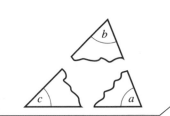 | Fit angles $a$, $b$ and $c$ together. They make a straight line. 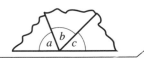 |

This *shows* that the angles in this triangle add up to 180° but it is not a **proof**.
That comes later in this chapter.

The angles on a straight line add up to 180° and so the angles in this triangle add up to 180°.

> **The angle sum of a triangle is 180°.**

### Example 1

Work out the size of angle $x$.

*Solution 1*

$72° + 57° = 129°$    | Add 72° and 57°

$180° − 129° = 51°$    | Take the result away from 180°, as the angle sum of a triangle is 180°.

$x = 51°$    | State the size of angle $x$.

Sometimes the fact that the angle sum of a triangle is 180° and other angle facts are needed.

### Example 2

Work out the size of
**a** angle $x$    **b** angle $y$.
**Give reasons for your answers.**

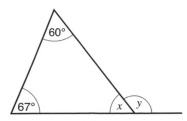

*Solution 2*

**a**   $67° + 60° = 127°$      **b**   $180° − 53° = 127°$
    $180° − 127° = 53°$             $y = 127°$
         $x = 53°$

| Angle sum of triangle is 180°. |    | Sum of angles on a straight line is 180°. |

**Exercise 7A**

In this exercise, the triangles are not accurately drawn.

In Questions **1–12**, find the size of each of the angles marked with letters and show your working.

**1**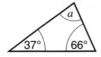

**2**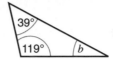

**3**

**4**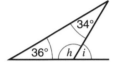

**5**

**6**

**7**

**8**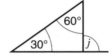

**9**

**10**

**11**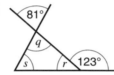

**12**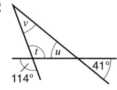

In Questions **13–15**, find the size of each of the angles marked with letters and show your working. **Give reasons for your answers.**

**13**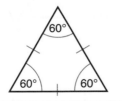

**14**

**15**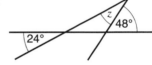

## 7.2 Equilateral triangles and isosceles triangles

An **equilateral** triangle has three equal sides and three equal angles.
As the angle sum of a triangle is 180°, the size of each angle is
180 ÷ 3 = 60°.

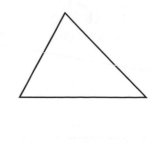

An **isosceles** triangle has two equal sides and the angles opposite the equal sides are equal.

A triangle whose sides are all different lengths is called a **scalene** triangle.

## Example 3

Work out the size of    **a** angle $x$    **b** angle $y$.
**Give reasons for your answers.**

*Solution 3*

**a**  $x = 41°$    | Isosceles triangle with equal angles opposite equal sides. |

**b**    $41° + 41° = 82°$

   $180° - 82° = 98°$    | Angle sum of triangle is 180°. |

      $y = 98°$

## Example 4

Work out the size of angle $x$.
**Give reasons for your answer.**

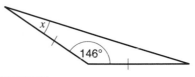

*Solution 4*

   $180° - 146° = 34°$    | Angle sum of triangle is 180°. |

   $34° \div 2 = 17°$    | Isosceles triangle with equal angles opposite equal sides. |

      $x = 17°$

## Exercise 7B

In this exercise, the triangles are not accurately drawn.

In Questions **1–12**, find the size of each of the angles marked with letters and show your working.

**1**

**2**
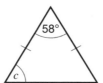

**3**
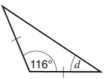

**4**
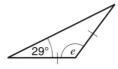

**5**

**6**
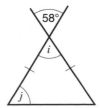

**7**
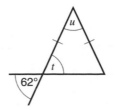

**8**
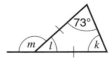

**9**
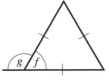

**10**
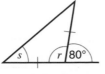

**11**
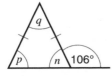

**12**
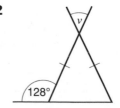

In Questions **13–15**, find the size of each of the angles marked with letters and show your working.
**Give reasons for your answers.**

**13**

**14**

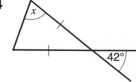

**15**

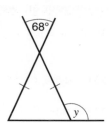

## 7.3 Corresponding angles and alternate angles

Parallel lines are always the same distance apart. They never meet.
In diagrams, arrows are used to show that lines are parallel.

In the diagram, a straight line crosses two parallel lines.
The shaded angles are called **corresponding angles** and are equal to each other.

The F shape formed by corresponding angles can be helpful in recognising them.

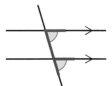

Other pairs of corresponding angles have been shaded in the diagrams below.

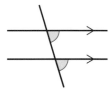

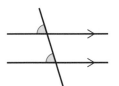

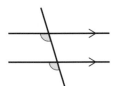

In the diagram, a straight line crosses two parallel lines.

The shaded angles are called **alternate angles** and are equal to each other.

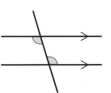

The Z shape formed by alternate angles can be helpful in recognising them.

Another pair of alternate angles has been shaded in this diagram.

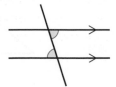

## Example 5

Write down the letter of the angle which is

**a** corresponding to the shaded angle,

**b** alternate to the shaded angle.

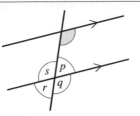

### Solution 5

**a** Angle $q$ is the corresponding angle to the shaded angle.

| Notice that they form an F shape. |

**b** Angle $s$ is the alternate angle to the shaded angle.

| Notice that they form a Z shape. |

## Example 6

**a** Find the size of angle $x$.

**b** Give a reason for your answer.

### Solution 6

**a** $x = 78°$

**b** Alternate angles.

## Example 7

**a** Find the size of angle $p$.

**b** Give a reason for your answer.

**c** Find the size of angle $q$.

**d** Give a reason for your answer.

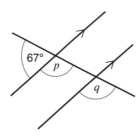

### Solution 7

**a** $180° - 67° = 113°$

$p = 113°$

**b** The sum of the angles on a straight line is 180°.

**c** $q = 113°$

**d** Corresponding angles.

## Exercise 7C

In this exercise, the diagrams are not accurately drawn.

**1** Write down the letter of the angle which is

  **a** corresponding to the shaded angle,

  **b** alternate to the shaded angle.

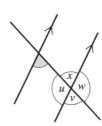

**2** Write down the letter of the angle which is

  **a** corresponding to the shaded angle,

  **b** alternate to the shaded angle.

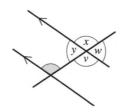

In Questions **3–5**, find the sizes of the angles marked with letters
and state whether the pairs of angles are corresponding or alternate.

**3**
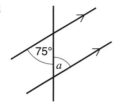

**4**
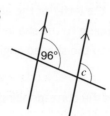

**5**
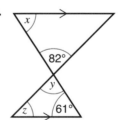

In Questions **6–20**, find the sizes of the angles marked with letters.
**Give reasons for your answers.**

**6**
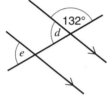

**7**
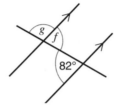

**8**
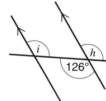

**9**
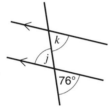

**10**
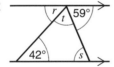

**11**
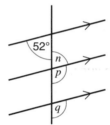

**12**
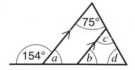

**13**

**14**
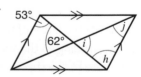

**15**
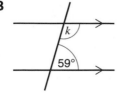

**16**
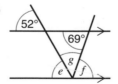

**17**
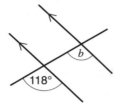

**18**
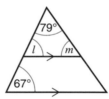

**19**
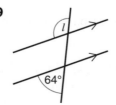

**20**
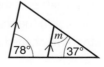

## 7.4 Proofs

In mathematics, a proof is a reasoned argument to show that a statement is always true. The proofs which follow make use of corresponding and alternate angles.

### Proof 1

**An exterior angle of a triangle is equal to the sum of the interior angles at the other two vertices**

The diagram shows a triangle $PQR$.

Extend the side $PQ$ to $S$.

At $Q$ draw a line $QT$ parallel to $PR$.

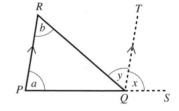

Then    angle $x$ = angle $a$ (corresponding angles)

and    angle $y$ = angle $b$ (alternate angles)

Adding,    $x + y = a + b$

$x + y$ is the exterior angle of the triangle and $a + b$ is the sum of the interior angles at the other two vertices and so the statement is true.

### Proof 2

**The angle sum of a triangle is 180°**

This proof starts in the same way as Proof 1.

The diagram shows a triangle $PQR$.

Extend the side $PQ$ to $S$.

At $Q$ draw a line $QT$ parallel to $PR$.

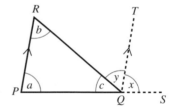

Then    angle $x$ = angle $a$ (corresponding angles)

and    angle $y$ = angle $b$ (alternate angles)

Adding,    $x + y = a + b$

As $x, y$ and $c$ are angles on a straight line, their angle sum is 180°, that is

$$x + y + c = 180°$$

So    $a + b + c = 180°$ which proves that the statement is true.

### Proof 3

**The opposite angles of a parallelogram are equal**

Draw a diagonal of the parallelogram.

angle $a$ = angle $c$ (alternate angles)

angle $b$ = angle $d$ (alternate angles)

Adding, $a + b = c + d$ which proves that the statement is true.

A parallelogram is a 4-sided shape with opposite sides parallel.

### Example 8

**a** Find the size of angle $w$.    **b** Give a reason for your answer.

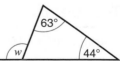

**Solution 8**

**a** $63° + 44° = 107°$

$w = 107°$

**b** Exterior angle of a triangle.

(As the full reason is long, it may be shortened to this.)

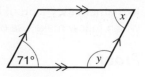

## Example 9

**a** Find the size of angle $x$.

**b** Give a reason for your answer.

**c** Find the size of angle $y$.

**d** Give reasons for your answer.

### Solution 9

**a** $x = 71°$

**b** Opposite angles of a parallelogram are equal.

**c** $\quad 2 \times 71° = 142°$

$360° - 142° = 218°$

$218° \div 2 = 109°$

$y = 109°$

**d** Angle sum of a quadrilateral is 360°.

Opposite angles of a parallelogram are equal.

### Exercise 7D

In this exercise, the diagrams are not accurately drawn.
Find the size of each of the angles marked with letters.
**Give reasons for your answers.**

**1**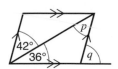

**2**

**3**

**4**

**5**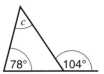

**6**

**7**

**8**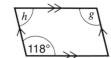

**9**

**10**
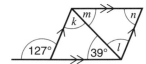

## 7.5 Bearings

**Bearings** are used to describe directions.

Bearings are measured **clockwise** ⟳ from **North**.

When the angle is less than 100°, one or two zeros are written in front of the angle, so that the bearing still has three figures.

## Example 10

Measure the bearing of $B$ from $A$.

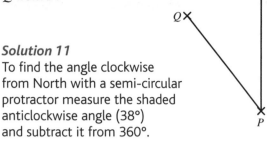

### Solution 10

From North, measure the angle clockwise.

The angle is 52°.

So the bearing is 052°.

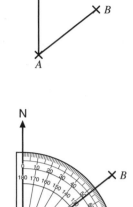

## Example 11

Measure the bearing of $Q$ from $P$.

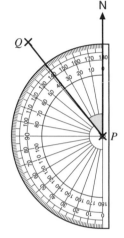

### Solution 11

To find the angle clockwise from North with a semi-circular protractor measure the shaded anticlockwise angle (38°) and subtract it from 360°.

$$360° - 38° = 322°$$

The bearing of $Q$ from $P$ is 322°.

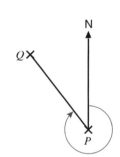

## Example 12

Folkestone and Dover are shown on the map.

The bearing of a ship from Folkestone is 117°.

The bearing of the ship from Dover is 209°.

Draw an accurate diagram to show the position of the ship.

Mark the position with a cross X. Label it $S$.

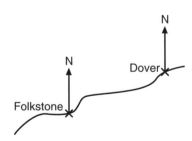

### Solution 12

Draw a line on a bearing of 117° from Folkestone.

Draw a line on a bearing of 209° from Dover by measuring an angle of 151° (360° − 209°) anticlockwise from North.

(Alternatively, measure an angle of 29° clockwise from *South*.)

Put a X where the lines cross.

Label the position $S$.

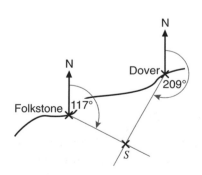

Sometimes, answers to questions have to be worked out, not found using a protractor.

### Example 13

The bearing of $B$ from $A$ is 061°.

Work out the bearing of $A$ from $B$.

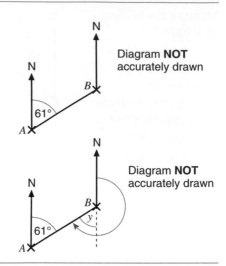

Diagram **NOT** accurately drawn

### Solution 13

The bearing of $A$ from $B$ is the reflex angle at $B$.

$\quad\quad y = 61°$ (alternate angles)

$\quad$ Bearing $= 180° + 61°$

$\quad\quad\quad\quad = 241°$

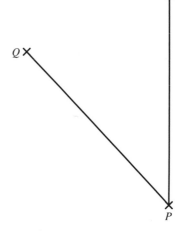

Diagram **NOT** accurately drawn

---

### Exercise 7E

In Questions **1–4**, measure the bearing of $Q$ from $P$.

**1**

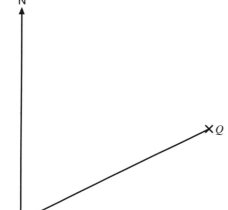

**2**

**3**

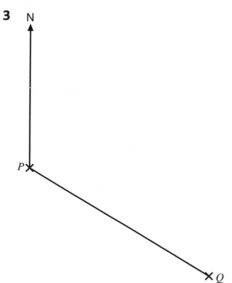

**4**

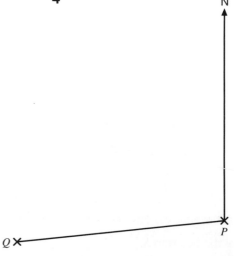

**5** Draw diagrams similar to those in Questions **1–4** to show the bearings

  **a**  026°    **b**  217°    **c**  109°    **d**  334°.

**6** The diagram shows two points, $A$ and $B$.
The bearing of a point $L$ from $A$ is 048°.
The bearing of $L$ from $B$ is 292°.
On the diagram on the resource sheet find the
position of $L$ by making an accurate drawing.

**7** The diagram shows two points, $P$ and $Q$.
The bearing of a point $M$ from $P$ is 114°.
The bearing of $M$ from $Q$ is 213°.
On the diagram on the resource sheet find the
position of $M$ by making an accurate drawing.

**8** Cromer and Great Yarmouth are shown on the map.
The bearing of a ship from Cromer is 052°.
The bearing of the ship from Great Yarmouth is 348°.
On the diagram on the resource sheet find the position
of the ship by making an accurate drawing.
Mark the position of the ship with a cross X.
Label it $S$.

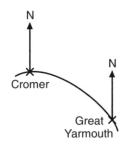

**9** The bearing of $Q$ from $P$ is 038°.
Work out the bearing of $P$ from $Q$.

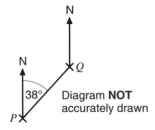

**10** The bearing of $T$ from $S$ is 146°.
Work out the bearing of $S$ from $T$.

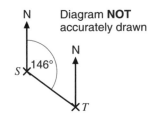

**11** The bearing of $B$ from $A$ is 074°.
The bearing of $C$ from $B$ is 180°.
$AB = AC$.
Work out the bearing of $C$ from $A$.

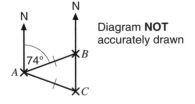

**12** The diagram shows the positions of
York, Scarborough and Hull.
The bearing of Scarborough from
York is 052°.
The bearing of Hull from York is 118°.
The distance between York and Scarborough is
the same as the distance between York and Hull.
Work out the bearing of Hull from Scarborough.

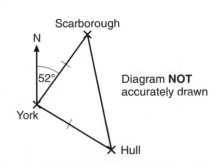

## Chapter summary

**You should know and be able to use these facts**

★ The angle sum of a triangle is 180°.

★ An **equilateral** triangle has three equal angles and three equal sides.

★ An **isosceles** triangle has two equal sides and the angles opposite the equal sides are equal.

★ A triangle whose sides are all different lengths is called a **scalene** triangle.

★ Where a straight line crosses two parallel lines, the corresponding angles are equal.

★ Where a straight line crosses two parallel lines, the **alternate angles** are equal.

★ **Bearings** are measured **clockwise** ⌒ from **North**.

**You should also know these proofs**

★ An exterior angle of a triangle is equal to the sume of the interior angles at the other two vertices.

★ The angle sum of a triangle is 180°.

## Chapter 7 review questions

In Questions **1–7**, find the size of each of the angles marked with a letter.
The diagrams are not accurately drawn.

**1**

**2**
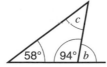

**3**

**4**
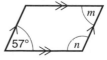

**5**
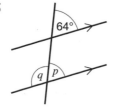

**6**

**7**
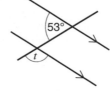

**6** In triangle $ABC$, $AB = AC$ and angle $C = 50°$.

   **a** Write down the special name of triangle $ABC$.

   **b** Work out the value of $y$.

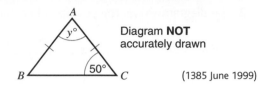

Diagram **NOT** accurately drawn

(1385 June 1999)

**7 a** **i** Write down the size of the angle marked $x$.

   **ii** Give a reason for your answer.

   **b** **i** Write down the size of the angle marked $y$.

   **ii** Give a reason for your answer.

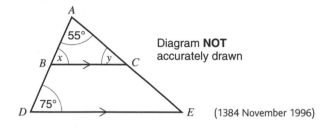

Diagram **NOT** accurately drawn

(1384 November 1996)

**8** $AC = BC$

   $AB$ is parallel to $DC$

   Angle $ABC = 52°$

   **a** Work out the value of    **i** $p$   **ii** $q$

   The angles marked $p°$ and $r°$ are equal.

   **b** What geometrical name is given to this type of equal angles?

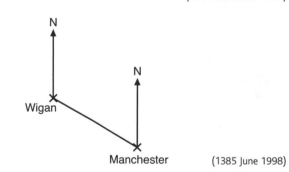

Diagram **NOT** accurately drawn

(1384 November 1997)

**9** The diagram represents the positions of Wigan and Manchester.

   **a** Measure and write down the bearing of Manchester from Wigan.

   **b** Find the bearing of Wigan from Manchester.

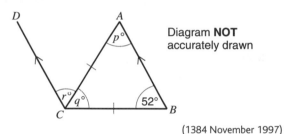

(1385 June 1998)

**10** Measure the bearing of $A$ from $B$.

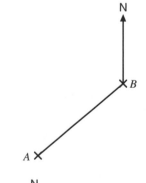

(1388 March 2004)

**11** Work out the bearing of

   **i** $B$ from $P$,

   **ii** $P$ from $A$.

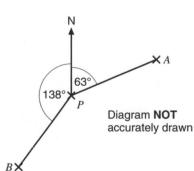

Diagram **NOT** accurately drawn

(1387 November 2004)

**12** The diagram shows the position of each of three buildings in a town.

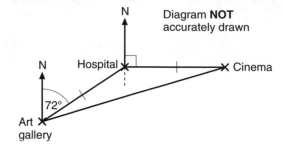

The bearing of the Hospital from the Art gallery is 072°.
The Cinema is due east of the Hospital.
The distance from the Hospital to the Art gallery is equal to the distance from the Hospital to the Cinema.
Work out the bearing of the Cinema from the Art gallery.

(1387 November 2003

# Expressions and sequences

## 8.1 Expressions and collecting like terms

There are three rows of students. Each row has the same number of students. We will use the letter $n$ to stand for the number of students in a row.

In three rows of students there are $3 \times n$ students.
This can be written as $3n$.
Five students are added. The total number of students is $3n + 5$

$3n$ and $3n + 5$ are called **algebraic expressions.**

An algebraic expression must contain at least one letter.
Each part of an expression is called a **term of the expression**.
For the expression $3n + 5$ the terms are $3n$ and $+5$

An expression can contain more than one letter.

---

### Example 1

There are $b$ large nails in each box and $c$ small nails in each packet.
Bob buys three boxes of large nails.

**a** Write down an expression, in terms of $b$, for the total number of large nails Bob buys.

Bob also buys two packets of small nails.
He uses 7 large nails and 5 small nails.

**b** Write down an expression, in terms of $b$ and $c$, for the total number of nails left.

#### Solution 1

**a** In the three boxes of large nails the total number of large nails is $3b$.

> $3 \times b$ is written as $3b$.

**b** In the two packets of small nails, there is a total of $2c$ small nails.

> Two packets of $c$ nails have $2 \times c$ nails.

Bob buys a total of $3b + 2c$ nails.

He uses a total of $7 + 5 = 12$ nails.

The total number of nails left is $3b + 2c - 12$

---

The value of an expression can be worked out if the value of each letter is known.

---

### Example 2

Work out the value of the expression $3p + 4q$ when $p = 10$ and $q = -2$

#### Solution 2

$3p + 4q = 3 \times 10 + 4 \times (-2)$
$\qquad\quad\;\; = 30 - 8$
$\qquad\quad\;\; = 22$

> positive $\times$ negative = negative.

### Example 3

Work out the value of the expression $2c - 5d - 15$ when $c = 3$ and $d = -1$

**Solution 3**

$$
\begin{aligned}
2c - 5d - 15 &= 2 \times 3 - 5 \times (-1) - 15 \\
&= 6 + 5 - 15 \\
&= 11 - 15 \\
&= -4
\end{aligned}
$$

> negative $\times$ negative $=$ positive.

In the expression $5p + 3q + 2p - 7q + 6$ the terms $5p$ and $+ 2p$ are called **like terms**. $+ 3q$ and $-7q$ are also like terms.

Like terms are terms which have the same letter. Like terms can be **collected**.

$5p$ and $+2p$ can be collected to give $7p$.
$+3q$ and $-7q$ can be collected to give $-4q$.

So the expression $5p + 3q + 2p - 7q + 6$ can be **simplified** to $7p - 4q + 6$

$7p$ and $-4q$ and $+6$ are not like terms so they cannot be collected.

### Example 4

Simplify $3x - 2y + 4 - 3y - 2x$

**Solution 4**

Collect like terms together to write
$3x - 2y + 4 - 3y - 2x$

$$= 3x - 2x - 2y - 3y + 4$$

> Collect the like terms.

$$= x - 5y + 4$$

> Since $3x - 2x = 1x = x$ and $-2y - 3y = -5y$.

### Exercise 8A

**1** Bill runs $x$ metres and Jane runs $y$ metres. Bill runs further than Jane. Write down an expression, in terms of $x$ and $y$, for the number of metres Bill has run further than Jane.

**2** Marbles are sold in packets and boxes.
There are $p$ marbles in each packet and $q$ marbles in each box.
Tim buys four packets and three boxes.

  **a** Write down an expression, in terms of $p$ and $q$, for the total number of marbles Tim buys.

  **b** Tim then gives 10 marbles to each of his two sisters. Write down an expression, in terms of $p$ and $q$, for the total number of marbles Tim has left.

**3** Each bicycle needs 2 wheels. Each tricycle needs 3 wheels. Each car needs 5 wheels.

  **a** What is the total number of wheels needed to make
    **i** 4 bicycles,
    **ii** 4 bicycles and 2 tricycles,
    **iii** 4 bicycles and 2 tricycles and 3 cars.

  **b** Write down an expression for the total number of wheels needed to make
    **i** $m$ bicycles, 2 tricycles and 3 cars,
    **ii** $m$ bicycles, $n$ tricycles and 3 cars,
    **iii** 4 bicycles, $n$ tricycles and $p$ cars,
    **iv** $m$ bicycles, 2 tricycles and $p$ cars,
    **v** $m$ bicycles, $n$ tricycles and $p$ cars.

**4** Work out the value of each of these expressions when $x = 5$ and $y = -4$

    **a** $2x + y$　　　　　　**b** $x + 2y$　　　　　　**c** $y - x$

    **d** $2x + 2y + 6$　　　　**e** $2x - 3y + 2$

**5** Work out the value of each of these expressions, when $x = 1$, $y = -4$ and $z = -3$

    **a** $x + y + z$　　　　　**b** $x + y - 2z$　　　　**c** $x + 3y - 2z$

    **d** $x - 2y - 3z$　　　　**e** $2x - 3y - 3z - 1$

**6** Simplify

    **a** $3x + x - 4y + 2y$　　　　　　**b** $6x - x - 5y - 2y$

    **c** $4x - 5y - 3x + 2y$　　　　　　**d** $x + 2x + 5y - 4x - 7y - 2y$

    **e** $5x - 3y - y - 2x$　　　　　　　**f** $x - 6y - 4x + 2y + 3x$

    **g** $4p + 2q - 7p - 3q$　　　　　　**h** $8m - n - 5n + 2m$

    **i** $2w + v - 8w + 4v + 2w$　　　　**j** $3c + 2d - 5d + c + d - 2c$

    **k** $5e - 3f - 2 - 5f - e - 1$　　　　**l** $4 - s - 3t - 7s + 8 - 2t$

## 8.2 Working with numbers and letters and using index notation

> also assessed in
> Module 4

The simplest way to write $4 \times a \times b$ is $4ab$, leaving out the multiplication signs.

So $2mn$ is the same as $2 \times m \times n$.

The simplest way to write $a \times b \times c$ is $abc$.

The simplest way to write $p \times q \div r$ is $\dfrac{pq}{r}$

### Example 5

Simplify $5x \times y \times 4z$

*Solution 5*

$$
\begin{aligned}
5x \times y \times 4z &= 5 \times x \times y \times 4 \times z \\
&= 5 \times 4 \times x \times y \times z \\
&= 20 \times xyz \\
&= 20\,xyz
\end{aligned}
$$

### Example 6

Work out the value of the expression $\dfrac{3ab + 6}{c}$ when $a = 2$, $b = 3$ and $c = 4$

*Solution 6*

$$\frac{3ab + 6}{c} = \frac{3 \times 2 \times 3 + 6}{4}$$

> Substitute the values of $a$, $b$ and $c$ into the expression.

$$= \frac{18 + 6}{4}$$

$$= \frac{24}{4}$$

> $\dfrac{24}{4}$ is $24 \div 4$

$$= 6$$

The simplest way to write $a \times a$ is $a^2$.
$a^2$ is read as '$a$ squared'
The number 2 in the expression $a^2$ is called the **index** or the power of $a$.

The simplest way to write $a \times a \times a$ is $a^3$.
$a^3$ is read as '$a$ cubed'

The index (or power) in the expression $a^3$ is 3.

This can be continued.
$a \times a \times a \times a = a^4$ read as    '$a$ times $a$ times $a$ times $a$ equals $a$ to the power 4'
$a \times a \times a \times a \times a = a^5$       $a^5$ is read as '$a$ to the power 5'
and so on.

### Example 7

Simplify $3p \times 2p \times 4p$.

*Solution 7*
$$\begin{aligned} 3p \times 2p \times 4p &= 3 \times p \times 2 \times p \times 4 \times p \\ &= 3 \times 2 \times 4 \times \boldsymbol{p} \times \boldsymbol{p} \times \boldsymbol{p} \\ &= 24 \times p^3 \\ &= 24p^3 \end{aligned}$$

### Example 8

Work out the value of the expression $3b^3$ when $b = 2$

*Solution 8*
$$\begin{aligned} 3b^3 &= 3 \times b^3 \\ &= 3 \times 2^3 \\ &= 3 \times 8 \\ &= 24 \end{aligned}$$

### Example 9

Work out the value of the expression $xy^2 + x^4$ when $x = 2$, $y = -3$

*Solution 9*
$$\begin{aligned} xy^2 + x^4 &= x \times y^2 + x^4 \\ &= 2 \times (-3)^2 + 2^4 \\ &= 2 \times 9 + 16 \\ &= 18 + 16 \\ &= 34 \end{aligned}$$

Only the $y$ is being squared in $xy^2$.

Negative squared = positive.

### Exercise 8B

**1** Write these expressions in their simplest form

   **a** $r \times s$              **b** $x \times y \times z$            **c** $m \times m$

   **d** $12 \times s \times t \times u$      **e** $6x \times y \times z$         **f** $a \times 24 \times b \times c$

   **g** $w \times w \times w$          **h** $3 \times x \times x$           **i** $a \times 4 \times a \times a$

   **j** $a \times a \times b \times b \times c \times c$      **k** $p \times p \times q \times q \times p$

**2** Simplify

   **a** $3x \times 2$       **b** $x \times 4q$        **c** $6y \times 4x$        **d** $4m \times 2p \times 2$

   **e** $3c \times 2d \times 5e$    **f** $5a \times a \times 4a$     **g** $2ef \times 3e$

**3** Work out the value of each of these expressions when $x = -1$

   **a** $x^2$          **b** $x^3$          **c** $x^4$          **d** $x^5$         **e** $x^6$

**4** Work out the value of each of these expressions, when $x = 10$

   **a** $x^2$          **b** $x^3$          **c** $2x^2$        **d** $\dfrac{x^2}{5}$       **e** $\dfrac{x^6 - x^5}{x^4}$

**5** Work out the value of the expression $2a^2 - a$ when $a = -4$

**6** Work out the value of the expression $3a^3 - 2a$ when $a = -2$

**7** Work out the value of each of these expression when $a = -2$

   **a** $\dfrac{3a^3}{a^2}$          **b** $3a^2 + 8$          **c** $\dfrac{3a^2 + 8}{a}$          **d** $\dfrac{2a^3 + 6}{a}$

**8** Work out the value of each of these expressions when $a = 2$, $b = 3$, $c = -1$

   **a** $abc$          **b** $4ab + c$          **c** $5bc - ac$          **d** $\dfrac{ab}{c}$          **e** $\dfrac{b + 2c^2}{a}$

**9** Work out the value of each of these expressions, when $p = 2$, $q = -3$, $r = 5$

   **a** $pq^2$          **b** $4rp^2 - 10q^2$          **c** $p^3 - q^3$          **d** $\dfrac{p^4 + 2r^2}{pq}$          **e** $20 - \dfrac{pq^2r}{p - q + r}$

**10** Work out the value of the expression $3(x + 1)^2$ when $x = -3$

## 8.3  Index laws

also assessed in
Module 4

Expressions involving powers of the same letter can be multiplied.

$$a^2 \times a^3 = (a \times a) \times (a \times a \times a)$$
$$= a \times a \times a \times a \times a$$
$$= a^5$$

Note that $a^2 \times a^3 = a^{2+3} = a^5$.

> **In general, $x^m \times x^n = x^{m+n}$ (add the indices).**

Expressions involving powers of the same letter can be divided.

$$a^5 \div a^2 = \frac{a^5}{a^2}$$
$$= \frac{\cancel{a} \times \cancel{a} \times a \times a \times a}{\cancel{a} \times \cancel{a}}$$
$$= a \times a \times a$$
$$= a^3$$

Note that $a^5 \div a^2 = a^{5-2} = a^3$

> **In general, $x^m \div x^n = x^{m-n}$ (subtract the indices).**

$x^m \times x^n = x^{m+n}$ and $x^m \div x^n = x^{m-n}$ are two of the **laws of indices** which can be used to simplify expressions.

### Example 10

Simplify $3b^4 \times 2b^6$.

*Solution 10*
$$3b^4 \times 2b^6 = 3 \times b^4 \times 2 \times b^6$$
$$= 3 \times 2 \times b^4 \times b^6$$
$$= 6 \times b^{4+6}$$
$$= 6 \times b^{10}$$
$$= 6b^{10}$$

## Example 11

Simplify $8p^7 \div 2p^6$.

**Solution 11**

$$8p^7 \div 2p^6 = \frac{\overset{4}{\cancel{8}} \times p^7}{\underset{}{\cancel{2}} \times p^6} = \frac{4p^7}{p^6}$$

$$= 4 \times p^7 \div p^6$$

$$= 4 \times p^{7-6}$$

$$= 4p^1$$

$$= 4p \quad \boxed{\text{Since } p^1 = p.}$$

## Example 12

Simplify $3ab^3 \times 5a^2b^4$.

**Solution 12**

$$3ab^3 \times 5a^2b^4 = 3 \times 5 \times a^1 \times a^2 \times b^3 \times b^4$$

$$= 15 \times a^{1+2} \times b^{3+4}$$

$$= 15a^3b^7$$

The two index laws found so far can be used to find some more laws.

When $m = n$, the law $x^m \div x^n = x^{m-n}$ becomes $x^n \div x^n = x^{n-n}$.

$$\text{So} \qquad\qquad 1 = x^0.$$

$x^n \div x^n = 1$ because anything divided by itself is 1

> **So for all non-zero values of $x$, $x^0 = 1$**

When $m = 0$ the law $x^m \div x^n = x^{m-n}$ becomes $x^0 \div x^n = x^{0-n}$.
But $x^0 = 1$, so $1 \div x^n = x^{-n}$.

> **This can be written as $x^{-n} = \dfrac{1}{x^n}$**    For example, $x^{-2} = \dfrac{1}{x^2}$.

## Example 13

$\dfrac{1}{x^3} = x^2 \times x^k$. Find the value of $k$.

**Solution 13**

$$\frac{1}{x^3} = x^2 \times x^k$$

$$x^{-3} = x^2 \times x^k \qquad \boxed{\text{Using } x^{-n} = \frac{1}{x^n} \text{ with } n = 3}$$

$$x^{-3} = x^{2+k} \qquad \boxed{\text{Using } x^m \times x^n = x^{m+n} \text{ with } m = 2 \text{ and } n = k.}$$

$$-3 = 2 + k \qquad \boxed{\text{Equating the powers of } x.}$$

$$k = -5$$

Expressions like $(x^3)^2$ and $(x^2)^3$ can be simplified using the law $x^m \times x^n = x^{m+n}$.

$(x^3)^2 = x^3 \times x^3 = x^{3+3} = x^6$

$(x^2)^3 = x^2 \times x^2 \times x^2 = x^{2+2+2} = x^6$

> **In general $\left(x^m\right)^n = x^{mn}$**

## Example 14

Simplify $(2a^3b^2)^4$.

**Solution 14**

$$(2a^3b^2)^4 = 2^4 \times (a^3)^4 \times (b^2)^4$$

$$= 16 \times a^{12} \times b^8 \qquad \boxed{\text{Using } (x^m)^n = x^{mn}}$$

$$= 16a^{12}b^8$$

## Example 15

Simplify $(2d^{-3})^{-2} \div d$.

**Solution 15**

$(2d^{-3})^{-2} \div d$

$= 2^{-2} \times (d^{-3})^{-2} \div d$

$= \dfrac{1}{4} \times d^6 \div d$

> $2^{-2} = \dfrac{1}{2^2}$ using the law $x^{-n} = \dfrac{1}{x^n}$ and $(d^{-3})^{-2} = d^6$
>
> using the law $(x^m)^n = x^{mn}$ and $(-3) \times (-2) = 6$

$= \dfrac{1}{4} d^5$

> $d = d^1$ and using law $x^m \div x^n = x^{m-n}$

## Exercise 8C

**1** Simplify

  **a** $x^3 \times x^2$
  **b** $y^5 \times y^3$
  **c** $n \times n^6$
  **d** $q^7 \times q$
  **e** $x^5 \div x^3$

  **f** $y^7 \div y^3$
  **g** $p^5 \div p^4$
  **h** $q^7 \div q$
  **i** $y \times y^4 \times y^3$
  **j** $q^4 \times q \div q^3$

**2** Simplify

  **a** $3x^2 \times x^5$
  **b** $4p \times 2p^4$
  **c** $4p \times 5p$
  **d** $2 \times 2r^8 \times 4r$

  **e** $6y^6 \div 2y^3$
  **f** $12q^2 \div 6q$
  **g** $8x^9 \div 2x^8$
  **h** $4q \div 2q$

  **i** $2y^2 \times 3y^3 \times y^3$
  **j** $6q \times 5q^4 \div 2q^5$

**3** Simplify

  **a i** $x^2 \times x$
  **ii** $x^5 \div x^2$
  **iii** $(x^2 \times x) + (x^5 \div x^2)$

  **b** $(8y^6 \div 2y^2) - (2y^2 \times y^2)$

**4** Simplify

  **a** $3a^4 \times a^3b^2$
  **b** $2ab^4 \times 4a^3b$
  **c** $5p^4q^3 \times 2q^3p^2$

  **d** $18x^8y^6 \div 6x^3y^2$
  **e** $12a^3b^5 \div 3a^3b$
  **f** $20p^4q \div 2p^3q^2$

**5** Find the value of

  **a** $4x^0$
  **b** $(xy)^0$

**6** Write as a power of $x$

  **a** $\dfrac{1}{x^4}$
  **b** $\dfrac{1}{x}$
  **c** $\dfrac{1}{x^4 \times x^3}$
  **d** $\dfrac{1}{x^4 \div x}$
  **e** $\dfrac{1}{x^5 \div x^7}$

**7** Simplify

  **a** $(x^5)^3$
  **b** $(2y^2)^4$
  **c** $(a^2b^4)^5$
  **d** $(3a^3b)^3$

  **e** $(x^2)^{-1}$
  **f** $(4y^{-2})^2$
  **g** $(a^{-4})^{-3}$
  **h** $(-2b^{-4})^{-2} \div b^2$

**8** Simplify

  **a** $(x^2)^0$
  **b** $(y^0)^4$
  **c** $(2a^3b^{-2})^{-3} \times (2a^{-3}b)^3$

## 8.4 Sequences

A **sequence** is a pattern of shapes or numbers which follow a rule.

Here is a sequence of patterns made up of yellow and blue triangles.

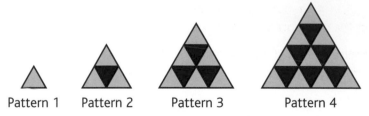

| | Pattern 1 | Pattern 2 | Pattern 3 | Pattern 4 |

| Pattern number | 1 | 2 | 3 | 4 |
|---|---|---|---|---|
| Number of yellow triangles | 1 | 3 | 6 | 10 |
| | | +2 | +3 | +4 |
| Number of **blue** triangles | 0 | 1 | 3 | 6 |
| | | +1 | +2 | +3 |
| Total number of triangles | 1 | 4 | 9 | 16 |
| | | +3 | +5 | +7 |

The number of yellow triangles form the number sequence 1, 3, 6, 10 ....

The differences between pairs of consecutive terms increase by 1. This is the **rule** which describes the sequence. Knowing the rule, all the terms in the sequence can be found.
The next term in the sequence is 15 ($=10 + 5$).
The numbers 1, 3, 6, 10, 15 ... are called **triangular numbers.**

The number of **blue** triangles form the number sequence 0, 1, 3, 6 ...

Again the rule is that the differences between pairs of consecutive terms increase by 1.
The next term in the sequence is 10 ($=6 + 4$).

The total number of triangles in each pattern form the number sequence 1, 4, 9, 16, 25 ...

1, 4, 9, 16 and 25 are **square numbers**.

$1 = 1^2$, $4 = 2^2$, $9 = 3^2$, $16 = 4^2$, $25 = 5^2$ so the next term in the sequence is 36 ($= 6^2$)

So in the sequence 1, 4, 9, 16, 25 ...

> **the 10th term is $10^2$ or 100.**
> **the 100th term is $100^2$ or 10 000**

In general, the $n$th term of the sequence of square numbers is given by the expression $n^2$.

### *Finding an expression for the nth term of an arithmetic sequence*

Here are some examples of sequences

> A  9, 10, 11, 12, 13 ...
> B  5, 8, 11, 14, 17 ...
> C  20, 18, 16, 14, 12 ...

These are all called **arithmetic sequences**. An arithmetic sequence is a sequence of numbers with the rule 'add a fixed number'.
The fixed number is the **difference**, or step, between consecutive terms.
In **A** the first term of the sequence is 9 and the second term is 10. So the difference is 1 since $10 - 9 = 1$, and also $11 - 10 = 1$ and so on.
In **B** the difference is 3 since $8 - 5 = 3$, $11 - 8 = 3$ and so on.
In **C** the terms decrease in value. The difference is $-2$ since $18 - 20 = -2$, $16 - 18 = -2$ and so on.

This machine shows the operation $\times 3$ followed by the operation $+2$

This is called a **two-stage input and output machine.** It is sometimes called a two-stage number machine.
The arrows show the direction through the machine.

> **When the input is 1, the output is $(1 \times 3 + 2) = 5$**
> **When the input is 2, the output is $(2 \times 3 + 2) = 8$**
> **When the input is 3, the output is $(3 \times 3 + 2) = 11$**
> **When the input is 4, the output is $(4 \times 3 + 2) = 14$**

This number machine, for inputs 1, 2, 3 and 4, gives the first four terms of arithmetic sequence **B** above: 5, 8, 11, 14 ...

To find the output, when the input is $n$, work forwards through the machine.

$n \times 3 + 2 = 3n + 2$ so the output is $3n + 2$

Here is a completed table for this input and output machine.

The $n$th term is the output when the input is $n$.

The $n$th term of the arithmetic sequence 5, 8, 11, 14, ... is $3n + 2$

The 3 in $3n + 2$ is the difference between consecutive terms.
The $+2$ in $3n + 2$ is the term before the 1st term, sometimes called the **zero term**.
The zero term is the output when the input is $0 (3 \times 0 + 2 = 2)$.

| ×3 | +2 | |
|---|---|---|
| **Input** | **Output** | |
| 0 | 2 | zero term |
| 1 | 5 | 1st term |
| 2 | 8 | 2nd term |
| 3 | 11 | 3rd term |
| 4 | 14 | 4th term |
| $n$ | $3n + 2$ | $n$th term |

In general, for an arithmetic sequence, the $n$th term is the

> **difference $\times n$ + zero term**

## Example 16

Here are the first five terms of an arithmetic sequence 7, 11, 15, 19, 23
Write down

**a** the difference between consecutive terms,

**b** the zero term,

**c** an expression, in terms of $n$, for the $n$th term of this sequence.

**Solution 16**

| Input | Output |            |
|-------|--------|------------|
| 0     |        | zero term  |
| 1     | 7      | 1st term   |
| 2     | 11     | 2nd term   |
| 3     | 15     | 3rd term   |
| 4     | 19     | 4th term   |
| 5     | 23     | 5th term   |
| $n$   |        | $n$th term |

**a** Difference $= 11 - 7 = 4$    | The difference between consecutive terms is 4, since $11 - 7 = 4$ |

**b** Zero term is $7 - 4 = \textbf{3}$    | To find the zero term, subtract from the first term the difference between consecutive terms. |

**c** $n$th term $= 4n + \textbf{3}$    | $n$th term $=$ **difference** $\times n +$ **zero term**. |

---

## Example 17

Here are the first five terms of an arithmetic sequence.    9, 7, 5, 3, 1
Write down, in terms of $n$, an expression for the $n$th term of this sequence.

**Solution 17**
Difference $= 7 - 9 = -2$

Zero term $= 9 - (-2) = 9 + 2 = 11$    | Subtract a negative $=$ add a positive. |

$n$th term $= -2n + 11$    | $n$th term $=$ difference $\times n +$ zero term. |

---

## Using the nth term of a sequence

## Example 18

The $n$th term of a number sequence is $\dfrac{n(n + 1)}{2}$.

Find the values of the first four terms of this sequence.

**Solution 18**

1st term $= \dfrac{1 \times (1 + 1)}{2} = \dfrac{1 \times 2}{2} = 1$    | To find the 1st term, substitute $n = 1$ |

2nd term $= \dfrac{2 \times (2 + 1)}{2} = 3$    | To find the 2nd term, substitute $n = 2$ |

3rd term $= \dfrac{3 \times (3 + 1)}{2} = 6$    | To find the 3rd term, substitute $n = 3$ |

4th term $= \dfrac{4 \times (4 + 1)}{2} = 10$    | To find the 4th term, substitute $n = 4$ |

| Since 1, 3, 6, 10 are the first four triangular numbers, $\dfrac{n(n + 1)}{2}$ is an expression, in terms of $n$, for the $n$th term of the sequence of triangular numbers. |

## Exercise 8D

Questions **1–5** each show an arithmetic sequence.
For each arithmetic sequence
**a** Write down the difference between consecutive terms
**b** Write down the zero term
**c** Write down, in terms of $n$, an expression for the $n$th term of the sequence.

**1** 5, 6, 7, 8 ...

**2** 5, 9, 13, 17, 21 ...

**3** 2, 5, 8, 11, 14 ...

**4** 22, 19, 16, 13, 10 ...

**5** −4, 0, 4, 8, 12 ...

**6** Here are the first five terms of an arithmetic sequence 6, 11, 16, 21 and 26
  **a** Write down, in terms of $n$, an expression for the $n$th term of this sequence.
  **b** Find the 100th term of the sequence.

**7** Here are the first five terms of an arithmetic sequence 1, 7, 13, 19 and 25
  **a** Write down, in terms of $n$, an expression for the $n$th term of this sequence.
  **b** Find the 50th term of the sequence.

**8** Here are the first five terms of an arithmetic sequence −4, −1, 2, 5 and 8
  **a** Write down, in terms of $n$, an expression for the $n$th term of this sequence.
  **b** Find the 1000th term of the sequence.

**9** Here are the first five terms of an arithmetic sequence 20, 10, 0, −10 and −20
  **a** Write down, in terms of $n$, an expression for the $n$th term of this sequence.
  **b** Find the 20th term of the sequence.

**10 a** Write down, in terms of $n$, an expression for the $n$th term of the arithmetic sequence
    1, 3, 5, 7 ...
  **b** Write down, in terms of $n$, an expression for the $n$th term of the arithmetic sequence
    2, 4, 6, 8 ...
  **c** Use your answers to parts **a** and **b** to find, in terms of $n$, an expression for the $n$th term of the
    arithmetic sequence 3, 7, 11, 15 ...

**11** If $n$ is a whole number, what type of whole number is $(2n - 1)$?

**12** The $n$th term of a number sequence is $5n$.
  **a** Find the values of the first four terms of this sequence.
  **b** Find  **i** the 20th term  **ii** the 100th term.

**13** The $n$th term of a number sequence is $(5 - n^2)$.
  **a** Find the values of the first five terms of this sequence.
  **b** Find  **i** the 10th term  **ii** the 12th term.

**14** The $n$th term of a number sequence is $(n^2 - n)$.

    **a** Find the values of the first four terms of this sequence.

    **b** Find the value of the 10th term of this sequence.

**15** An expression for the $n$th term of a number sequence is $3^n$.
Find the values of the first four terms of this sequence.

**16** Here are the first four terms of a number sequence 2, 4, 8 and 16.
Find an expression for the $n$th term of the number sequence.

## Chapter summary

**You should now know that:**

★ an **algebraic expression** must contain at least one letter. For example, $3n + 5$ is an algebraic expression

★ the $3n$ and the $+5$ are called terms of the expression $3n + 5$

★ to **collect like terms**, for example in $2a + 3b + a - 4b$, combine the terms which contain the same letter to give $3a - b$

★ to simplify expressions which have a multiplication sign, remove the multiplication sign, for example $4 \times a = 4a$

★ when the same letter is multiplied you use indices, for example $a \times a = a^2$

★ to find the value of an expression, substitute the value of each letter and work out the result

★ $x^m \times x^n = x^{m+n}$, $x^m \div x^n = x^{m-n}$ and $(x^m)^n = x^{mn}$ are three laws of indices which can be used to simplify expressions

★ $x^0 = 1$ and $x^{-n} = \dfrac{1}{x^n}$ for all non-zero values of $x$

★ a **sequence** is a pattern of shapes or numbers which follow a rule

★ the **terms** of a sequence are the numbers in the sequence

★ an **arithmetic sequence** is a sequence of numbers with the rule, 'add a fixed number'

★ the **difference** between consecutive terms in an arithmetic sequence is the increase or decrease in value from one term to the next

★ the **zero term** of a sequence is the output when the input is 0

★ to find the $n$th term of an arithmetic sequence you use

    $n$th term = the difference $\times n$ + zero term

# Chapter 8 review questions

1 Eggs are sold in boxes. There are 6 eggs in each box. Mrs Smith buys $p$ boxes of eggs.
   a Write down an expression, in terms of $p$, for the number of eggs that Mrs Smith buys.
   She uses 5 of these eggs.
   b Write down an expression, in terms of $p$, for the number of eggs left.

2 A bicycle has 2 wheels.
   A tricycle has 3 wheels.
   In a shop there are $p$ bicycles and $q$ tricycles.
   Write down an expression, in terms of $p$ and $q$, for the total number of wheels on the bicycles and the tricycles in the shop.

3 Simplify   a $6a - b - 2a - 2b$   b $5x^2 + y^3 + 4x^2 - 3y^3$   c $6xy + 2wz + 3xy - 5wz$

4 $P = a + bc$
   Work out the value of $P$ when $a = -5, b = 6$ and $c = 2$

5 a Simplify $x^2 + 2x^2 + 3x^2$.
   b Work out the value of these expressions when $x = -10$
     i $x^2$   ii $x^2 + 2x^2 + 3x^2$.

6 $p = 3, q = -2$
   Work out the value of   a $p^2 + q^2$   b $2p + q^3$

7 Work out the value of the expression $4mn^2$ when $m = -2, n = 3$

8 Simplify   a $d^4 \times d^6$   b $e^9 \div e^4$   c $6p^5 \times 3p^2$   d $6q^9 \div 2q^3$

9 Here are the first five terms of a number sequence.
        16   12   8   4   0
   a Write down the next two terms of the sequence.

   Here are the first six terms of a different number sequence.
        7   12   17   22   27   32
   b Write down an expression, in terms of $n$, for the $n$th term of this sequence.    (1385 November 2001)

10 The $n$th term of an arithmetic sequence is $5n - 6$
   Find the value of
   a the 1st term of this sequence        b the 2nd term of this sequence,
   c the 3rd term of this sequence        d the 10th term of this sequence.

11 Here is a table for an input and output machine.
   Copy and complete the table.

| +1 | ×3 |
|---|---|
| Input | Output |
| 1 | |
| 2 | |
| 4 | |
| | 21 |
| | 30 |
| $n$ | |

**12** Here are the first 5 terms of an arithmetic sequence.

$$7 \quad 10 \quad 13 \quad 16 \quad 19$$

Find an expression, in terms of $n$, for the $n$th term of this sequence.

**13** Here are the first four terms of an arithmetic sequence.

$$3 \quad 7 \quad 11 \quad 15$$

Write down, in terms of $n$, an expression for the $n$th term of the sequence.          (1388 January 2005)

**14** Here are the first five terms of an arithmetic sequence.

$$24 \quad 20 \quad 16 \quad 12 \quad 8$$

**a** Write down, in terms of $n$, an expression for the $n$th term of this sequence.
**b** Find the 20th term of the sequence.

**15** Here are the first five terms of an arithmetic sequence.

$$-1 \quad 3 \quad 7 \quad 11 \quad 15$$

**a** Find, in terms of $n$, an expression for the $n$th term of this sequence.

In another arithmetic sequence the $n$th term is $8n - 16$
John says that there is a number that is in both sequences.
**b** Explain why John is wrong.                                              (1388 March 2004)

**16** Here are the first 5 terms of an arithmetic sequence.

$$8 \quad 11 \quad 14 \quad 17 \quad 20$$

**a** **i** Find an expression, in terms of $n$, for the $n$th term of this sequence.
   **ii** Joan says the number 100 is a term of this sequence. Joan is wrong. Explain why.

The $n$th term of another sequence is given by $2n^2 - 16n$.
**b** Work out the 8th term of this sequence.

**17** **a** Simplify $x^5 \div x^2$          **b** Simplify $2w^4y \times 3w^3y^2$          (1388 March 2003)

**18** **a** Simplify $12y^3 \div 3y^5$          **b** Simplify $2w^3x^2 \times 3w^4x$          (1388 January 2003)

**19** **a** $7^5 \times 7^6 = 7^3 \times 7^k$   Find the value of $k$.          **b** Simplify $\dfrac{15a^3b^7}{3a^2b^3}$          (1388 March 2004)

**20** Simplify fully
   **a** $(p^3)^3$          **b** $\dfrac{3q^4 \times 2q^5}{q^3}$          (1387 June 2003)

**21** Simplify fully
   **a** $(t^{-1})^{-3}$          **b** $(2x^2y^3)^4$          **c** $\dfrac{(a^{-2})^{-4}}{a^2 \times a^6}$

# Measure (1)

## 9.1 Compound measures – speed and density

### Speed

A car travelled 90 kilometres in 3 hours. If it had travelled the 90 km at the same speed for the whole 3 hours, then the car would travel $\frac{90}{3} = 30$ km each hour. This means that 30 kilometres per hour is the **average speed** of the car. The word 'per' here means 'each' or 'for every'. Notice that the distance travelled was divided by the time taken so

$$\text{average speed} = \frac{\text{total distance travelled}}{\text{total time taken}}$$

Speed can be measured, for example, in kilometres per hour, miles per hour and metres per second. Speed is called a **compound measure** because it involves more than one unit of measure. It involves a unit of length and a unit of time.

We write 30 kilometres per hour as 30 km/h, where the '/' is a sort of division sign showing that speed is distance divided by time.

If a car travels at an average speed of 40 km/h, the car travels 40 km in 1 hour,

$40 \times 2 = 80$ km in 2 hours
$40 \times 3 = 120$ km in 3 hours
and so on.

So $\qquad$ **distance = average speed × time**

The time the car takes to travel 120 km at 40 km/h is $\frac{120}{40} = 3$ hours.

So $\qquad$ $$\text{time} = \frac{\text{distance}}{\text{average speed}}$$

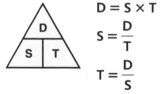

$D = S \times T$

$S = \dfrac{D}{T}$

$T = \dfrac{D}{S}$

Using D to stand for distance, S to stand for average speed and T to stand for time, learning this diagram is a way to remember these results.

---

### Example 1

The distance from Birmingham to Swansea is 155 km. Nitesh drove from Birmingham to Swansea in $2\frac{1}{2}$ hours. Work out Nitesh's average speed for this journey.

**Solution 1**

Average speed $= \dfrac{155}{2.5} = 62$ km/h

$$\text{average speed} = \frac{\text{total distance travelled}}{\text{total time taken}}$$

The distance is in km and the time is in hours, so the speed is in km/h.

---

### Example 2

The distance from Glasgow to Liverpool is 348 km.
Susan drove from Glasgow to Liverpool at an average speed of 40 km/h.
Work out the time, in hours and minutes, her journey took.

**Solution 2**

$$\text{Time} = \frac{348}{40}$$

$time = \dfrac{distance}{average\ speed}$

$$= 8.7\,\text{h}$$

Speed is in km/h and distance is in km so the time is in hours.

$$0.7 \times 60 = 42$$

$8.7\,\text{h} = 8\,\text{h} + 0.7\,\text{h}$

$$\text{Time} = 8\,\text{h }42\text{ minutes}$$

To change 0.7 h to minutes, multiply 0.7 by 60, as there are 60 minutes in an hour.

## Exercise 9A

**1** Paul went for a walk. He walked 21 km in $3\frac{1}{2}$ hours. Work out his average speed on the walk.

**2** Greg went for a cycle ride. He rode for 4 hours 15 minutes at an average speed of 20 km/h. Work out the distance that Greg rode.

**3** June drove the 275 km from London to Hull at an average speed of 50 km/h. Work out the time her journey took.

**4** Janet ran 3.8 km in 40 minutes. Work out her average speed in km/h.

**5** Sven drove the 119 km from Bournemouth to Bristol at an average speed of 68 km/h. Work out the time, in hours and minutes, his journey took.

**6** A car is travelling at an average speed of 85 m/s. Work out the distance the car travels in 0.4 seconds.

**7** A horse ran 12 km at an average speed of 10 km/h. How long, in hours and minutes, did this take?

**8** Change a speed of 85 m/s to km/h.

**9** John drove from his home to visit a friend.
John drove the first 3 hours at an average speed of 40 km/h.
John then drove the remaining 60 km to his friend's house at an average speed of 30 km/h.
Work out John's average speed for his whole journey from his home to his friend's house.

**10** In an athletics match the 100 m was won in a time of 9.91 s and the 200 m was won in a time of 19.79 s. Which race was won with the faster average speed? You must give a reason for your answer.

## *Density*

To work out the density of a substance, divide its mass by its volume

$$density = \frac{mass}{volume}$$

Density is also a compound measure. It involves a unit of mass and a unit of volume.
In the diagram M stands for mass, D stands for density and V stands for volume.
From the diagram

$$mass = density \times volume \qquad M = D \times V$$

$$density = \frac{mass}{volume} \qquad D = \frac{M}{V}$$

$$volume = \frac{mass}{density} \qquad V = \frac{M}{D}$$

When the mass is measured in kg and the volume in cubic metres or m³, then density is measured in kg per m³ or kg/m³. Density can also be measured in g/cm³.

## Example 3

A piece of silver has a mass of 42 g and a volume of 4 cm³. Work out the density of silver.

**Solution 3**

Density $= \dfrac{\text{mass}}{\text{volume}}$

Density $= \dfrac{42}{4}$

   | Divide the mass by the volume. |

  $= 10.5 \text{ g/cm}^3$

   | As the mass is in g and the volume is in cm³, the density is in g/cm³. |

## Example 4

The density of steel is 7700 kg/m³.

**a** A steel bar has a volume of 2.5 m³. Work out the mass of the bar.
**b** A block of steel has a mass of 1540 kg. Work out the volume of the block.

**Solution 4**

**a** Mass = density × volume

 Mass $= 7700 \times 2.5$

   | Multiply the density by the volume. |

  $= 19\,250 \text{ kg}$

   | As the density is in kg/m³ and the volume is in m³, the mass is in kg. |

**b** Volume $= \dfrac{\text{mass}}{\text{density}}$

 Volume $= \dfrac{1540}{7700}$

   | Divide the mass by the density. |

  $= 0.2 \text{ m}^3$

   | As the mass is in kg and the density is in kg/m³ the volume is in m³. |

### Exercise 9B

**1** The density of iron is 7.86 g/cm³. The volume of an iron block is 100 cm³.
Work out the mass of the iron block.

**2** A slab of concrete has a volume of 60 cm³ and a mass of 15 000 g.
Work out the density of the concrete.

**3** Gold has a density of 19.3 g/cm³. The gold in a ring has a mass of 15 g.
Work out the volume of the gold in the ring.

**4** The density of balsa wood is 0.2 g/cm³. The volume of a model made of balsa wood is 150 cm³.
Work out the mass of the model.

**5** 14.7 g of sulphur has a volume of 7.5 cm³. Work out the density of sulphur.

## 9.2 Converting between metric and imperial units

When converting between the metric and imperial systems, methods of proportion can be used.
These approximate conversions should be learnt.

| Metric | Imperial |
|---|---|
| 2.5 cm | 1 inch |
| 8 km | 5 miles |
| 1 m | 39 inches |
| 30 cm | 1 foot |
| 1 kg | 2.2 pounds |
| 4.5 litres | 1 gallon |
| 1 litre | 1.75 pints |

### Example 5

**a** Change 6 litres to pints.

**b** Change 11 pounds to kilograms.

*Solution 5*

**a** 1 litre is approximately $1\frac{3}{4}$ pints, or 1.75 pints

$6 \times 1.75 = 10.5$

6 litres is approximately 10.5 pints

There will be more pints than litres so *multiply* the number of litres by the number of pints in 1 litre.

**b** 1 kg is approximately 2.2 pounds
$11 \div 2.2 = 5$
11 pounds is approximately 5 kg

There will be fewer kg than pounds so divide the number of pounds by 2.2

### Exercise 9C

**1** Change 10 miles to kilometres.

**2** Change 3 kilograms to pounds.

**3** Change 7 pints to litres.

**4** Change 6 gallons to litres.

**5** Change 8 inches to centimetres.

**6** Change 32 kilometres to miles.

**7** Change 90 centimetres to feet.

**8** Change 22 pounds to kilograms.

**9** Change 2 litres to pints.

**10** Change 36 litres to gallons.

**11** Changes 25 centimetres to inches.

**12** Alan fills his car up with petrol. He buys 45 litres. Change 45 litres to gallons.

**13** Oliver buys 4 pounds of apples. Change 4 pounds to kilograms.

**14** The distance between two towns is 300 miles. Change 300 miles to kilometres.

**15** A car is travelling at a speed of 80 kilometres per hour.
Change 80 kilometres per hour to miles per hour.

# Chapter summary

**You should now:**

★ understand and use average speed

★ understand and use density

★ be able to convert between metric and imperial units.

## Chapter 9 review questions

**1** Kim drove the 90 km from London to Oxford in 1 hour 15 minutes.
Work out her average speed in km/h.

**2** Tom cycled from Birmingham to Leicester at an average speed of 24 km/h.
His journey time was 2 hours 45 minutes.
Work out the distance from Birmingham to Leicester.

**3** The mass of 5 m³ of copper is 44 800 kg.

**a** Work out the density of copper.

The density of zinc is 7130 kg/m³

**b** Work out the mass of 5 m³ of zinc. *(1387 November 2003)*

**4** These two metal blocks each have a volume of 0.5 m³.

The density of the copper block is 8900 kg per m³.
The density of the nickel block is 8800 kg per m³.
Calculate the difference in the masses of the blocks. *(1385 June 2001)*

**5** Daniel leaves his home at 07 00
He drives 87 miles to work.
He drives at an average speed of 36 miles per hour.
At what time does Daniel arrive at work? *(1387 November 2003)*

**6** Ann drives 210 km in 2 hours 40 minutes.
Work out Ann's average speed. *(1387 November 2005)*

**7** Fred runs 200 metres in 21.2 seconds.

**a** Work out Fred's average speed.

Write down all the figures on your calculator display.

**b** Round off your answer to part **a** to an appropriate degree of accuracy. *(1387 November 2004)*

**8** The crosses on the diagram show the positions of three places $A$, $B$ and $C$.
The scale of the diagram is 1 cm to 5 km.

Tariq cycled in a straight line from $A$ to $C$.
He left $A$ at 1.30 pm.
He cycled at an average speed of 10 kilometres per hour.

**a** Find the time he arrived at $C$.

**b** Find the bearing of

  **i** $B$ from $A$      **ii** $A$ from $C$.                                (1385 November 2002)

**9** Change 4 kg to pounds.

**10** Change 28 miles to kilometres.                                         (1387 November 2003)

# Decimals and fractions 10

## 10.1 Fractions revision

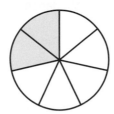

$\frac{2}{7}$ of this circle is shaded.

$\frac{2}{7}$ is a **fraction**.

| | | |
|---|---|---|
| The top number shows that 2 parts of the circle are shaded | $\dfrac{2}{7}$ | The top number of the fraction is called the **numerator** |
| The bottom number shows that the circle is divided into 7 equal parts | | The bottom number of the fraction is called the **denominator** |

**Equivalent fractions** are fractions that are equal.
If both the numerator and the denominator of a fraction are multiplied by the same number then an equivalent fraction is obtained.

$\frac{1}{3}$ $\frac{2}{6}$ and $\frac{3}{9}$ are all equivalent fractions.

A fraction can be simplified if the numerator and denominator can both be divided by the same number.

This process is called **cancelling**.

When a fraction cannot be simplified, it is in its **simplest form** or in its **lowest terms**.

---

### Example 1

Find the simplest form of $\frac{18}{30}$

**Solution 1**
6 is the highest common factor of 18 and 30
Divide both 18 and 30 by 6
The simplest form of $\frac{18}{30}$ is $\frac{3}{5}$

---

Fractions sometimes have to be put in order of size.
To do this when fractions have the same denominator, compare the numerators.

When fractions have different denominators:

- find the **lowest common denominator** (lowest common multiple of the denominators)
- change each fraction to its equivalent fraction with this denominator
- compare the numerators to put the fractions in order.

## Example 2

Write the fractions $\frac{1}{4}$ $\frac{2}{10}$ and $\frac{2}{5}$ in order of size.

Start with the smallest fraction.

*Solution 2*

The lowest common denominator for the three fractions is 20

Equivalent fractions for each of $\frac{1}{4}$ $\frac{2}{10}$ and $\frac{2}{5}$ with a denominator of 20 are

Starting with the smallest fraction, the order is $\quad\frac{4}{20}\quad\frac{5}{20}\quad\frac{8}{20}$

that is $\quad\frac{2}{10}\quad\frac{1}{4}\quad\frac{2}{5}$

---

An **improper fraction** is one in which the numerator is greater than the denominator.

For example, $\frac{5}{4}$ $\frac{12}{5}$ and $\frac{14}{8}$ are improper fractions.

The improper fraction $\frac{5}{4}$ can be thought of as '5 over 4' or as 5 quarters.

Similarly, the improper fraction $\frac{12}{5}$ can be thought of as '12 over 5' or as 12 fifths.

A **mixed number** consists of a whole number and a fraction.

For example, $2\frac{3}{4}$ $3\frac{1}{5}$ and $6\frac{7}{8}$ are mixed numbers.

Mixed numbers can be changed to improper fractions and vice versa.

For example, to change $2\frac{3}{4}$ to an improper fraction, work out how many quarters there are in $2\frac{3}{4}$

There are 4 quarters in 1 and so there are 8 ($2 \times 4$) quarters in 2
Add the extra 3 quarters to get 11 quarters.

So $\quad 2\frac{3}{4}=\frac{11}{4}$ $\qquad\boxed{2\frac{3}{4}=\frac{(2\times4)+3}{4}=\frac{11}{4}}$

To change the improper fraction $\frac{17}{6}$ to a mixed number, firstly work out how many whole ones there are. 6 sixths is 1; 12 sixths is 2 and so 17 sixths is 2 whole ones and 5 sixths.

$\frac{17}{6}=2\frac{5}{6}$ $\qquad\boxed{\begin{array}{l}17\div6=2\text{ remainder }5\\2\text{ is the whole number}\\\frac{5}{6}\text{ is the fraction}\end{array}}$

## Exercise 10A

**1** Copy the fractions and fill in the missing number to make the fractions equivalent.

**a** $\frac{1}{5}=\frac{}{10}$      **b** $\frac{2}{7}=\frac{4}{}$      **c** $\frac{2}{3}=\frac{}{15}$      **d** $\frac{3}{5}=\frac{12}{}$

**e** $\frac{3}{8}=\frac{}{32}$      **f** $\frac{4}{9}=\frac{}{45}$      **g** $\frac{5}{6}=\frac{15}{}$      **h** $\frac{7}{8}=\frac{28}{}$

**2** Copy the fractions and fill in the missing number to make the fractions equivalent.

**a** $\frac{4}{7}=\frac{}{28}$      **b** $\frac{5}{8}=\frac{}{48}$      **c** $\frac{2}{9}=\frac{}{54}$      **d** $\frac{3}{5}=\frac{24}{}$

**e** $\frac{5}{12}=\frac{}{36}$      **f** $\frac{7}{20}=\frac{21}{}$      **g** $\frac{3}{16}=\frac{15}{}$      **h** $\frac{7}{25}=\frac{}{100}$

**3** Write each fraction in its simplest form.

**a** $\frac{3}{6}$    **b** $\frac{6}{8}$    **c** $\frac{7}{21}$    **d** $\frac{10}{40}$    **e** $\frac{20}{25}$    **f** $\frac{12}{16}$    **g** $\frac{42}{48}$    **h** $\frac{36}{40}$

**4** Write each fraction in its simplest form.

**a** $\frac{25}{100}$    **b** $\frac{24}{50}$    **c** $\frac{100}{150}$    **d** $\frac{75}{200}$    **e** $\frac{48}{60}$    **f** $\frac{72}{90}$    **g** $\frac{85}{100}$    **h** $\frac{125}{1000}$

**5** Write each set of fractions in order. Start with the smallest fraction.

**a** $\frac{3}{5}$ $\frac{7}{10}$    **b** $\frac{5}{6}$ $\frac{3}{4}$    **c** $\frac{2}{3}$ $\frac{5}{6}$ $\frac{7}{12}$    **d** $\frac{9}{20}$ $\frac{4}{5}$ $\frac{3}{4}$

**e** $\frac{4}{15}$ $\frac{1}{3}$ $\frac{3}{10}$    **f** $\frac{3}{4}$ $\frac{9}{16}$ $\frac{5}{8}$    **g** $\frac{23}{40}$ $\frac{7}{10}$ $\frac{3}{5}$ $\frac{13}{20}$    **h** $\frac{1}{2}$ $\frac{3}{5}$ $\frac{5}{12}$ $\frac{11}{30}$ $\frac{7}{15}$

**6** Julie and Susan have identical chocolate bars.
Julie eats $\frac{3}{4}$ of her chocolate bar.
Susan eats $\frac{7}{8}$ of her chocolate bar.
Who eats more chocolate?
Give a reason for your answer.

**7** Ahmid says that $\frac{7}{12}$ is bigger than $\frac{5}{6}$ because 7 is bigger than 5
Is Ahmid correct? You must give a reason for your answer.

**8** Change these mixed numbers to improper fractions.

**a** $1\frac{1}{2}$    **b** $1\frac{3}{5}$    **c** $2\frac{1}{6}$    **d** $2\frac{7}{8}$    **e** $4\frac{3}{4}$    **f** $6\frac{2}{5}$

**9** Change these improper fractions to mixed numbers.

**a** $\frac{7}{5}$    **b** $\frac{4}{3}$    **c** $\frac{19}{4}$    **d** $\frac{14}{5}$    **e** $\frac{27}{5}$    **f** $\frac{30}{7}$

## 10.2 Arithmetic of decimals

In a decimal number the decimal point separates the whole number part from the part that is less than 1
When adding or subtracting decimals line up the decimal points first.
To multiply by a decimal ignore the decimal point and do the multiplication with whole numbers.
Then decide on the position of the decimal point.

**Example 3**

**a** Multiply 5.12 by 4.6     **b** Multiply 3.4 by 0.2

*Solution 3*

**a**
```
    5¹1 2
  ×   4 6
  -------
  3 0 7 2
2 0¹4 8 0
  -------
2 3 5 5 2
```
Ignore the decimal points and do the multiplication with whole numbers.

Estimate $= 5 \times 5$
$= 25$

Round 5.12 to 5 and round 4.6 to 5
An estimate for the answer is $5 \times 5$

$5.12 \times 4.6 = 23.552$

This means that the decimal point will go between the 3 and the 5 as 23.552 is close to 25

$5.12 \times 4.6 = 23.552$

The number of decimal places in the answer is 3 which is the same as the total number of decimal places in the question.
This rule is another way of finding the position of the decimal point in the answer.

**b**
```
   3 4
 ×   2
 -----
   6 8
```
Do the multiplication with whole numbers.

$3.4 \times 0.2 = 0.68$

The total number of decimal places in the question is 2 so there must be 2 decimal places in the answer.

To divide a number by a decimal multiply both the number and the decimal by a power of 10 (10, 100, 1000 ...) to make the decimal a whole number. It is much easier to divide by a whole number than by a decimal.

## Example 4

Divide 20 by 0.4

*Solution 4*

$$\frac{20}{0.4} = \frac{200}{4}$$

To make 0.4 a whole number multiply it by 10
So multiply both 20 and 0.4 by 10

Then divide in the usual way.

$$\begin{array}{r} 5\,0 \\ 4\overline{)2\,0\,0} \end{array}$$

So $20 \div 0.4 = 50$

## Example 5

Divide 58.2 by 0.03

*Solution 5*

$$\frac{58.2}{0.03} = \frac{5820}{3}$$

To make 0.03 a whole number multiply it by 100
So multiply both 58.2 and 0.03 by 100

Then divide in the usual way.

$$\begin{array}{r} 1\,9\,4\,0 \\ 3\overline{)5\,8^2\,2^1\,0} \end{array}$$

So $58.2 \div 0.03 = 1940$

## Exercise 10B

**1** Write each of the following sets of numbers in order of size. Start with the smallest number each time.

    **a** 0.373  0.37  0.73  0.333  0.733      **b** 15.8  15.38  15.3  15.833  15.803

    **c** 0.045  0.05  0.0545  0.055  0.0454    **d** 6.067  6.006  6.07  6.06  6.077  6.076

    **e** 8.092  8.9  8.02  8.09  8.2  8.29  8.92

**2** Work out

    **a** 8.2 + 9.7      **b** 5.67 + 0.94      **c** 12.45 + 3.49      **d** 76.29 + 64.67

    **e** 13.1 + 5.69    **f** 87.34 + 45.9      **g** 345.06 + 24.8     **h** 15.2 + 8.953

    **i** 4 + 6.2 + 8.77   **j** 23 + 15.6 + 3.45

**3** Work out

    **a** 8.57 − 3.21     **b** 19.31 − 7.16     **c** 56.43 − 12.56     **d** 67.65 − 45.8

    **e** 8.6 − 3.42      **f** 14.6 − 4.31      **g** 9 − 3.4        **h** 17 − 5.43

    **i** 8.72 − 6.04     **j** 7.34 − 3.286

**4** Work out

    **a** 9.62 × 10      **b** 67.231 × 100    **c** 0.83 × 10      **d** 0.0065 × 1000

    **e** 8.2 × 100      **f** 0.9 × 100      **g** 8.41 × 100     **h** 43.2 × 1000

    **i** 0.21 × 1000    **j** 6.08 × 100     **k** 0.0134 × 10    **l** 56.1 × 100

**5** Work out

    **a** 6.34 × 0.4      **b** 4.21 × 0.3      **c** 0.02 × 0.4      **d** 0.08 × 0.3

    **e** 2.16 × 0.3      **f** 0.54 × 0.8      **g** 0.723 × 0.06    **h** 3.15 × 0.8

**6** Work out

   **a** 3.1 × 4.2      **b** 0.36 × 1.4      **c** 3.6 × 2.3      **d** 7.4 × 0.53

   **e** 8.6 × 2.4      **f** 9.2 × 0.15      **g** 0.064 × 0.73      **h** 0.095 × 3.4

**7** Work out the cost of 0.6 kg of carrots at 25p per kilogram. Give your answer in pounds.

**8** Work out the cost of 1.6 m of material at £4.20 per metre.

**9** Work out

   **a** 456 ÷ 100      **b** 72.3 ÷ 10      **c** 0.76 ÷ 10      **d** 53 ÷ 100

   **e** 0.9 ÷ 10      **f** 67.2 ÷ 100      **g** 7 ÷ 1000      **h** 4 ÷ 100

   **i** 0.054 ÷ 10      **j** 2.31 ÷ 100      **k** 45 ÷ 1000      **l** 6.01 ÷ 100

**10** Work out

   **a** 12 ÷ 0.2      **b** 5 ÷ 0.2      **c** 26 ÷ 0.4      **d** 9 ÷ 0.04

   **e** 4.2 ÷ 0.3      **f** 0.72 ÷ 0.03      **g** 0.145 ÷ 0.5      **h** 19.2 ÷ 0.03

   **i** 6.12 ÷ 0.003      **j** 0.035 ÷ 0.7      **k** 0.048 ÷ 0.6      **l** 0.00828 ÷ 0.09

**11** Five people share £130.65 equally. Work out how much each person will get.

**12** A bottle of lemonade holds 1.5 litres. A glass will hold 0.3 litres.
How many glasses can be filled from the bottle of lemonade?

## 10.3 Manipulation of decimals

Using a calculator $\frac{8.4}{0.2} = 42$

$\frac{84}{0.2} = 420$ so **multiplying the numerator by 10** without altering the denominator **multiplies the answer by 10**

$\frac{8.4}{2} = 4.2$ so **multiplying the denominator by 10** without altering the numerator **divides the answer by 10**.

$\frac{0.84}{0.2} = 4.2$ so **dividing the numerator by 10** without altering the denominator **divides the answer by 10**.

$\frac{8.4}{0.02} = 420$ so **dividing the denominator by 10** without altering the numerator **multiplies the answer by 10**.

Similar results are obtained by using other powers of 10

For example $\frac{8400}{0.2} = 42\,000$ (the numerator 8.4 has been multiplied by 1000 without altering the denominator, so the answer has been multiplied by 1000).

Sometimes the numerator and denominator are both changed.

For example starting with the result $\frac{8.4}{0.2} = 42$ it is possible to write down the value of $\frac{84}{0.002}$

The numerator 8.4 has been multiplied by 10 and the denominator has been divided by 100 so the answer has been multiplied by 1000, that is $\frac{84}{0.002} = 42\,000$

## Example 6

Given that $\dfrac{16.3}{2.5} = 6.52$ work out the value of $\dfrac{16.3}{25}$

### Solution 6

$\dfrac{16.3}{25} = \dfrac{16.3}{2.5 \times 10}$ | Starting with $\dfrac{16.3}{2.5}$ multiply the denominator by 10 to get $\dfrac{16.3}{25}$

$\dfrac{16.3}{25} = 6.52 \div 10$ | Multiplying the denominator by 10 without altering the numerator divides the answer by 10

$\dfrac{16.3}{25} = 0.652$ | To get the answer divide 6.52 by 10

## Example 7

Given that $\dfrac{3.46 \times 25.5}{3.4} = 25.95$ find the value of each of the following.

**a** $\dfrac{34.6 \times 2.55}{0.34}$          **b** $\dfrac{2.595 \times 0.34}{25.5}$

### Solution 7

**a** $\dfrac{34.6 \times 2.55}{0.34} = \dfrac{3.46 \times 10 \times 25.5 \div 10}{3.4 \div 10}$ | Starting with $\dfrac{3.46 \times 25.5}{3.4}$ multiply 3.46 by 10 and divide 25.5 by 10 (so the value of the numerator is not altered) and divide the denominator by 10 to get $\dfrac{34.6 \times 2.55}{3.4}$

$\dfrac{34.6 \times 2.55}{0.34} = 25.95 \times 10$ | Dividing the denominator by 10 without altering the numerator multiplies the answer by 10

$\dfrac{34.6 \times 2.55}{0.34} = 259.5$ | To get the answer multiply 25.95 by 10

**b** $3.46 \times 25.5 = 25.95 \times 3.4$ | Multiply both sides of $\dfrac{3.46 \times 25.5}{3.4} = 25.95$ by 3.4

$\dfrac{25.95 \times 3.4}{25.5} = 3.46$ | Divide both sides of $3.46 \times 25.5 = 25.95 \times 3.4$ by 25.5

$\dfrac{2.595 \times 0.34}{25.5} = \dfrac{25.95 \div 10 \times 3.4 \div 10}{25.5}$ | Starting with $\dfrac{25.95 \times 3.4}{25.5}$ divide 25.95 by 10 and divide 3.4 by 10 (so the value of the numerator is divided by 100 and the denominator is not altered) to get $\dfrac{2.595 \times 0.34}{25.5}$

$\dfrac{2.595 \times 0.34}{25.5} = 3.46 \div 100$ | Dividing the numerator by 100 without altering the denominator divides the answer by 100

$\dfrac{2.595 \times 0.34}{25.5} = 0.0346$ | To get the answer divide 3.46 by 100

**Exercise 10C**

**1** Given that $6.4 \times 2.8 = 17.92$ work out

   **a** $64 \times 28$    **b** $640 \times 2.8$    **c** $0.64 \times 28$    **d** $0.64 \times 0.028$

**2** Given that $18.3 \div 1.25 = 14.64$ work out

   **a** $183 \div 1.25$    **b** $1.83 \div 1.25$    **c** $0.183 \div 1.25$    **d** $0.183 \div 12.5$

**3** Given that $13.2 \times 5.5 = 72.6$ work out

   **a** $132 \times 5.5$    **b** $1.32 \times 0.55$    **c** $0.132 \times 55$    **d** $0.0132 \times 550$

**4** Given that $30.4 \div 4.75 = 6.4$ work out

   **a** $30.4 \div 47.5$    **b** $3.04 \div 4.75$    **c** $304 \div 4.75$    **d** $3.04 \div 0.475$

**5** Given that $\dfrac{23.2 \times 5.1}{3.4} = 34.8$ work out

   **a** $\dfrac{23.2 \times 51}{3.4}$    **b** $\dfrac{232 \times 51}{3.4}$    **c** $\dfrac{23.2 \times 5.1}{34}$    **d** $\dfrac{232 \times 51}{34}$

**6** Given that $\dfrac{17.2 \times 4.5}{2.4} = 32.25$ work out

   **a** $\dfrac{172 \times 45}{2.4}$    **b** $\dfrac{17.2 \times 4.5}{240}$    **c** $\dfrac{17.2 \times 45}{240}$    **d** $\dfrac{1.72 \times 0.45}{0.24}$

**7** Given that $23 \times 56 = 1288$ work out

   **a** $0.23 \times 560$    **b** $1288 \div 5.6$    **c** $12.88 \div 0.23$    **d** $1288 \div (23 \times 28)$

**8** Given that $52 \times 32 = 1664$ work out

   **a** $0.52 \times 0.32$    **b** $1664 \div 5.2$    **c** $16.64 \div 0.32$    **d** $166.4 \div 0.64$

**9** Given that $884 \div 34 = 26$ work out

   **a** $8.84 \div 340$    **b** $884 \div 2.6$    **c** $8.84 \div 260$    **d** $884 \div (3.4 \times 2.6)$

**10** Given that $1512 \div 36 = 42$ work out

   **a** $15.12 \div 3.6$    **b** $1.512 \div 0.036$    **c** $15.12 \div 420$    **d** $1.512 \div 0.84$

**11** Given that $\dfrac{144 \times 28}{42} = 96$ work out

   **a** $\dfrac{14.4 \times 28}{0.42}$    **b** $\dfrac{1.44 \times 2.8}{420}$    **c** $\dfrac{14.4 \times 2.8}{9.6}$    **d** $\dfrac{4.2 \times 9.6}{0.028}$

**12** Given that $\dfrac{84 \times 45}{35} = 108$ work out

   **a** $\dfrac{8.4 \times 4.5}{350}$    **b** $\dfrac{0.84 \times 4.5}{0.035}$    **c** $\dfrac{8.4 \times 0.45}{10.8}$    **d** $\dfrac{10.8 \times 0.35}{840}$

**13** Given that $\dfrac{1872}{1.2^2} = 1300$ work out

   **a** $\dfrac{1872}{12^2}$    **b** $\dfrac{18.72}{1.2^2}$    **c** $\dfrac{187.2}{0.12^2}$    **d** $\dfrac{936}{120^2}$

## 10.4 Conversion between decimals and fractions

also assessed in Module 4

A **terminating decimal** is a decimal which ends.
For example 0.56, 0.0004 and 4.57 are terminating decimals.

All terminating decimals can be converted to fractions using place value.

$0.7 = \frac{7}{10}$        $0.06 = \frac{6}{100}$

$0.76 = \frac{7}{10} + \frac{6}{100}$
$= \frac{70}{100} + \frac{6}{100}$
$= \frac{76}{100}$

### Example 8

Write 0.024 as a fraction.
Give your fraction in its simplest form.

*Solution 8*

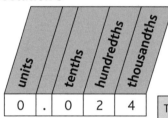

$0.024 = \frac{24}{1000}$

The heading of the last column with a figure in it is thousandths, so the denominator is 1000

$\frac{24}{1000} = \frac{12}{500} = \frac{6}{250} = \frac{3}{125}$

### Example 9

Write 3.7 as a fraction.

*Solution 9*

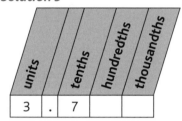

$3.7 = 3\frac{7}{10}$

The 3 is the whole number part, the .7 is $\frac{7}{10}$

All fractions can be changed into decimals.

### Example 10

Write the following fractions as decimals.    **a** $\frac{9}{10}$    **b** $\frac{23}{100}$

*Solution 10*

**a** $\frac{9}{10} = 0.9$                        **b** $\frac{23}{100} = 0.23$

### Example 11

Write the following fractions as decimals.    **a** $\frac{2}{5}$    **b** $\frac{11}{25}$

*Solution 11*

**Method 1** (non-calculator using equivalent fractions)

**a** $\frac{2}{5} = \frac{4}{10} = 0.4$                  **b** $\frac{11}{25} = \frac{44}{100} = 0.44$

**Method 2** (calculator)

**a** $\frac{2}{5}$ means $2 \div 5$
Using a calculator
2 $\div$ 5 $=$ 0.4
$\frac{2}{5} = 0.4$

**b** $\frac{11}{25}$ means $11 \div 25$
Using a calculator
11 $\div$ 25 $=$ 0.44
$\frac{11}{25} = 0.44$

Short division is suitable for changing $\frac{2}{5}$ to a decimal because the denominator is small.

$\frac{2}{5}$ means $2 \div 5$

| 2.0 is the same as 2 so divide 2.0 by 5 |

$$5\overline{)2.^20}^{\quad 0.\ 4}$$

| 5 does not divide into 2 so put down a zero and carry. |

$\frac{2}{5} = 0.4$

| 5 divides into 20 four times. |

Not all fractions can be written as terminating decimals.

For example $\frac{1}{3} = 1 \div 3 = 0.333\ 33 \ldots$ which is a recurring decimal.

A **recurring decimal** is a decimal in which one or more figures repeat.

$0.11111111 \ldots$, $0.565\ 656\ 56 \ldots$ and $9.762\ 333\ 33 \ldots$ are also recurring decimals.

To show that a figure recurs put a dot above the figure.

So $0.333\ 33 \ldots$ is written as $0.\dot{3}$ and $\frac{1}{3} = 0.\dot{3}$

Sometimes more than one figure recurs.

$\frac{3}{11} = 3 \div 11 = 0.272\ 727 \ldots$

A dot is placed above each recurring figure.

So $\frac{3}{11} = 0.\dot{2}\dot{7}$

## Example 12

Write the following fractions as decimals.

  **a** $\frac{7}{9}$        **b** $\frac{13}{22}$        **c** $\frac{5}{7}$

### Solution 12

**a** $\frac{7}{9}$ means $7 \div 9$

| Work out $7 \div 9$ on a calculator. |

Using a calculator

$7 \boxed{\div} 9 \boxed{=} 0.777\ 777 \ldots = 0.\dot{7}$

| The 7 recurs so put a dot above the 7 |

**b** $\frac{13}{22}$ means $13 \div 22$

| Work out $13 \div 22$ on a calculator. |

Using a calculator

$13 \boxed{\div} 22 \boxed{=} 0.590\ 909\ 0 \ldots = 0.5\dot{9}\dot{0}$

| The 90 recurs so put a dot above each of these figures. Do not put a dot above the 5 as it does not recur. |

**c** $\frac{5}{7}$ means $5 \div 7$

| Work out $5 \div 7$ on a calculator. |

Using a calculator

$5 \boxed{\div} 7 \boxed{=} 0.714\ 285\ 714 \ldots = 0.\dot{7}14\ 28\dot{5}$

| A group of six figures recurs. There isn't enough room to see all the figures recurring but you can see that the same pattern of figures is starting again. When more than two figures recur just two dots are used, one above the first figure in the recurring group and one above the last figure in the group. |

Fractions written in their simplest form with denominators 2, 4, 5, 8, 10, 16, 20, ... will convert to terminating decimals.

Fractions written in their simplest form with denominators 3, 6, 7, 9, 11, 12, 13, 14, 15, 17, 18, 19, ... will convert to recurring decimals.

In general if a fraction written in its simplest form has a denominator with a prime factor other than 2 or 5, it will convert to a recurring decimal.

**Example 13**

Work out whether the following fractions will convert to terminating or recurring decimals.

**a** $\frac{8}{30}$          **b** $3\frac{7}{20}$

*Solution 13*

**a** $\frac{8}{30} = \frac{4}{15}$

> Write the fraction in its simplest form.

$15 = 3 \times 5$

> Write the denominator as the product of its prime factors.

3 is a prime factor as well as 5

$\frac{8}{30}$ will convert to a recurring decimal.

**b** $\frac{7}{20}$

> Just consider the fraction part.

$20 = 2 \times 2 \times 5$

> Write the denominator as the product of its prime factors.

The only prime factors are 2 and 5

$\frac{7}{20}$ will convert to a terminating decimal.

---

The fractions and decimals in the table are some that are used frequently and should be learnt.

| Fraction | Decimal |
|---|---|
| $\frac{1}{100}$ | 0.01 |
| $\frac{1}{10}$ | 0.1 |
| $\frac{1}{5}$ | 0.2 |
| $\frac{1}{4}$ | 0.25 |
| $\frac{1}{3}$ | $0.\dot{3}$ |
| $\frac{1}{2}$ | 0.5 |
| $\frac{2}{3}$ | $0.\dot{6}$ |
| $\frac{3}{4}$ | 0.75 |

**Exercise 10D**

**1** Write each of the decimals as a fraction in its simplest form.

    **a** 0.7        **b** 0.14        **c** 0.093        **d** 0.006

    **e** 0.2        **f** 2.5         **g** 25.08        **h** 2.84

**2** Write the following fractions as decimals.

    **a** $\frac{9}{10}$        **b** $\frac{37}{100}$        **c** $\frac{3}{100}$        **d** $\frac{561}{1000}$        **e** $\frac{8}{1000}$

**3** Write the following as equivalent fractions and then as decimals.

    **a** $\frac{4}{5} = \frac{}{10}$    **b** $\frac{7}{50} = \frac{}{100}$    **c** $\frac{8}{25} = \frac{}{100}$    **d** $\frac{9}{500} = \frac{}{1000}$    **e** $\frac{3}{20} = \frac{}{100}$

**4** Write down the following fractions as decimals.

    **a** $\frac{1}{2}$        **b** $\frac{1}{4}$        **c** $\frac{2}{3}$        **d** $\frac{1}{5}$        **e** $\frac{3}{4}$

**5** Use short division to change these fractions to decimals.

    **a** $\frac{3}{5}$        **b** $\frac{3}{8}$        **c** $\frac{1}{6}$

**6** Use a calculator to change these fractions to decimals.

**a** $\frac{1}{32}$    **b** $\frac{9}{40}$    **c** $\frac{23}{125}$    **d** $\frac{5}{8}$    **e** $\frac{11}{16}$

**7** By writing the denominator in terms of its prime factors, decide whether the following fractions will convert to recurring or terminating decimals.

**a** $\frac{9}{40}$    **b** $\frac{17}{32}$    **c** $\frac{8}{45}$    **d** $\frac{13}{42}$    **e** $\frac{6}{125}$    **f** $\frac{37}{60}$

**8** Use a calculator to change these fractions to decimals.

**a** $\frac{5}{6}$    **b** $\frac{8}{9}$    **c** $\frac{5}{11}$    **d** $\frac{7}{12}$    **e** $\frac{2}{7}$

## 10.5 Converting recurring decimals to fractions

All recurring decimals can be converted to fractions.

To convert a recurring decimal to a fraction:

● introduce a letter, usually $x$
● form an equation by putting $x$ equal to the recurring decimal
● multiply both sides of the equation by 10 if 1 digit recurs, by 100 if 2 digits recur, by 1000 if 3 digits recur and so on
● subtract the original equation from the new equation
● rearrange to find $x$ as a fraction.

### Example 14

Convert the recurring decimal $0.\dot{2}$ to a fraction.

**Solution 14**

Let    $x = 0.2222\ldots$       Put $x$ equal to the recurring decimal.

$\quad 10x = 2.222\ldots$       Multiply both sides of the equation by *10* as *1* digit recurs.
$-\quad x = 0.222\ldots$       Subtract the equations.
$\quad\ \ 9x = 2$

$\qquad x = \frac{2}{9}$       Divide both sides by 9

$0.\dot{2} = \frac{2}{9}$

### Example 15

Convert the recurring decimal $0.2\dot{3}7\dot{1}$ to a fraction.

**Solution 15**

Let        $x = 0.237\,137\,1\ldots$       Put $x$ equal to the recurring decimal.
              Care is needed here, the 2 does *not* recur.

$\quad 1000x = 237.1371\ldots$       Multiply both sides of the equation by 1000 $(= 10^3)$ as 3 digits recur.
$-\qquad x = \quad 0.2371\ldots$       Subtract the equations.
$\quad\ \ 999x = 236.9$

$\qquad x = \dfrac{236.9}{999}$       Divide both sides by 999

$\qquad\ \ = \dfrac{2369}{9990}$       Multiply both the numerator and denominator by 10 to change the decimal in the numerator to an integer.

$0.2\dot{3}7\dot{1} = \dfrac{2369}{9990}$

## Example 16

Convert the recurring decimal $3.08\dot{6}$ to a fraction.

### Solution 16

$$3.08\dot{6} = 3 + 0.086\,86...$$

Consider the decimal part 0.086 86 ...

Let $\quad x = 0.086\,86 ...$

Put $x$ equal to the recurring decimal.
Care is needed here. The 0 does *not* recur.

$$100x = 8.686 ....$$
$$-\quad \underline{x = 0.086\,86...}$$
$$99x = 8.6$$

Multiply both sides of the equation by $100(= 10^2)$ as 2 digits recur.
Subtract the equations.

$$x = \frac{8.6}{99}$$

Divide both sides by 99

$$= \frac{86}{990}$$

Multiply both the numerator and denominator by 10 to change the decimal in the numerator to an integer.

$$= \frac{43}{495}$$

Simplify the fraction.

$$3.08\dot{6} = 3\tfrac{43}{495}$$

Include the whole number part in the answer.

## Example 17

Given that $0.\dot{2} = \frac{2}{9}$

Express the recurring decimal $0.3\dot{2}$ as a fraction.

### Solution 17

$$0.3\dot{2} = 0.3 + 0.0\dot{2}$$

To use the information given in the question split up the decimal 0.32222 ... = 0.3 + 0.02222 ...

$$= 0.3 + 0.\dot{2} \div 10$$

Rewrite the recurring part of the decimal using the given result.

$$= \tfrac{3}{10} + \tfrac{2}{9} \times \tfrac{1}{10}$$

Change all decimals to fractions.

$$= \tfrac{3}{10} + \tfrac{2}{90}$$

$$= \tfrac{27}{90} + \tfrac{2}{90}$$

Write fractions with a common denominator so they can be added.

$$= \tfrac{29}{90}$$

$$0.3\dot{2} = \tfrac{29}{90}$$

## Exercise 10E

Convert each recurring decimal to a fraction. Give each fraction in its simplest form.

**1**   0.777 77 ...        **2**   0.343 434 ...       **3**   0.915 915 ...      **4**   $0.\dot{1}\dot{8}$

**5**   $0.3\dot{1}\dot{7}$              **6**   $0.0\dot{5}$                **7**   $0.3\dot{2}\dot{6}$            **8**   $0.7\dot{0}\dot{1}$

**9**   $0.2\dot{3}$               **10**   $6.8\dot{3}$            **11**   $2.10\dot{6}$          **12**   $7.35\dot{2}$

**13** Given that $\frac{6}{11} = 0.\dot{5}\dot{4}$ write the recurring decimal $0.5\dot{5}\dot{4}$ as a fraction.

**14** Given that $\frac{5}{33} = 0.\dot{1}\dot{5}$ write the recurring decimal $0.2\dot{1}\dot{5}$ as a fraction.

**15** Given that $\frac{1}{6} = 0.1\dot{6}$ write the recurring decimal $0.401\dot{6}$ as a fraction.

## 10.6 Rounding to significant figures

A number rounded to **one significant figure** has only one figure that is not zero.

5937 rounded to one significant figure is 6000
0.006 183 rounded to one significant figure is 0.006

A number rounded to two significant figures is more accurate than a number rounded to one significant figure.

To round 5937 to *two* significant figures look at the *third* figure (3). As this is less than 5, do not change the previous figure (9) and write zeros in the tens column and the units column.

So 5937 rounded to two significant figures is 5900

To round 0.006 183 to *two* significant figures, look at the *third* figure (8) after the zeros at the beginning. As this is more than 5, increase the previous figure (1) by 1. Remember to include the zeros at the beginning in your answer.

So 0.006 183 rounded to two significant figures is 0.006 2

To round whole numbers greater than one to *three* significant figures, look at the *fourth* figure.

To round decimals to *three* significant figures, look at the *fourth* figure *after the zeros* at the beginning.

5937 rounded to three significant figures is 5940
0.006 183 rounded to three significant figures is 0.006 18

### Example 18

Round

**a** 3462 to one significant figure
**c** 0.3469 to three significant figures

**b** 7.38 to two significant figures
**d** 0.0201 to two significant figures

*Solution 18*

**a** 3462
3462 rounds to 3000 to one significant figure.

> The *second* figure is **4**. As this is less than 5, the 3 stays as it is and a zero goes in all the other places.

**b** 7.3**8**
7.38 rounds to 7.4 to two significant figures.

> The *third* figure is **8**
> As this is more than 5, increase the 3 by 1

**c** 0.346**9**
0.3469 rounds to 0.347 to three significant figures.

> The *fourth* figure after the zero at the beginning is **9**
> As this is more than 5, increase the 6 by 1

**d** 0.020**1**
0.0201 rounds to 0.020 to two significant figures.

> The *third* figure after the zeros at the beginning is **1**
> As this is less than 5, the zero before the 1 stays as it is.
> The zero at the end is needed as it is the second significant figure.

### Example 19

Use your calculator to work out the value of $\dfrac{6.73 + 4.5}{12.03 - 9.73}$

Give your answer correct to two significant figures.

*Solution 19*

$6.73 + 4.5 = 11.23$

> Use a calculator to work out the value of the numerator.

$12.03 - 9.73 = 2.3$

> Use a calculator to work out the value of the denominator.

$11.23 \div 2.3 = 4.882\,608\,696\,...$

> The line in a fraction means divide.
> Now use a calculator to work out $11.23 \div 2.3$
> Write down all the figures shown on your calculator.

$4.882\,608\,696 = 4.9$ correct to two significant figures

> To give the answer to two significant figures, look at the third figure (**8**).
> As this is more than 5, increase the figure before it by 1

### Exercise 10F

**1** Round these numbers to one significant figure

| | | | |
|---|---|---|---|
| **a** 8234 | **b** 76 420 | **c** 453 | **d** 72 |
| **e** 0.381 | **f** 0.004 56 | **g** 0.109 | **h** 532.4 |

**2** Round these numbers to two significant figures

| | | | |
|---|---|---|---|
| **a** 4263 | **b** 8719 | **c** 685 | **d** 3.84 |
| **e** 798 | **f** 0.005 62 | **g** 703 | **h** 0.4032 |

**3** Round these numbers to three significant figures

| | | | |
|---|---|---|---|
| **a** 8736 | **b** 56.24 | **c** 27.839 | **d** 0.786 21 |
| **e** 0.030 56 | **f** 87.98 | **g** 6 735 412 | **h** 907.189 |

**4** Round these to the number of significant figures given in the brackets

| | | | |
|---|---|---|---|
| **a** 6712 (1) | **b** 8614 (3) | **c** 6926 (2) | **d** 82.14 (2) |
| **e** 876.3 (3) | **f** 12.52 (3) | **g** 0.0426 (1) | **h** 0.002 345 1 (2) |
| **i** 7.6024 (3) | **j** 8.795 (2) | **k** 508 342 (3) | **l** 0.000 481 6 (3) |

**5** Use your calculator to work out the value of the following.
Give each answer correct to three significant figures.

**a** $5421 \div 23$      **b** $423 \times 871$     **c** $0.0562 \times 0.041$

**d** $\dfrac{3250 \times 720}{0.32}$     **e** $\dfrac{9.6}{13.21 - 9.1}$     **f** $\dfrac{27.31 - 8.96}{4.56 + 9.8}$

# Chapter summary

**You should now know that:**

★ that equivalent fractions are fractions that are equal

★ how to find an equivalent fraction by multiplying both the numerator and denominator by the same number

★ how to cancel a fraction to obtain its simplest form

★ how to order fractions by writing each fraction with the same denominator

★ that an improper fraction is one in which the numerator is greater than the denominator

★ that a mixed number consists of a whole number and a fraction

★ how to convert between mixed numbers and improper fractions

★ in a decimal number the decimal point separates the whole number part from the part that is less than one

★ to multiply by a decimal ignore the decimal point and do the multiplication with whole numbers. Then decide on the position of the decimal point

★ to divide by a decimal write the division as a fraction then multiply numerator and denominator by a power of 10 to find an equivalent fraction with an integer as the denominator

★ if a number in the numerator of an expression is multiplied by a power of 10 (or a number in the denominator is divided by a power of 10) then the value of the expression is multiplied by the same power of 10

★ if a number in the numerator of an expression is divided by a power of 10 (or a number in the denominator is multiplied by a power of 10) then the value of the expression is divided by the same power of 10

★ terminating decimals can be converted to fractions by using place value

★ fractions can be converted to decimals by using equivalent fractions or division

★ some fractions convert to recurring decimals

★ when the denominator of a fraction written in its simplest form has prime factors containing only 2s and/or 5s then the fraction will convert to a terminating decimal; otherwise the fraction will convert to a recurring decimal

★ every recurring decimal can be converted to a fraction

★ numbers can be rounded to significant figures.

# Chapter 10 review questions

1 Work out

   **a** $5.6 \times 10$      **b** $76.2 \div 100$    **c** $9 \div 100$    **d** $0.0062 \times 100$   **e** $0.87 \times 1000$

2 Work out

   **a** $0.2 \times 0.3$           **b** $1.2 \times 0.6$           **c** $0.37 \times 0.5$

   **d** $0.4 \times 0.08$        **e** $6.1 \times 4.2$           **f** $0.32 \times 5.6$

3 Work out

   **a** $6.25 \div 0.5$       **b** $75.6 \div 0.3$        **c** $47.7 \div 0.09$

   **d** $56 \div 0.2$          **e** $46.2 \div 0.03$      **f** $0.84 \div 0.004$

**4**   0.3      0.06      0.058      0.26

     **a**   Write these four decimals in order of size. Start with the smallest decimal.

     **b**   Write 0.3 as a fraction.     **c**   Work out $0.3 - 0.26$        **d**   Work out $0.058 \times 100$

                                                             (1388 January 2002)

**5**   **a**   Work out $41.3 \times 100$      **b**   Work out $0.4 \times 0.6$          **c**   Work out $5.2 - 1.37$

                                                             (1388 March 2003)

**6**   Change $\frac{7}{8}$ to a decimal.

**7**   Karen says that $\frac{13}{24}$ can be converted into a terminating decimal. Lucy says that the fraction converts to a recurring decimal. Who is correct? You must give a reason for your answer.

**8**   By writing the denominator as the product of its prime factors, decide whether the following fractions will convert to a recurring decimal or a terminating decimal.

     **a**   $\frac{9}{16}$               **b**   $\frac{37}{55}$                  **c**   $\frac{2}{75}$                    **d**   $\frac{19}{96}$

**9**   **a**   Write 0.35 as a fraction. Give your answer in its simplest form.

     **b**   Write $\frac{3}{8}$ as a decimal.                                             (1387 June 2002)

**10**   Use your calculator to write each fraction as a decimal.

     **a**   $\frac{23}{40}$               **b**   $\frac{11}{16}$                  **c**   $\frac{9}{11}$                    **d**   $\frac{7}{90}$

**11**   $1.54 \times 450 = 693$

     Use this result to write down the answer to

     **a**   $1.54 \times 45$            **b**   $1.54 \times 4.5$             **c**   $0.154 \times 0.45$        (1387 May 2002)

**12**   Using the information that $97 \times 123 = 11\,931$ write down the value of

     **a**   $9.7 \times 12.3$          **b**   $0.97 \times 123\,000$        **c**   $11.931 \div 9.7$        (1387 June 2003)

**13**   **a**   Express $\frac{4}{9}$ as a recurring decimal.

     **b**   Convert the recurring decimal $0.1\dot{3}\dot{6}$ to a fraction in its simplest form.

**14**   **a**   Change $\frac{3}{11}$ to a decimal.      **b**   Prove that the recurring decimal $0.\dot{3}\dot{9} = \frac{13}{33}$       (1387 June 2005)

**15**   Express the recurring decimal $2.0\dot{6}$ as a fraction.

     Write your answer in its simplest form.                                            (1388 March 2005)

**16**   Change to a single fraction

     **a**   the recurring decimal $0.\dot{1}\dot{3}$                 **b**   the recurring decimal $0.5\dot{1}\dot{3}$       (1388 March 2002)

**17**   $a$ is an integer such that $1 \leqslant a \leqslant 9$

     $b$ is an integer such that $1 \leqslant b \leqslant 9$

     Prove that $0.0\dot{a}\dot{b} = \dfrac{ab}{990}$

**18**   **a**   Express $0.\dot{2}\dot{7}$ as a fraction in its simplest form.

     **b**   $x$ is an integer such that $1 \leqslant x \leqslant 9$

     Prove that $0.0\dot{x} = \dfrac{x}{99}$                                                (1388 January 2003)

**19 a** Convert the recurring decimal $0.3\dot{6}$ to a fraction.

    **b** Convert the recurring decimal $2.1\dot{3}\dot{6}$ to a mixed number.
       Give your answer in its simplest form.               (1388 March 2004)

**20** The recurring decimal $0.\dot{7}\dot{2}$ can be written as the fraction $\frac{8}{11}$
    Write the recurring decimal $0.5\dot{7}\dot{2}$ as a fraction.         (1387 November 2005)

**21** Round the following to the number of significant figures given in the brackets

    **a** 3546 (1)          **b** 3546 (2)          **c** 0.005 62 (1)

    **d** 23.76 (2)        **e** 2.4387 (3)       **f** 696 213 (2)

# Expanding brackets and factorising

## 11.1 Expanding brackets

There are three rows. Each row has $n$ students.

The number of students is $3 \times n = 3n$.

Two students are added to each of the three rows, so $3 \times 2 = 6$ students are added.

There are now $n + 2$ students in each row.

$$\begin{array}{r} n + 2 \\ n + 2 \\ +\ n + 2 \\ \hline \end{array}$$

Total number of students $= 3(n + 2) = 3 \times n + 3 \times 2 = 3n + 6$

The simplest way of writing this is

$$3(n + 2) = 3n + 6$$

Writing $3(n + 2)$ as $3n + 6$ is called **expanding brackets**. This is also known as multiplying out brackets.

---

### *Example 1*

Expand $5(2x + 3)$.

**Solution 1**

$5(2x + 3) = 5 \times 2x + 5 \times 3$

$\qquad\quad = 10x + 15$

> Multiply each term inside the bracket by the term outside the bracket.

---

### *Example 2*

Expand $p(p - 3)$.

**Solution 2**

$p(p - 3) = p \times p + p \times - 3$

$\qquad\quad = p^2 - 3p$

> Multiply each term inside the bracket by the term outside the bracket.

> Positive $(+)\times$ negative $(-)$ = negative $(-)$.

---

### *Example 3*

Expand and simplify $5p - 3(p - q)$.

**Solution 3**

$5p - 3(p - q) = 5p - 3 \times p - 3 \times -q$

$\qquad\qquad\quad = 5p - 3p + 3q$

$\qquad\qquad\quad = 2p + 3q$

> Multiply each term inside the bracket by the $-3$ outside the bracket.

> Negative $(-3) \times$ negative $(-q)$ is positive $(+3q)$.

> Collect like terms.

### Example 4

Expand $2y(y^2 - 4y + 3)$.

**Solution 4**

$2y(y^2 - 4y + 3) = 2y \times y^2 + 2y \times -4y + 2y \times 3$

$\qquad\qquad\qquad = 2y^3 - 8y^2 + 6y$

> Multiply each term inside the bracket by the term outside the bracket.

### Exercise 11A

**1** Expand

| | | | |
|---|---|---|---|
| **a** $2(n + 1)$ | **b** $2(2 + n)$ | **c** $4(p - 2)$ | **d** $2(3 - n)$ |
| **e** $5(c + d)$ | **f** $2(c - d)$ | **g** $4(m - n)$ | **h** $2(2x + 3y)$ |
| **i** $5(3g - 4)$ | **j** $2(3g - 4h)$ | **k** $7(x + 2y + 3)$ | **l** $3(p - 2q - 1)$ |

**2** Expand

| | | | |
|---|---|---|---|
| **a** $n(n + 1)$ | **b** $b(b^2 + 3b)$ | **c** $2n(n - 3)$ | **d** $a(4 - a)$ |
| **e** $y(3y - y^2)$ | **f** $-3(5n + 4)$ | **g** $-3(2a - 1)$ | **h** $-2(n - 4)$ |
| **i** $-(y - 1)$ | **j** $-5(d^2 - d)$ | **k** $a(b + c)$ | **l** $2x(3x^2 + 4x + 1)$ |

**3** Expand and simplify

| | | |
|---|---|---|
| **a** $2(p + 1) + 5$ | **b** $2(t + 3) - t$ | **c** $2(n + 1) + 3(n + 2)$ |
| **d** $2(2c + 1) + 3(3c + 2)$ | **e** $2(m + 3) + 3(2m - 1)$ | **f** $2(3d + 4) + 3(1 - 2d)$ |
| **g** $2(2p - 3q) + 2(2p - q)$ | **h** $4p - 2(p + 1)$ | **i** $5y - 3(2y + 1)$ |
| **j** $3 - (n + 3)$ | **k** $2q - 3(q - 1)$ | **l** $4x - 3(2x - 1)$ |

**4** Expand and simplify

| | | |
|---|---|---|
| **a** $x(x + 1) + 1(x + 1)$ | **b** $q(q + 1) + 3(q + 1)$ | **c** $s(s + 4) - 2(s + 4)$ |
| **d** $t(t - 3) - 2(t - 5)$ | **e** $a(a - 3) - 2(a - 1)$ | **f** $n(n + 2) - 4(n - 2)$ |

## 11.2 Factorising by taking out common factors

Expanding $3(2b + 5)$ gives $6b + 15$

**Factorising** is the reverse process to expanding brackets.
So factorising $6b + 15$ gives $3(2b + 5)$.

3 is the highest common factor of 6 and 15

The two factors of the expression $6b + 15$ are 3 and $2b + 5$

### Example 5

Factorise $8c + 12s$.

**Solution 5**

$8c + 12s = 4 \times 2c + 4 \times 3s$

> The common factors of 8 and 12 are 1, 2 and 4
> The HCF of 8 and 12 is 4

$\qquad\quad = 4(2c + 3s)$

> Write the HCF outside the bracket and the other factor inside the bracket.

## Example 6

Factorise $ab + ac$.

**Solution 6**

$ab + ac = a \times b + a \times c$ — The common factor of $ab$ and $ac$ is $a$.

$\quad = a(b + c)$ — Write the $a$ outside the bracket.

## Example 7

Factorise $y^2 - y$.

**Solution 7**

$y^2 - y = y \times y - 1 \times y$ — The common factor of $y^2$ and $y$ is $y$.

$\quad = y \times y - y \times 1$ — Re-writing $1 \times y$ as $y \times 1$

$\quad = y(y - 1)$ — Write the $y$ outside the bracket.

## Example 8

Factorise completely $4ax - 8xb + 6xc$.

**Solution 8**

$4ax - 8xb + 6xc = 2x \times 2a - 2x \times 4b + 2x \times 3c$ — The common factor of $4ax$, $8xb$ and $6xc$ is $2x$.

$\quad = 2x(2a - 4b + 3c)$ — Write the $2x$ outside the bracket.

In questions where you are asked to 'Factorise *completely*' check that the terms of the expression inside the bracket do not have a common factor.

## Example 9

Factorise completely $9a^2b - 12ab^2c$.

**Solution 9**

$9a^2b - 12ab^2c = 3 \times 3 \times a \times a \times b - 3 \times 4 \times a \times b \times b \times c$ — The common factor is $3 \times a \times b$.

$\quad = 3ab(3 \times a - 4 \times b \times c)$ — Write the $3ab$ outside the bracket.

$\quad = 3ab(3a - 4bc)$ — Note that $3a$ and $4bc$ do not have a common factor.

### Exercise 11B

**1** Factorise these expressions.

a $3y + 6$    b $2x + 2$    c $5p - 10$    d $3d + 6e$

e $6x + 2y$    f $3a - 12b$    g $12p - 6q$    h $4d + 6e$

i $10c - 6d$    j $8x + 12y$    k $10m - 15n + 5$    l $8p - 6q + 4$

m $9d - 6e - 12f$    n $pq + pr$    o $xy + zy$    p $ab - 7b$

q $db - b$    r $pq + py + p$    s $y^2 + yz$    t $y^2 + y$

u $4y^2 - 3y$    v $y^3 + 2y$    w $y^3 - 5y$    x $p^2 + pq + pr$

y $ay^3 - by^2 + cy$

**2** Factorise completely

    **a** $2xy + 4x$          **b** $3pq + 9qr + 6pqr$     **c** $4xy - 8y^2$

    **d** $5ac - abc$          **e** $4pq - 6pqr$        **f** $rt + rst$

    **g** $14y^4 + 7y^2$        **h** $9d^3 + 6d^2$         **i** $ab^3 + a^2b$

    **j** $10x^2y + 15y^2$      **k** $3xy^2 - 8x^2y^5$     **l** $12pq^2r - 16qp^3$

    **m** $8a^2b + 16a^3 + 12ab^2$   **n** $4a^4b^2 + 6a^3b^3 + 12a^2b^2$   **o** $(p^2q)^2 + p^3q$

# 11.3 Expanding the product of two brackets

Consider the areas of these two rectangles.

Area $= x(x + 2)$

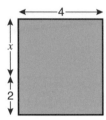

Area $= 4(x + 2)$

Combining these two rectangles gives a single rectangle with length $(x + 4)$ and width $(x + 2)$

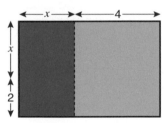

Area $= (x + 4)(x + 2)$

The area of the large rectangle is equal to the sum of the areas of the two smaller rectangles so

$$(x + 4)(x + 2) = x(x + 2) + 4(x + 2)$$

To expand two brackets, multiply each term in the first bracket by the second bracket so

$$(x + p)(x + q) = x(x + q) + p(x + q)$$
$$= x^2 + qx + px + pq$$

Expanding $(x + p)(x + q)$ gives $x^2 + qx + px + pq$.

---

**Example 10**

Expand and simplify $(x + 2)(x - 3)$.

**Solution 10**

$(x + 2)(x - 3) = x(x - 3) + 2(x - 3)$
                                                | Multiply each term, $x$ and $+2$ in the first bracket by the second bracket, $(x - 3)$. |

        $= x^2 - 3x + 2x - 6$

        $= x^2 - x - 6$               | Collect like terms. |

**Example 11**

Expand and simplify $(n - 2)^2$.

**Solution 11**

$(n - 2)^2 = (n - 2)(n - 2)$

> $(n - 2)^2 = (n - 2) \times (n - 2)$.

$\quad\quad = n(n - 2) - 2(n - 2)$

> Multiply each term in the first bracket by the second bracket.

$\quad\quad = n^2 - 2n - 2n + 4$

$\quad\quad = n^2 - 4n + 4$

> Collect like terms.

**Example 12**

Expand and simplify $(4p - 5q)(3p + 2q)$.

**Solution 12**

$(4p - 5q)(3p + 2q) = 4p(3p + 2q) - 5q(3p + 2q)$

> Multiply each term in the first bracket by the second bracket.

$\quad\quad\quad = 12p^2 + 8pq - 15qp - 10q^2$

$\quad\quad\quad = 12p^2 - 7pq - 10q^2$

> Collecting like terms; note $-15qp$ is the same as $-15pq$.

**Exercise 11C**

1 Expand and simplify
   a $(x + 1)(x + 2)$      b $(q + 3)(q + 1)$      c $(r + 1)(r - 3)$      d $(x + 4)(x + 3)$
   e $(x + 2)(x - 1)$      f $(x + 3)(x - 3)$      g $(r - 1)(r + 3)$      h $(x - 2)(x + 4)$
   i $(x - 2)(x - 1)$      j $(x - 3)(x - 4)$

2 Expand and simplify
   a $(x + 1)^2$           b $(x + 3)^2$           c $(x + 7)^2$
   d $(x - 5)^2$           e $(x - 8)^2$           f $(x - 10)^2$

3 Expand and simplify
   a $(x + y)(x + 2y)$     b $(2x + y)(x + y)$     c $(p + q)(3p - q)$
   d $(3p + q)(p - q)$     e $(x + 3y)(2x - 5y)$   f $(a - b)(a + b)$
   g $(p - q)(p + 3q)$     h $(2x + 3y)(3x - 4y)$  i $(3x - 2y)(x - y)$
   j $(4x - 5y)(2x - 3y)$  k $(x + y)^2$           l $(2x + y)^2$
   m $(3p - 5q)^2$         n $(6a + 5b)(6a - 5b)$

4 Expand
   a $(a + b)(c + d)$      b $(e + f)(g - h)$

## 11.4 Factorising by grouping

In Section 11.2, an expression was factorised by taking out the common factor of the terms in the expression. The common factor was a single term, for example 3, $4a$, $ab$. For some expressions the common factor can involve the sum or difference of terms, for example $x + 3$ or $2a - 4b$.

## Example 13

Factorise completely $12(x + 2)^2 - 9(x + 2)$.

**Solution 13**

$12(x + 2)^2 - 9(x + 2)$

$= \mathbf{3} \times 4 \times (\mathbf{x + 2}) \times (x + 2) - \mathbf{3} \times 3 \times (\mathbf{x + 2})$ | 3 is the HCF of 12 and 9

$= \mathbf{3} \times (\mathbf{x + 2}) \times 4 \times (x + 2) - \mathbf{3} \times (\mathbf{x + 2}) \times 3$ | 3 and $(\mathbf{x + 2})$ are both common factors.

$= 3(\mathbf{x + 2})[4(x + 2) - 3]$ | So write $3(\mathbf{x + 2})$ outside the square bracket.

$= 3(\mathbf{x + 2})[4x + 8 - 3]$ | Simplify the terms inside the square bracket.

$= 3(x + 2)(4x + 5)$

Expanding $(a + b)(c + d)$ gives $ac + ad + bc + bd$.
Since factorising is the reverse process to expanding brackets,
factorising $ac + ad + bc + bd$ gives $(a + b)(c + d)$.

In the expression $ac + ad + \mathbf{bc + bd}$ there is no common factor of all four terms.
Pairs of terms with a common factor can be grouped together and factorised to give
$a(c + d) + \mathbf{b(c + d)}$. These two terms have a common bracketed factor, $(c + d)$.

Using this common factor gives $(c + d)(a + b)$ which can also be written as $(a + b)(c + d)$.
This is called **factorising by grouping**.

## Example 14

Factorise $pr + qs - ps - qr$.

**Solution 14**

$pr + qs - ps - qr = pr - ps + qs - qr$ | Group the terms in pairs so that each pair has a common factor – in this case $p$ and $q$.

$= p(r - s) - q(r - s)$ | Factorise each pair. The **bracketed term** must be the same.

$= (r - s)(p - q)$ | $(r - s)$ is a common factor. This answer could be written as $(p - q)(r - s)$.

## Exercise 11D

**1** Factorise completely

    **a** $y(x + 3) + 2(x + 3)$          **b** $a(x - y) + b(x - y)$

    **c** $p(x + 2y) - q(x + 2y)$      **d** $2p(x + 4) + 3q(x + 4)$

    **e** $(x + 1)^2 - 4(x + 1)$         **f** $(x - y)^2 - b(x - y)$

    **g** $(x + 5) + 3(x + 5)^2$         **h** $(x + 3y) + 2(x + 3y)^2$

    **i** $4(x + 2)^2 - 2(x + 2)$       **j** $6(x - y)^2 - 3(x - y)$

    **k** $6(x + 4) + 4(x + 4)^2$      **l** $6y(x + 3y) + 9(x + 3y)^2$

**2** Factorise completely

    **a** $ab + ac + db + dc$           **b** $pq + 2r + pr + 2q$

    **c** $x^2 + ax + 2x + 2a$           **d** $ps + qr - pr - qs$

    **e** $x^2 - 3x + 2x - 6$            **f** $x^2 - 3x - x + 3$

    **g** $2x^2 - 8x + 3x - 12$         **h** $2x^2 - 3x - 2x + 3$

## 11.5 Factorising expressions of the form $x^2 + bx + c$

Expanding $(x + 4)(x + 2)$ gives

$$(x + 4)(x + 2) = x(x + 2) + 4(x + 2)$$
$$= x^2 + 2x + 4x + 8$$
$$= x^2 + 6x + 8$$

Since factorising is the reverse process to expanding brackets, factorising $x^2 + 6x + 8$ gives $(x + 4)(x + 2)$.

In the expression $x^2 + 6x + 8$ there is no common factor of all three terms.

> Taking out a common factor of just two of the three terms is not factorising the expression. An answer $x(x + 6) + 8$ is *incorrect*.

From Section 11.3      $(x + p)(x + q) = x^2 + qx + px + pq$.

So                  $(x + p)(x + q) = x^2 + (p + q)x + pq$.

> $(p + q)x$ is the same as $x(p + q)$.

The diagrams show the expansions of

$$(x + p)(x + q) \qquad \text{and} \qquad (x + 4)(x + 2)$$

|   | $x$ | $+p$ |
|---|---|---|
| $x$ | $x^2$ | $+px$ |
| $+q$ | $+qx$ | $+pq$ |

|   | $x$ | $+4$ |
|---|---|---|
| $x$ | $x^2$ | $+4x$ |
| $+2$ | $+2x$ | $+8$ |

For these expansions to be the same, obviously $x^2 = x^2$ but also

$$+pq = +8 \qquad \text{and} \qquad + px + qx = +4x + 2x = +6x$$

that is, the product of $p$ and $q$ is $+8$

and the sum of $p$ and $q$ is $+6$, since $px + qx = (p + q)x$.

To factorise $x^2 + \mathbf{6}x + \mathbf{8}$
find two numbers whose product is $+\mathbf{8}$ and whose sum is $+\mathbf{6}$.
The two numbers are $+2$ and $+4$.

So    $x^2 + 6x + 8 = x^2 + 2x + 4x + 8$
$$= x(x + 2) + 4(x + 2)$$
$$= (x + 2)(x + 4)$$

### Example 15

Factorise $x^2 - 7x + 12$

**Solution 15**

$x^2 - 7x + \mathbf{12}$      $-4 \times -3 = +12$
                     $-4 + -3 = -7$

> Find two numbers whose product is $+\mathbf{12}$ and whose sum is $-7$

$x^2 - 7x + 12 = x^2 - 4x - 3x + 12$

> Write $-7x$ as $-4x - 3x$.

$\qquad\qquad\quad = x(x - 4) - 3(x - 4)$

> Factorise by grouping. The bracketed term must be the same.

$\qquad\qquad\quad = (x - 4)(x - 3)$

> $(x - 4)$ is a common factor.

> This answer could also be written as $(x - 3)(x - 4)$.

### Example 16

Factorise $x^2 - x - 6$

### Solution 16

$x^2 - 1x - 6$     $+2 \times -3 = -6$    | Find two numbers whose product is **−6** and whose sum is **−1** |

               $+2 + -3 = -1$

$x^2 - x - 6 = x^2 + 2x - 3x - 6$    | Write $-x$ as $+2x - 3x$. |

         $= x(x + 2) - 3(x + 2)$    | Factorise by grouping. The bracketed term must be the same. |

         $= (x + 2)(x - 3)$    | $(x + 2)$ is a common factor. |

| This answer could also be written as $(x - 3)(x + 2)$. |

### Exercise 11E

**1** Write down the two numbers
  **a** whose product is $+10$ and whose sum is $+7$
  **b** whose product is $+7$ and whose sum is $-8$
  **c** whose product is $+24$ and whose sum is $+11$
  **d** whose product is $-6$ and whose sum is $-5$
  **e** whose product is $-8$ and whose sum is $+2$
  **f** whose product is $-12$ and whose sum is $-1$
  **g** whose product is $-12$ and whose sum is $+4$
  **h** whose product is $+9$ and whose sum is $+6$
  **i** whose product is $+20$ and whose sum is $-9$
  **j** whose product is $-16$ and whose sum is $0$

**2** Factorise these expressions.
  **a** $x^2 + 3x + 2$       **b** $x^2 + 4x + 3$       **c** $x^2 + 6x + 5$
  **d** $x^2 + 5x + 4$       **e** $x^2 + 7x + 10$      **f** $x^2 - 8x + 7$
  **g** $x^2 + 11x + 24$     **h** $x^2 - 5x - 6$       **i** $x^2 + 2x - 8$
  **j** $x^2 + 7x - 8$       **k** $x^2 + 17x - 18$     **l** $x^2 - 3x - 18$
  **m** $x^2 + 2x + 1$       **n** $x^2 + 6x + 9$       **o** $x^2 + 10x + 25$
  **p** $x^2 - 5x - 36$      **q** $x^2 + 16x - 36$

## 11.6 Factorising the difference of two squares

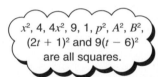

$x^2$, 4, $4x^2$, 9, 1, $p^2$, $A^2$, $B^2$, $(2t + 1)^2$ and $9(t - 6)^2$ are all squares.

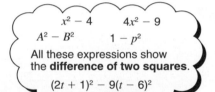

$x^2 - 4$      $4x^2 - 9$
$A^2 - B^2$     $1 - p^2$
All these expressions show
the **difference of two squares**.
$(2t + 1)^2 - 9(t - 6)^2$

An expression that is the difference of two squares can be factorised using the method in Section 11.5

## Example 17

Factorise $x^2 - 4$

### Solution 17

$x^2 - 4 = x^2 + 0x - 4$

| Write $x^2 - 4$ in the form $x^2 + bx + c$. |

$x^2 + 0x - 4$      $-2 \times +2 = -4$
                         $-2 + +2 = +0$

| Find two numbers whose product is $-4$ and whose sum is $+0$ |

$x^2 - 4 = x^2 - 2x + 2x - 4$

| Write $+0x$ as $-2x + 2x$. |

$\quad = x(x - 2) + 2(x - 2)$

| Factorise by grouping. The bracketed term must be the same. |

$\quad = (x - 2)(x + 2)$

| $(x - 2)$ is a common factor. |

| This answer could also be written as $(x + 2)(x - 2)$. |

## Example 18

Factorise $x^2 - n^2$.

### Solution 18

$x^2 - n^2 = x^2 + 0x - n^2$

$x^2 + 0x - n^2$      $+n \times -n = -n^2$
                         $+n + -n = +0$

| Find two numbers whose product is $-n^2$ and whose sum is $+0$ |

$x^2 - n^2 = x^2 + nx - nx - n^2$

| Write $+0x$ as $+nx - nx$. |

$\quad = x(x + n) - n(x + n)$

| Factorise by grouping. The bracketed term must be the same. |

$\quad = (x + n)(x - n)$

| $(x + n)$ is a common factor. |

| This answer could also be written as $(x - n)(x + n)$. |

To factorise the difference of the squares of two terms, multiply the sum of the two terms by the difference of the two terms.

So, for any two terms $A$ and $B$,

$$A^2 - B^2 = (A + B)(A - B)$$

## Example 19

Factorise $4x^2 - 9$

### Solution 19

$4x^2 - 9 = (2x)^2 - 3^2$

| $4x^2$ and 9 are squares since $4x^2 = (2x)^2$ and $9 = 3^2$, so this expression is the difference of two squares. Write $4x^2 - 9$ in the form $A^2 - B^2$. |

$\quad = (2x + 3)(2x - 3)$

| Use $A^2 - B^2 = (A + B)(A - B)$ with $A = 2x$ and $B = 3$ |

## Example 20

Factorise $18x^2 - 50y^2$

### Solution 20

$18x^2 - 50y^2 = 2(9x^2 - 25y^2)$

> Take out the common factor, 2, and $9x^2$ and $25y^2$ are squares since $9x^2 = (3x)^2$ and $25y^2 = (5y)^2$

$= 2[(3x)^2 - (5y)^2]$

> Write $9x^2 - 25y^2$ in the form $A^2 - B^2$.

$= 2[(3x + 5y)(3x - 5y)]$

> Use $A^2 - B^2 = (A + B)(A - B)$ with $A = 3x$ and $B = 5y$.

$= 2(3x + 5y)(3x - 5y)$

## Example 21

Factorise $(2t + 1)^2 - 9(t - 6)^2$

### Solution 21

$(2t + 1)^2 - 9(t - 6)^2$

$= (2t + 1)^2 - [3(t - 6)]^2$

> Write $(2t + 1)^2 - 9(t - 6)^2$ in the form $A^2 - B^2$.

$= [(2t + 1) + 3(t - 6)][(2t + 1) - 3(t - 6)]$

> Use $A^2 - B^2 = (A + B)(A - B)$ with $A = 2t + 1$ and $B = 3(t - 6)$.

$= [2t + 1 + 3t - 18][2t + 1 - 3t + 18]$
$= [5t - 17][-t + 19]$

> Expand and simplify the terms in the squared brackets.

$= (5t - 17)(-t + 19)$

## Exercise 11F

**1** Factorise these expressions.

  **a** $p^2 - 1$              **b** $y^2 - 9$

  **c** $x^2 - 36$           **d** $a^2 - 100$

  **e** $144 - b^2$         **f** $1 - p^2$

  **g** $4p^2 - 1$           **h** $9x^2 - 1$

  **i** $25y^2 - 4$         **j** $(x + 1)^2 - 25$

  **k** $49 - (1 - x)^2$

**2 a** Factorise $x^2 - p^2$.

  **b** Hence find the value of

    **i** $51^2 - 49^2$     **ii** $7.64^2 - 2.36^2$

**3** Find the value of $1006^2 - 994^2$

**4** Factorise these expressions completely.

  **a** $2p^2 - 32$           **b** $27a^2 - 48$

  **c** $3y^2 - 75x^2$         **d** $4a^2 - 64b^2$

  **e** $12p^2 - 27q^2$      **f** $9(p + 1)^2 - 4p^2$

  **g** $8(x + 2)^2 - 2(x - 1)^2$      **h** $50(2x + 1)^2 - 18(1 - x)^2$

## Chapter summary

**You should now be able to:**

★ **expand** (multiply out) **brackets** by multiplying each term inside the bracket by the term outside the bracket, for example $3x(x + 2) = 3x^2 + 6x$

★ **factorise** by taking out a common factor, for example $4c + 6cs = 2c(2 + 3s)$; factorising is the opposite process to expanding brackets

★ expand two brackets by multiplying each term in the first bracket by the second bracket, for example $(x + p)(x + q) = x(x + q) + p(x + q) = x^2 + qx + px + pq$ and then simplify if possible

★ factorise by grouping

★ factorise an expression of the form $x^2 + bx + c$

★ factorise the **difference of two squares** using the general rule $A^2 - B^2 = (A + B)(A - B)$.

## Chapter 11 review questions

**1 a** Multiply out $4(3x + 2)$        **b** Simplify $2(3x + 1) + 3(x - 2)$

**2** Simplify

    **a** $3(a + 2) - 5$      **b** $4(b - 3) - 3b$      **c** $2(3c + d) + 3(c + d)$

**3** Simplify

    **a** $5(2n + 3m) + 3(n - 4m)$      **b** $4(2n + 2m + 1) + 3(2n - 4p - 3)$

**4** Expand and simplify $3(5x - 2) - 2(2x - 5)$                    (1388 March 2002)

**5 a** Simplify   **i** $p^2 \times p^7$     **ii** $x^8 \div x^3$     **iii** $\dfrac{y^4 \times y^3}{y^5}$

   **b** Expand $t(3t^2 + 4)$                                    (1387 November 2003)

**6 a** Expand the brackets $p(q - p^2)$

   **b** Expand and simplify $5(3p + 2) - 2(5p - 3)$            (1387 November 2004)

**7** Expand $y(3y^2 + 5y)$

**8** Factorise

    **a** $2r + 6$           **b** $4s - 10t$

**9** Factorise

    **a** $6a - 12b + 30$      **b** $8x + 12y - 16z$

**10 a** Expand and simplify $3(2x - 1) - 2(2x - 3)$

     **b** Factorise $y^2 + y$                                      (1387 November 2003)

**11** Factorise $k^2 + k$

**12** Factorise

    **a** $ab + 2bc$        **b** $2x + 3ax^3$

**13 a** Factorise $x^2 - 3x$  **b** Simplify $k^5 \div k^2$  (1387 June 2004)

**14** Expand and simplify $(y + 5)(y + 3)$  (1388 January 2005)

**15** Expand and simplify $(x - 9)(x + 4)$  (1388 March 2005)

**16 a** Simplify $5p - 4q + 3p + q$  **b** Simplify $\dfrac{x^7}{x^2}$

  **c** Factorise $4x + 6$  **d** Multiply out and simplify $(x + 3)(x - 2)$

  **e** Simplify $2x^3 \times x^5$  (1386 November 2002)

**17 a** Expand and simplify $(x + 7)(x - 4)$  **b** Expand $y(y^3 + 2y)$

  **c** Factorise $p^2 + 6p$  **d** Factorise completely $6x^2 - 9xy$  (1387 June 2005)

**18 a** Simplify $k^5 \div k^2$

  **b** Expand and simplify
   **i** $4(x + 5) + 3(x - 7)$   **ii** $(x + 3y)(x + 2y)$

  **c** Factorise $(p + q)^2 + 5(p + q)$

  **d** Simplify $(m^{-4})^{-2}$

  **e** Simplify $2t^2 \times 3r^3t^4$  (1387 June 2004)

**19** Expand and simplify
  **a** $3b + 1 - 4(b - 2)$  **b** $y(2y - y^3)$
  **c** $(x + 2)(x + 7)$  **d** $4 + (m + 1)^2$

**20 a** Expand $(x + 5)(x + 8)$  **b** Factorise $x^2 - 5x - 14$  (1388 March 2004)

**21 a** Expand and simplify $(x + 1)(x - 7)$  **b** Factorise $y^2 + 3y - 10$  (1388 November 2005)

**22 a** Factorise $x^2 + 4x - 21$

  **b** Factorise $4x^2 - 25$

  **c** Factorise $ab - 2ay + bx - 2xy$

**23 a** Factorise $m^2 - n^2$  **b** Hence find the value of $198^2 - 2^2$

**24** Factorise completely $6p^2 - 4p - 3pq + 2q$  (1385 June 1998)

**25** Factorise completely
  **a** $3a^2 - 12b^2$  **b** $8(n + 1)^2 - 2(n - 3)^2$

**26 a** Factorise $x^2 + 4x + 4$

  **b** Hence, factorise completely $(3x + 4)^2 - (x^2 + 4x + 4)$

# Two-dimensional shapes (1)

This chapter is about two-dimensional shapes. Two-dimensional shapes are flat. Two-dimensional is often written as 2-D.

## 12.1  Special quadrilaterals

A **quadrilateral** has 4 sides. Some quadrilaterals have special names.
The table shows some of the properties of special quadrilaterals.

| Quadrilateral | Properties |
|---|---|
| **Square** | All sides equal in length<br>All angles are 90° |
| **Rectangle** | Opposite sides equal in length<br>All angles are 90° |
| **Rhombus** | All sides equal in length<br>Opposite sides parallel<br>Opposite angles equal |
| **Parallelogram** | Opposite sides equal in length and parallel<br>Opposite angles equal |
| **Trapezium** | One pair of parallel sides |
| **Isosceles trapezium** | One pair of parallel sides<br>Non-parallel sides equal in length |
| **Kite** | Two pairs of adjacent sides equal in length<br>(**adjacent** means 'next to') |

# 12.2 Perimeter and area of rectangles

The **perimeter** of a two-dimensional shape is the total distance around the edge or boundary of the shape.

As a perimeter is a distance the units of perimeter are the units of length. So a perimeter can be measured in millimetres (mm), centimetres (cm), metres (m) or kilometres (km) for example.

So the perimeter of a rectangle which is 4 cm long and 3 cm wide is $(4 + 3 + 4 + 3)$ cm $= 14$ cm.

A formula for the perimeter, $P$, of a rectangle with length, $l$, and width, $w$, is

$$P = l + w + l + w$$

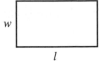

which simplifies to $\boxed{P = 2l + 2w \quad \text{or} \quad P = 2(l + w)}$

The **area** of a two-dimensional shape is a measure of the amount of space inside the shape.

The area of a centimetre square is 1 square centimetre. This is written as 1 cm²
The area of a square with sides of length 1 m (a metre square) is 1 square metre or 1 m²

The diagram shows a rectangle. The length of the rectangle is 4 cm and its width is 3 cm.
The rectangle is divided up into centimetre squares.
There are 12 centimetre squares inside the rectangle so that the area of the rectangle is 12 cm².

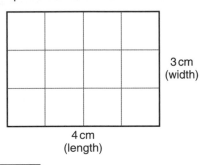

3 cm
(width)

4 cm
(length)

The number of squares inside the rectangle is $4 \times 3 = 12$

So to find the area of a rectangle multiply its length by its width.

$\boxed{\textbf{Area of a rectangle} = \textbf{length} \times \textbf{width}}$

width

length

The area, $A$, of a rectangle with length, $l$, and width, $w$, is given by the formula

$\boxed{A = lw}$

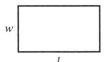

$w$

$l$

For a square the width is equal to the length and so to find the area of a square, multiply the length of the side of the square by itself, that is, square it.

$\boxed{\textbf{Area of a square} = \textbf{length} \times \textbf{length}}$

The area, $A$, of a square of side, $l$, is given by the formula $\boxed{A = l \times l \quad \text{or} \quad A = l^2}$

$l$

$l$

---

## Example 1

A football pitch is a rectangle with a length of 120 m and a width of 75 m.
Work out  **a** its perimeter  **b** its area.

### Solution 1
**a** Perimeter $= 2 \times 120 + 2 \times 75$

$= 240 + 150$

$= 390$ m

> Perimeter of a rectangle $= 2 \times$ length $+ 2 \times$ width.

> As the lengths are in metres the units of the perimeter are m.

**b** Area = 120 × 75

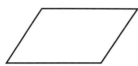

Area of a rectangle = length × width.

= 9000 m²

As the lengths are in metres the units of the area are m².

## 12.3 Area of a parallelogram

To find the area of this parallelogram          remove the orange triangle

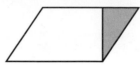

and replace it at the other end of the parallelogram to make a rectangle.

The area of the parallelogram is equal to the area of a rectangle with the same base and the same height as the parallelogram.

**Area of a parallelogram = base × height**

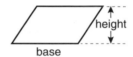

The 'height' of the parallelogram is the **vertical** or **perpendicular** height.

The area, $A$, of a parallelogram with base $b$ and height $h$ is given by the formula

$A = bh$

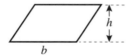

### Example 2

Work out the area of the parallelogram.

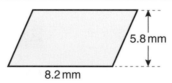

5.8 mm

8.2 mm

### Solution 2

Area = 8.2 × 5.8

Area of a parallelogram = base × height.

Area = 47.56 mm²

As the lengths are in millimetres the units of the area are mm².

## 12.4 Area of a triangle

Start with this triangle.          Join a congruent triangle to it as shown to make a parallelogram.

 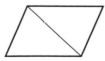

The area of the triangle is half the area of the parallelogram.

**Area of a triangle = $\frac{1}{2}$ × base × height**

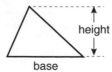

height

base

The 'height' of the triangle is the **vertical** or **perpendicular** height.

The area, $A$, of a triangle with base, $b$, and height, $h$, is given by the formula

$A = \frac{1}{2}bh$

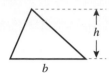

---

## Example 3

Work out the area of the triangle.

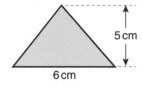
5 cm
6 cm

**Solution 3**

Area $= \frac{1}{2} \times 6 \times 5$ | Area of a triangle $= \frac{1}{2} \times$ base $\times$ height.

Area $= 15$ cm$^2$ | Either multiply the base by the height and then halve the result that is, $6 \times 5 = 30$ and $\frac{1}{2} \times 30 = 15$ or halve the base and multiply by the height that is $\frac{1}{2} \times 6 = 3$ and $3 \times 5 = 15$

---

## 12.5 Area of a trapezium

Start with this trapezium.          Join a congruent trapezium to it as shown to make a parallelogram.

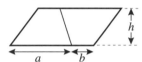

The area of the trapezium is half the area of the parallelogram.
The base of the parallelogram is the sum of the parallel sides of the trapezium.
$h$ is the **perpendicular** distance between the parallel sides.

**Area of a trapezium $= \frac{1}{2} \times$ sum of parallel sides $\times$ distance between them**

The area, $A$, of a trapezium with parallel sides of length $a$ and $b$ and a distance, $h$, between the parallel sides is given by the formula

$A = \frac{1}{2}(a + b)h$

---

## Example 4

Work out the area of the trapezium.

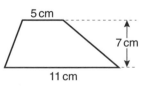
5 cm
7 cm
11 cm

**Solution 4**

Area $= \frac{1}{2} \times (11 + 5) \times 7$ | Area of a trapezium $= \frac{1}{2} \times$ sum of parallel sides $\times$ distance between them.

$= \frac{1}{2} \times 16 \times 7$ | Work out the brackets first.

Area $= 56$ cm$^2$

**Exercise 12A**

**1** Write down the names of the quadrilaterals with
  **a** all sides the same length          **b** all angles equal
  **c** both pairs of opposite sides parallel    **d** opposite angles equal

**2** The length of a rectangle is 9 cm and its width is 4 cm.
   Work out **a** the perimeter and **b** the area.

**3** The length of each side of a square is 7 cm.
   Work out **a** the perimeter and **b** the area.

**4** Work out the areas of these triangles and parallelograms.

**a**
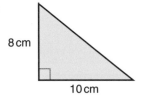
8 cm  10 cm

**b**
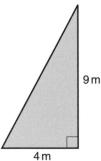
9 m  4 m

**c**
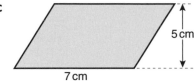
5 cm  7 cm

**d**
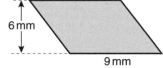
6 mm  9 mm

**e**
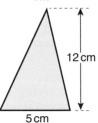
12 cm  5 cm

**5** Copy and complete this table.

| Shape | Length | Width | Area |
|---|---|---|---|
| Rectangle | 7 cm | 9 cm | |
| Rectangle | 10 cm | | 40 cm² |
| Rectangle | | 5 cm | 30 cm² |

**6** Copy and complete this table.

| Shape | Base | Height | Area |
|---|---|---|---|
| Triangle | 5 cm | 10 cm | |
| Parallelogram | 8 cm | 4 cm | |
| Parallelogram | 7 cm | | 56 cm² |
| Triangle | | 8 cm | 24 cm² |

**7** Work out the areas of these trapezia.

**a**
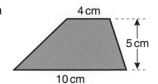
4 cm  5 cm  10 cm

**b**
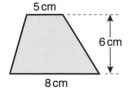
5 cm  6 cm  8 cm

**c**
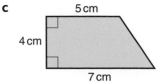
5 cm  4 cm  7 cm

**8** Work out the areas of these rectangles. Give answers correct to the nearest whole number.

**a**

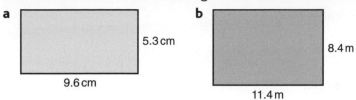

5.3 cm

9.6 cm

**b**

8.4 m

11.4 m

**9** Work out the areas of these triangles and parallelograms.
Give answers correct to the nearest whole number.

**a**

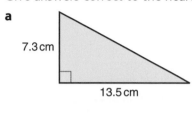

7.3 cm

13.5 cm

**b**

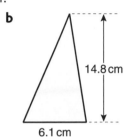

14.8 cm

6.1 cm

**c**

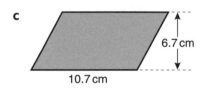

6.7 cm

10.7 cm

**d**

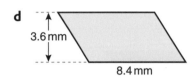

3.6 mm

8.4 mm

**10** Work out the areas of these trapezia.
Give answers correct to the nearest whole number.

**a**

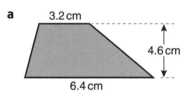

3.2 cm

4.6 cm

6.4 cm

**b**

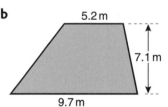

5.2 m

7.1 m

9.7 m

**c**

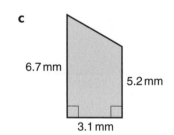

6.7 mm

5.2 mm

3.1 mm

# 12.6 Problems involving areas

Questions sometimes involve using the areas of rectangles, squares, triangles and parallelograms.

## Example 5

Work out the area of the shape.

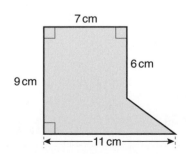

7 cm

6 cm

9 cm

11 cm

## Solution 5

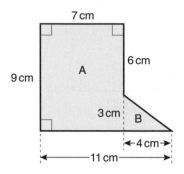

Split the shape up into a rectangle A and a right-angled triangle B.

The base of the triangle is 11 − 7 = 4 cm.
The height of the triangle is 9 − 6 = 3 cm.

Area of rectangle A $= 9 \times 7$
$= 63 \text{ cm}^2$

Area of a rectangle = length × width.

Area of triangle B $= \frac{1}{2} \times 4 \times 3$
$= 6 \text{ cm}^2$

Area of a triangle $= \frac{1}{2} \times$ base × height.

Area of shape $= 63 + 6$
$= 69 \text{ cm}^2$

Add the area of rectangle A and the area of triangle B to find the area of the shape.

## Example 6

A rectangular wall is 450 cm long and 300 cm high. The wall is to be tiled. The tiles are squares of side 50 cm. How many tiles are needed?

## Solution 6

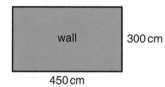

There is no diagram given with this question so it is a good idea to draw one.

### Method 1

Number of tiles needed for the length $= \dfrac{450}{50} = 9$

One way to answer questions like these is to work out how many tiles are needed for the length and how many are needed for the height.

Number of tiles needed for the height $= \dfrac{300}{50} = 6$

Number of tiles needed = 9 x 6

So there are 6 rows each with 9 tiles.

Number of tiles needed = 54

Multiply 9 by 6 to get the total number of tiles needed.

### Method 2

Area of wall $= 450 \times 300$
$= 135\,000 \text{ cm}^2$

The other method is to divide the area of the wall by the area of one tile.

Area of a tile $= 50 \times 50$
$= 2500 \text{ cm}^2$

Number of tiles $= \dfrac{135\,000}{2500} = 54$

**Exercise 12B**

1 Work out **a** the area and **b** the perimeter of this shape.
All the corners are right angles.

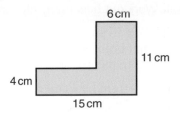

2 Work out the areas of these shapes.

a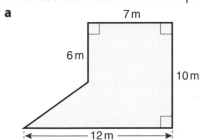

b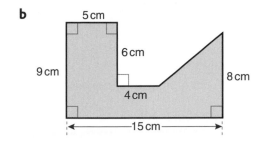

3 Work out the area of this shape.

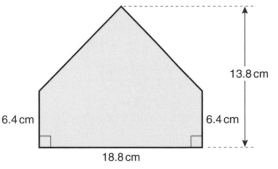

4 The diagram shows the floor plan of a room.

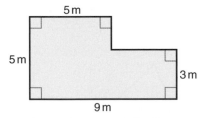

Work out the area of the floor.
Give the units of your answer.

5 The diagram shows a rectangular piece
of yellow paper with a corner removed.

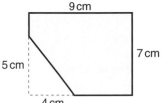

Work out the area of the yellow paper
that is left.

6 The floor of a room is a 5 m by 3 m
rectangle. The carpet used to cover the
floor completely costs £8.95 a square
metre. Work out the cost of the carpet
used.

7 Karl wants to make a rectangular lawn in his
garden. He wants the lawn to be 30 m by
10 m. Karl buys rectangular strips of turf
5 m long and 1 m wide. Work out how many
strips of turf Karl needs to buy.

8 Work out the area of the shaded region in this diagram.

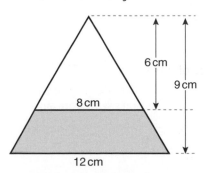

**9** A wall is a 300 cm by 250 cm rectangle. Tiles, which are squares of side 50 cm, are used to tile the wall. Work out how many tiles are needed.

**10** Trevor is going to paint some doors in his house. Each door is a 2 m by 0.85 m rectangle and he is going to paint both sides of each door. Each tin of paint that he is going to use covers 8 m². Trevor wants to paint 20 doors. How many tins of paint does he need to buy?

**11** A rectangle is 9 cm by 4 cm. A square has the same area as the rectangle.
Work out the length of each side of the square.

## Chapter summary

**You should now:**

★ be able to draw triangles and quadrilaterals accurately using ruler, protractor and compasses.

**You should also know:**

★ the names and properties of special quadrilaterals

★ the **perimeter** of a two-dimensional shape is the total distance around the edge or boundary of the shape

★ the perimeter, $P$, of a rectangle with length, $l$, and width, $w$, is given by the formula

$$P = 2l + 2w \quad \text{or} \quad P = 2(l + w)$$

★ how to find the area of a rectangle using
area of a rectangle = length × width or using the formula

$$A = lw$$

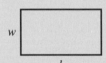

★ how to find the area of a parallelogram using
area of a parallelogram = base × height
(where the height is the vertical or perpendicular height)
or using the formula

$$A = bh$$

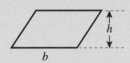

★ how to find the area of a triangle using
area of a triangle = $\frac{1}{2}$ × base × height or using the formula

$$A = \frac{1}{2}bh$$

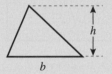

★ how to find the area of a trapezium using
area of a trapezium = $\frac{1}{2}$ × sum of parallel sides × distance between them
or using the formula

$$A = \frac{1}{2}(a + b)h$$

★ how to find the area and perimeter of a shape made from triangles and rectangles
★ how to solve problems involving areas

## Chapter 12 review questions

**1** Work out the area of the triangle.
Give the units with your answer.

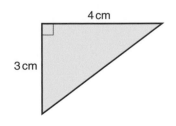

Diagram NOT
accurately drawn

4 cm

3 cm

(1385 May 2002)

**2** This diagram shows the plan of a floor.
   **a** Work out the perimeter of the floor.
   **b** Work out the area of the floor.

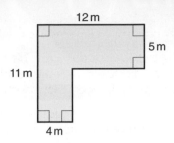

**Diagram NOT accurately drawn**

(1385 June 2001)

**3** Work out the area of the trapezium.

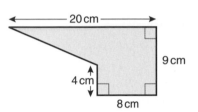

**Diagram NOT accurately drawn**

**4** The diagram shows a shape.
   Work out the area of the shape.

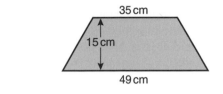

**Diagram NOT accurately drawn**

(1387 November 2003)

**5** The diagram shows a wall with a door in it.
   Work out the grey area.

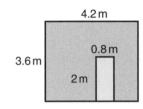

**6** Mary's floor is a rectangle 8 m long and 5 m wide.
   She wants to cover the floor completely with carpet tiles.
   Each carpet tile is square with sides of length 50 cm.
   Each carpet tile costs £4.19
   Work out the cost of covering Mary's floor completely with carpet tiles.

(1387 November 2004)

# Graphs (1)

## 13.1 Coordinates and line segments

The diagram shows the straight line joining the points $A(2, 1)$ and $B(8, 5)$. A line joining two points is called a **line segment**. So in the diagram, $AB$ is the line segment joining the points $A$ and $B$.

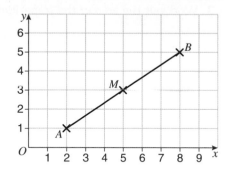

### *Midpoint of a line segment*

The midpoint $M$ of the line segment $AB$ has coordinates $(5, 3)$.

Notice that the $x$-coordinate of $A$ is 2, the $x$-coordinate of $B$ is 8 and $\dfrac{2 + 8}{2} = 5$, the $x$-coordinate of $M$.

Similarly the $y$-coordinate of $A$ is 1, the $y$-coordinate of $B$ is 5 and $\dfrac{1 + 5}{2} = 3$, the $y$-coordinate of $M$.

In general the $x$-coordinate of the midpoint of a line segment is the mean of the $x$-coordinates of its endpoints and the $y$-coordinate of its midpoint is the mean of the $y$-coordinates of its endpoints.

That is the midpoint of the line joining $(a, b)$ and $(p, q)$ is the point $\left( \dfrac{a + p}{2}, \dfrac{b + q}{2} \right)$

---

### *Example 1*

Find the midpoint of the line joining
**a** $(3, 5)$ and $(13, 7)$        **b** $(-5, 8)$ and $(9, -13)$.

*Solution 1*

**a** $3 + 13 = 16$

> The $x$-coordinates are 3 and 13

$\dfrac{16}{2} = 8$

$5 + 7 = 12$

> The $y$-coordinates are 5 and 7

$\dfrac{12}{2} = 6$

Midpoint is $(8, 6)$

**b** $-5 + 9 = 4$

> The $x$-coordinates are $-5$ and 9

$\dfrac{4}{2} = 2$

$8 + -13 = -5$

> The $y$-coordinates are 8 and $-13$

$\dfrac{-5}{2} = -2.5$

Midpoint is $(2, -2.5)$

---

## 13.2 Straight line graphs

In the diagram,

> all points through which the red vertical line passes have an $x$-coordinate equal to $-2$

$x = -2$ is called the **equation of the line**

> all points through which the blue horizontal line passes have a $y$-coordinate equal to 1. $y = 1$ is called the equation of this line.

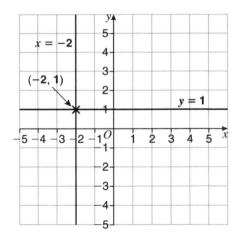

The lines $x = -2$ and $y = 1$ intersect at the point with coordinates $(-2, 1)$.

The $x$-coordinate of all points on the $y$-axis is 0. $x = 0$ is the equation of the $y$-axis.
The $y$-coordinate of all points on the $x$-axis is 0. $y = 0$ is the equation of the $x$-axis.

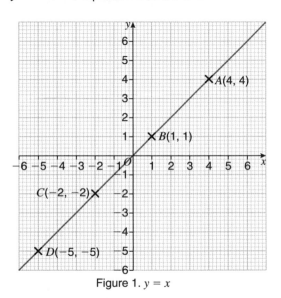

Figure 1. $y = x$

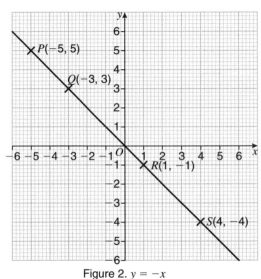

Figure 2. $y = -x$

Figure 1 shows a line drawn through the points

$A(4, 4)$ $B(1, 1)$ $C(-2, -2)$ $D(-5, -5)$ and $O(0, 0)$ (the origin)

For each of these five points the $y$-coordinate is equal to the $x$-coordinate. All other points on this line also have the $y$-coordinate equal to the $x$-coordinate.

**The equation of this line is $y = x$**

Figure 2 shows a line drawn through the points

$P(-5, 5)$ $Q(-3, 3)$ $R(1, -1)$ $S(4, -4)$ and $O(0, 0)$ (the origin)

For each of these five points the $y$-coordinate is equal to minus the $x$-coordinate. For example, for the point $P$ the $x$-coordinate is $-5$ and the $y$-coordinate is $-(-5) = 5$. All other points on this line also have $y$-coordinate equal to minus the $x$-coordinate.

**The equation of this line is $y = -x$**

## Example 2

a  Find the equation of the line which passes through the points $E(3, 6)$ $F(1, 2)$ and $H(-2, -4)$.

b  Find the equation of the line which passes through the points $T(3, 9)$ $U(2, 6)$ $V(-1, -3)$ and $W(-3, -9)$.

### Solution 2

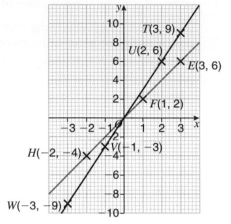

a
$$\overset{\times 2}{E(3, 6)} \quad \overset{\times 2}{F(1, 2)} \quad \overset{\times 2}{H(-2, -4)}$$

For all points on this line the $y$-coordinate is twice the $x$-coordinate, so the equation of this line is $y = 2x$.

b
$$\overset{\times 3}{T(3, 9)} \quad \overset{\times 3}{U(2, 6)} \quad \overset{\times 3}{V(-1, -3)} \quad \overset{\times 3}{W(-3, -9)}$$

For all points on this line the $y$-coordinate is three times the $x$-coordinate, so the equation of this line is $y = 3x$.

---

## Example 3

Draw the graph of $y = 5x - 4$. Use values of $x$ from $x = -1$ to $x = 4$

### Solution 3

$y = 5x - 4$

| | |
|---|---|
| When $x = 4$, | $y = 5 \times 4 - 4 = 16$ |
| When $x = 3$, | $y = 5 \times 3 - 4 = 11$ |
| When $x = 2$, | $y = 5 \times 2 - 4 = 6$ |
| When $x = 1$, | $y = 5 \times 1 - 4 = 1$ |
| When $x = 0$, | $y = 5 \times 0 - 4 = -4$ |
| When $x = -1$, | $y = 5 \times (-1) - 4 = -9$ |

> Substitute integer values of $x$ from $x = -1$ to $x = 4$ into $y = 5x - 4$

> Alternatively these points can be found and shown in a **table of values**

| $x$ | $-1$ | 0 | 1 | 2 | 3 | 4 |
|---|---|---|---|---|---|---|
| $y$ | $-9$ | $-4$ | 1 | 6 | 11 | 16 |

Plot each of the points

(4, 16) (3, 11) (2, 6)
(1, 1) (0, −4) (−1, −9)

and join them.

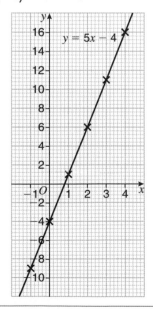

All of the graphs of the equations considered so far have been straight-line graphs.
Straight-line graphs are also called **linear graphs**.

The equations have been of the form $y = mx + c$, where $m$ and $c$ are numbers.
($y = x$, $y = -x$, $y = 2x$ and $y = 3x$ are also equations of the form $y = mx + c$, where $c = 0$)

To draw a straight-line graph only two points need to be plotted but it is safer to find and plot three points.

## Example 4

Draw the graph of $y = 2x + 3$ from $x = -3$ to $x = 3$

### Solution 4

$y = 2x + 3$

When $x = 3$, $y = 9$

When $x = -3$, $y = -3$

When $x = 0$, $y = 3$

| $y = 2x + 3$ is of the form $y = mx + c$ so the graph will be a straight line. |
| :--- |

| Substitute $x = 3$ and $x = -3$ into $y = 2x + 3$ |
| :--- |

| Use another value of $x$ as a check. |
| :--- |

| Plot each of the points (3, 9) (0, 3) and (−3, −3) and join them. |
| :--- |

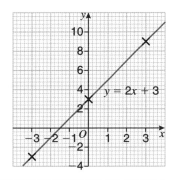

Equations of the form $x + y = k$, where $k$ is a number, also always give straight line graphs, since $x + y = k$ can be rearranged to give $y = -x + k$ which is of the form $y = mx + c$ with $m = -1$
To draw these straight-line graphs, tables of values are not necessary.

## Example 5

Draw the graph of $x + y = 5$

### Solution 5

$x + y = 5$

When $x = 0$, $y = 5$
When $y = 0$, $x = 5$

When $x = 2$, $y = 3$

| $x + y = 5$ is of the form $x + y = k$ so the graph will be a straight line. |
| :--- |

| Substitute $x = 0$ and $y = 0$ into $x + y = 5$ |
| :--- |

| Use another value of $x$ as a check. |
| :--- |

| Plot each of the points (0, 5) (5, 0) and (2, 3) and join them. |
| :--- |

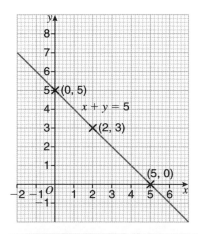

## Exercise 13A

**1** Work out the coordinates of the midpoint of the line joining

  **a** (3, 1) and (11, 7)      **b** (2, 5) and (12, 29)      **c** (−6, 9) and (8, 13)

  **d** (−4, −6) and (6, 12)      **e** (9, −15) and (−11, 6)      **f** (0, −5) and (9, −11)

**2** Write down the equation of the lines through these points.

  **a** (−4, −3) and (−4, 1)      **b** (0, −8) and (0, 3)      **c** (−1, −8) and (7, −8)

**3 a** Plot the points (−2, 4) (0, 0) (1, −2) and (3, −6) and join them with a straight line.

  **b** Find the equation of the line.

**4 a** Draw the graph of $y = 5 - 2x$. Use values of $x$ from −2 to 4

  **b** Write down the coordinates of the point where your graph crosses

    **i** the $y$-axis,   **ii** the $x$-axis.

  **c** Use your graph to find the value of $x$ when $y = 6$

**5 a** Draw the graph of $y = 4x - 7$. Use values of $x$ from 0 to 5

  **b** Write down the coordinates of the point where your graph crosses the $y$-axis.

  **c** Estimate and write down the $x$-coordinate of the point where your graph crosses the $x$-axis.

  **d**  **i** On the same axes, draw the graph of $x = 2.5$

    **ii** Write down the coordinates of the point where the two graphs cross.

  **e**  **i** On the same axes, draw the graph of $y = -5$

    **ii** Write down the coordinates of the point where the graph of $y = -5$ crosses the graph of $y = 4x - 7$

**6 a** Draw the graph of $y = 2 - x$. Use values of $x$ from $x = -3$ to $x = 4$

  **b** The point with coordinates $(k, 3.5)$ lies on the graph of $y = 2 - x$.
    Use your graph to find the value of $k$.

**7 a** On the same axes, draw the graphs of

    **i** $x + y = 2$      **ii** $x + y = 7$      **iii** $x + y = 10$

    **iv** $x + y = 13$      **v** $x + y = -10$      **vi** $x + y = -8$

  **b** What do you notice about the six graphs you have drawn?

**8** The diagram shows the graphs of three straight lines.
Write down the equation of these lines.

  **a** $AB$,      **b** $CD$,      **c** $EF$.

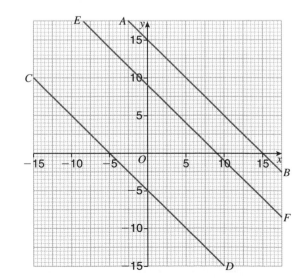

# Chapter summary

**You should be able to:**

★ find the coordinates of the midpoint if a line segment, given the coordinates of the endpoints if the segment

★ recognise the equations of vertical and horizontal lines, for example knowing that:

- $x = 5$ is the **equation of the vertical line** passing through the point $(5, 0)$ on the $x$-axis
- $y = 3$ is the **equation of the horizontal line** passing through the point $(0, 3)$ on the $y$-axis

★ recognise the graphs of equations $y = x$ and $y = -x$

★ draw a **straight-line graph** from a given equation and given values of $x$ by

- plotting three points
- joining the points with a straight line

★ draw a straight-line graph from an equation in the form $x + y =$ constant by

- plotting the points where the line crosses the $x$-axis and the $y$-axis
- joining these points with a straight line
- for example, $x + y = 2$ crosses the axes at $(0, 2)$ and $(2, 0)$

# Chapter 13 review questions

**1** The point $T$ is $(2, -3)$.

  **a** Write down the equation of the horizontal line through $T$.

  **b** Write down the equation of the vertical line through $T$.

**2** Write down the equations of the lines through these points.

  **a** $(-4, 0)$ and $(4, 0)$       **b** $(-7, -8)$ and $(-7, 8)$

**3 a** Draw the graph of $y = 6 - 2x$. Use values of $x$ from $-1$ to $5$

  **b** Write down the coordinates of the point where your graph crosses

    **i** the $y$-axis   **ii** the $x$-axis.

  **c** Use your graph to find the value of $x$ when $y = 5$

## 14.1 Significant figures

In the number 48.76, the 4 stands for 40 and is the most **significant** digit or figure in the number as it represents the largest part of the number.
The number 8 (8 units) is the second most significant digit.

To round a number correct to 1 significant figure, look at the second significant figure in the number. If that number is 5 or more, then increase the first figure by 1 If the second significant figure is less than 5, then do not change the first significant figure. The process is similar for rounding to other significant figures.

48.76 rounded correct to 1 significant figure is 50 as it is nearer to 50 than it is to 40

48.76 rounded correct to 2 significant figures is 49, as it is nearer to 49 than it is to 48

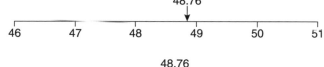

48.76 rounded correct to 3 significant figures is 48.8 as it is nearer to 48.8 than it is to 48.7

In the number 0.04268, the 4 stands for 4 hundredths and is the most significant figure in the number as it represents the largest part of the number.

0.04268 rounded correct to 1 significant figure is 0.04, as it is nearer to 0.04 than it is to 0.05

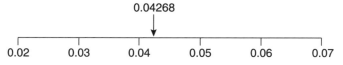

0.04268 rounded correct to 2 significant figures is 0.043 as it is nearer 0.043 than 0.042

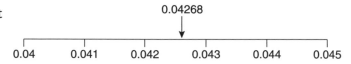

0.04268 rounded correct to 3 significant figures is 0.0427 as it is nearer 0.0427 than 0.0426

Zeros at the front of a number are not counted as significant figures.
The first 4 significant figures of 0.057328 are 5, 7, 3 and 2.

Zeros within a number are counted as significant figures.
The first 4 significant figures of 5.0487 are 5, 0, 4 and 8

---

### Example 1

Round each of these numbers correct to 3 significant figures.

**a** 6547      **b** 12.48      **c** 0.4763      **d** 3.0486

**Solution 1**

**a** 6550 | The 4th significant figure is 7 so increase the 3rd significant figure by 1

**b** 12.5 | The 4th significant figure is 8 so increase the 3rd significant figure by 1

**c** 0.476 | The 0 is not a significant figure, so the first three significant figures in the number are 4, 7 and 6. Since 3 is less than 5, do not change the 6

**d** 3.05 | The 0 between the 3 and the 4 *is* a significant figure, so 8 is the fourth significant figure.

Estimates for the values of expressions can be found by first writing each number correct to 1 significant figure and then carrying out the calculation.

## Example 2

Work out an estimate for the value of each of these expressions.

**a**  $17.2 \times 52.3$      **b**  $18.3 \div 3.9$      **c**  $\dfrac{6.4 \times 179.8}{0.52}$

**Solution 2**

**a**  $20 \times 50$ | Write each number correct to 1 significant figure.

  $= 1000$ | 1000 is an estimate for the value of $17.2 \times 52.3$

**b**  $20 \div 4$ | Write each number correct to 1 significant figure.

  $= 5$ | 5 is an estimate for the value of $18.3 \div 3.9$

**c**  $\dfrac{6 \times 200}{0.5}$ | Write each number correct to 1 significant figure.

  $= \dfrac{1200}{0.5}$ | Multiply 6 by 200

  $= 2400$ | $0.5 = \frac{1}{2}$ and dividing by $\frac{1}{2}$ is the same as multiplying by 2

2400 is an estimate for the value of $\dfrac{6.4 \times 179.8}{0.52}$

When working out answers to calculations, do not give too many figures in the answer.
For example, if the diameter of a circle is 6.4 centimetres, then the circumference of the circle is $\pi \times 6.4$ cm. Using $\pi = 3.14$ leads to the value 20.096 cm for the circumference. This answer has too many figures as a length could not be measured to such accuracy under normal circumstances. A more sensible answer is 20.1 cm, as it is possible to measure lengths correct to the nearest mm. This value 20.1 is correct to 3 significant figures.

Generally, when answers are not exact, 2 or 3 significant figures is an appropriate degree of accuracy for the answer.

## Example 3

In each case work out the value of the expression.
  **i** Write down all the figures on your calculator display
  **ii** write your answer to an appropriate degree of accuracy.

**a**  $24.6 \div 18.2$      **b**  $\dfrac{6.46 \times \sqrt{4.95}}{2.3^2 - 1.6^2}$

**Solution 3**

**a  i** 1.351 648 | Your calculator may display more figures than this.
**b  i** 5.264 686

**a  ii** 1.4 or 1.35 | 2 or 3 significant figures is an appropriate degree of accuracy.
**b  ii** 5.3 or 5.26

## Exercise 14A

**1** Write these numbers correct to 2 significant figures.

   **a** 648       **b** 34.78       **c** 7.082       **d** 0.0557       **e** 36 578

**2** Write these numbers correct to 3 significant figures.

   **a** 4879       **b** 21.036       **c** 345.88       **d** 21.675       **e** 0.004 776

**3** Write these numbers correct to 2 significant figures.

   **a** 399       **b** 0.205       **c** 7577       **d** 2.963       **e** 15.09

**4** Write these numbers correct to 3 significant figures.

   **a** 4599       **b** 0.005 996       **c** 2.000 47       **d** 1.9956       **e** 21.97

**5** Write these numbers correct to 1 significant figure.

   **a** 16.3       **b** 4.7       **c** 24.9       **d** 7.7       **e** 8.5

**6** Write these numbers correct to 1 significant figure.

   **a** 0.61       **b** 0.77       **c** 0.45       **d** 0.056       **e** 0.0079

**7** Work out an estimate for the value of each of these expressions by first rounding the numbers to 1 significant figure.

   **a** $3.4 \times 4.9$       **b** $5.4 \times 3.8$       **c** $9.4 \times 9.9$       **d** $13.4 \times 3.9$       **e** $24.4 \times 4.7$

**8** Work out an estimate for the value of each of these expressions by first writing the numbers correct to 1 significant figure.

   **a** $211 \div 12$       **b** $346 \div 3.2$       **c** $985 \div 478$       **d** $377 \div 48$       **e** $2334 \div 12.9$

**9** Work out an estimate for the value of each of these expressions by first rounding the numbers to 1 significant figure.

   **a** $\dfrac{5.4 \times 9.9}{10.1}$       **b** $\dfrac{9.8 \times 4.9}{5.1}$       **c** $\dfrac{16.4 \times 9.9}{16.8}$       **d** $\dfrac{6.8 \times 19.9}{12.1}$       **e** $\dfrac{17.4 \times 20.3}{38.7}$

**10** Calculate the value of each of the following.

   **i** Write down all the figures on your calculator display.

   **ii** Write down the answer correct to a sensible degree of accuracy.

   **a** $\dfrac{3.4^3}{5.66 \times 8.44}$       **b** $\dfrac{2.46 \times 4.98}{4.50^2}$       **c** $\dfrac{8.1^2 + \sqrt{298}}{41}$       **d** $\dfrac{\sqrt[3]{645} \times 3.8^2}{45}$

## 14.2  Accuracy of measurements

With a centimetre stick, lengths can be measured correct to the nearest centimetre.

Measured with a centimetre stick, the length of a piece of A4 paper is 30 cm.

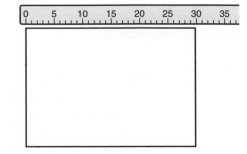

This does not mean that the length of the A4 paper is exactly 30 centimetres.

It is 30 centimetres correct to the nearest centimetre.

The exact length of the piece of A4 paper is somewhere between 29.5 cm and 30.5 cm.

So 30 cm to the nearest centimetre means that the minimum (least) possible length is 29.5 cm and the maximum (greatest) possible length is 30.5 cm.

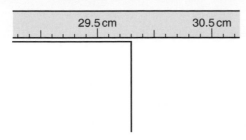

With a ruler, lengths can be measured correct to the nearest millimetre.
Measured with a ruler, the length of a piece of A4 paper is 298 mm.

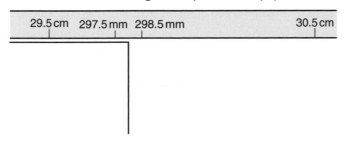

The exact length of the piece of A4 paper is between 297.5 mm and 298.5 mm.
So 298 mm to the nearest millimetre means that the minimum (least) possible length is 297.5 mm and the maximum (greatest) possible length is 298.5 mm.

Measurements given to the nearest whole unit may be inaccurate by up to one half of a unit below and one half of a unit above.
For example, '8 kg to the nearest kilogram' has a least possible weight of 7.5 kg and a greatest possible weight of 8.5 kg.

## Example 4

Write down the smallest possible volume and the greatest possible volume of
**a** a large tank which holds 52 litres of water correct to the nearest litre.
**b** a small bowl which holds 124 millilitres of water correct to the nearest millilitre.

### Solution 4

| | | | |
|---|---|---|---|
| **a** | 52 litres correct to the nearest litre | 51.5 litres | 52.5 litres |
| **b** | 124 millilitres correct to the nearest millilitre | 123.5 millilitres | 124.5 millilitres |

## Example 5

The length of a calculator is 12.8 cm correct to the nearest **millimetre**. Write down
**a** the minimum possible length it could be     **b** the maximum possible length it could be.

### Solution 5
Write 12.8 cm as 128 mm, as the measurement is accurate to the nearest millimetre.

| | minimum possible length | maximum possible length |
|---|---|---|
| 128 mm correct to the nearest millimetre | 127.5 mm or 12.75 cm | 128.5 mm or 12.85 cm |

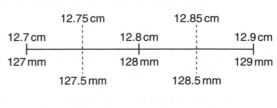

### Exercise 14B

**1** The length of a pencil is 12 cm correct to the nearest cm. Write down the maximum length it could be.

**2** The weight of an envelope is 45 grams correct to the nearest gram. Write down the minimum weight it could be.

**3** The capacity of a jug is 4 litres correct to the nearest litre. Write down the minimum capacity of the jug.

**4** The radius of a plate is 9.7 cm correct to the nearest millimetre.
Write down **a** the least possible length it could be **b** the greatest possible length it could be.

**5** Magda's height is 1.59 m correct to the nearest centimetre. Write down in **metres**
**a** the minimum possible height she could be **b** the maximum possible height she could be.

**6** The length of a pen is 10 cm correct to the nearest cm. The length of a pencil case is 102 mm correct to the nearest mm. Explain why the pen might not fit in the case.

**7** The width of a cupboard is measured to be 82 cm correct to the nearest cm. There is a gap of 822 mm correct to the nearest mm in the wall. Explain how the cupboard might fit in the wall.

## Chapter summary

**You should now be able to:**

★  write a number correct to a given number of significant figures

★  write the answer to a calculation to a sensible degree of accuracy

★  work out an estimate for the value of an expression by writing all the numbers correct to 1 significant figure

## Chapter 14 review questions

**1** Write these numbers correct to 3 significant figures.
  **a** 6482      **b** 35.795      **c** 0.052 81      **d** 1.6040      **e** 2.998      **f** 1.0992

**2** Work out the approximate value of these expressions by rounding the values to 1 significant figure.
  **a** $4.2 \times 7.8$      **b** $19.2 \times 31.8$      **c** $42 \times 78$      **d** $231 \times 47.8$

**3** Work out the approximate value of these expressions by rounding the values to 1 significant figure.
  **a** $\dfrac{101}{9.9}$   **b** $\dfrac{51.2}{4.9}$   **c** $\dfrac{798}{19.9}$   **d** $\dfrac{6.1 \times 19.2}{9.9}$   **e** $\dfrac{2.4 \times 0.388}{4.2}$

**4** Estimate the answers to these expressions by first writing all values correct to 1 significant figure.
  **a** $\dfrac{4.2 \times 3.6}{0.52}$   **b** $\dfrac{6.2 \times 7.2}{0.21}$   **c** $\dfrac{8.9^2}{0.49}$   **d** $\dfrac{6.3 \times 0.48}{0.022}$   **e** $\dfrac{18.4 \times 0.036}{0.38}$

**5** Work out an estimate for the value of $\dfrac{296}{1.84 \times 0.32}$

**6** Work out an estimate for the value of $\dfrac{637}{3.2 \times 9.8}$

(1387 June 2005)

# Three-dimensional shapes (1)

All the photographs show examples of **three-dimensional shapes**.
Three-dimensional shapes have length, breadth and height. They are not flat, like squares and circles.

Three-dimensional is often written as **3-D**.

## 15.1 Volume of three-dimensional shapes

The **volume** of a 3-D shape is the amount of space it takes up.

The diagram shows a cube with edges 1 cm long.
It is called a centimetre cube and its volume is 1 cubic centimetre (written 1 cm³).

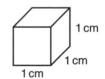

The volume of a 3-D shape is the number of centimetre cubes it contains.

### *Volume of a cuboid*

The diagram shows a cuboid made from centimetre cubes.

There are 12 centimetre cubes in each layer and there are two layers.
So the number of centimetre cubes is $12 \times 2 = 24$ and the volume of
the cuboid is 24 cm³.

$4 \times 3$ (length $\times$ width) gives 12, the number of centimetre cubes in
each layer.
$12 \times 2$ (the height) gives 24, the number of centimetre cubes.
So the number of centimetre cubes is $4 \times 3 \times 2 = 24$
This shows that

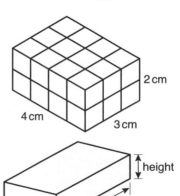

> **volume of a cuboid = length × width × height**

The volume, $V$, of a cuboid with length $l$, width $w$ and height $h$ is given by the formula

$$V = lwh$$

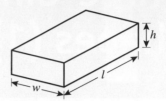

When the length, width and height are measured in centimetres, the volume is in $cm^3$.
When they are measured in metres, the volume is in $m^3$, and so on.
The length, width and height must all be measured in the *same* units.

## Example 1

Work out the volume of the cuboid shown in the diagram.

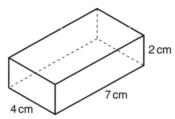

*Solution 1*
Volume $= 7 \times 4 \times 2$
$\qquad = 56\ cm^3.$

The volume of the cuboid might be given and one of its dimensions has to be found.

## Example 2

The volume of a cuboid is $216\ cm^3$.
Its length is 9 cm and its width is 6 cm.
Work out its height.

*Solution 2*
Let the height of the cuboid be $h$ cm.

$$216 = 9 \times 6 \times h$$

$$54h = 216$$

$$h = \frac{216}{54} = 4$$

The height of the cuboid is 4 cm.

| Use a letter to stand for the height. |
| Substitute the given values for $V$, $l$ and $w$ in $V = lwh$. |
| Simplify $9 \times 6 \times h$ as $54h$ and write it on the left-hand side. |
| Divide both sides by 54 to find the value of $h$. |
| State the answer. |

Sometimes the number of small cuboids which will fit inside a large cuboid has to be found.

## Example 3

A packet is a cuboid which is 10 cm by 5 cm by 4 cm.
A box is a cuboid which is 80 cm by 60 cm by 40 cm.

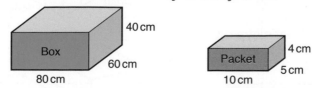

Work out how many packets will fit exactly into the box.

## Solution 3

### Method 1

$$\frac{80}{10} = 8$$    Work out how many times the length of a packet will fit into the length of the box.

$$\frac{60}{5} = 12$$    Work out how many times the width of a packet will fit into the width of the box.

$$\frac{40}{4} = 10$$    Work out how many times the height of a packet will fit into the height of the box.

$8 \times 12 \times 10 = 960$    *Multiply* the three results to find the number of packets which will fit exactly into the box.

### Method 2

$80 \times 60 \times 40 = 192\,000 \text{ cm}^3$    Work out the volume of the box.

$10 \times 5 \times 4 = 200 \text{ cm}^3$    Work out the volume of a packet.

$$\frac{192\,000}{200} = 960$$    *Divide* the volume of the box by the volume of a packet to find the number of packets which will fit exactly into the box.

## Exercise 15A

**1** Work out the volume of a cuboid which is 40 cm by 30 cm by 20 cm.

**2** Work out the volume of a cuboid which is 20 m by 7 m by 5 m.

**3** Work out the volume of a cuboid which is 25 mm by 10 mm by 8 mm.

**4** A cuboid measures 4 m by 2 m by 50 cm.
   **a** Explain why its volume is *not* 400 m³.     **b** Work out its volume in m³.

**5** Work out the volume, in m³, of a cuboid which is 5 m by 2 m by 40 cm.

**6** A cuboid measures 10 cm by 4 cm by 5 mm.
   Work out its volume
   **a** in cm³       **b** in mm³

**7** The volume of a cuboid is 400 cm³. Its length is 10 cm and its width is 8 cm.
   Work out its height.

**8** The volume of a cuboid is 180 cm³. Its width is 6 cm and its height is 2 cm.
   Work out its length.

**9** The volume of a cuboid is 120 m³. Its length is 8 m and its width is 6 m.
   Work out its height.

**10** Copy and complete the table.

| Length (cm) | Width (cm) | Height (cm) | Volume (cm³) |
|---|---|---|---|
| 10 | | 4 | 280 |
| 50 | 20 | | 7000 |
| | 6 | 5 | 270 |
| 5 | 4 | | 70 |
| 25 | | 4 | 650 |

**11** A packet is a cuboid which is 5 cm by 3 cm by 2 cm.
A box is a cuboid which is 20 cm by 18 cm by 10 cm.
Work out how many packets will fit exactly into the box.

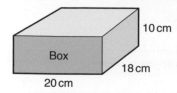
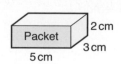

**12** A box of chocolates measures 20 cm by 20 cm by 5 cm.
A container is a cuboid which measures 80 cm by 60 cm by 30 cm.
Work out the greatest number of boxes which can be packed in the container.

**13** A crate is full of boxes.
The crate is a cube with side 1 m.
Each box is a cuboid which is 25 cm by 20 cm by 10 cm.
Work out the number of boxes in the crate.

**14** A packet of butter measures 11 cm by 6.5 cm by 4 cm.
A box measures 55 cm by 26 cm by 24 cm.
How many packets of butter are needed to fill the box?

**15** Kate fills a container with boxes. Each box is a cube of side 0.5 m.
The container is a cuboid of length 6 m, width 5 m and height 3 m.
Work out how many boxes will fit exactly into the container.

## Volume of a prism

A 3-D shape which is the same shape all along its length is called a **prism**. When you cut through a prism so that the cut is parallel to an end, you always get the same shape, called the **cross-section** of the prism.

The diagram shows a cuboid.
A cuboid is a prism with a rectangle as its cross-section.

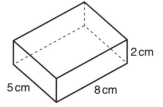

Volume of the cuboid = $5 \times 2 \times 8$
$= 80 \text{ cm}^3$

$5 \times 2 = 10 \text{ cm}^2$ is the area of cross-section of the prism.
8 cm is its length.

This shows that

> **volume of a prism = area of cross-section × length**

---

**Example 4**

The diagram shows a prism with a right-angled triangle as its cross-section.

Work out the volume of the prism.

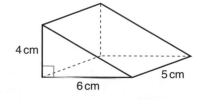

**Solution 4**
Area of cross-section $= \frac{1}{2} \times 6 \times 4$
$= 12 \text{ cm}^2$

Volume $= 12 \times 5 = 60 \text{ cm}^3$

> Work out the area of the triangular cross-section using area $= \frac{1}{2} \times$ base $\times$ vertical height.

> Multiply the area of cross-section by the length to get the volume of the prism.

## Example 5

The diagram shows a trapezium.
The trapezium is the cross-section of a prism.
The length of the prism is 8 cm.

Work out the volume of the prism.

## Solution 5

### Method 1

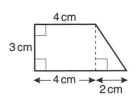

Split the trapezium into a rectangle and a triangle.

Area of rectangle = $4 \times 3 = 12$ cm²

Area of triangle = $\frac{1}{2} \times 2 \times 3 = 3$ cm²

Work out the area of the rectangle and the area of the triangle.

Area of cross-section = 12 + 3

$\qquad\qquad\qquad\quad = 15$ cm²

Add the area of the rectangle and the area of the triangle to get the area of the trapezium cross-section.

Volume = $15 \times 8 = 120$ cm³

Multiply the area of cross-section by the length to get the volume of the prism.

### Method 2

Area of cross-section = $\frac{1}{2} \times (6 + 4) \times 3$

$\qquad\qquad\qquad\quad = \frac{1}{2} \times 10 \times 3$

$\qquad\qquad\qquad\quad = 15$ cm²

Work out the area of the trapezium cross-section using area = $\frac{1}{2}(a + b)h$

Volume = $15 \times 8 = 120$ cm³

Multiply the area of cross-section by the length to get the volume of the prism.

## Exercise 15B

**1** The area of the cross-section of a prism is 14 cm². The length of the prism is 6 cm.
Work out the volume of the prism.

**2** The volume of a prism is 63 cm³. The area of its cross-section is 9 cm².
Work out the length of the prism.

**3** The diagram shows a prism with a right-angled triangle
as its cross-section.
Work out the volume of the prism.

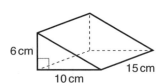

**4** The diagram shows a parallelogram.
The parallelogram is the cross-section of a prism.
The length of the prism is 7 cm.
Work out the volume of the prism.

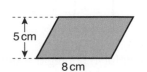

**5** The diagram shows a trapezium.
The trapezium is the cross-section of a prism.
The length of the prism is 5 cm.
Work out the volume of the prism.

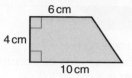

**6** The diagram shows a prism.
The cross-section of the prism is a trapezium.
The lengths of the parallel sides of the trapezium are 12 cm and 8 cm.
The distance between the parallel sides of the trapezium is 6 cm.
The length of the prism is 9 cm.
Work out the volume of the prism.

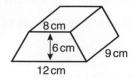

**7** The diagram shows the cross-section of a prism.
All the corners are right angles.
The length of the prism is 8 cm.
Work out the volume of the prism.

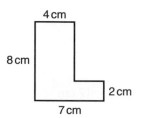

**8** The diagram shows the cross-section of a barn.
The length of the barn is 15 m.
Find the volume of the barn.

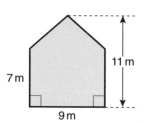

**9** The diagram shows a prism with a right-angled
triangle as its cross-section.
The volume of the prism is 126 cm³.
Work out the length of the prism.

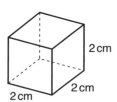

## 15.2 Surface area of three-dimensional shapes

### *Prisms and pyramids*

The diagram shows a cube of side 2 cm.
Each of its six faces is a square.
The area of each square face is 2 × 2 = 4 cm².
So the **surface area** of the cube is 6 × 4 = 24 cm².

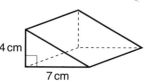

The surface area of the cube could be found by drawing an accurate, full-size
net of the cube and counting the number of centimetre squares inside it.

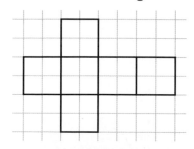

Nets are helpful in finding the surface areas of 3-D shapes – a sketch of a net of the 3-D shape with
lengths shown on it is usually used.

## Example 6

The diagram shows a prism with a right-angled triangle as its cross-section.
Work out the surface area of the prism.

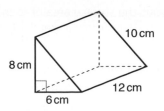

## Solution 6

Draw a sketch of a net of the prism. There is no need to show the length of every edge but show all those which will be used in the working.

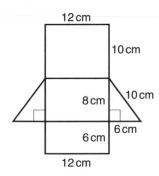

*Method 1*

$(12 \times 10) + (12 \times 8) + (12 \times 6)$ | Work out the areas of the three rectangles.

$(\frac{1}{2} \times 6 \times 8) + (\frac{1}{2} \times 6 \times 8)$ | Work out the areas of the two triangles.

$120 + 96 + 72 + 24 + 24$ | Add the results to find the surface area.
$= 336 \text{ cm}^2$

*Method 2*

Work out the total area of the three rectangles in one step by treating them as a single rectangle which is 12 cm by $(10 + 8 + 6)$ cm that is 12 cm by 24 cm.

Total area of three rectangles $= 12 \times 24$
$= 288 \text{ cm}^2$.

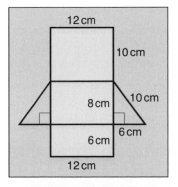

The two triangles are congruent.
So find the area of one triangle and multiply it by 2

Total area of two triangles $= 2 \times \frac{1}{2} \times 6 \times 8$
$= 48 \text{ cm}^2$

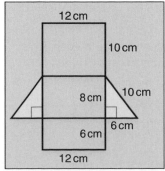

$288 + 48 = 336 \text{ cm}^2$ | Add the two results to find the surface area.

**Exercise 15C**

**1** Work out the surface area of each of these 3-D shapes.
For each shape, draw a sketch of the net.

   **a** A cube of side 5 cm.

   **b** A cuboid which is 4 cm by 4 cm by 8 cm.

   **c** A cuboid which is 7 cm by 3 cm by 10 cm.

   **d** A prism with a right-angled triangle as its cross-section.

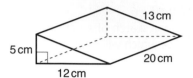

   **e** A prism with an isosceles triangle as its cross-section.

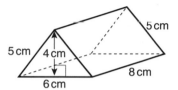

   **f** A square-based pyramid.
   The *vertical* height, $v$, of each of the triangular faces is 10 cm.

   **g** A prism of length 20 cm with this trapezium as its cross-section.

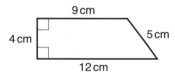

   **h** A prism of length 15 cm with this isosceles trapezium as its cross-section.

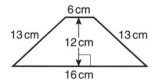

## 15.3 Coordinates in three dimensions

To locate a point in two dimensions, two perpendicular axes are used, the $x$-axis and the $y$-axis, and two coordinates are given, the $x$-coordinate and the $y$-coordinate.

In three dimensions, an extra axis is needed (the $z$-axis).
The three axes are perpendicular to each other.
The position of a point is given by three coordinates

● the $x$-coordinate     ● the $y$-coordinate     ● the $z$-coordinate.

The coordinates of a point are written $(x, y, z)$.
For example:

● the coordinates of $P$ are $(0, 1, 0)$

● the coordinates of $Q$ are $(0, 3, 2)$

● the coordinates of $R$ are $(1, 3, 2)$.

As with coordinates in two dimensions, negative coordinates are needed to locate some points.

Example 7

Write down the coordinates of $S$, $T$ and $U$.

*Solution 7*

The coordinates are $S(0, 0, -2)$, $T(0, -3, 2)$
and $U(-1, -3, 2)$

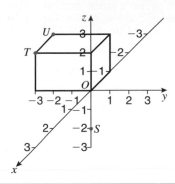

## Exercise 15D

**1** Write down the coordinates of $P$, $Q$ and $R$.

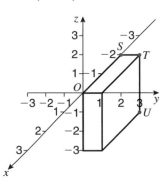

**2** Draw a diagram to show the points $A(1, 0, 0)$, $B(1, 0, 3)$ and $C(1, 2, 3)$.

**3** Write down the coordinates of $S$, $T$ and $U$.

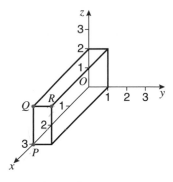

**4** Draw a diagram to show the points $D(0, 0, -1)$, $E(3, 0, -1)$ and $F(3, -2, -1)$.

**5** The coordinates of five of the corners of a cuboid are $(1, 0, 0)$, $(1, -3, 0)$, $(1, -3, -1)$, $(1, 0, -1)$ and $(-2, 0, 0)$. Find the coordinates of the other three corners.

# Chapter summary

**You should now know:**

★ that **three-dimensional** (3-D) shapes have length, breadth and height

★ how to find the **volume** of a cuboid and solve related problems

  volume of a cuboid = length × width × height

$$V = lwh$$

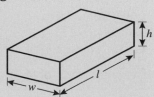

★ that a **prism's** cross-section is the same all along its length
★ how to find the volume of a prism
   **volume of a prism = area of cross-section × length**
★ how to find the **surface area** of cubes, cuboids, prisms and pyramids, including the use of nets
★ how to use coordinates in three dimensions

## Chapter 15 review questions

Where necessary, give answers correct to three significant figures.

**1** The diagram represents a large tank in the shape of a cuboid.
The tank has a base. It does not have a top.
The width of the tank is 2.8 metres.
The length of the tank is 3.2 metres.
The height of the tank is 4.5 metres.

The outside of the tank is going to be painted.
1 litre of paint will cover 2.5 m² of the tank.
The cost of paint is £2.99 per litre.

Calculate the cost of paint needed to paint the outside of the tank.

4.5 m

Diagram **NOT** accurately drawn

2.8 m

3.2 m

(1387 June 2003)

**2** The diagram shows a box in the shape of a cuboid.

**a** Work out the volume of the box.

The box is full of sugar lumps.
Each sugar lump is in the shape of a cuboid.
Each sugar lump is 1 cm by 1 cm by 2 cm.

**b** Work out the number of sugar lumps in the box.

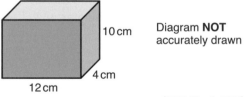

10 cm

Diagram **NOT** accurately drawn

4 cm

12 cm

(1388 March 2003)

**3** A cuboid *OABCDEFG* is drawn on 3-dimensional axes.

Point *P* is the midpoint of the edge *EF*.
Write down the coordinates of point *P*.

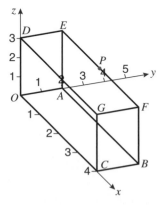

(1384 November 1994)

# Indices and standard form

# 16

**CHAPTER**

## 16.1  Zero and negative powers

> *also assessed in*
> *Module 4*

Positive powers of a number are introduced in Chapter 6

Zero and negative powers appear in Chapter 8

For non-zero values of $a$

$$a^0 = 1 \text{ and } a^{-1} = \frac{1}{a} \text{ (the reciprocal of } a)$$

and

for any number $n$,  $a^{-n} = \dfrac{1}{a^n}$

---

### Example 1

Work out the value of

**a**  $4^{-1}$　　　　**b**  $\left(\dfrac{2}{5}\right)^{-1}$　　　　**c**  $\dfrac{2^3 \times 2^5}{2 \times 2^{10}}$

### Solution 1

**a**  $\dfrac{1}{4}$

**b**  $\dfrac{5}{2} = 2\frac{1}{2}$

> The reciprocal of the fraction $\dfrac{a}{b}$ is the fraction $\dfrac{b}{a}$

**c**  $\dfrac{2^3 \times 2^5}{2^1 \times 2^{10}} = \dfrac{2^8}{2^{11}}$

> Use $x^m \times x^n = x^{m+n}$
> Use $x^m \div x^n = x^{m-n}$　　(See Section 8.3)

$= 2^{-3}$

> Use $a^{-n} = \dfrac{1}{a^n}$

$= \dfrac{1}{2^3}$

$= \dfrac{1}{8}$

---

The values of negative powers of numbers can be found using a calculator.

For example to find $2.5^{-3}$ a common key sequence is $\boxed{2}\ \boxed{.}\ \boxed{5}\ \boxed{y^x}\ \boxed{-}\ \boxed{3}\ \boxed{=}$ giving the value 0.064 but you should make sure that you know how to use your own calculator to find powers of numbers.

---

### Example 2

Use a calculator to work out

**a**  $1.1^{-2}$　　　　**b**  $(-2.4)^{-3}$

Give your answers correct to 3 significant figures.

**Solution 2**

**a**   0.826 446...

so $1.1^{-2} = 0.826$ (to 3 s.f.)

Key in $\boxed{1}\boxed{.}\boxed{1}\boxed{y^x}\boxed{-}\boxed{2}\boxed{=}$

Give the answer correct to 3 significant figures.

**b**   −0.072 337...

so $(-2.4)^{-3} = -0.0723$ (to 3 s.f.)

Key in $\boxed{-}\boxed{2}\boxed{.}\boxed{4}\boxed{y^x}\boxed{-}\boxed{3}\boxed{=}$

Give the answer correct to 3 significant figures.

### Exercise 16A

**1**  Work out the value of the following.

   **a** $2^{-1}$        **b** $3^{-2}$        **c** $5^{-1}$        **d** $10^{-3}$        **e** $2^0$

   **f** $2.5^{-1}$        **g** $\left(\dfrac{1}{3}\right)^{-1}$        **h** $\left(\dfrac{2}{3}\right)^{-2}$

**2**  Simplify the following.

   **a** $3^2 \times 3^{-3}$        **b** $4^{-2} \times 4$        **c** $5^4 \times 5^{-2}$        **d** $6^2 \times 6^{-4}$        **e** $2^2 \times 2^{-5}$

**3**  Simplify the following.

   **a** $4^{-2} \div 4^{-1}$        **b** $3^2 \div 3^{-1}$        **c** $2^{-2} \div 2^{-4}$        **d** $10^{-4} \div 10^{-3}$        **e** $5^{-3} \div 5^{-1}$

**4**  Simplify the following.

   **a** $\dfrac{2^4 \times 2^2}{2^7}$        **b** $\dfrac{3^4 \times 3^{-2}}{3^3}$        **c** $\dfrac{5^{-2} \times 5^2}{5}$        **d** $\dfrac{4^{-3} \times 4^3}{4^{-2}}$        **e** $\dfrac{2^{-4} \times 2^2}{2^{-7}}$

**5**  Simplify the following.

   **a** $\dfrac{2^4}{2^7 \times 2^{-2}}$        **b** $\dfrac{3^4 \times 3^2}{3 \times 3^7}$        **c** $\dfrac{5^4 \times 5^{-2}}{5^2 \times 5^{-1}}$        **d** $\dfrac{4^4 \times 4^{-2}}{4^{-1}}$        **e** $\dfrac{2^4 \times 2^2}{2^7 \times 2^{-1}}$

**6**  Find the value of $n$ in each of the following.

   **a** $2^n = \dfrac{2^2}{2^5}$        **b** $3 \times 3^n = \dfrac{3^3}{3^5}$        **c** $\dfrac{5^n}{5} = \dfrac{5^2}{5^5}$        **d** $4^2 \times 4^n = \dfrac{4^2}{4^6}$

**7**  Work out the value of

   **a** $7 \times 10^2$        **b** $2 \times 10^5$        **c** $4.9 \times 10^3$        **d** $8 \times 10^{-1}$        **e** $6.9 \times 10^{-3}$

**8**  Find the value of $n$ in each of the following.

   **a** $9000 = 9 \times 10^n$        **b** $72\,000 = 7.2 \times 10^n$   **c** $0.4 = 4 \times 10^n$        **d** $0.082 = 8.2 \times 10^n$

**9**  Work out

   **a** $1.25^{-1}$        **b** $0.16^{-1}$        **c** $0.4^{-2}$        **d** $2.5^{-3}$        **e** $12.5^{-2}$

**10**  Work out each of the following. Give your answer correct to 3 significant figures.

   **a** $0.45^{-1}$        **b** $2.25^{-2}$        **c** $0.96^{-3}$        **d** $0.0057^{-2}$

## 16.2  Standard form

*also assessed in Module 4*

There is often a need especially in science to carry out calculations with large numbers. For example the age of the Earth is about 4.6 thousand million years.
Written out in full this is 4 600 000 000 years.
Large numbers can be written in **standard form**.

A number written in standard form has the form $a \times 10^n$ where $1 \leqslant a < 10$, that is, $a$ is a number between 1 and 10 and $n$ is an integer.

For example $9000 = 9 \times 1000 = 9 \times 10^3$ in standard form. 9 is the number between 1 and 10, that is, $a = 9$ and $n = 3$

The 9 has moved 3 places to the *left* because it has been multiplied by $10^3$

| Th | H | T | U |
|----|---|---|---|
|    |   |   | 9 |
| 9  | 0 | 0 | 0 |

## Example 3

Write in standard form     **a** 312     **b** 590 000

*Solution 3*

**a**

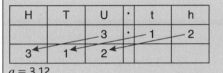

| H | T | U | . | t | h |
|---|---|---|---|---|---|
|   |   | 3 | . | 1 | 2 |
| 3 | 1 | 2 |   |   |   |

$a = 3.12$
The digits have moved 2 places to the left so $n = 2$

$312 = 3.12 \times 10^2$

**b** $590\,000 = 5.9 \times 10^5$

$a = 5.9$
The digits move 5 places to the left so $n = 5$

Numbers in standard form can be written as ordinary numbers.

## Example 4

Write as an ordinary number     **a** $8 \times 10^6$     **b** $7.3 \times 10^4$

*Solution 4*
**a** $8 \times 10^6 = 8\,000\,000$     The 8 moves 6 places to the left from the units column to the millions column.

**b** $7.3 \times 10^4 = 73\,000$     The digits move 4 places to the left.

## Example 5

Write in standard form     **a** $30 \times 10^5$     **b** $164 \times 10^8$

*Solution 5*
**a** $30 \times 10^5 = 3 \times 10^1 \times 10^5$     Write 30 in standard form.
$\quad\quad\quad\quad\ = 3 \times 10^{1+5}$
$\quad 30 \times 10^5 = 3 \times 10^6$     Add the indices.

**b** $164 \times 10^8 = 1.64 \times 10^2 \times 10^8$     Write 164 in standard form.
$\quad\quad\quad\quad\quad\ = 1.64 \times 10^{2+8}$
$\quad 164 \times 10^8 = 1.64 \times 10^{10}$     Add the indices.

Numbers in standard form can be multiplied or divided.

For example $(3 \times 10^{15}) \times (2 \times 10^7) = 3 \times 2 \times 10^{15} \times 10^7 = 6 \times 10^{22}$

Similarly $(3 \times 10^{15}) \div (2 \times 10^7) = \dfrac{3 \times 10^{15}}{2 \times 10^7} = \dfrac{3}{2} \times \dfrac{10^{15}}{10^7} = 1.5 \times 10^{15-7} = 1.5 \times 10^8$

## Example 6

Express as a number in standard form

**a** $(4 \times 10^5) \times (5 \times 10^8)$     **b** $(6 \times 10^{15}) \div (5 \times 10^9)$

Do not use a calculator.

### Solution 6

**a** $(4 \times 10^5) \times (5 \times 10^8) = 4 \times 5 \times 10^5 \times 10^8$
$$= 20 \times 10^{5+8}$$
$$= 2 \times 10^1 \times 10^{13}$$
$$= 2 \times 10^{14}$$

Write 20 in standard form.

**b** $(6 \times 10^{15}) \div (5 \times 10^9) = \dfrac{6 \times 10^{15}}{5 \times 10^9}$
$$= \frac{6}{5} \times \frac{10^{15}}{10^9}$$
$$= 1.2 \times 10^{15-9}$$
$$(6 \times 10^{15}) \div (5 \times 10^9) = 1.2 \times 10^6$$

It is possible to estimate the value of an expression in which the numbers are in standard form by rounding each number to 1 significant figure. For example

$3.48 \times 10^{21} = 3 \times 10^{21}$ correct to 1 significant figure.

## Example 7

Work out an estimate in standard form for the value of each of these expressions.

**a** $(4.04 \times 10^{12}) \times (5.3 \times 10^{14})$     **b** $(8.97 \times 10^{11}) \div (1.987 \times 10^7)$

### Solution 7

**a** $(4 \times 10^{12}) \times (5 \times 10^{14})$

Write each number correct to 1 significant figure.

$$= 20 \times 10^{26}$$

$4 \times 5 = 20$ and $10^{12} \times 10^{14} = 10^{26}$

Estimate $= 2 \times 10^{27}$

Write the estimate in standard form.

**b** $(9 \times 10^{11}) \div (2 \times 10^7)$

Write each number correct to 1 significant figure.

$$= \frac{9 \times 10^{11}}{2 \times 10^7} = \frac{9}{2} \times \frac{10^{11}}{10^7}$$

$9 \div 2 = 4.5$ and $10^{11} \div 10^7 = 10^4$

Estimate $= 4.5 \times 10^4$

Write the estimate in standard form.

A scientific calculator can be used to evaluate expressions involving numbers in standard form. There are two main ways of entering a number in standard form. Make sure that you know how to use your calculator for standard form.

The first method uses the (EXP) key. For example $2.4 \times 10^5$ is entered by keying in

(2)(.)(4)(EXP)(5)  Possible displays are 2.4E5, 2.4 05, 2.4$^{05}$ and 2.4$^{\times 10^5}$

The second method uses the (∧) key, the number $2.4 \times 10^5$ being entered as

(2)(.)(4)(×)(1)(0)(∧)(5)

## Example 8

Use a calculator to work out

**a** $(2.4 \times 10^5) \times (5.6 \times 10^8)$     **b** $(4.86 \times 10^{21}) \div (2.88 \times 10^9)$

**Solution 8**

**a** $1.344 \times 10^{14}$

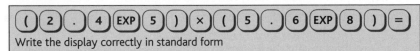

Write the display correctly in standard form

**b** $1.6875 \times 10^{12}$

In both of the cases above the brackets need not be used because there is only a single mathematical operation to carry out. In more complex expressions the brackets must be used.

## Example 9

$x = 2.4 \times 10^{15}, y = 9.5 \times 10^{14}$

Use a calculator to work out the value of $\dfrac{xy}{x + y}$

Give your answer in standard form correct to 3 significant figures.

**Solution 9**

$\dfrac{2.4 \times 10^{15} \times 9.5 \times 10^{14}}{(2.4 \times 10^{15} + 9.5 \times 10^{14})}$

Here the brackets are essential as the division is by the sum of both numbers.

$= 6.80590... \times 10^{14}$

Write the display correctly in standard form.

$= 6.81 \times 10^{14}$

Give your answer correct to 3 significant figures.

### Exercise 16B

**1** Write these numbers in standard form.

    **a** 40 000     **b** 530 000     **c** 56     **d** 487.5     **e** 60     **f** 5

**2** Write these as ordinary numbers.

    **a** $3 \times 10^6$     **b** $5.2 \times 10^5$     **c** $6.03 \times 10^4$     **d** $4 \times 10^1$     **e** $8.3 \times 10^0$

**3** Write these in standard form.

    **a** $50 \times 10^8$     **b** $47 \times 10^4$     **c** $187 \times 10^{10}$     **d** $3400 \times 10^2$     **e** $500 \times 10^{16}$

**4** Express as a number in standard form.

    **a** $(4 \times 10^8) \times (2 \times 10^3)$     **b** $(6 \times 10^5) \times (1.5 \times 10^3)$     **c** $(4 \times 10^8) \times (3 \times 10^5)$

    **d** $(6 \times 10^7) \times (3 \times 10^6)$     **e** $(6 \times 10^9) \times (5 \times 10^3)$     **f** $(5 \times 10^8) \times (2 \times 10^6)$

**5** Express as a number in standard form.

    **a** $(4 \times 10^8) \div (2 \times 10^3)$     **b** $(6 \times 10^5) \div (1.5 \times 10^3)$     **c** $(8 \times 10^8) \div (5 \times 10^5)$

    **d** $(5 \times 10^8) \div (5 \times 10^6)$

**6** Express as a number in standard form.

    **a** $(3 \times 10^4)^2$     **b** $(5 \times 10^4)^2$     **c** $(2 \times 10^5)^3$     **d** $(3 \times 10^4)^3$

**7** Express as a number in standard form.

**a** $(4 \times 10^8) + (2 \times 10^8)$      **b** $(5 \times 10^8) - (2.3 \times 10^8)$      **c** $(4 \times 10^8) - (9 \times 10^7)$

**d** $(8 \times 10^8) + (4 \times 10^8)$      **e** $(4 \times 10^8) + (6 \times 10^8)$      **f** $(4 \times 10^8) - (4 \times 10^7)$

**8** Write these numbers in standard form correct to 2 significant figures.

**a** $3.78 \times 10^7$      **b** $5.08 \times 10^9$      **c** $9.78 \times 10^{12}$      **d** $7.42 \times 10^6$      **e** $6.449 \times 10^6$

**9** By writing the numbers in standard form correct to one significant figure, work out an estimate of the value of these expressions. Give your answers in standard form.

**a** $600\,008 \times 598$      **b** $78\,018 \times 4880$      **c** $699\,008 \div 198$      **d** $8\,104\,660\,000 \div 78$

**10** Evaluate these expressions. Give your answers in standard form.

**a** $(3.2 \times 10^{10}) \times (6.5 \times 10^6)$      **b** $(1.3 \times 10^7) \times (4.5 \times 10^6)$

**c** $(2.46 \times 10^{10}) \div (2.5 \times 10^6)$      **d** $(3.6 \times 10^{20}) \div (3.75 \times 10^6)$

**11** Evaluate these expressions. Give your answers in standard form correct to 3 significant figures.

**a** $(3.5 \times 10^{11}) \div (6.5 \times 10^6)$      **b** $(1.33 \times 10^{10}) \times (4.66 \times 10^4)$

**c** $(5.3 \times 10^8) \times (6.45 \times 10^6)$      **d** $(3.24 \times 10^8) \div (6.4 \times 10^4)$

**12** $x = 3.5 \times 10^9$, $y = 4.7 \times 10^5$

Work out the value of the following. Give your answer in standard form correct to 3 significant figures.

**a** $\dfrac{x}{y}$      **b** $x(x + 800y)$      **c** $\dfrac{xy}{x + 800y}$      **d** $\left(\dfrac{x}{2000}\right)^2 + y^2$

Numbers less than 1 can also be written in standard form.

For example $0.009 = 9 \times \dfrac{1}{1000} = 9 \times \dfrac{1}{10^3} = 9 \times 10^{-3}$ in standard form.

9 is the number between 1 and 10, that is, $a = 9$, and $n = -3$

| U | . | t | h | th |
|---|---|---|---|---|
| 9 | . |   |   |    |
| 0 | . | 0 | 0 | 9  |

The 9 has moved 3 places to the *right* because it has been multiplied by $10^{-3}$.

## *Example 10*

Write, in standard form    **a** 0.043     **b** 0.000 756

*Solution 10*

**a**

| H | T | U | . | t | h | th |
|---|---|---|---|---|---|----|
|   |   | 4 | . | 3 |   |    |
|   |   | 0 | . | 0 | 4 | 3  |

$a = 4.3$
The digits have moved 2 places to the right so $n = -2$

$0.043 = 4.3 \times 10^{-2}$

**b**

$a = 7.56$
The digits move 4 places to the right so $n = -4$

$0.000\,756 = 7.56 \times 10^{-4}$

## Example 11

Change $3 \times 10^{-4}$ to an ordinary number.

**Solution 11**

$3 \times 10^{-4} = 0.0003$

> The **3** moves 4 places to the right.

## Example 12

Express as a number in standard form

**a** $(6 \times 10^5) \times (2.4 \times 10^{-9})$     **b** $(2 \times 10^3) \div (8 \times 10^{-5})$

Do not use a calculator.

**Solution 12**

**a** $(6 \times 10^5) \times (2.4 \times 10^{-9}) = 6 \times 2.4 \times 10^5 \times 10^{-9}$

$\qquad\qquad\qquad\qquad = 14.4 \times 10^{5 + (-9)}$

$\qquad\qquad\qquad\qquad = 14.4 \times 10^{-4}$

$\qquad\qquad\qquad\qquad = 1.44 \times 10^1 \times 10^{-4}$

$(6 \times 10^5) \times (2.4 \times 10^{-9}) = 1.44 \times 10^{-3}$

> Write 14.4 in standard form.

**b** $(2 \times 10^3) \div (8 \times 10^{-5}) = \dfrac{2 \times 10^3}{8 \times 10^{-5}}$

$\qquad\qquad\qquad\qquad = \dfrac{2}{8} \times \dfrac{10^3}{10^{-5}}$

$\qquad\qquad\qquad\qquad = 0.25 \times 10^{3 - (-5)}$

$\qquad\qquad\qquad\qquad = 0.25 \times 10^8$

$\qquad\qquad\qquad\qquad = 2.5 \times 10^{-1} \times 10^8$

$(2 \times 10^3) \div (8 \times 10^{-5}) = 2.5 \times 10^7$

> Write 0.25 in standard form.

## Example 13

$a = 2.4 \times 10^{-8}$, $b = 4.47 \times 10^{-7}$

Evaluate $(a + 2b)^2$

Give your answer in standard form correct to 3 significant figures.

**Solution 13**

$(2.4 \times 10^{-8} + 2 \times 4.47 \times 10^{-7})^2$

$= 8.427\,24 \times 10^{-13}$

$= 8.43 \times 10^{-13}$

> Give your answer correct to 3 significant figures.

## Example 14

Light travels at $3 \times 10^8$ metres per second.

Work out the time it takes light to travel 1.5 centimetres.

Give your answer in seconds in standard form.

**Solution 14**

$\dfrac{1.5 \times 10^{-2}}{3 \times 10^8}$

$= 5 \times 10^{-11}$ seconds

> $1.5\ \text{cm} = 0.015\ \text{m} = 1.5 \times 10^{-2}\ \text{m}$
> Time = distance ÷ speed

**Exercise 16C**

**1** Write these numbers in standard form.
  **a** 0.004    **b** 0.000 53    **c** 0.0056    **d** 0.4875    **e** 0.6    **f** 0.000 007 5

**2** Write these as ordinary numbers.
  **a** $3 \times 10^{-4}$    **b** $5.2 \times 10^{-5}$    **c** $6.03 \times 10^{-3}$    **d** $4 \times 10^{-1}$    **e** $7.3 \times 10^{-2}$

**3** Write these in standard form.
  **a** $50 \times 10^{-4}$    **b** $47 \times 10^{-3}$    **c** $187 \times 10^{-5}$    **d** $3400 \times 10^{-1}$    **e** $0.5 \times 10^{-5}$
  **f** $0.06 \times 10^{-6}$    **g** $0.0075 \times 10^{-15}$

**4** Express as a number in standard form.
  **a** $(4 \times 10^{-8}) \times (2 \times 10^{-5})$    **b** $(6 \times 10^{-5}) \times (1.5 \times 10^{-4})$    **c** $(4 \times 10^{-7}) \times (3 \times 10^{-6})$
  **d** $(6 \times 10^{-7}) \times (3 \times 10^{-6})$    **e** $(6 \times 10^{-9}) \times (4 \times 10^{3})$    **f** $(5 \times 10^{-10}) \times (2 \times 10^{6})$

**5** Express as a number in standard form.
  **a** $(4 \times 10^{-8}) \div (2 \times 10^{3})$    **b** $(6 \times 10^{5}) \div (1.5 \times 10^{-3})$    **c** $(8 \times 10^{-8}) \div (5 \times 10^{-5})$
  **d** $(3 \times 10^{-10}) \div (6 \times 10^{-6})$    **e** $(4 \times 10^{-9}) \div (5 \times 10^{-13})$    **f** $(5 \times 10^{-8}) \div (5 \times 10^{6})$

**6** Express as a number in standard form.
  **a** $(3 \times 10^{-4})^2$    **b** $(5 \times 10^{-5})^2$    **c** $(2 \times 10^{-5})^3$    **d** $(3 \times 10^{-4})^3$

**7** Express as a number in standard form.
  **a** $(4 \times 10^{-4}) + (2 \times 10^{-4})$    **b** $(5 \times 10^{-3}) - (2.3 \times 10^{-3})$    **c** $(4 \times 10^{-3}) - (9 \times 10^{-4})$
  **d** $(8 \times 10^{-5}) + (4 \times 10^{-5})$    **e** $(4 \times 10^{-6}) + (6 \times 10^{-6})$    **f** $(4 \times 10^{-3}) - (4 \times 10^{-4})$

**8** Express as a number in standard form.
  **a** $(3.2 \times 10^{-8}) \times (6.5 \times 10^{6})$    **b** $(1.3 \times 10^{-7}) \times (4.5 \times 10^{-6})$
  **c** $(2.46 \times 10^{-10}) \div (2.5 \times 10^{6})$    **d** $(3.6 \times 10^{-20}) \div (3.75 \times 10^{-6})$

**9** Express as a number in standard form correct to 3 significant figures.
  **a** $(3.5 \times 10^{11}) \div (6.5 \times 10^{-6})$    **b** $(1.33 \times 10^{-10}) \times (4.66 \times 10^{4})$
  **c** $(5.3 \times 10^{-8}) \times (6.45 \times 10^{-6})$    **d** $(3.24 \times 10^{-8}) \div (6.4 \times 10^{-4})$

**10** $x = 2.4 \times 10^{-5}$, $y = 9.6 \times 10^{-6}$
  Evaluate these expressions. Give your answers in standard form correct to 3 significant figures where necessary.
  **a** $\dfrac{x^2}{y}$    **b** $\dfrac{x^2 + y^2}{x + y}$    **c** $\dfrac{xy}{x - y}$

**11** The base of a microchip is in the shape of a rectangle. Its length is $2 \times 10^{-3}$ mm and its width is $1.55 \times 10^{-3}$ mm. Find the area of the base. Give your answer in mm² in standard form.

**12** An atomic particle has a lifetime of $3.86 \times 10^{-5}$ seconds. It travels at a speed of $4.2 \times 10^{6}$ metres per second. Calculate the distance it travels in its lifetime.

**13** The surface area of a sphere, radius $r$, is given by the formula $S = 4\pi r^2$
  Use this formula to find the surface area of a sphere with a radius of $6.4 \times 10^{6}$ m
  Give your answer in standard form correct to 3 significant figures.

**14** The distance of the Earth from the Sun is $1.5 \times 10^8$ km.
The distance of the planet Neptune from the Sun is 4510 million km.
Write in the form $1 : n$ the ratio

distance of the Earth from the Sun : distance of Neptune from the Sun

**15** The mass of a uranium atom is $3.98 \times 10^{-22}$ grams. Work out the number of uranium atoms in 2.5 kilograms of uranium.

## 16.3 Fractional indices

Using $x^m \times x^n = x^{m+n}$

$$x^{\frac{1}{2}} \times x^{\frac{1}{2}} = x^{\frac{1}{2} + \frac{1}{2}} = x^1 = x$$

This means that $x^{\frac{1}{2}}$ multiplied by itself gives $x$

So $x^{\frac{1}{2}}$ is the same as the square root of $x$, that is, $x^{\frac{1}{2}} = \sqrt{x}$

Similarly $x^{\frac{1}{3}} \times x^{\frac{1}{3}} \times x^{\frac{1}{3}} = x$ so $x^{\frac{1}{3}}$ is the same as the cube root of $x$

that is, $x^{\frac{1}{3}} = \sqrt[3]{x}$

In general $x^{\frac{1}{n}} = \sqrt[n]{x}$ where $\sqrt[n]{x}$ means the $n$th root of $x$

---

### Example 15

Find the values of the following.

**a** $4^{\frac{1}{2}}$ **b** $(-1000)^{\frac{1}{3}}$ **c** $16^{0.25}$

*Solution 15*

**a** $4^{\frac{1}{2}} = \sqrt{4} = 2$    $\boxed{x^{\frac{1}{2}} = \sqrt{x}}$

**b** $(-1000)^{\frac{1}{3}} = \sqrt[3]{-1000} = -10$    $\boxed{x^{\frac{1}{3}} = \sqrt[3]{x}}$

**c** $16^{0.25} = 16^{\frac{1}{4}} = \sqrt[4]{16} = 2$    $\boxed{x^{\frac{1}{4}} = \sqrt[4]{x}}$

---

In Section 8.3 the law of indices $(x^p)^q = x^{pq}$ was used for integer values of $p$ and $q$.
This law can also be used when $p$ and $q$ are not integers.

For example $x^{\frac{2}{3}} = x^{2 \times \frac{1}{3}} = (x^2)^{\frac{1}{3}}$ also $x^{\frac{2}{3}} = x^{\frac{1}{3} \times 2} = (x^{\frac{1}{3}})^2$

In general $x^{\frac{n}{m}} = (x^n)^{\frac{1}{m}} = (x^{\frac{1}{m}})^n$

---

### Example 16

Write as a single power    **a** $\sqrt{3}$    **b** $(\sqrt{5})^3$    **c** $(\sqrt[3]{7})^2$

*Solution 16*

**a** $\sqrt{3} = 3^{\frac{1}{2}}$    $\boxed{x^{\frac{1}{2}} = \sqrt{x}}$

**b** $(\sqrt{5})^3 = (5^{\frac{1}{2}})^3 = 5^{\frac{3}{2}}$    $\boxed{(x^{\frac{1}{m}})^n = x^{\frac{m}{n}}}$

**c** $(\sqrt[3]{7})^2 = (7^{\frac{1}{3}})^2 = 7^{\frac{2}{3}}$    $\boxed{(x^{\frac{1}{m}})^n = x^{\frac{m}{n}}}$

---

### Example 17

Work out the values of     **a** $8^{\frac{2}{3}}$     **b** $16^{-\frac{3}{4}}$

*Solution 17*

**a** $8^{\frac{2}{3}} = (8^{\frac{1}{3}})^2 = 2^2 = 4$

$$\boxed{x^{\frac{n}{m}} = (x^{\frac{1}{m}})^n}$$

**b** $16^{-\frac{3}{4}} = \dfrac{1}{16^{\frac{3}{4}}}$

$$\boxed{a^{-n} = \dfrac{1}{a^n}}$$

$\dfrac{1}{16^{\frac{3}{4}}} = \left(\dfrac{1}{16^{\frac{1}{4}}}\right)^3 = \dfrac{1}{2^3}$

$$\boxed{x^{\frac{n}{m}} = (x^{\frac{1}{m}})^n}$$

$16^{-\frac{3}{4}} = \dfrac{1}{8}$

---

A fraction can be raised to a power, for example $\left(\dfrac{2}{5}\right)^3 = \dfrac{2}{5} \times \dfrac{2}{5} \times \dfrac{2}{5} = \dfrac{2^3}{5^3}$

In general $\left(\dfrac{a}{b}\right)^n = \dfrac{a^n}{b^n}$

---

### Example 18

Work out the value of $\left(\dfrac{8}{27}\right)^{-\frac{1}{3}}$

*Solution 18*

$\left(\dfrac{8}{27}\right)^{-\frac{1}{3}} = \dfrac{1}{\left(\dfrac{8}{27}\right)^{\frac{1}{3}}}$

$$\boxed{a^{-n} = \dfrac{1}{a^n}}$$

$\left(\dfrac{8}{27}\right)^{\frac{1}{3}} = \dfrac{8^{\frac{1}{3}}}{27^{\frac{1}{3}}} = \dfrac{2}{3}$

$$\boxed{\left(\dfrac{a}{b}\right)^n = \dfrac{a^n}{b^n} \text{ and } x^{\frac{1}{3}} = \sqrt[3]{x}}$$

$\left(\dfrac{8}{27}\right)^{-\frac{1}{3}} = \dfrac{1}{\left(\dfrac{2}{3}\right)} = \dfrac{3}{2}$

$\left(\dfrac{8}{27}\right)^{-\frac{1}{3}} = 1\dfrac{1}{2}$

---

### Exercise 16D

**1** Work out the value of

    **a** $9^{\frac{1}{2}}$          **b** $25^{\frac{1}{2}}$          **c** $100^{\frac{1}{2}}$          **d** $4^{\frac{1}{2}}$          **e** $\left(\dfrac{1}{4}\right)^{\frac{1}{2}}$

**2** Work out the value of

    **a** $27^{\frac{1}{3}}$          **b** $1000^{\frac{1}{3}}$          **c** $-64^{\frac{1}{3}}$          **d** $125^{\frac{1}{3}}$          **e** $\left(\dfrac{1}{125}\right)^{\frac{1}{3}}$

**3** Work out as a single fraction the value of

   **a** $\left(\dfrac{1}{2}\right)^4$      **b** $\left(\dfrac{1}{3}\right)^2$      **c** $\left(\dfrac{2}{3}\right)^2$      **d** $\left(\dfrac{2}{5}\right)^2$      **e** $\left(\dfrac{3}{4}\right)^3$

**4** Work out the value of

   **a** $27^{\frac{2}{3}}$      **b** $1000^{\frac{2}{3}}$      **c** $64^{\frac{2}{3}}$      **d** $16^{\frac{3}{4}}$      **e** $25^{\frac{3}{2}}$

**5** Work out as a single fraction the value of

   **a** $25^{-\frac{1}{2}}$      **b** $9^{-\frac{1}{2}}$      **c** $27^{-\frac{1}{3}}$      **d** $8^{-\frac{2}{3}}$      **e** $64^{-\frac{3}{2}}$

**6** Find the value of $n$.

   **a** $\dfrac{1}{\sqrt{5}} = 5^n$      **b** $(\sqrt{7})^5 = 7^n$      **c** $(\sqrt[3]{2})^{11} = 2^n$

## Chapter summary

> **You should now know that:**
>
> ★   for non-zero values of $a$
>
>      $a^0 = 1$         $a^{-1} = \dfrac{1}{a}$
>
>      $a^{\frac{1}{2}} = \sqrt{a}$         $a^{\frac{1}{3}} = \sqrt[3]{a}$
>
> ★   and for any numbers $m$ and $n$
>
>      $a^{-m} = \dfrac{1}{a^m}$         $a^{\frac{1}{n}} = \sqrt[n]{a}$         $a^{\frac{n}{m}} = (a^n)^{\frac{1}{m}} = (a^{\frac{1}{m}})^n$
>
> **You should also know how to:**
>
> ★   solve problems which involve powers of several numbers
>
> ★   work out the value of a number raised to a fractional power
>
> ★   write numbers in standard form
>
> ★   solve problems in standard form with or without a calculator

## Chapter 16 review questions

**1** Write these numbers in standard form.

   **a** 6000      **b** 75 000      **c** 375 500      **d** 8

**2** Write these numbers in standard form.

   **a** 0.005      **b** 0.000 34      **c** 0.000 456      **d** 0.7

**3** Write as ordinary numbers.

   **a** $2 \times 10^3$      **b** $4.2 \times 10^5$      **c** $6.442 \times 10^4$      **d** $6.2 \times 10^0$

   **e** $7.3 \times 10^{-4}$      **f** $4.52 \times 10^{-3}$      **g** $1.2 \times 10^{-1}$

**4** Express as a number in standard form.

  **a** $(2 \times 10^4) \times (3 \times 10^6)$          **b** $(6 \times 10^4) \times (4 \times 10^6)$          **c** $(3 \times 10^{-3}) \times (3 \times 10^6)$
  **d** $(1.5 \times 10^{-4}) \times (4 \times 10^{-6})$     **e** $(8 \times 10^{-4}) \times (3 \times 10^{-7})$     **f** $(4 \times 10^5)^2$

**5** The population of an island at the start of 2004 was $1.3 \times 10^5$

  **a** Write $1.3 \times 10^5$ as an ordinary number.

  At the end of 2004 the population had decreased by 35 000

  **b** Find the population of this island at the end of 2004
  Give your answer in standard form.

**6** Express as a number in standard form.

  **a** $(6 \times 10^{14}) \div (3 \times 10^6)$          **b** $(4 \times 10^7) \div (8 \times 10^3)$          **c** $(5 \times 10^{14}) \div (2 \times 10^{16})$
  **d** $(8 \times 10^{-8}) \div (4 \times 10^6)$          **e** $(9 \times 10^{-6}) \div (3 \times 10^{-5})$

**7 a  i** Write 40 000 000 in standard form.
     **ii** Write $3 \times 10^{-5}$ as an ordinary number.

  **b** Work out the value of $3 \times 10^{-5} \times 40\,000\,000$
  Give your answer in standard form.                                         (1387 November 2004)

**8** Write as a power of 2

  **a** $2^5 \times 2^4 \times 2^2$     **b** $\dfrac{2^4 \times 2^7}{2^8}$          **c** $\dfrac{2^5 \times 2^7}{2^3 \times 2^4}$          **d** $(2^3)^4$          **e** $\sqrt{2^4}$

**9** Write down the value of

  **a** $8^0$               **b** $4^{-1}$               **c** $3^{-2}$               **d** $25^{\frac{1}{2}}$               **e** $1000^{-\frac{1}{3}}$

**10** Work out

  **a** $27^{\frac{2}{3}}$               **b** $16^{\frac{3}{4}}$               **c** $100^{\frac{3}{2}}$               **d** $25^{-\frac{3}{2}}$

**11 a** Write down the value of $36^{\frac{1}{2}}$

  **b** $4n^{\frac{3}{2}} = 8^{-\frac{1}{3}}$

  Find the value of $n$.                                                      (1388 January 2004)

**12** Use a calculator to work out the values of these expressions. Give your answers in standard form
correct to 3 significant figures where appropriate.

  **a** $(6.4 \times 10^8) \div (1.25 \times 10^5)$               **b** $\dfrac{8.2 \times 10^9}{2.3 \times 10^4 + 9.7 \times 10^4}$

  **c** $(1.4 \times 10^{-3}) \times (5.8 \times 10^6)$          **d** $\sqrt{3.2 \times 10^8}$

**13** $y = \dfrac{pq}{p + q}$

  $p = 2.4 \times 10^5$ and $q = 3.7 \times 10^5$  Work out the value of $y$. Give your answer in standard form
  correct to 3 significant figures.

**14** A floppy disk can store 1 440 000 bytes of data.

  **a** Write the number 1 440 000 in standard form.

  A hard disk can store $2.4 \times 10^9$ bytes of data.

  **b** Calculate the number of floppy disks needed to store the $2.4 \times 10^9$ bytes of data.
                                                                             (1387 November 2003)

**15** The Andromeda Galaxy is 21 900 000 000 000 000 000 km from Earth.

   **a** Write 21 900 000 000 000 000 000 in standard form.

Light travels $9.46 \times 10^{12}$ km in one year.

   **b** Calculate the number of years that light takes to travel from the Andromeda Galaxy to Earth. Give your answer in standard form correct to 2 significant figures.

**16** $v = \sqrt{\dfrac{GM}{R}}$     $G = 6.6 \times 10^{-11}$    $M = 6 \times 10^{24}$    $R = 5\,600\,000$

Calculate the value of $v$. Give your answer in standard form correct to 3 significant figures.

**17** $y^2 = \dfrac{ab}{a + b}$

$a = 3 \times 10^8$
$b = 2 \times 10^7$
Find $y$.
Give your answer in standard form correct to 2 significant figures.       (1387 June 2003)

# Further factorising, simplifying and algebraic proof

## 17.1 Further factorising

Quadratic expressions of the form $x^2 + bx + c$ were factorised in Section 11.5 by finding two numbers whose product is $+c$ and whose sum is $+b$.

For example factorising $x^2 + 7x + 12$ gives $(x + 4)(x + 3)$ since the two numbers which have a product of $+12$ and a sum of $+7$ are $+4$ and $+3$

This method is now extended to factorise more general quadratic expressions of the form $ax^2 + bx + c$ where $a$, $b$ and $c$ are numbers and $a \neq 1$

$a$ is the **coefficient of $x^2$**, $b$ is the **coefficient of $x$** and $c$ is the **constant term**.

So for the quadratic expression $6x^2 + 7x + 2$

the coefficient of $x^2$ is 6, the coefficient of $x$ is $+7$ and the constant term is $+2$

To factorise $6x^2 + 7x + 2$:

- multiply the coefficient of $x^2$ (6) by the constant term ($+2$) which gives $+12$

- find two numbers whose product is $+12$ and whose sum is the coefficient of $x$ ($+7$). The two numbers are $+4$ and $+3$

- split the $x$ term using these two numbers then factorise by grouping

  So $6x^2 + 7x + 2 = 6x^2 + 4x + 3x + 2$
  $= 2x(3x + 2) + 1(3x + 2)$
  $= (3x + 2)(2x + 1)$

> The order of $4x$ and $3x$ can be reversed.
> $6x^2 + 3x + 4x + 2$
> $= 3x(2x + 1) + 2(2x + 1)$
> $= (2x + 1)(3x + 2)$

So $6x^2 + 7x + 2$ when factorised gives $(3x + 2)(2x + 1)$

To check this answer expand $(3x + 2)(2x + 1)$ as in Section 11.3 to get $6x^2 + 7x + 2$

---

### Example 1

Factorise $2x^2 + 11x + 12$

**Solution 1**

| | | |
|---|---|---|
| $2x^2 + 11x + 12$ | $2 \times +12 = +24$ | Multiply the coefficient of $x^2$ by the constant term. |
| $2x^2 + \mathbf{11}x + 12$ | $+8 \times +3 = +24$ $+8 + +3 = \mathbf{+11}$ | Find two numbers whose product is $+24$ and whose sum is $\mathbf{+11}$. |
| $2x^2 + 11x + 12 = 2x^2 + 8x + 3x + 12$ | | Write $+11x$ as $+8x + 3x$ or as $+3x + 8x$ |
| $= 2x(x + 4) + 3(x + 4)$ | | Factorise by grouping. The bracketed term must be the same. |
| $= (x + 4)(2x + 3)$ | | $(x + 4)$ is a common factor. |
| $2x^2 + 11x + 12 = (x + 4)(2x + 3)$ | | This answer could also be written as $(2x + 3)(x + 4)$ |

## Example 2

Factorise $6x^2 - 7x - 3$

**Solution 2**

| | | |
|---|---|---|
| $6x^2 - 7x - 3$ | $6 \times -3 = -18$ | Multiply the coefficient of $x^2$ by the constant term. |

| | | |
|---|---|---|
| $6x^2 - \mathbf{7x} - 3$ | $-9 \times +2 = -18$ | Find two numbers whose product is $-18$ and whose sum is $-7$ |
| | $-9 + +2 = -7$ | |

| | |
|---|---|
| $6x^2 - 7x - 3 = 6x^2 - 9x + 2x - 3$ | Write $-7x$ as $-9x + 2x$ |
| $= 3x(2x - 3) + 1(2x - 3)$ | Factorise by grouping. The bracketed term must be the same. |
| $= (2x - 3)(3x + 1)$ | $(2x - 3)$ is a common factor. |
| $6x^2 - 7x - 3 = (2x - 3)(3x + 1)$ | This answer could also be written as $(3x + 1)(2x - 3)$ |

Expressions of the form $px^2 + qxy + ry^2$ where $p$, $q$ and $r$ are numbers can be factorised in a similar way.

## Example 3

Factorise $2x^2 - 5xy + 3y^2$

**Solution 3**

$2x^2 - 5xy + 3y^2$

| | | |
|---|---|---|
| $2x^2 - 5xy + 3y^2$ | $2 \times +3 = +6$ | Multiply the coefficient of $x^2$ by the coefficient of $y^2$ |

| | | |
|---|---|---|
| $2x^2 - \mathbf{5xy} + 3y^2$ | $-2 \times -3 = +6$ | Find two numbers whose product is $+6$ and whose sum is $-5$ |
| | $-2 + -3 = -5$ | |

| | |
|---|---|
| $2x^2 - 5xy + 3y^2 = 2x^2 - 2xy - 3xy + 3y^2$ | Write $-5xy$ as $-2xy - 3xy$. |
| $= 2x(x - y) - 3y(x - y)$ | Factorise by grouping. Note that $-3xy$ is the same as $-3yx$. The bracketed term must be the same. |
| $2x^2 - 5xy + 3y^2 = (x - y)(2x - 3y)$ | $(x - y)$ is a common factor. |
| | Check by expanding the brackets $(x - y)(2x - 3y) = 2x^2 - \mathbf{3xy} - \mathbf{2yx} + 3y^2$ $= 2x^2 - \mathbf{5xy} + 3y^2$ |

---

**Example 4**

Factorise completely $6x^2 - 3x - 3$

*Solution 4*

$6x^2 - 3x - 3 = 3(2x^2 - x - 1)$    | Take out the common factor 3

Consider $2x^2 - x - 1$    | Factorise $2x^2 - x - 1$

$2x^2 - 1x - 1$    $2 \times -1 = -2$    | Multiply the coefficient of $x^2$ by the constant term.

$2x^2 - \mathbf{1}x - 1$    $-2 \times +1 = -2$ | Find two numbers whose product is $-2$ and whose sum is $-1$
                $-2 + +1 = -1$

$2x^2 - x - 1 = 2x^2 - 2x + 1x - 1$    | Write $-x$ as $-2x + x$.

$\qquad\qquad = 2x(x - 1) + 1(x - 1)$    | Factorise by grouping. The bracketed term must be the same.

$2x^2 - x - 1 = (x - 1)(2x + 1)$    | $(x - 1)$ is a common factor.

$6x^2 - 3x - 3 = 3(2x^2 - x - 1)$
$\qquad\qquad = 3[(x - 1)(2x + 1)]$

$6x^2 - 3x - 3 = 3(x - 1)(2x + 1)$

---

**Exercise 17A**

In questions **1–23** factorise the expressions.

**1** $6x^2 + 5x + 1$     **2** $6x^2 + 7x + 1$     **3** $12x^2 - 13x + 1$

**4** $12x^2 - 7x + 1$     **5** $12x^2 - 8x + 1$     **6** $4x^2 + 4x + 1$

**7** $6x^2 + 7x + 2$     **8** $2x^2 - 5x + 3$     **9** $4x^2 + 3x - 1$

**10** $20x^2 - x - 1$     **11** $6x^2 + 5x - 1$     **12** $3x^2 - 5x - 2$

**13** $5x^2 + 14x - 3$     **14** $7x^2 + 5x - 2$     **15** $3x^2 + 10x + 8$

**16** $2x^2 + x - 15$     **17** $8x^2 - 10x - 3$     **18** $3x^2 - 10x - 8$

**19** $2y^2 + 7y - 15$     **20** $4y^2 + 5y - 6$     **21** $4y^2 + 19y + 12$

**22** $6x^2 + 5xy + y^2$     **23** $5x^2 - 9xy - 2y^2$

In questions **24–26** factorise the expressions completely.

**24** $6x^2 + 4x - 2$     **25** $16y^2 - 16y + 4$     **26** $12x^3 + 10x^2 - 12x$

## 17.2 Simplifying rational expressions

Algebraic expressions in the form of a fraction are called **rational expressions.**

Each of these rational expressions can be simplified by factorising the numerator and denominator and then cancelling any expression which is common.

The methods of factorising used in Sections 11.2, 11.5, 11.6 and 17.1 are required in this section.

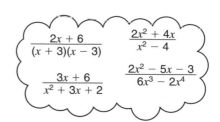

## Example 5

Simplify fully $\dfrac{2x^2 + 4x}{x^2 - 4}$

### Solution 5

$2x^2 + 4x = 2x(x + 2)$

> Factorise the numerator by taking out the common factor $2x$.

$x^2 - 4 = (x + 2)(x - 2)$

> $x^2 - 4$ is the difference of two squares.
> Factorise it by using $A^2 - B^2 = (A + B)(A - B)$.

$\dfrac{2x^2 + 4x}{x^2 - 4} = \dfrac{2x(x + 2)}{(x + 2)(x - 2)}$

> Write $\dfrac{2x^2 + 4x}{x^2 - 4}$ in a fully factorised form.

$= \dfrac{2x\cancel{(x + 2)}}{\cancel{(x + 2)}(x - 2)}$

> Cancel the common factor $(x + 2)$.

$\dfrac{2x^2 + 4x}{x^2 - 4} = \dfrac{2x}{x - 2}$

> $\dfrac{2x}{(x - 2)}$ is usually written as $\dfrac{2x}{x - 2}$.
>
> It is not possible to simplify $\dfrac{2x}{x - 2}$ further.

## Example 6

Simplify fully $\dfrac{3x + 3}{x^2 + 3x + 2}$

### Solution 6

$3x + 3 = 3(x + 1)$

> Factorise the numerator by taking out the common factor.

$x^2 + 3x + 2 = (x + 2)(x + 1)$

> Factorise the denominator.
> Find two numbers whose product is $+2$ and whose sum is $+3$ The numbers are $+2$ and $+1$

$\dfrac{3x + 3}{x^2 + 3x + 2} = \dfrac{3(x + 1)}{(x + 2)(x + 1)}$

> Write $\dfrac{3x + 3}{x^2 + 3x + 2}$ in a fully factorised form.

$= \dfrac{3\cancel{(x + 1)}}{(x + 2)\cancel{(x + 1)}}$

> Cancel the common factor $(x + 1)$.

$\dfrac{3x + 3}{x^2 + 3x + 2} = \dfrac{3}{x + 2}$

> It is not possible to simplify $\dfrac{3}{x + 2}$ further.

## Example 7

Simplify fully $\dfrac{2x^2 - 5x - 3}{6x^3 - 2x^4}$

*Solution 7*

$2x^2 - 5x - 3$

$\qquad = 2x^2 - 6x + 1x - 3$

$\qquad = 2x(x - 3) + 1(x - 3)$

$\qquad = (x - 3)(2x + 1)$

> Factorise the numerator.
> Multiply the coefficient of $x^2$ by the constant term to get $-6$
> Find two numbers whose product is $-6$ and whose sum is $-5$
> The numbers are $-6$ and $+1$
> Write $-5x$ as $-6x + 1x$.

$6x^3 - 2x^4 = 2x^3(3 - x)$

> Factorise the denominator by taking out the common factors.

$\dfrac{2x^2 - 5x - 3}{6x^3 - 2x^4} = \dfrac{(x - 3)(2x + 1)}{2x^3(3 - x)}$

> Write $\dfrac{2x^2 - 5x - 3}{6x^3 - 2x^4}$ in a fully factorised form.

$\qquad = \dfrac{-1(3 - x)(2x + 1)}{2x^3(3 - x)}$

> $(x - 3) = -1(3 - x).$

$\qquad = \dfrac{-1(3 - x)(2x + 1)}{2x^3(3 - x)}$

> Cancel the common factor $(3 - x)$.

$\dfrac{2x^2 - 5x - 3}{6x^3 - 2x^4} = \dfrac{-(2x + 1)}{2x^3}$

## Exercise 17B

**1** Simplify

**a** $\dfrac{4k^3}{2k^2}$

**b** $\dfrac{2(y + 1)}{(y + 1)^2}$

**c** $\dfrac{(p + 1)(p + 2)}{p(p + 2)}$

**d** $\dfrac{(2 + x)}{(x + 2)(x - 2)}$

**e** $\dfrac{1 - x}{x - 1}$

**f** $\dfrac{(3 - 2x)(2 - x)}{(2x - 3)(x + 1)}$

In Questions **2–25** simplify the expressions fully.

**2** $\dfrac{2x + 6}{2x}$

**3** $\dfrac{x^2 + 5x}{2x}$

**4** $\dfrac{4x + 8}{3x + 6}$

**5** $\dfrac{x^2 + 2x}{8x + 16}$

**6** $\dfrac{9 - 3x}{2x^2 - 6x}$

**7** $\dfrac{2x + 10}{x^2 - 25}$

**8** $\dfrac{x^2 - 16}{x^2 - 4x}$

**9** $\dfrac{(x + 3)}{x^2 - 9}$

**10** $\dfrac{x^2 - 4}{2x^2 + 4x}$

**11** $\dfrac{2x^2 - 18}{2x^2 + 6x}$

**12** $\dfrac{x^2 + 3x + 2}{x^2 - 1}$

**13** $\dfrac{x^2 + x - 2}{x^2 - 4}$

**14** $\dfrac{x^2 - 36}{x^2 - 7x + 6}$

**15** $\dfrac{2x^2 - 8}{x^2 + 6x + 8}$

**16** $\dfrac{x^2 + 7x + 10}{2x^2 - 50}$

**17** $\dfrac{x^2 + 5x + 6}{x^2 + x - 6}$

**18** $\dfrac{x^2 - 7x + 12}{x^2 - x - 6}$

**19** $\dfrac{x^2 - 2x - 8}{x^2 + x - 20}$

**20** $\dfrac{2x^2 + 3x + 1}{x^2 - 2x - 3}$

**21** $\dfrac{x^2 - 9}{2x^2 - 7x + 3}$

**22** $\dfrac{6x^2 - x - 1}{4x^2 - 1}$

**23** $\dfrac{6x^2 + 5x - 6}{9x^2 - 4}$

**24** $\dfrac{2x^2 + 3x - 2}{x^2 + 7x} \times \dfrac{x}{x + 2}$

**25** $\dfrac{2y^2 + 4y}{3y^2 + 7y + 2} \times \dfrac{9y^2 - 1}{3y^2 - y}$

## 17.3 Adding and subtracting rational expressions

Numerical fractions are added and subtracted in Section 20.1
Similar methods are used to add and subtract algebraic fractions.

To add fractions with the same denominator, add the numerators but do not change the denominator. For example $\dfrac{4}{x} + \dfrac{3}{x} = \dfrac{7}{x}$

To subtract fractions with the same denominator, subtract the numerators but do not change the denominator. For example $\dfrac{7x}{9} - \dfrac{2x}{9} = \dfrac{5x}{9}$

To add or subtract fractions with different denominators, each fraction must be written with a common denominator.

For example $\frac{2}{5} + \frac{1}{6} = \frac{12}{30} + \frac{5}{30} = \frac{17}{30}$  | The LCM of 5 and 6 is $5 \times 6 = 30$ |

To find $\dfrac{2}{5x} + \dfrac{1}{6x}$ firstly write each fraction with a common denominator of $30x$

so $\dfrac{2}{5x} = \dfrac{12}{30x}$ and $\dfrac{1}{6x} = \dfrac{5}{30x}$     $\dfrac{2}{5x} + \dfrac{1}{6x} = \dfrac{12}{30x} + \dfrac{5}{30x} = \dfrac{17}{30x}$

In general, to add or subtract algebraic fractions with different denominators
- factorise the denominators if possible
- write each fraction as a fraction with a common denominator
- add or subtract the fractions and factorise the numerator if possible
- simplify the algebraic fraction if possible as in Section 17.2

### Example 8

Write $\dfrac{1}{x} - \dfrac{2}{x^2 + 2x}$ as a single fraction in its simplest form.

### Solution 8

$\dfrac{1}{x} - \dfrac{2}{x^2 + 2x} = \dfrac{1}{x} - \dfrac{2}{x(x + 2)}$   | Factorise the denominator $x^2 + 2x$ |

Common denominator is $x(x + 2)$   | Since $x$ and $x(x + 2)$ divide exactly into $x(x + 2)$. |

$\dfrac{1}{x} = \dfrac{x + 2}{x(x + 2)}$

| Write $\dfrac{1}{x}$ as a fraction with denominator $x(x + 2)$. The other fraction $\dfrac{2}{x(x + 2)}$ has denominator $x(x + 2)$ so does not change. |

$\dfrac{1}{x} - \dfrac{2}{x(x + 2)} = \dfrac{x + 2}{x(x + 2)} - \dfrac{2}{x(x + 2)}$

$= \dfrac{x + 2 - 2}{x(x + 2)}$   | Since the denominators are the same just subtract the numerators. |

$= \dfrac{x}{x(x + 2)}$

$= \dfrac{\overset{1}{\cancel{x}}}{\cancel{x}(x + 2)}$   | Cancel the common factor $x$. |

| The single fraction $\dfrac{1}{x + 2}$ cannot be simplified further. |

$\dfrac{1}{x} - \dfrac{2}{x^2 + 2x} = \dfrac{1}{x + 2}$

## Example 9

Simplify $\dfrac{1}{3x+6} + \dfrac{1}{5x+10} - \dfrac{2}{15x+30}$

### Solution 9

$\dfrac{1}{3x+6} + \dfrac{1}{5x+10} - \dfrac{2}{15x+30} = \dfrac{1}{3(x+2)} + \dfrac{1}{5(x+2)} - \dfrac{2}{15(x+2)}$

> Factorise the denominators.

Common denominator is $15(x+2)$

> All denominators divide exactly into $15(x+2)$.

$\dfrac{1}{3(x+2)} = \dfrac{5}{15(x+2)}$　　$\dfrac{1}{5(x+2)} = \dfrac{3}{15(x+2)}$

> Write each fraction as a fraction with denominator $15(x+2)$.

$\dfrac{1}{3(x+2)} + \dfrac{1}{5(x+2)} - \dfrac{2}{15(x+2)} = \dfrac{5}{15(x+2)} + \dfrac{3}{15(x+2)} - \dfrac{2}{15(x+2)}$

$= \dfrac{5+3-2}{15(x+2)}$

> Since the denominators are the same just combine the numerators.

$= \dfrac{6}{15(x+2)}$

$= \dfrac{\overset{2}{6}}{\underset{5}{15}(x+2)}$

> Divide the numerator and the denominator by 3

$\dfrac{1}{3x+6} + \dfrac{1}{5x+10} - \dfrac{2}{15x+30} = \dfrac{2}{5(x+2)}$

## Example 10

Write $\dfrac{5}{2x+1} - \dfrac{3}{2x+3}$ as a single fraction.

### Solution 10

$\dfrac{5}{2x+1} - \dfrac{3}{2x+3}$

> The denominators do not factorise.

Common denominator is $(2x+1)(2x+3)$

> Since the denominators have no common factor.

$\dfrac{5}{(2x+1)} = \dfrac{5(2x+3)}{(2x+1)(2x+3)}$　　$\dfrac{3}{(2x+3)} = \dfrac{3(2x+1)}{(2x+1)(2x+3)}$

> Write each fraction as a fraction with denominator $(2x+1)(2x+3)$.

$\dfrac{5}{2x+1} - \dfrac{3}{2x+3} = \dfrac{5(2x+3)}{(2x+1)(2x+3)} - \dfrac{3(2x+1)}{(2x+1)(2x+3)}$

$= \dfrac{5(2x+3) - 3(2x+1)}{(2x+1)(2x+3)}$

> Since the denominators are the same just subtract the numerators.

$= \dfrac{10x+15-6x-3}{(2x+1)(2x+3)}$

> Simplify the numerator.

$= \dfrac{4x+12}{(2x+1)(2x+3)}$

$= \dfrac{4(x+3)}{(2x+1)(2x+3)}$

> Factorise the numerator.

$\dfrac{5}{2x+1} - \dfrac{3}{2x+3} = \dfrac{4(x+3)}{(2x+1)(2x+3)}$

> The single fraction $\dfrac{4(x+3)}{(2x+1)(2x+3)}$ cannot be simplified further.

**Exercise 17C**

1 Write $\dfrac{2}{5x} + \dfrac{3}{5x}$
  as a single fraction in its simplest form.

2 Write $\dfrac{1}{4x} + \dfrac{2}{3x}$ as a single fraction.

3 Write $\dfrac{3}{4x} - \dfrac{1}{6x}$ as a single fraction.

4 Write $\dfrac{2}{3x} + \dfrac{5}{6x} - \dfrac{7}{9x}$ as a single fraction.

5 Write $\dfrac{3}{5(x+1)} - \dfrac{1}{2(x+1)}$
  as a single fraction.

6 a Factorise $x^2 - 3x$

  b Simplify $\dfrac{2}{x^2 - 3x} + \dfrac{3}{x}$

7 a Factorise $x^2 - 5x + 6$

  b Write $\dfrac{1}{x-3} - \dfrac{1}{x^2 - 5x + 6}$
    as a single fraction in its simplest form.

8 Write $\dfrac{6}{x^2 - 9} + \dfrac{1}{x+3}$ as a single fraction
  in its simplest form.

9 Write $\dfrac{2}{x-1} - \dfrac{4}{x^2 - 1}$
  as a single fraction in its simplest form.

10 Write $\dfrac{3}{x+1} + \dfrac{4}{(x+1)^2}$ as a single fraction.

11 Simplify $\dfrac{1}{x+1} + \dfrac{1}{x-1}$

12 Simplify $\dfrac{4}{x+3} + \dfrac{2}{x-2}$

13 Simplify $\dfrac{1}{2x-5} - \dfrac{1}{2x-3}$

14 Express $\dfrac{1}{2} + \dfrac{2}{3x-1}$ as a single fraction.

15 Express $\dfrac{1}{x-3} - \dfrac{2}{3x-1}$ as a single fraction.

16 Express $\dfrac{2}{2-x} - \dfrac{4}{4-x}$ as a single fraction.

17 a Factorise $2x^2 + 5x + 3$

  b Write as a single fraction in its simplest form $\dfrac{5}{x+1} - \dfrac{4x}{2x^2 + 5x + 3}$

18 a Factorise $2x^2 - 7x - 4$

  b Write $\dfrac{2}{x-4} - \dfrac{2x+10}{2x^2 - 7x - 4}$ as a single fraction in its simplest form.

19 a Factorise  i $x^2 + 5x + 6$   ii $x^2 + 7x + 12$

  b Write $\dfrac{1}{x^2 + 5x + 6} + \dfrac{1}{x^2 + 7x + 12}$ as a single fraction in its simplest form.

20 a Factorise $4x^2 - 4x - 3$

  b Write $\dfrac{2}{4x^2 - 4x - 3} + \dfrac{1}{4x^2 + 8x + 3}$ as a single fraction in its simplest form.

## 17.4 Algebraic proof

Jane finds this puzzle.

- Write down an odd integer.
- Write down the next integer (which will be an even number).
- Square your odd number and double your even number and add your two results.
- The answer is always 1 more than the square of your even number.

Jane chooses 7 as her odd integer.
She writes the numbers 7 and 8 (7 and 8 are **consecutive integers**.)
Jane squares 7 to get 49 and doubles 8 to get 16
She then adds her two results to get 65 ($= 49 + 16$).

Jane says that she has proved the puzzle because 65 is 1 more than the square of 8 (her even number).

Jane is wrong because she has only shown that the puzzle works for the odd number 7
She has not proved that it works for all odd numbers.

Here are examples of some important facts including algebraic expressions for odd numbers, which will help when writing algebraic proofs

- three consecutive integers can be written in the form $n, n + 1, n + 2$ where $n$ is an integer. In some questions it is more useful to write three consecutive integers in the form $n - 1, n, n + 1$ so that the middle term is the simplest
- any even number can be written in the form $2n$ where $n$ is an integer
- the next even number after $2n$ is $2n + 2$ so $2n$ and $2n + 2$ are consecutive even integers
- any odd number can be written in the form $2n - 1$ where $n$ is an integer
- $2n - 1$ and $2n + 1$ are consecutive odd integers.

The next example shows an algebraic proof of Jane's puzzle.

### Example 11

Prove algebraically that for any odd number and the even number after it, the square of the odd number added to twice the even number is always one more than the square of the even number.

> Understand the problem by first writing some numerical examples
> e.g. **3, 4** gives $3^2 + 2 \times \textbf{4} = 17 = \textbf{4}^2 + 1$

*Solution 11*
For any integer $n$
Odd number $= 2n - 1$
*Next* (even) number $= 2n$

> Set up the problem by using algebraic expressions.
> Remember to state what $n$ stands for.
> The number after $2n - 1$ is $2n - 1 + 1$

(odd number)$^2$ + twice the even number
$= (2n - 1)^2 + \textbf{2} \times \textbf{2n}$

$= 4n^2 - 4n + 1 + \textbf{4n}$

> $(2n - 1)^2 = (2n - 1)(2n - 1)$

$= 4n^2 + 1$

> $-4n + \textbf{4n} = 0$

$= (2n)^2 + 1$

> Interpret the result by linking the expression $4n^2 + 1$ to the problem.
> $4n^2 = 2n \times 2n$

$= $ (the even number)$^2$ + 1

> The even number was $2n$

So the square of the odd number added to twice the even number is always one more than the square of the even number.

## Example 12

**a** Factorise fully $A^2 - B^2 + A - B$.

**b** Prove that the difference between the squares of any two integers added to the difference between the two integers is an even number.

### Solution 12

**a** $A^2 - B^2 + A - B = (A + B)(A - B) + 1(A - B)$

> Factorise $A^2 - B^2$ as the difference of two squares and write $A - B$ as $1(A - B)$.

$\qquad\qquad = (A - B)(A + B + 1)$

> $(A - B)$ is a common factor.

**b** Let $m$ and $n$ be any two integers with $m > n$.
The difference between the integers is $m - n$

> *Note* that the question refers to any two integers so two different letters are used.

$$m^2 - n^2 + m - n$$

> Add the **difference between the squares of the integers** to the difference between the integers.

$$= (m - n)(m + n + 1)$$

> Use part **a**

If $m$ and $n$ are both even or both odd,
$m - n$ is even and $m + n + 1$ is odd
so $(m - n)(m + n + 1)$ is even.

> even − even = even, odd − odd = even
> even + even + 1 = odd, odd + odd + 1 = odd
> even × odd = even

If one of $m$ and $n$ is even and the other odd, then $m - n$ is odd and $m + n + 1$ is even so again $(m - n)(m + n + 1)$ is even.

For any integers, $m$ and $n$, $m^2 - n^2 + m - n$ is even.

So the difference between the squares of any two integers added to the difference between the two integers is an even number.

---

## Exercise 17D

In questions **1–7** prove the result algebraically.

**1** The sum of any three consecutive integers is a multiple of 3

**2** The sum of any two consecutive odd numbers is a multiple of 4

**3** The sum of any two odd numbers is an even number.

**4** The sum of any four consecutive odd numbers is a multiple of 8

**5** The sum of any three consecutive odd numbers is never a multiple of 6

**6** The difference between the squares of any two consecutive even numbers is twice the sum of the two even numbers.

**7** The sum of the squares of any three consecutive integers is never a multiple of 3

**8** Prove algebraically that for any even number and the odd number after it, the square of the even number added to four times the odd number is always a square number.

**9** Prove that the difference between the squares of any two integers added to the sum of the two integers is an even number.

**10** The expression $\dfrac{(n+3)(n+4)}{2}$ is the $n$th term of the sequence of numbers 10, 15, 21, 28, …

   **a** Write down an expression in terms of $n$ for the $(n+1)$th term of the sequence 10, 15, 21, 28, …

   **b** By finding an expression for the sum of the $n$th term and the $(n+1)$th term of the sequence, prove that the sum of any two consecutive terms in the sequence is a square number.

## Chapter summary

> **You should now know:**
>
> ★ how to factorise quadratic expressions of the form $ax^2 + bx + c$ and $ax^2 + bxy + cy^2$ where $a$, $b$ and $c$ are numbers with $a \neq 1$
>
> ★ that in the expression $ax^2 + bx + c$, $a$ is the **coefficient of $x^2$**, $b$ is the **coefficient of $x$** and $c$ is the **constant term**
>
> ★ how to simplify **rational expressions** by factorising both the numerator and denominator and cancelling any common factors
>
> ★ how to add or subtract algebraic fractions with different denominators by applying these steps
> - factorise the denominators if possible
> - write each fraction as a fraction with a common denominator
> - add or subtract the fractions and factorise the numerator if possible
> - simplify the algebraic fraction if possible as in Section 17.2
>
> ★ how to prove algebraically a given result.

## Chapter 17 review questions

**1** Simplify fully   **i** $m^4 \times m^5$     **ii** $p^6 \div p^2$     **iii** $\dfrac{q^2 \times q^6}{q^3}$     **iv** $\dfrac{4(k+8)^2}{k+8}$

**2** Simplify fully

   **a** $2(3x+4) - 3(4x-5)$     **b** $(2xy^3)^5$     **c** $\dfrac{n^2-1}{n+1} \times \dfrac{2}{n-2}$     (1387 June 2003)

**3** Write as a single fraction in its simplest form $\dfrac{1}{3x} + \dfrac{1}{2x} - \dfrac{1}{6x}$     (1385 November 2002)

**4** Simplify fully $\dfrac{25 - x^2}{25 + 5x}$     (1387 November 2005)

**5** Simplify fully

   **a** $(2x^3y)^5$                               **b** $\dfrac{x^2 - 4x}{x^2 - 6x + 8}$     (5540 June 2005)

**6** Simplify $\dfrac{x^2(5+x)}{x^2 - 25}$     (1388 March 2004)

**7 a** Factorise $9x^2 - 6x + 1$

   **b** Simplify $\dfrac{6x^2 + 7x - 3}{9x^2 - 6x + 1}$

**8 a** Solve $\dfrac{40 - x}{3} = 4 + x$

  **b** Simplify fully $\dfrac{4x^2 - 6x}{4x^2 - 9}$  (1387 June 2004)

**9** Simplify $\dfrac{1}{2x + 3} + \dfrac{1}{2x - 1}$  (1385 November 1998)

**10 a** Factorise $x^2 + 3x + 2$

  **b** Write as a single fraction in its simplest form $\dfrac{3}{x + 1} + \dfrac{3x}{x^2 + 3x + 2}$  (1385 June 2000)

**11** Simplify $\dfrac{2x^2 - x - 10}{4x^2 - 25}$

**12 a** Factorise completely $2(x - 5)^2 + 3(x - 5)$

  **b** Simplify $\dfrac{3(y - 4)}{(y - 4)^2}$  (1388 January 2005)

**13** Show that $\dfrac{x + 2}{x} - \dfrac{x - 3}{x + 1} = \dfrac{2(3x + 1)}{x(x + 1)}$

**14 a** Factorise $5x^2 + 6xy + y^2$

  **b** Hence, write 50 601 as a product of two integers both of which are greater than 5

**15 a** Factorise $x^2 - 4x - 21$

  **b** Simplify $\dfrac{1}{(x + 2)} - \dfrac{3}{(x - 1)}$  (1384 November 1997)

**16** Prove algebraically that the sum of the squares of any two consecutive even integers is never a multiple of 8

**17** Prove algebraically that the sum of the squares of any two odd numbers leaves a remainder of 2 when divided by 4  (1387 November 2005)

**18 a** Show that $(2a - 1)^2 - (2b - 1)^2 = 4(a - b)(a + b - 1)$

  **b** Prove that the difference between the squares of any two odd numbers is a multiple of 8 (You may assume that any odd number can be written in the form $2r - 1$ where $r$ is an integer.)  (1387 June 2003)

**19** Humera wrote down three consecutive square numbers in order of size.
She then added the smallest and the largest of these numbers together and finally subtracted the middle number of the three numbers.
Prove algebraically that Humera's answer should always be two more than the middle number.

# Circle geometry (1)

## 18.1 Parts of a circle

The diagrams show the mathematical names for some parts of a circle.

- The **circumference** is the distance around the edge of a circle.
- A **chord** is a straight line segment joining two points on a circle.
- A **diameter** is a chord that passes through the centre of a circle.
- A **radius** is the distance from the centre of a circle to a point on the circle.
- A **tangent** is a line that touches the circle at only one point.

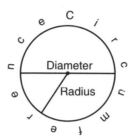

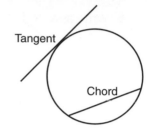

## 18.2 Isosceles triangles

Triangles formed by two radii and a chord are **isosceles** because they have two sides of equal length (the two sides that are radii). In an isosceles triangle, the angles opposite the equal sides are also equal.

### Example 1

$A$ and $B$ are points on the circumference of a circle, centre $O$.
Angle $OAB = 40°$.

Calculate the size of angle $AOB$.
Give reasons for your answer.

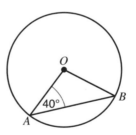

### Solution 1

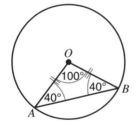

> At each step, mark the new information on the diagram.

$OA = OB$

> $OA$ and $OB$ are radii.

Angle $OBA = 40°$

> Triangle $OAB$ is isosceles.

In an isosceles triangle, the angles opposite the equal sides are equal.

> Give the reason.

Angle $AOB = 180° - (40° + 40°)$
$= 100°$

> Add the equal angles and subtract the sum from 180°.

The angle sum of a triangle is 180°.

> Give the reason.

**Exercise 18A**

In questions **1–9** each diagram shows a circle, centre $O$.
Calculate the size of each of the angles marked with a letter.
The diagrams are **NOT** accurately drawn.

**1**
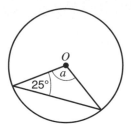

**2**
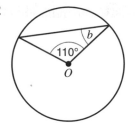

**3**
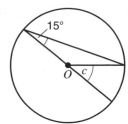

**4**

**5**
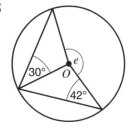

**6**
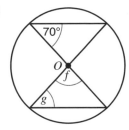

In Questions **7–9**, give reasons for your answers.

**7**
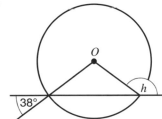

**8**
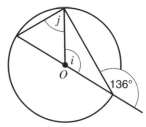

**9**
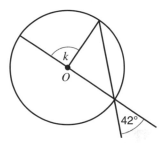

## 18.3  Tangents and chords

Here are four geometric facts which involve tangents or chords.

- **A tangent is perpendicular to the radius at the point of contact.**
  Angle $OTP = 90°$
  Angle $OTQ = 90°$

- **Tangents from an external point to a circle are equal in length.**
  $PA = PB$

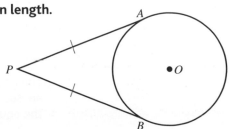

- **A line drawn from the centre of a circle perpendicular to a chord bisects the chord.**

  $AM = BM$

  The **converse** (opposite) of this statement is also true.

- **A line drawn from the centre of a circle to the midpoint of a chord is perpendicular to the chord.**

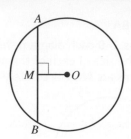

## Example 2

$PT$ is a tangent at $T$ to a circle, centre $O$.
$TU$ is a chord of the circle.
Angle $PTU = 54°$.

Find the size of angle $TOU$.
Give reasons for your answer.

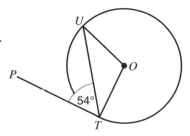

### Solution 2

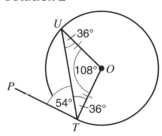

Angle $OTU = 90° - 54°$

$\quad\quad\quad\quad = 36°$

| Subtract 54° from 90°. |

Tangent is perpendicular to the radius.

| Give the reason. |

Angle $OUT = 36°$

| $OT = OU$. |

In an isosceles triangle, the angles opposite the equal sides are equal.

| Give the reasons. |

Angle $TOU = 180° - (36° + 36°)$

$\quad\quad\quad\quad = 108°$

| Add the equal angles and subtract the sum from 180°. |

Angle sum of a triangle is 180°.

| Give the reason. |

## Example 3

$PA$ and $PB$ are tangents to a circle.
Angle $APB = 68°$.

Calculate the size of angle $PAB$.
Give reasons for your answer.

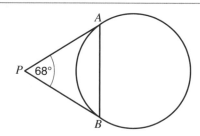

### Solution 3

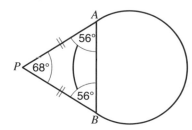

$PA = PB$

Tangents from an external point to a circle are equal in length.

| Give the reason. |

Angle $PAB = \dfrac{180° - 68°}{2}$

$\quad\quad\quad\quad = 56°$

| Subtract 68° from 180° and divide the result by 2. Triangle $PAB$ is isosceles. |

The angle sum of a triangle is 180° and in an isosceles triangle the angles opposite the equal sides are equal.

| Give the reasons. |

## Exercise 18B

The diagrams are **NOT** accurately drawn.

**1** $PT$ is a tangent at $T$ to a circle, centre $O$.
Angle $POT = 37°$.
Find the size of angle $a$.
Give reasons for your answer.

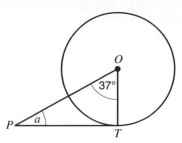

**2** $PA$ is a tangent at $A$ to a circle, centre $O$.
$B$ is a point on the circumference of the circle.
$POB$ is a straight line.
Find the size of each of the angles marked with letters.

**a**

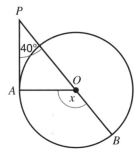

**b**

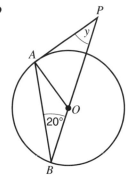

**3** $PA$ is a tangent at $A$ to a circle, centre $O$.
$AB$ is a chord of the circle.
Calculate the size of angles $x$ and $y$.

**a**

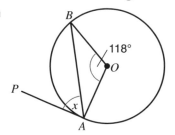

**b**

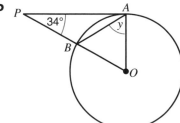

**4** $AB$ is a chord of a circle, centre $O$.
$M$ is the midpoint of $AB$.
Angle $BAO = 64°$.
Find the size of angle $AOM$.
Give reasons for your answer.

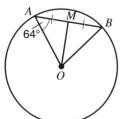

**5** $PA$ and $PB$ are tangents.
Angle $ABP = 61°$.
Calculate the size of angle $APB$.
Give reasons for your answer.

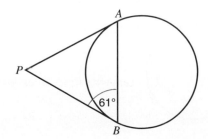

**6** *PA* and *PB* are tangents to a circle, centre *O*.
Find the size of angles *x* and *y*.

**a**
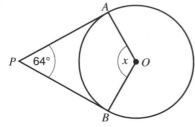

**b**
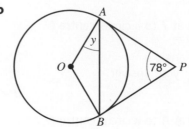

**7** *PA* is a tangent to the circle at *A*.
*AB* is a diameter of the circle.
*D* is a point on *PB* such that angle *BAD* = 72°.
*AP* = *AB*.
Calculate the size of angle *PDA*.

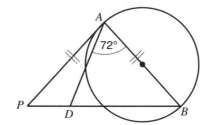

**8** *PA* is a tangent to the circle, centre *O*.
*AB* is a chord of the circle.
Angle *AOB* = 152°.
Angle *APB* = 71°.
Find the size of angle *PBA*.

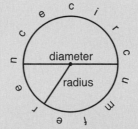

## Chapter summary

**You should know the meaning of:**

- **circumference**
- **chord**
- **diameter**
- **radius**
- **tangent**

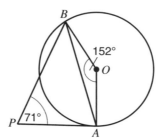

**You should now know these geometric facts and be able to use them:**

★　a tangent is perpendicular to the radius at the point of contact

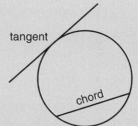

★　tangents from an external point to a circle are equal in length

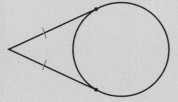

★ a line drawn from the centre of a circle perpendicular to a chord bisects the chord

# Chapter 18 review questions

The diagrams are **NOT** accurately drawn.

**1** *P* and *Q* are points on a circle, centre *O*.
Angle *POQ* = 116°.

Work out the size of angle *OPQ*.
Give reasons for your answer.

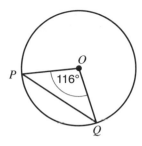

**2** *A* and *B* are points on a circle, centre *O*.
*SBO* and *TBA* are straight lines.
Angle *SBT* = 47°.

Work out the size of angle *AOB*.
Give reasons for your answer.

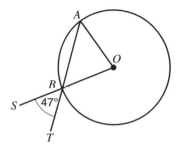

**3** *PT* is a tangent at *T* to a circle, centre *O*.
Angle *OPT* = 39°.

Work out the size of angle *POT*.
Give reasons for your answer.

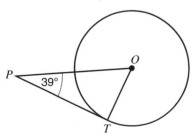

**4** *A* and *B* are points on a circle.
*PA* and *PB* are tangents to the circle.
Angle *APB* = 54°.

Work out the size of angle *PAB*.
Give reasons for your answer.

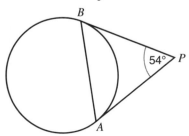

**5** *PQ* is a chord of a circle, centre *O*.
*M* is the midpoint of *PQ*.
Angle *POM* = 57°.

Work out the size of angle *OPM*.
Give reasons for your answer.

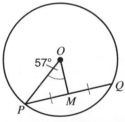

# Angles (2)

## 19

**CHAPTER**

## 19.1 Quadrilaterals

A **quadrilateral** is a shape with four straight sides and four angles.

| To find the angle sum of a quadrilateral, draw a quadrilateral on paper and label its angles $a, b, c$ and $d$. 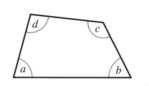 | Tear off its corners. 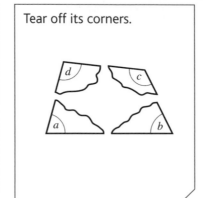 | Fit angles $a, b, c$ and $d$ together at a point. 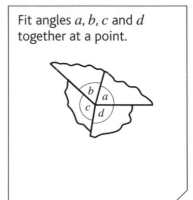 |
|---|---|---|

The angles at a point add up to 360° and so this shows that the angles in this quadrilateral add up to 360°.

> **The angle sum of a quadrilateral is 360°.**

To *prove* this result, draw a **diagonal** of the quadrilateral.
The diagonal splits the quadrilateral into two triangles.
The angle sum of each triangle is 180°.
So the angle sum of the quadrilateral is 2 × 180° = 360°.

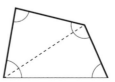

---

### Example 1

Work out the size of angle $x$.

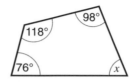

*Solution 1*

$76° + 118° + 98° = 292°$    | Add 76, 118 and 98 |

$360° - 292° = 68°$    | Take the result away from 360, as the angle sum of a quadrilateral is 360°. |

$x = 68°$    | State the size of angle $x$. |

---

## Example 2

**a** Write down the size of angle $x$.

**b** Work out the size of angle $y$.

   **Give a reason for each answer.**

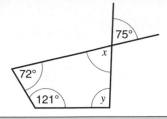

*Solution 2*

**a** $x = 75°$

| Where two straight lines cross, the opposite angles are equal. |

**b** $121° + 72° + 75° = 268°$

$$360° - 268° = 92°$$

| Angle sum of a quadrilateral is 360° |

$$y = 92°$$

## Example 3

The diagram shows a kite.

**a** Write down the size of angle $x$.

**b** Work out the size of angle $y$.  **Give a reason for each answer.**

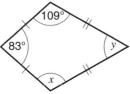

*Solution 3*

**a** $x = 109°$

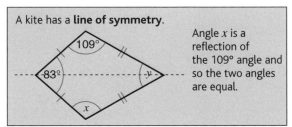

A kite has a **line of symmetry**.

Angle $x$ is a reflection of the 109° angle and so the two angles are equal.

**b** $83° + 109° + 109° = 301°$

$$360° - 301° = 59°$$

$$y = 59°$$

| Angle sum of a quadrilateral is 360°. |

## Exercise 19A

In this exercise, the quadrilaterals are not accurately drawn.

In Questions **1–12**, find the size of each of the angles marked with letters and show your working.

**1**

**2**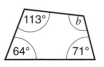

**3**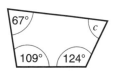

**4**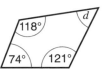

**5**

**6**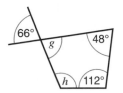

**7**

**8**

**9**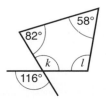

**10** 4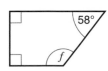

**11**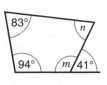

**12**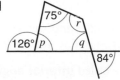

**13** The diagram shows a kite.

  **a** Write down the size of angle $v$.

  **b** Work out the size of angle $w$.

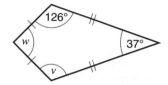

**14** The diagram shows a kite.
    Work out the value of $x$.

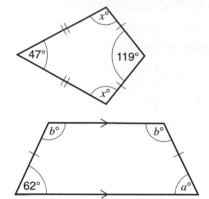

**15** The diagram shows an isosceles trapezium.

  **a** Write down the value of $a$.

  **b** Work out the value of $b$.

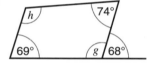

In Questions **16–20,** find the sizes of the angles marked with letters and show your working.
**Give reasons for your answers.**

**16**

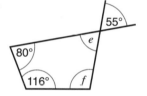

**17**

**18**

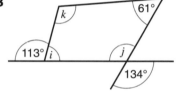

**19**

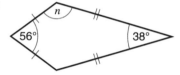

**20**

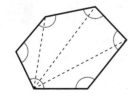

## 19.2 Polygons

A **polygon** is a shape with three or more straight sides.

Some polygons have special names.

A 3-sided polygon is called a **triangle**.
A 4-sided polygon is called a **quadrilateral**.
A 5-sided polygon is called a **pentagon**.
A 6-sided polygon is called a **hexagon**.
An 8-sided polygon is called an **octagon**.
A 10-sided polygon is called a **decagon**.

To find the sum of the angles of a polygon, split it into triangles.

For example, for this hexagon, draw as many diagonals as possible from one corner.

This splits the hexagon into four triangles.

The angle sum of a triangle is 180° and so the sum of the angles of a hexagon is 4 × 180° = 720°.

Sometimes, these angles are called **interior angles** to emphasise that they are *inside* the polygon.

Using this method, the sum of the interior angles of any polygon can be found.

| Number of sides | Number of triangles | Sum of the interior angles |
|---|---|---|
| 4 | 2 | 360° |
| 5 | 3 | 540° |
| 6 | 4 | 720° |
| 7 | 5 | 900° |
| 8 | 6 | 1080° |
| 9 | 7 | 1260° |
| 10 | 8 | 1440° |

The number of triangles into which the polygon can be split up is always two less than the number of sides.

## Example 4

Find the sum of the angles of a 12-sided polygon (**dodecagon**).

**Solution 4**

$$12 - 2 = 10$$ Subtract 2 from the number of sides to find the number of triangles.

$$10 \times 180 = 1800$$ Multiply the number of triangles by 180.

The sum of the angles $= 1800°$ State the sum of the angles in degrees.

A polygon with all its sides the same length and all its angles the same size is called a **regular** polygon.

So a square *is* a regular polygon, because all its sides are the same length and all its angles are 90°, but a rhombus is *not* a regular polygon.

Although its sides are all the same length, its angles are not all the same size.

Here are three more regular polygons.

a regular pentagon     a regular hexagon     a regular octagon

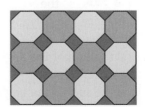

The Pentagon in Washington DC is the headquarters of the US Department of Defence.

Bees' honeycomb is made up of regular hexagons.

Regular octagons tessellate with squares.

Example 5

Find the size of each interior angle of a regular decagon.

**Solution 5**

| | |
|---|---|
| $10 - 2 = 8$ | Subtract 2 from the number of sides to find the number of triangles. |
| $8 \times 180 = 1440$ | Multiply the number of triangles by 180 to find the sum of all 10 interior angles. |
| $1440 \div 10 = 144$ | All 10 interior angles are the same size. So divide 1440 by 10 |
| Each interior angle is 144° | State the size of each interior angle. |

**Example 6**

The diagram shows a regular 9-sided polygon (nonagon) with centre $O$.

**a** Work out the size of
   **i** angle $x$   **ii** angle $y$.

**b** Use your answer to part **a ii** to work out the size of each interior angle of the polygon.

**Solution 6**

**a i** $x = 360° \div 9$

Each corner of the polygon could be joined to the centre O to make 9 equal angles at O. The total of all 9 angles is 360°, as altogether they make a complete turn.

$x = 40°$

State the size of angle $x$.

(40° is the angle at the centre of a regular 9-sided polygon.)

**ii** $180° - 40° = 140°$

The angle sum of a triangle is 180° and so the sum of the two base angles is 140°.

$140° \div 2 = 70°$

The triangle is isosceles and so the two base angles are equal.

$y = 70°$

State the size of angle $y$.

**b** $2 \times 70° = 140°$

Because the polygon is regular, it has nine lines of symmetry and each interior angle is twice the size of each base angle of the triangle.

Each interior angle is 140°.

State the size of each interior angle.

### Exercise 19B

In this exercise, the polygons are not accurately drawn.

**1** Find the sum of the angles of a 15-sided polygon.

**2** Find the sum of the angles of a 20-sided polygon.

**3** A polygon can be split into 17 triangles by drawing diagonals from one corner. How many sides has the polygon?

In Questions **4–9**, find the size of each of the angles marked with letters and show your working.

**4**

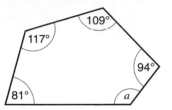

**5**

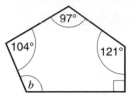

**6**

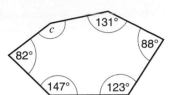

**7**

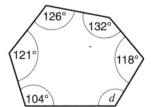

**8**

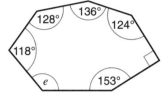

**9**

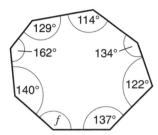

**10** The diagram shows a pentagon.
All its sides are the same length.

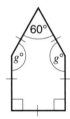

  **a** Work out the value of $g$.

  **b** Is the pentagon a regular polygon?
Explain your answer.

**11** Work out the size of each interior angle of

  **a** a regular pentagon       **b** a regular hexagon       **c** a regular octagon.

**12** Work out the size of each interior angle of a regular 15-sided polygon.

**13** Work out the size of each interior angle of a regular 20-sided polygon.

**14** Work out the size of the angle at the centre of a regular pentagon.

**15** Work out the size of the angle at the centre of a regular 12-sided polygon.

Australia's 50 cent coin is a regular
12-sided polygon (dodecagon)

**16** The angle at the centre of a regular polygon is 60°.
How many sides has the polygon?

**17** The angle at the centre of a regular polygon is 20°.

  **a** How many sides has the polygon?

  **b** Work out the size of each interior angle of the polygon.

**18 a** Work out the angle at the centre of a regular octagon.

  **b** Draw a circle with a radius of 5 cm and, using your answer to part **a** , draw a regular octagon
inside the circle.

**19 a** Work out the angle at the centre of a regular 10-sided polygon.

  **b** Draw a circle with a radius of 5 cm and, using your answer to part **a** , draw a regular 10-sided polygon inside the circle.

**20** The diagram shows a pentagon. Work out the size of

  **a** angle $h$   **b** angle $i$.

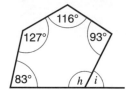

**21** The diagram shows a hexagon. Work out the size of

  **a** angle $j$   **b** angle $k$.

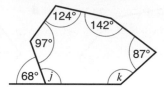

**22** Craig says, 'The sum of the interior angles of this polygon is 1000°'. Explain why he must be wrong.

**23** The diagram shows a quadrilateral.

  **a** Work out the size of each of the angles marked with letters.

  **b** Work out $l + m + n + p$

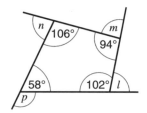

**24** The diagram shows a pentagon.

  **a** Work out the size of each of the angles marked with letters.

  **b** Work out $q + r + s + t + u$

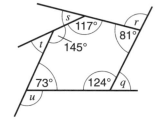

**25** The diagram shows a hexagon.

  **a** Work out the size of each of the angles marked with letters.

  **b** Work out $u + v + w + x + y + z$

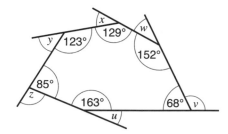

# 19.3 Exterior angles

A polygon's interior angles are the angles *inside* the polygon.

Extend a side to make an **exterior angle**, which is *outside* the polygon.

At each **vertex** (corner), the interior angle and the exterior angle are on a straight line and so their sum is 180°.

  interior angle + exterior angle = 180°

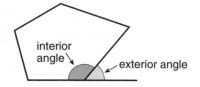

**The sum of the exterior angles of any polygon is 360°.**

To show this, imagine someone standing at $P$ on this quadrilateral, facing in the direction of the arrow.

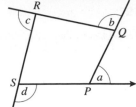

They turn through angle $a$, so that they are facing in the direction $PQ$, and then walk to $Q$.

At $Q$, they turn through angle $b$, so that they are facing in the direction $QR$, and then walk to $R$.

At $R$, they turn through angle $c$, so that they are facing in the direction $RS$, and then walk to $S$.

At $S$, they turn through angle $d$.

They are now facing in the direction of the arrow again and so they have turned through 360°.

The total angle they have turned through is also the sum of the exterior angles of the quadrilateral.

So  $a + b + c + d = 360°$

The same argument can be used with *any* polygon, not just a quadrilateral.

## Example 7

The sizes of four of the exterior angles of a pentagon are 67°, 114°, 58° and 73°.
Work out the size of the other exterior angle.

*Solution 7*

$67° + 114° + 58° + 73° = 312°$      | Add the four given exterior angles.

$360° - 312° = 48°$      | Subtract the result from 360

Exterior angle $= 48°$      | State the size of the exterior angle.

## Example 8

For a regular 18-sided polygon, work out
**a** the size of each exterior angle,
**b** the size of each interior angle.

*Solution 8*

**a**  $360° \div 18 = 20°$      | Because the polygon is regular, all 18 exterior angles are equal. Their sum is 360° and so divide 360° by 18

**b**  $180° - 20° = 160°$      | At a corner, the sum of the interior angle and the exterior angle is 180°. So subtract 20° from 180°.

## Example 9

The size of each interior angle of a regular polygon is 150°. Work out
**a** the size of each exterior angle,
**b** the number of sides the polygon has.

*Solution 9*

**a**  $180° - 150° = 30°$      | At a corner, the sum of the interior angle and the exterior angle is 180°. So subtract 150° from 180°.

**b**  $360 \div 30 = 12$      | Because the polygon is regular, all the exterior angles are 30°. Their sum is 360° and so divide 360 by 30

**Exercise 19C**

1 At a vertex (corner) of a polygon, the size of the interior angle is 134°.
Work out the size of the exterior angle.

2 At a vertex of a polygon, the size of the exterior angle is 67°.
Work out the size of the interior angle.

3 The sizes of three of the exterior angles of a quadrilateral are 72°, 119° and 107°.
Work out the size of the other exterior angle.

4 The sizes of five of the exterior angles of a hexagon are 43°, 109°, 58°, 74° and 49°.
Work out the size of the other exterior angle.

5 Work out the size of each exterior angle of a regular octagon.

6 Work out the size of each exterior angle of a regular 9-sided polygon.

7 For a regular 24-sided polygon, work out
  a the size of each exterior angle,
  b the size of each interior angle.

8 For a regular 40-sided polygon, work out
  a the size of each exterior angle,
  b the size of each interior angle.

9 The size of each interior angle of a regular polygon is 168°. Work out
  a the size of each exterior angle,
  b the number of sides the polygon has.

10 The size of each interior angle of a regular polygon is 170°.
Work out the number of sides the polygon has.

# Chapter summary

**You should know and be able to use these facts**

★ A **quadrilateral** is a shape with four straight sides and four angles.

★ The angle sum of a quadrilateral is 360°.

★ A **polygon** is a shape with three or more straight sides.

★ A 5-sided polygon is called a **pentagon**.

★ A 6-sided polygon is called a **hexagon**.

★ An 8-sided polygon is called an **octagon**.

★ A 10-sided polygon is called a **decagon**.

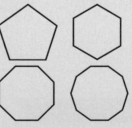

★ The angle sum of a polygon can be found by subtracting 2 from the number of sides and multiplying the result by 180°.

★ A polygon with all its sides the same length and all its angles the same size is called a **regular** polygon.

★ At a vertex, interior angle + exterior angle = 180°.

★ The sum of the exterior angles of any polygon is 360°.
To find the size of each exterior angle of a regular
polygon, divide 360° by the number of sides.

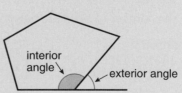

interior angle

exterior angle

# Chapter 19 review questions

In Questions **1–3**, find the size of each of the angles marked with a letter.
The diagrams are not accurately drawn.

**1**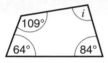

**2**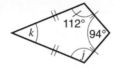

**3**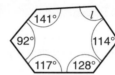

**4** Calculate the value of $x$.

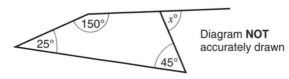

Diagram **NOT** accurately drawn

(4400 November 2004)

**5** Work out the value of $a$.

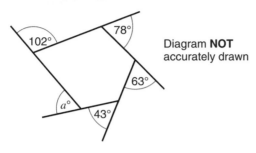

Diagram **NOT** accurately drawn

(1388 March 2002)

**6** Work out the size of each exterior angle of a regular 10-sided polygon.

**7 a** Work out the sum of the interior angles of a 9-sided polygon.

The size of each exterior angle of a regular polygon is 20°.
**b** Work out how many sides the polygon has.

**8** The diagram shows a regular hexagon.
**a** Work out the value of $x$.
**b** Work out the value of $y$.

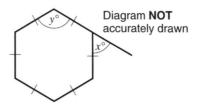

Diagram **NOT** accurately drawn

(1385 June 2001)

**9** *ABCDE* is a regular pentagon.
*AEF* and *CDF* are straight lines.
Work out the size of angle *DFE*.
**Give reasons for your answer.**

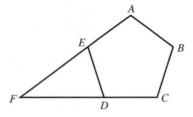

Diagram **NOT** accurately drawn

(1388 March 2004)

# Fractions

## 20.1 Addition and subtraction of fractions

$\frac{3}{5}$ of the rectangle is shaded red and $\frac{1}{5}$ of the rectangle is shaded green.

$\frac{4}{5}$ of the rectangle is shaded.

So $\quad \frac{3}{5} + \frac{1}{5} = \frac{4}{5}$

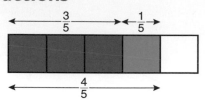

**To add fractions with the same denominator, add the numerators but do not change the denominator.**

For example, $\frac{4}{9} + \frac{2}{9} = \frac{6}{9}$ which in its simplest form is $\frac{2}{3}$

To add fractions with different denominators, firstly find the **lowest common denominator** and then change each fraction to its equivalent fraction with this denominator.

### Example 1

Work out $\frac{2}{3} + \frac{1}{5}$

**Solution 1**

$$\frac{2}{3} = \frac{10}{15} \quad \text{and} \quad \frac{1}{5} = \frac{3}{15}$$

> The lowest common denominator is 15
> Change each fraction to its equivalent fraction with a denominator of 15

So $\quad \frac{2}{3} + \frac{1}{5} = \frac{10}{15} + \frac{3}{15}$

$$\frac{2}{3} + \frac{1}{5} = \frac{13}{15}$$

> Add the numerators but do not change the denominator.

### Example 2

Work out $\frac{7}{8} + \frac{1}{2}$   Give your answer as a mixed number.

**Solution 2**

$$\frac{7}{8} + \frac{1}{2} = \frac{7}{8} + \frac{4}{8}$$

> The lowest common denominator is 8
> Change $\frac{1}{2}$ to the equivalent fraction with a denominator of 8

$$\frac{7}{8} + \frac{4}{8} = \frac{11}{8}$$

> Add the numerators but do not change the denominator.

$$\frac{11}{8} = 1\frac{3}{8}$$

> Change the improper fraction to a mixed number.

$$\frac{7}{8} + \frac{1}{2} = 1\frac{3}{8}$$

Fractions can be subtracted in a similar way.

## Example 3

Work out $\frac{4}{5} - \frac{3}{4}$

### Solution 3

$\frac{4}{5} = \frac{16}{20}$  and  $\frac{3}{4} = \frac{15}{20}$    | The lowest common denominator is 20<br>Change each fraction to its equivalent fraction with a denominator of 20

$\frac{4}{5} - \frac{3}{4} = \frac{16}{20} - \frac{15}{20}$    | Subtract the numerators but do not change the denominator.

$\frac{4}{5} - \frac{3}{4} = \frac{1}{20}$

## Exercise 20A

In questions **1–5** give each answer as a mixed number or a fraction in its simplest form.

**1** Work out

a $\frac{1}{3} + \frac{1}{2}$    b $\frac{1}{3} + \frac{1}{4}$    c $\frac{1}{5} + \frac{1}{4}$    d $\frac{1}{8} + \frac{1}{4}$

e $\frac{2}{3} + \frac{1}{4}$    f $\frac{1}{2} + \frac{2}{5}$    g $\frac{1}{5} + \frac{2}{3}$    h $\frac{1}{20} + \frac{4}{5}$

**2** Work out

a $\frac{2}{5} + \frac{3}{8}$    b $\frac{3}{4} + \frac{2}{5}$    c $\frac{2}{3} + \frac{1}{6}$    d $\frac{2}{7} + \frac{3}{5}$

e $\frac{4}{5} + \frac{7}{10}$    f $\frac{5}{6} + \frac{3}{8}$    g $\frac{9}{10} + \frac{3}{4}$    h $\frac{7}{20} + \frac{5}{8}$

**3** Work out

a $\frac{1}{2} - \frac{1}{4}$    b $\frac{1}{3} - \frac{1}{4}$    c $\frac{1}{3} - \frac{1}{9}$    d $\frac{1}{5} - \frac{1}{20}$

e $\frac{3}{4} - \frac{1}{2}$    f $\frac{2}{3} - \frac{1}{4}$    g $\frac{1}{2} - \frac{2}{5}$    h $\frac{11}{12} - \frac{1}{6}$

**4** Work out

a $\frac{3}{4} - \frac{2}{3}$    b $\frac{9}{10} - \frac{3}{5}$    c $\frac{7}{8} - \frac{2}{5}$    d $\frac{8}{9} - \frac{2}{3}$

e $\frac{5}{6} - \frac{3}{5}$    f $\frac{3}{4} - \frac{2}{7}$    g $\frac{17}{20} - \frac{3}{4}$    h $\frac{5}{6} - \frac{7}{9}$

**5** Work out

a $\frac{3}{7} + \frac{2}{5}$    b $\frac{4}{9} - \frac{1}{5}$    c $\frac{5}{6} - \frac{2}{3}$    d $\frac{3}{10} + \frac{1}{4}$

e $\frac{7}{8} - \frac{2}{3}$    f $\frac{7}{12} + \frac{3}{4} - \frac{5}{6}$    g $\frac{4}{5} - \frac{1}{2} + \frac{7}{10}$    h $\frac{4}{9} + \frac{2}{3} - \frac{1}{6}$

# 20.2 Addition and subtraction of mixed numbers

**When adding mixed numbers, add the whole numbers and the fractions separately.**

## Example 4

Work out $2\frac{1}{3} + 4\frac{1}{2}$

### Solution 4

$2 + 4 = 6$    | Add the whole numbers.

$\frac{1}{3} + \frac{1}{2} = \frac{2}{6} + \frac{3}{6} = \frac{5}{6}$    | Add the fractions.

$2\frac{1}{3} + 4\frac{1}{2} = 6 + \frac{5}{6} = 6\frac{5}{6}$    | Add the two results.

Sometimes adding the fractions gives an improper fraction.

For example, adding the fraction parts of $2\frac{2}{3}$ and $4\frac{1}{2}$ gives $\frac{2}{3} + \frac{1}{2} = \frac{4}{6} + \frac{3}{6} = \frac{7}{6}$

$\frac{7}{6}$ is an improper fraction. As a mixed number, $\frac{7}{6} = 1\frac{1}{6}$

So $2\frac{2}{3} + 4\frac{1}{2} = 6 + 1\frac{1}{6} = 7\frac{1}{6}$

Mixed numbers can be subtracted in a similar way.

## Example 5

Work out $3\frac{2}{3} - 1\frac{1}{2}$

### Solution 5

$3 - 1 = 2$ | Subtract the whole numbers.

$\frac{2}{3} - \frac{1}{2} = \frac{4}{6} - \frac{3}{6} = \frac{1}{6}$ | Subtract the fractions.

$2 + \frac{1}{6} = 2\frac{1}{6}$ | Add the two results.

$3\frac{2}{3} - 1\frac{1}{2} = 2\frac{1}{6}$

## Example 6

Work out $4\frac{1}{4} - 2\frac{7}{10}$

### Solution 6

**Method 1**

$4\frac{1}{4} = \frac{17}{4}$ and $2\frac{7}{10} = \frac{27}{10}$ | Change mixed numbers to improper fractions.

 and  | The lowest common denominator is 20. Change each fraction to its equivalent fraction with a denominator of 20

$\frac{17}{4} - \frac{27}{10} = \frac{85}{20} - \frac{54}{20} = \frac{31}{20}$ | Subtract the numerators but do not change the denominator.

$\frac{31}{20} = 1\frac{11}{20}$ | Give the answer as a mixed number.

$4\frac{1}{4} - 2\frac{7}{10} = 1\frac{11}{20}$

**Method 2**

$4 - 2 = 2$ | Subtract the whole numbers.

$\frac{1}{4} - \frac{7}{10} = \frac{5}{20} - \frac{14}{20} = -\frac{9}{20}$ | Subtract the fractions.

$2 - \frac{9}{20} = 1 + 1 - \frac{9}{20} = 1 + \frac{20}{20} - \frac{9}{20} = 1 + \frac{11}{20} = 1\frac{11}{20}$ | Add the two results.

$4\frac{1}{4} - 2\frac{7}{10} = 1\frac{11}{20}$

## Exercise 20B

**1** Work out

a $2\frac{4}{5} + 1\frac{3}{5}$    b $3\frac{5}{9} + \frac{8}{9}$    c $6\frac{3}{4} + 1\frac{1}{2}$    d $4\frac{4}{5} + \frac{1}{2}$

e $2\frac{3}{5} + 1\frac{3}{4}$    f $1\frac{2}{3} + 1\frac{1}{2}$    g $4\frac{3}{4} + 1\frac{2}{3}$    h $3\frac{2}{3} + 1\frac{4}{9}$

**2** Work out

a $2\frac{1}{2} - 1\frac{1}{4}$    b $3\frac{7}{8} - 1\frac{1}{2}$    c $3\frac{2}{3} - 1\frac{1}{2}$    d $4\frac{4}{5} - 1\frac{2}{3}$

**3** Work out

a $1 - \frac{1}{3}$    b $3 - \frac{2}{5}$    c $6 - 5\frac{1}{4}$    d $8 - 4\frac{2}{3}$

**4** Work out

a $2\frac{1}{4} - 1\frac{1}{2}$    b $3\frac{1}{4} - 1\frac{2}{3}$    c $4\frac{1}{2} - 2\frac{3}{4}$    d $3\frac{1}{6} - 1\frac{1}{2}$

e $5\frac{1}{3} - 1\frac{3}{4}$    f $6\frac{3}{5} - 2\frac{2}{3}$    g $4\frac{2}{7} - 1\frac{3}{5}$    h $7\frac{1}{9} - 3\frac{2}{3}$

**5** Work out

a $3\frac{1}{5} + 2\frac{4}{15}$    b $8\frac{3}{8} - 3\frac{3}{4}$    c $6\frac{2}{3} - \frac{8}{9}$    d $1\frac{5}{6} + 4\frac{2}{7}$

e $12\frac{5}{8} - 3\frac{2}{3}$    f $3\frac{5}{6} + 3\frac{2}{5}$    g $8\frac{7}{10} + 3\frac{1}{3}$    h $9\frac{1}{6} - 1\frac{7}{9}$

## 20.3 Multiplication of fractions and mixed numbers

Multiplication by an integer is the same as repeated addition.

So $2 \times \frac{4}{9}$ is the same as $\frac{4}{9} + \frac{4}{9} = \frac{8}{9}$

**To multiply a fraction by an integer, multiply the numerator of the fraction by the integer. Do not change the denominator of the fraction.**

---

### Example 7

Work out $6 \times \frac{2}{3}$

**Solution 7**

$6 \times \frac{2}{3} = \frac{12}{3}$    Multiply the numerator of the fraction by the integer. Do not change the denominator.

$\frac{12}{3} = 4$    Simplify the fraction.

$6 \times \frac{2}{3} = 4$    The answer is an integer in this case.

---

To multiply $\frac{3}{4}$ by $\frac{2}{3}$

$\frac{3}{4} \times \frac{2}{3} = \frac{6}{12} = \frac{1}{2}$    Multiply the numerators $3 \times 2 = 6$

Multiply the denominators $4 \times 3 = 12$

Simplify the fraction.

$\frac{6}{12} (= \frac{1}{2})$ of the area of the square is shaded.

**To multiply two fractions, multiply the numerators and then multiply the denominators.**

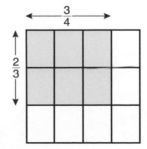

### Example 8

Work out $\frac{2}{3} \times \frac{2}{5}$

### Solution 8

$\frac{2}{3} \times \frac{2}{5} = \frac{4}{15}$

Multiply the numerators $2 \times 2 = 4$
Multiply the denominators $3 \times 5 = 15$
$\frac{4}{15}$ is in its simplest form.

When multiplying fractions, it is sometimes possible to simplify the multiplication by cancelling.

### Example 9

Work out $\dfrac{5}{14} \times \dfrac{7}{10}$

### Solution 9

$\dfrac{5}{14} \times \dfrac{7}{10} = \dfrac{5 \times 7}{14 \times 10}$

$\dfrac{\overset{1}{\cancel{5}} \times 7}{14 \times \underset{2}{\cancel{10}}} = \dfrac{1 \times 7}{14 \times 2}$

Cancel the 5 and the 10

$\dfrac{1 \times \overset{1}{\cancel{7}}}{\underset{2}{\cancel{14}} \times 2} = \dfrac{1 \times 1}{2 \times 2}$

Cancel the 7 and the 14

$\dfrac{1 \times 1}{2 \times 2} = \dfrac{1}{4}$

$\dfrac{5}{14} \times \dfrac{7}{10} = \dfrac{1}{4}$

When multiplying mixed numbers, first write the mixed numbers as improper fractions.

### Example 10

Work out $2\frac{2}{3} \times 1\frac{4}{5}$

### Solution 10

$2\frac{2}{3} \times 1\frac{4}{5} = \frac{8}{3} \times \frac{9}{5}$

Change each mixed number into an improper fraction.

$= \dfrac{8 \times \overset{3}{\cancel{9}}}{\underset{1}{\cancel{3}} \times 5}$

Cancel the 9 and the 3

$= \dfrac{24}{5} = 4\frac{4}{5}$

Change the improper fraction into a mixed number.

## Exercise 20C

Give each answer in its simplest form.

**1** Work out

**a** $2 \times \frac{1}{3}$        **b** $3 \times \frac{1}{4}$        **c** $2 \times \frac{2}{5}$        **d** $3 \times \frac{2}{7}$

**e** $\frac{3}{8} \times 2$        **f** $\frac{5}{12} \times 4$        **g** $\frac{9}{20} \times 8$        **h** $\frac{3}{5} \times 25$

**2** Work out

**a** $\frac{3}{5} \times \frac{1}{2}$      **b** $\frac{1}{4} \times \frac{3}{5}$      **c** $\frac{1}{3} \times \frac{2}{5}$      **d** $\frac{1}{2} \times \frac{1}{5}$

**e** $\frac{2}{3} \times \frac{2}{7}$      **f** $\frac{3}{4} \times \frac{3}{5}$      **g** $\frac{7}{10} \times \frac{3}{5}$      **h** $\frac{3}{4} \times \frac{5}{7}$

**3** Work out

**a** $\frac{3}{4} \times \frac{2}{5}$      **b** $\frac{2}{3} \times \frac{3}{8}$      **c** $\frac{10}{11} \times \frac{3}{5}$      **d** $\frac{5}{6} \times \frac{4}{15}$

**e** $\frac{3}{14} \times \frac{8}{9}$      **f** $\frac{16}{21} \times \frac{9}{40}$      **g** $\frac{9}{28} \times \frac{14}{15}$      **h** $\frac{25}{36} \times \frac{27}{40}$

**4** Work out

**a** $1\frac{1}{4} \times \frac{1}{3}$      **b** $1\frac{3}{5} \times \frac{1}{2}$      **c** $2\frac{2}{3} \times \frac{1}{5}$      **d** $\frac{3}{7} \times 3\frac{1}{2}$

**e** $1\frac{1}{3} \times 2\frac{1}{4}$      **f** $3\frac{1}{2} \times 1\frac{1}{4}$      **g** $3\frac{1}{3} \times 1\frac{4}{5}$      **h** $2\frac{1}{7} \times 2\frac{4}{5}$

**5** Work out

**a** $1\frac{1}{3} \times 1\frac{1}{4}$      **b** $2\frac{1}{2} \times 1\frac{3}{5}$      **c** $3\frac{3}{4} \times 1\frac{1}{10}$      **d** $1\frac{3}{5} \times 4\frac{1}{5}$

**e** $2\frac{2}{7} \times 4\frac{3}{8}$      **f** $4\frac{1}{6} \times 4\frac{4}{5}$      **g** $6\frac{3}{7} \times 1\frac{5}{9}$      **h** $8\frac{1}{3} \times 2\frac{7}{10}$

**6** Work out

**a** $\frac{3}{4} \times \frac{5}{6} + \frac{3}{16}$      **b** $\frac{17}{20} - \frac{4}{5} \times \frac{3}{8}$      **c** $\left(1\frac{2}{3} + \frac{5}{6}\right) \times \frac{8}{9}$      **d** $\left(2\frac{7}{9} - 1\frac{1}{3}\right) \times 4\frac{1}{2}$

## 20.4 Division of fractions and mixed numbers

$\frac{3}{4}$ of this rectangle is shaded red.

Divide the red area by 2

Now $\frac{3}{8}$ of the rectangle is shaded.

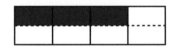

So $\frac{3}{4} \div 2 = \frac{3}{8}$    also    $\frac{3}{4} \times \frac{1}{2} = \frac{3}{8}$

So dividing by 2 is the same as multiplying by $\frac{1}{2}$

$\frac{1}{2}$ is the reciprocal of 2

To work out $3 \div \frac{3}{4}$ consider how many times $\frac{3}{4}$ goes into 3

There are $\frac{12}{4}$ in 3 whole squares, this is 4 lots of $\frac{3}{4}$

So $3 \div \frac{3}{4} = 4$    also    $3 \times \frac{4}{3} = 4$

So dividing by $\frac{3}{4}$ is the same as multiplying by $\frac{4}{3}$

$\frac{4}{3}$ is the **reciprocal** of $\frac{3}{4}$

Similarly, the reciprocal of 3 (or $\frac{3}{1}$) is $\frac{1}{3}$ and the reciprocal of $\frac{2}{7}$ is $\frac{7}{2}$

To divide by a fraction

- change the division sign into a multiplication sign
- write down the reciprocal of the second fraction.

**Example 11**

Work out $\frac{4}{5} \div 3$

**Solution 11**

$\frac{4}{5} \div 3 = \frac{4}{5} \times \frac{1}{3}$  | The reciprocal of 3 is $\frac{1}{3}$   Multiply $\frac{4}{5}$ by $\frac{1}{3}$ |

$\frac{4}{5} \div 3 = \frac{4}{15}$

**Example 12**

Work out $\frac{5}{6} \div \frac{3}{4}$
Give your fraction in its simplest form.

**Solution 12**

$\dfrac{5}{6} \div \dfrac{3}{4} = \dfrac{5}{\overset{}{\underset{3}{6}}} \times \dfrac{\overset{2}{4}}{3}$  | The reciprocal of $\frac{3}{4}$ is $\frac{4}{3}$ So multiply $\frac{5}{6}$ by $\frac{4}{3}$ |

$= \dfrac{10}{9}$

$\dfrac{5}{6} \div \dfrac{3}{4} = 1\frac{1}{9}$  | Write the improper fraction as a mixed number. |

When dividing mixed numbers, first write the mixed numbers as improper fractions.

**Example 13**

Work out $2\frac{4}{5} \div 2\frac{1}{10}$

**Solution 13**

$2\frac{4}{5} \div 2\frac{1}{10} = \frac{14}{5} \div \frac{21}{10}$

$= \dfrac{\overset{2}{14}}{\underset{1}{5}} \times \dfrac{\overset{2}{10}}{\underset{3}{21}}$  | Write the mixed numbers as improper fractions. The reciprocal of $\frac{21}{10}$ is $\frac{10}{21}$ |

$= \dfrac{4}{3}$

$2\frac{4}{5} \div 2\frac{1}{10} = 1\frac{1}{3}$  | Write the improper fraction as a mixed number. |

## Exercise 20D

Give each answer in its simplest form.

**1** Work out

  **a** $\frac{5}{6} \div 2$         **b** $\frac{3}{8} \div 2$         **c** $\frac{5}{8} \div 3$         **d** $\frac{5}{8} \div 10$
  **e** $\frac{1}{4} \div \frac{1}{3}$         **f** $\frac{3}{5} \div \frac{1}{2}$         **g** $\frac{4}{9} \div \frac{1}{3}$         **h** $\frac{5}{6} \div \frac{1}{4}$

**2** Work out

  **a** $\frac{2}{3} \div \frac{3}{5}$         **b** $\frac{3}{7} \div \frac{4}{5}$         **c** $\frac{4}{5} \div \frac{3}{10}$         **d** $\frac{9}{16} \div \frac{3}{8}$
  **e** $\frac{5}{8} \div \frac{15}{32}$         **f** $\frac{7}{10} \div \frac{14}{25}$         **g** $\frac{20}{21} \div \frac{8}{15}$         **h** $\frac{25}{32} \div \frac{15}{16}$

**3** Work out

  **a** $3\frac{1}{2} \div 7$         **b** $2\frac{4}{5} \div \frac{1}{10}$         **c** $2\frac{1}{3} \div \frac{1}{9}$         **d** $3\frac{3}{4} \div 5$
  **e** $1\frac{1}{2} \div \frac{3}{4}$         **f** $2\frac{4}{9} \div \frac{2}{3}$         **g** $7\frac{1}{5} \div \frac{9}{10}$         **h** $5\frac{1}{3} \div \frac{4}{9}$

**4** Work out

  **a** $2\frac{1}{3} \div 2\frac{4}{5}$         **b** $3\frac{1}{4} \div 1\frac{4}{9}$         **c** $3\frac{3}{4} \div 1\frac{4}{5}$         **d** $6\frac{2}{3} \div 2\frac{8}{9}$
  **e** $4\frac{9}{10} \div 2\frac{4}{5}$         **f** $5\frac{5}{6} \div 1\frac{1}{9}$         **g** $7\frac{1}{2} \div 1\frac{1}{4}$         **h** $2\frac{1}{12} \div 1\frac{1}{9}$

**5** Work out

  **a** $\left(\frac{1}{2} + \frac{5}{12}\right) \div \frac{3}{4}$     **b** $3 - \frac{3}{4} \div \frac{9}{10}$     **c** $3\frac{1}{8} \div \left(2\frac{1}{2} - 1\frac{3}{8}\right)$     **d** $3\frac{3}{5} \times 1\frac{1}{4} \div 2\frac{2}{5}$

## 20.5 Fractions of quantities

A **unit fraction** has a numerator of 1 and the denominator is a non-zero positive integer. Examples of unit fractions are $\frac{1}{2}$ $\frac{1}{4}$ $\frac{1}{5}$ and $\frac{1}{10}$

To find a unit fraction of an amount, think of that amount divided into equal parts.

---

### Example 14

Find $\frac{1}{4}$ of £24

**Solution 14**

$24 \div 4 = 6$ | Finding $\frac{1}{4}$ of an amount is the same as dividing the amount into 4 equal parts.

$\frac{1}{4}$ of £24 = £6

---

To find a fraction of an amount where the numerator is more than 1, think of the calculation in two stages. Firstly, divide the amount by the denominator. Then multiply the result by the numerator.

---

### Example 15

Find $\frac{2}{3}$ of £24

**Solution 15**

$24 \div 3 = 8$ | Divide 24 by 3    The result is 8

$8 \times 2 = 16$ | Multiply 8 by 2

$\frac{2}{3}$ of £24 = £16

---

Another way to find a fraction of a quantity is to multiply the quantity by the fraction. For example

$$\frac{2}{\cancel{3}_1} \times \cancel{24}^8 = 16$$

In mathematics, the word 'of' means the same as '$\times$'

---

### Example 16

Find $\frac{3}{7}$ of 10

**Solution 16**

$\frac{3}{7}$ of 10 $= \frac{3}{7} \times 10$ | 'of' means the same as '$\times$'

$= \frac{30}{7}$

$\frac{3}{7}$ of 10 $= 4\frac{2}{7}$ | Change the improper fraction into a mixed number.

---

### Exercise 20E

**1** Find

  **a** $\frac{1}{4}$ of 12      **b** $\frac{1}{3}$ of 18      **c** $\frac{1}{5}$ of 35      **d** $\frac{1}{6}$ of 24

  **e** $\frac{1}{10}$ of 70      **f** $\frac{1}{8}$ of 32      **g** $\frac{1}{12}$ of 48      **h** $\frac{1}{20}$ of 100

**2** Find

  **a** $\frac{2}{3}$ of 15      **b** $\frac{3}{4}$ of 12      **c** $\frac{2}{5}$ of 20      **d** $\frac{3}{7}$ of 14

  **e** $\frac{5}{6}$ of 24      **f** $\frac{4}{9}$ of 27      **g** $\frac{7}{10}$ of 50      **h** $\frac{7}{8}$ of 48

**3** Find

**a** $\frac{2}{5}$ of £150     **b** $\frac{3}{4}$ of 120 grams     **c** $\frac{2}{3}$ of 48 cm     **d** $\frac{4}{5}$ of £230

**e** $\frac{5}{6}$ of 240 m     **f** $\frac{7}{10}$ of 120 cm     **g** $\frac{3}{7}$ of 280 kg     **h** $\frac{2}{9}$ of 6120 km

**4** Find

**a** $\frac{2}{3}$ of 5     **b** $\frac{2}{5}$ of 4     **c** $\frac{2}{9}$ of 6     **d** $\frac{5}{12}$ of 16

**e** $\frac{7}{8}$ of 12     **f** $\frac{3}{7}$ of 8     **g** $\frac{9}{10}$ of 8     **h** $\frac{5}{9}$ of 24

**5** Find

**a** $\frac{1}{3}$ of $\frac{2}{5}$     **b** $\frac{8}{9}$ of $\frac{3}{7}$     **c** $\frac{6}{7}$ of $\frac{14}{15}$     **d** $\frac{5}{8}$ of $\frac{24}{25}$

**6** Find

**a** $\frac{3}{4}$ of £184     **b** $\frac{5}{8}$ of £496     **c** $\frac{5}{6}$ of £318     **d** $\frac{17}{20}$ of £460

**e** $\frac{15}{16}$ of 336 m     **f** $\frac{39}{50}$ of £1750     **g** $\frac{7}{40}$ of 660 kg     **h** $\frac{29}{100}$ of 40 km

**7** Find

**a** $\frac{2}{5}$ of £4     **b** $\frac{3}{8}$ of 12 m     **c** $\frac{7}{16}$ of 6 km     **d** $\frac{5}{12}$ of 78 cm

**e** $\frac{17}{32}$ of 20 kg     **f** $\frac{9}{10}$ of 175 grams     **g** $\frac{3}{40}$ of £1420     **h** $\frac{3}{4}$ of £68.40

## 20.6 Fraction problems

Problems can involve fractions.

### Example 17

In a cinema

$\frac{2}{5}$ of the audience are women.

$\frac{1}{8}$ of the audience are men.

All the rest of the audience are children.

What fraction of the audience are children?

**Solution 17**

$\frac{2}{5} + \frac{1}{8} = \frac{16}{40} + \frac{5}{40} = \frac{21}{40}$     | Add $\frac{2}{5}$ and $\frac{1}{8}$ to find the fraction of the audience who are women or men. |

$1 - \frac{21}{40} = \frac{40}{40} - \frac{21}{40} = \frac{19}{40}$     | Subtract $\frac{21}{40}$ from 1 to find the fraction of the audience who are children. |

$\frac{19}{40}$ of the audience are children.

### Example 18

A school has 1800 pupils.

860 of these pupils are girls.

$\frac{3}{4}$ of the girls like swimming.

$\frac{2}{5}$ of the boys like swimming.

Work out the total number of pupils in the school who like swimming.

**Solution 18**

$\frac{3}{4} \times 860 = 645$     | Work out the number of girls who like swimming. |

$1800 - 860 = 940$     | Work out the number of boys in the school. |

$\frac{2}{5} \times 940 = 376$     | Work out the number of boys who like swimming. |

$645 + 376 = 1021$     | Work out the total number of pupils who like swimming. |

1021 pupils like swimming.

**Exercise 20F**

1 Simon spends $\frac{1}{2}$ of his money on rent and $\frac{1}{3}$ of his money on transport.

   a  What fraction of his money does he spend on rent and transport altogether?

   b  What fraction of his money is left?

2 Dawn drives for $\frac{3}{4}$ of a journey. The journey lasts for 148 minutes.
   For how many minutes does Dawn drive?

3 There are 800 students in a school. $\frac{3}{5}$ of the students are boys.
   Work out the number of boys in the school.

4 Last season, Pearson Athletic won $\frac{7}{10}$ of its matches, drew $\frac{1}{5}$ and lost the rest.
   What fraction of its matches did it lose?

5 $\frac{1}{2}$ of a garden is lawn. $\frac{2}{5}$ of the garden is a vegetable patch. The rest of the garden is a flower bed.
   What fraction of the garden is a flower bed?

6 $\frac{8}{9}$ of an iceberg lies below the surface of the water. The total volume of an iceberg is 990 m³.
   What volume of this iceberg is below the surface?

7 There are 36 students in a class. Javed says that $\frac{3}{8}$ of these students are boys.
   Explain why Javed cannot be right.

8 John walks $2\frac{1}{2}$ miles to the next village. He then walks a further $2\frac{2}{3}$ miles to the river.
   How far has he walked altogether?

9 Tammy watches 2 films. The first film is $1\frac{3}{4}$ hours long and the second one is $2\frac{1}{3}$ hours long.
   Work out the total length of the two films.

10 Two sticks are $2\frac{1}{2}$ metres and $1\frac{1}{3}$ metres long.
   Work out the difference between the lengths of the two sticks.

11 $\frac{2}{3}$ of a square is shaded. $\frac{3}{4}$ of the shaded part is shaded blue.
   What fraction of the whole square is shaded blue?

12 In a crowd, $\frac{2}{5}$ of the people are female. $\frac{7}{10}$ of the females are girls.
   What fraction of the crowd is girls?

13 DVDs are sold for £14 each. $\frac{2}{5}$ of the £14 goes to the DVD company.
   How much of the £14 goes to the DVD company?

14 A school buys some textbooks. The total price of the textbooks is £2400
   The school gets a discount of $\frac{1}{8}$ off the price of the textbooks.
   Work out how much the school pays for the textbooks.

15 Alison, Becky and Carol take part in a charity relay race. The race is over a total distance
   of $2\frac{5}{8}$ kilometres. Each girl runs an equal distance. Work out how far each girl runs.

## Chapter summary

**You should now know:**

★ that to add (or subtract) fractions with the same denominator, add (or subtract) the numerators but do not change the denominator

★ that to add or subtract fractions with different denominators, firstly find the **lowest common denominator** and then change each fraction to its equivalent fraction with this denominator

★ how to add (or subtract) mixed numbers by adding (or subtracting) the whole numbers and the fractions separately

★ how to multiply fractions by multiplying the numerators and then multiplying the denominators

★ how to multiply or divide mixed numbers by firstly writing the mixed numbers as improper fractions

★ how to divide by a fraction by
  ● changing the division sign into a multiplication sign
  ● writing down the reciprocal of the second fraction

★ how to find a fraction of an amount by dividing the amount by the denominator and then multiplying the result by the numerator.

## Chapter 20 review questions

**1** Work out
  **a** $\frac{1}{4}$ of £48    **b** $\frac{1}{5}$ of £50    **c** $\frac{1}{3}$ of £60    **d** $\frac{1}{10}$ of £40    **e** $\frac{1}{6}$ of £30

**2** Work out
  **a** $\frac{3}{4}$ of £44    **b** $\frac{2}{3}$ of £18    **c** $\frac{2}{5}$ of £35    **d** $\frac{3}{5}$ of £100    **e** $\frac{4}{5}$ of £45

**3** Change to improper fractions
  **a** $2\frac{2}{3}$        **b** $8\frac{4}{5}$

**4** Work out
  **a** $\frac{1}{2} - \frac{1}{3}$    **b** $\frac{2}{3} - \frac{1}{4}$    **c** $\frac{1}{3} + \frac{1}{4}$    **d** $\frac{1}{4} + \frac{3}{5}$    **e** $\frac{4}{5} - \frac{2}{3}$

**5** Simon spent $\frac{1}{3}$ of his pocket money on a computer game.
  He spent $\frac{1}{4}$ of his pocket money on a ticket for a football match.
  Work out the fraction of his pocket money that he had left.                    (1387 June 2003)

**6** Asif, Curtly and Barbara share some money.
  Asif receives $\frac{3}{8}$ of the money.
  Barbara receives $\frac{1}{3}$ of the money.
  What fraction of the money does Curtly receive?                    (1388 March 2004)

**7** Work out $1 - (\frac{1}{2} + \frac{1}{6})$                    (1387 November 2004)

**8** Work out, giving your answers as mixed numbers
  **a** $3\frac{4}{5} + 2\frac{2}{3}$        **b** $4\frac{3}{4} - 2\frac{4}{5}$

**9** Work out and simplify where possible

**a** $\frac{1}{7} \times 4$      **b** $\frac{5}{12} \times 2$      **c** $\frac{1}{8} \times 4$      **d** $\frac{3}{10} \times 2$      **e** $\frac{3}{8} \times 2$

**10** Work out

**a** $\frac{2}{5} \div 2$      **b** $\frac{4}{7} \div 2$      **c** $\frac{6}{7} \div 3$      **d** $\frac{8}{9} \div 3$      **e** $\frac{4}{9} \div 5$

**11** Work out and simplify where possible

**a** $\frac{4}{5} \times \frac{1}{4}$      **b** $\frac{1}{3} \times \frac{4}{5}$      **c** $\frac{1}{4} \times \frac{3}{4}$      **d** $\frac{1}{5} \times \frac{10}{21}$      **e** $\frac{6}{11} \times \frac{1}{3}$

**12** Work out and simplify where possible

**a** $\frac{1}{3}$ of $\frac{3}{5}$      **b** $\frac{1}{4}$ of $\frac{3}{5}$      **c** $\frac{1}{2}$ of $\frac{6}{7}$      **d** $\frac{1}{4}$ of $\frac{8}{11}$      **e** $\frac{1}{5}$ of $\frac{10}{13}$

**13** Work out, giving each answer in its simplest form

**a** $\frac{3}{8} \times \frac{2}{5}$      **b** $\frac{5}{6} \times \frac{2}{5}$      **c** $\frac{7}{10} \times \frac{5}{6}$      **d** $\frac{5}{6} \times \frac{9}{10}$      **e** $\frac{5}{9} \times \frac{6}{25}$

**14** Work out, giving each answer in its simplest form

**a** $\frac{3}{4} \div \frac{1}{5}$      **b** $\frac{5}{7} \div \frac{10}{11}$      **c** $\frac{2}{9} \div \frac{2}{30}$      **d** $\frac{18}{35} \div \frac{6}{7}$      **e** $\frac{21}{40} \div \frac{3}{8}$

**15** Work out

**a** $\frac{1}{6} + \frac{4}{9}$      **b** $\frac{3}{7} \div 8$                              (1388 March 2002)

**16** Work out $\frac{2}{3} \times \frac{5}{6}$   Give your fraction in its simplest form.         (1388 January 2004)

**17** Work out, giving your answers as mixed numbers

**a** $1\frac{2}{3} \times 2\frac{3}{10}$      **b** $4\frac{2}{3} \div 1\frac{2}{5}$

**18 a** Work out the value of $\frac{2}{3} \times \frac{3}{4}$
Give your answer as a fraction in its simplest form.

**b** Work out the value of $1\frac{2}{3} + 2\frac{3}{4}$
Give your answer as a fraction in its simplest form.         (1387 June 2005)

**19** A school has 1200 pupils.

575 of these pupils are girls.

$\frac{2}{5}$ of the girls like sport.

$\frac{3}{5}$ of the boys like sport.

Work out the total number of pupils in the school who like sport.     (1387 November 2003)

**20** Work out $3\frac{3}{4} \times 2\frac{2}{3}$                                  (1388 March 2003)

**21** Work out $3\frac{1}{2} - 1\frac{2}{3}$   Give your answer as a mixed number in its simplest form.

# Scale drawings and dimensions

## 21.1 Scale drawings and maps

Scale drawings and scale models are drawings and models of places and objects. The lengths and distances in scale drawings and models are shorter than in the actual places and objects but the proportions and angles stay the same.

The scale of drawings and models may be given in various ways, for example '1 cm represents 2 m' and '1 cm to 1 km'. Scales are sometimes given as ratios.

1 : 200
(1 cm represents 2 m)

1 : 100 000
(1 cm to 1 km)

1 : 72
($\frac{1}{72}$ of full size)

---

### Example 1

Tom uses a scale of 1 : 250 to make a model of an aeroplane.

**a** The wing length of the model is 6 cm.
Work out the wing length of the real aeroplane.

**b** The length of the real aeroplane is 40 m.
Work out the length of the model.

*Solution 1*

**a**     $6 \times 250 = 1500$ cm

> The scale is 1 : 250, so every 1 cm on the model represents 250 cm on the real aeroplane. To find lengths on the real aeroplane, **multiply** lengths on the model by 250

   $1500 \div 100 = 15$ m.

> Change 1500 cm to metres.

The wing length of the real aeroplane is 15 m.

**b**     $40 \times 100 = 4000$ cm

> The model will be smaller than the aeroplane, so change 40 m to centimetres.

   $4000 \div 250 = 16$ cm

> To find lengths on the model, **divide** lengths on the real aeroplane by 250

The length of the model is 16 cm.

---

Maps are scale drawings. Map scales may also be given in various ways including ratios.

### Example 2

The scale of a map is 1 : 100 000
Work out the real distance that 6.4 cm on the map represents.

*Solution 2*

1 : 100 000

| | |
|---|---|
| | 1 cm on the map represents a real distance of 100 000 cm. |

6.4 × 100 000 = 640 000
The real distance is 640 000 cm

| | |
|---|---|
| | To find real distances, **multiply** lengths on the map by 100 000 |

640 000 ÷ 100 = 6400 m

| | |
|---|---|
| | Change 640 000 cm to kilometres using 1 m = 100 cm and 1 km = 1000 m. |

6400 ÷ 1000 = 6.4 km

### Exercise 21A

1 Jim uses a scale of 1 : 100 to draw a plan of a room to scale. On the scale drawing, the length of the room is 5.6 cm. What is the real length of the room?

2 Kylie makes a scale model of a rocking horse. She uses a scale of 1 : 5
The rocking horse is 125 cm high. How high will her scale model be?

3 On a map, 1 cm represents 2 km. What distance on the map will represent a real distance of
  **a**  10 km        **b**  22 km        **c**  7 km?

4 On a map, 1 cm represents 5 km. Work out the real distance between two towns, if their distance apart on the map is
  **a**  2 cm        **b**  3.1 cm        **c**  8.4 cm.

5 The scale of a map is 1 : 50 000    On the map, the distance between two towns is 4.2 cm. Work out the real distance between the towns. Give your answer in kilometres.

6 The scale of a map is 1 : 100 000    Work out the distance on the map between two towns, if the real distance between the towns is
  **a**  6 km        **b**  10.5 km

7 The scale of a model aeroplane is 1 : 72
  **a**  The length of the model aeroplane is 93 cm. Find, in metres, the real length of the aeroplane.
  **b**  The wingspan of the real aeroplane is 32.4 m. Find, in centimetres, the wingspan of the model.

8 A scale model of a car is 12 cm long. The length of the real car is 4.8 m. Find, as a ratio, the scale of the model.

9 A plan of a house is drawn to scale. A room with a real length of 6 m is 24 cm long on the plan. Find, as a ratio, the scale of the plan.

10 The distance between two towns is 6 km. On a map, the distance between the towns is 30 cm. Find, as a ratio, the scale of the map.

## 21.2 Dimensions

The perimeter of a rectangle is given by the expression $2l + 2b$. The circumference of a circle is given by $\pi d$. Perimeter and circumference are both *lengths*.

Each term in the expressions $2l + 2b$ and $\pi d$ consists of a number $\times$ a length. Therefore, we say that the expressions have the **dimension** *length*.

(Numbers, such as 2 and $\pi$, have no dimensions.)

The area of a rectangle is given by the expression $lb$. The area of a circle is given by $\pi r^2$.
Each term consists of a number $\times$ a length $\times$ a length.
The expressions have the dimensions *length $\times$ length*.

The volume of a cuboid is given by the expression $lwh$. The volume of a cylinder is given by $\pi r^2 h$.
Each term consists of a number $\times$ a length $\times$ a length $\times$ a length.
The expressions have the dimensions *length $\times$ length $\times$ length*.

So, for example, if $a$, $b$ and $c$ represent lengths,

$a + b$, $2a + 3b$, $\pi(b + c)$ and $\frac{1}{2}b + c$ represent lengths because each expression has dimension *length*.

$ab + bc$, $a(b + c)$, $\pi ab$, $\dfrac{b^2 c}{b + c}$ and $\frac{1}{2}bc$ represent areas because each expression has dimension *length $\times$ length*.

$abc$, $\frac{1}{3}b^2 c$, $4a(ab + c^2)$ and $\pi a^2(b + c)$ represent volumes because each expression has dimension *length $\times$ length $\times$ length*.

### Example 3

$x$, $y$ and $z$ represent lengths.
For each of these expressions, state whether it could represent a length, an area, a volume or none of these.
(Numbers have no dimensions.)

**a** $5xy$    **b** $\pi y(x^2 + z^2)$    **c** $2x(y + 3)$    **d** $\dfrac{3xy}{(x + y)}$

*Solution 3*

**a** An area

The expression has dimensions length $\times$ length.

**c** None of these

Multiply out the brackets, obtaining $2xy + 6x$.
The first term has the dimensions length $\times$ length but the second term has the dimension length.

**b** A volume

The dimension of $\pi y$ is length. Inside the brackets, both $x^2$ and $z^2$ have dimensions length $\times$ length.
So the dimensions of the expression are length $\times$ length $\times$ length.

**d** A length

The top term has the dimensions length $\times$ length.

The bottom term has the dimension length.

$\left(\dfrac{\text{length} \times \text{length}}{\text{length}}\right) = \text{length}$

## Exercise 21B

Throughout this exercise, the letters $b$, $h$ and $r$ represent lengths.
Numbers such as $3$, $\frac{1}{2}$ and $\pi$ are numbers which have no dimensions.

**1** Here are some expressions.

$$\tfrac{1}{2}bh \qquad 4h + 3r \qquad \pi(r + h) \qquad bhr \qquad 5h^2 \qquad bh^2$$

  **a** Write down the expressions which could represent a length.

  **b** Write down the expressions which could represent an area.

**2** Here are some expressions.

$$2bh \qquad r^2h \qquad 3b + 2r \qquad b(h + r) \qquad \pi r^2 b \qquad \frac{r^3}{b + h}$$

  **a** Write down the expressions which could represent an area.

  **b** Write down the expressions which could represent a volume.

**3** Here are some expressions.

$$b + rh \qquad b(3 + r) \qquad \pi b^3 + 4h^2 \qquad bh + \pi r^2 \qquad \frac{3b + r^2}{h}$$

Write down the **one** expression which could represent an area.

**4** Here are some expressions.

$$4b^2h \qquad \pi r^2(h + 2) \qquad \frac{5h^4}{b + r} \qquad (h + 2r)^3 \qquad \tfrac{1}{3}b^2h$$

Write down the **one** expression which could **not** represent a volume.

**5** Copy the table and complete it by putting a tick (✓) in the correct column to show whether the expression could be used for length, area, volume or none of these.

| Expression | Length | Area | Volume | None of these |
|---|---|---|---|---|
| $2rh$ | | | | |
| $\pi r + 4h$ | | | | |
| $\dfrac{(r + h)^2}{3b}$ | | | | |
| $b^3 + rh$ | | | | |
| $\pi r^2(h + r)$ | | | | |
| $\dfrac{bhr}{(b + h)}$ | | | | |

# Chapter summary

**You should now:**

★ be able to use and interpret scale drawings

★ understand dimensions and recognise whether an expression could represent a length, an area, a volume or none of these

# Chapter 21 review questions

**1** The length of a coach is 15 metres.
John makes a model of the coach.
He uses a scale of $1:24$
Work out the length, in centimetres, of the model coach.                    (1387 June 2005)

**2** The scale of a map is $1:50\,000$
Work out the real distance 6 cm represents.
Give your answer in kilometres.                    (1387 November 2005)

**3** A model is made of an aeroplane.
The length of the model is 18 centimetres.
The length of the real aeroplane is 45 metres.
Work out the ratio of the length of the model to the length of the real aeroplane.
Write your answer in the form $1:n$.                    (1388 March 2003)

**4** The expressions below can be used to calculate lengths, area or volumes of some shapes.
The letters $p, q$ and $r$ represent lengths.
$\pi$ and 2 are numbers which have no dimension.
Which **three** of these expressions can be used to calculate an area?

$$\pi(p+q) \qquad \frac{pq}{r} \qquad rq(p+q) \qquad \pi pq \qquad \frac{p^2 r}{2}$$

$$2r \qquad \frac{qr}{2} \qquad r(p+q) \qquad \frac{p^2\pi}{r} \qquad \frac{\pi pqr}{2}$$

(1385 June 2001)

**5** Here are three expressions.

| Expression | Length | Area | Volume | None of these |
|---|---|---|---|---|
| $\pi a^2 b$ | | | | |
| $\pi b^2 + 2h$ | | | | |
| $2ah$ | | | | |

$a, b$ and $h$ are lengths.
$\pi$ and 2 are numbers which have no dimensions.
Copy the table and put a tick ($\checkmark$) in the correct column to show whether the
expression can be used for length, area, volume or none of these.                    (1385 May 2002)

**6** The table shows six expressions.
$a, b$ and $c$ are lengths.
2 and 3 are numbers and have no dimension.

| $2a + 3b$ | $3ab$ | $a + b + c$ | $2a^2c$ | $2a^2 + bc$ | $ab(b + 2c)$ |
|---|---|---|---|---|---|
| | | | | | |

Copy the table and

**a** put the letter $A$ in the box underneath each of the **two** expressions that could
represent an **area**.

**b** put the letter $V$ in the box underneath each of the **two** expressions that could
represent a **volume**.                    (1385 November 2002)

**7** The crosses on the diagram show the positions of three places $A$, $B$ and $C$.
  The scale of the diagram is 1 cm to 5 km.

 $C$

$A$

$B$

Tariq cycled in a straight line from $A$ to $C$.
He left $A$ at 1.30 pm.
He cycled at an average speed of 10 kilometres per hour.

**a** Find the time he arrived at $C$.

**b** Find the bearing of
  **i** $B$ from $A$      **ii** $A$ from $C$.

(1385 November 2002)

# Two-dimensional shapes (2)

This chapter is about two-dimensional shapes. Two-dimensional shapes are flat. Two-dimensional is often written as 2-D.

## 22.1 Drawing shapes

Here are some examples of making accurate drawings of shapes using a ruler, compasses and in some cases a protractor.

### Example 1

Here is a sketch of a triangle.

Make an accurate drawing of the triangle.

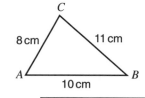

*Solution 1*

An accurate drawing made with a ruler and compasses but not a protractor is called a **construction**. So Example 1 is a construction.

*A* ——————— 10 cm ——————— *B*

**Step 1**
Draw a line 10 cm long using a ruler.

This is the base, *AB*, of the triangle.

Sometimes in exam questions this base line will be drawn on the question paper.

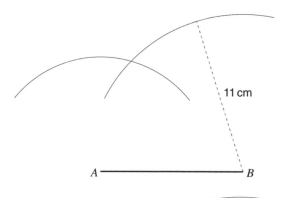

**Step 2**
Using a ruler set your compasses to 8 cm. Put the point of the compasses on *A* and draw an arc of a circle.

**Step 3**
Using a ruler set your compasses to 11 cm. Put the point of the compasses on *B* and draw an arc of a circle.

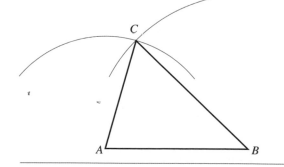

**Step 4**
The point where the two arcs cross is the third vertex (corner), *C*, of the triangle.

*CB* is 11 cm long and *CA* is 8 cm long so Join *C* to *A* and *C* to *B* to complete triangle *ABC*.

It is important that all construction lines can be seen. They should not be rubbed out.

## Exercise 22A

**1** Here is a sketch of triangle *ABC*.
Use ruler and compasses to construct the triangle when

   **a** *AB* = 10 cm, *CA* = 8 cm and *CB* = 9 cm

   **b** *AB* = 8.7 cm, *CA* = 9.4 cm and *CB* = 8.1 cm

   **c** *AB* = 4.6 cm, *CA* = 10.4 cm and *CB* = 7.9 cm

   **d** *AB* = 3.5 cm, *CA* = 12 cm and *CB* = 12.5 cm.

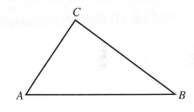

In each case measure the size of the largest angle of the triangle.

**2** Use ruler and compasses to construct an equilateral triangle with sides of length 8 cm.

**3** Here is a sketch of triangle *LMN*.

   **a** Make an accurate drawing of triangle *LMN*.

   **b** Measure the length of the side *LN*.

Give your answer to the nearest 0.1 cm.

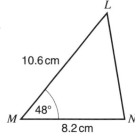

**4** Here is a sketch of a shape (rhombus).
Make an accurate drawing of the rhombus.

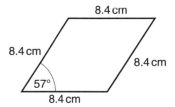

**5** Here is a sketch of a shape (kite) *ABCD*.
*AC* = 10.8 cm, *AB* = *AD* = 8.2 cm, *BC* = *DC* = 4.7 cm.
Use ruler and compasses to construct the kite.

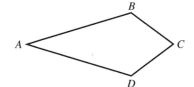

**6** Here are sketches of three triangles.
Make an accurate drawing of each triangle.

   **a** 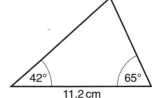    **b** 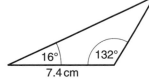    **c**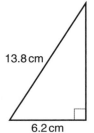

**7** Triangle *ABC* is isosceles. The sides *AB* and *AC* are both 5.6 cm long.
The angles at *B* and *C* are both 38°.

   **a** Draw a sketch of triangle *ABC* showing the lengths of sides *AB* and *AC* and the size of the angles at *B* and *C* on the sketch.

   **b** Make an accurate drawing of triangle *ABC*.

   **c** Measure the length of *BC*. Give your answer to the nearest 0.1 cm.

## 22.2 Circumference of a circle

The **circumference** is the special name of the perimeter of a circle, that is, the distance all around it.
Measure the circumference and diameter of some circular objects.

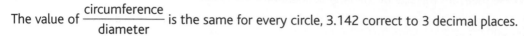

For each one, work out the value of $\dfrac{\text{circumference}}{\text{diameter}}$.

The answer is always just over 3

The value of $\dfrac{\text{circumference}}{\text{diameter}}$ is the same for every circle, 3.142 correct to 3 decimal places.

The value cannot be found exactly and the Greek letter $\pi$ is used to represent it.

So for all circles $\quad \dfrac{\textbf{circumference}}{\textbf{diameter}} = \boldsymbol{\pi}$

and $\quad$ **circumference $= \pi \times$ diameter**

Using $C$ to stand for the circumference of a circle with diameter $d$

$$\dfrac{C}{d} = \pi \quad \text{and} \quad C = \pi d$$

**To find the circumference of a circle, multiply its diameter by $\pi$**

---

### Example 2

Work out the circumference of a circle with a diameter of 6.8 cm.
Give your answer correct to 3 significant figures.

**Solution 2**

$\pi \times 6.8$

> Multiply the diameter by $\pi$.
> Use the calculator's $\pi$ button, if it has one.
> Otherwise use 3.142

$= 21.3628 \ldots$

> Write down at least 4 figures of the calculator display.

Circumference $= 21.4$ cm

> Round the circumference to 3 significant figures.
> The units (cm) are the same as the diameter's.

---

When the radius rather than the diameter is given in a question, one way of finding the circumference is to double the radius to obtain the diameter and then multiply the diameter by $\pi$.

Alternatively use the fact that a circle's diameter $d$ is twice its radius $r$, that is, $d = 2r$.

Replace $d$ by $2r$ in the formula $\boxed{C = \pi d}$ giving $C = \pi \times 2r$ which can be written as $\boxed{C = 2\pi r}$.

Sometimes the circumference is given and the diameter or radius has to be found.

---

### Example 3

The circumference of a circle is 29.4 cm.
Work out its diameter.
Give your answer correct to 3 significant figures.

*Solution 3*

**Method 1**

$29.4 = \pi d$    | Substitute 29.4 for $C$ in the formula $C = \pi d$. |

$d = 29.4 \div \pi$    | Divide both sides by $\pi$. |

$= 9.3583...$    | Divide 29.4 by $\pi$ and write down at least 4 figures of the calculator display. |

Diameter $= 9.36$ cm    | Round the diameter to 3 significant figures. The units are cm. |

The formula $C = \pi d$ can be rearranged with $d$ as the subject and used to find the diameter of a circle if its circumference is given.

Dividing both sides of $C = \pi d$ by $\pi$ gives $d = \dfrac{C}{\pi}$

> **To find the diameter of a circle, divide its circumference by $\pi$**

**Method 2**

$29.4 \div \pi$    | Divide the circumference by $\pi$. |

$d = 9.3583 ...$    | Write down at least 4 figures of the calculator display. |

Diameter $= 9.36$ cm    | Round the diameter to 3 significant figures. The units are cm. |

## Exercise 22B

If your calculator does not have a $\pi$ button, take the value of $\pi$ to be 3.142
Give answers correct to 3 significant figures.

1 Work out the circumferences of circles with these diameters.

   **a** 4.2 cm       **b** 9.7 m       **c** 29 cm       **d** 12.7 cm       **e** 17 m

2 Work out the circumferences of circles with these radii.

   **a** 3.9 cm       **b** 13 cm       **c** 6.3 m       **d** 29 m       **e** 19.4 cm

3 Work out the diameters of circles with these circumferences.

   **a** 17 cm       **b** 25 m       **c** 23.8 cm       **d** 32.1 cm       **e** 76.3 m

4 The circumference of a circle is 28.7 cm. Work out its radius.

5 The diameter of the London Eye is 135 m. Work out its circumference.

6 The tree with the greatest circumference in the world is a Montezuma cypress tree in Mexico. Its circumference is 35.8 m. Work out its diameter.

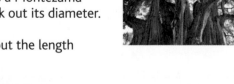

7 Taking the equator as a circle of radius 6370 km, work out the length of the equator.

8 The circumference of a football is 70 cm. Work out its radius.

9 A semicircle has a diameter of 25 cm.
   Work out its perimeter.
   (Hint: the perimeter includes the diameter)

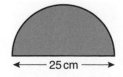

← 25 cm →

**10** A semicircle has a radius of 19 m.
Work out its perimeter.

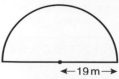

←19m→

**11** The diagram shows a running track.
The ends are semicircles of diameter 57.3 m and
the straights are 110 m long.
Work out the total perimeter of the track.

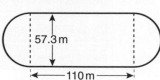

57.3 m

←110 m→

**12** A reel of cotton has a radius of 1.3 cm. The cotton is wrapped round it 500 times.
Work out the total length of cotton. Give your answer in metres.

**13** The radius of a cylindrical tin of soup is 3.8 cm.
Work out the length of the label.
(Ignore the overlap.)

**14** The diameter of a car wheel is 52 cm.
   **a** Work out the circumference of the wheel.
   **b** Work out the distance the car travels when the wheel makes 400 complete turns.
      Give your distance in metres.

**15** The big wheel of a 'penny farthing' bicycle has a radius of 0.75 m.
Work out the number of complete turns the big wheel makes
when the bicycle travels 1 kilometre.

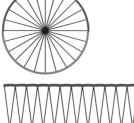

## 22.3 Area of a circle

The diagram shows a circle which has been
split up into equal 'slices' called sectors.

The sectors can be rearranged to make this new shape.

Splitting the circle up into more and more sectors and rearranging
them, the new shape becomes very nearly a rectangle.
The length of the rectangle is half the circumference of the circle.
The width of the rectangle is equal to the radius of the circle.
The area of the rectangle is equal to the area of the circle.

radius

½ circumference

Area of circle = ½ circumference × radius

$= \frac{1}{2} \times 2\pi r \times r$

$$A = \pi r^2$$

**To find the area of a circle multiply $\pi$ by the square of the radius**

or     **Area of a circle = $\pi$ × radius × radius**

If the diameter rather than the radius is given in a question, the first step is to halve the diameter to
get the radius.

## Example 4

The diameter of a circle is 9.6 m.
Work out its area.
Give your answer correct to 3 significant figures.

*Solution 4*

$9.6 \div 2 = 4.8$ — Divide the diameter by 2 to get the radius.

$\pi \times 4.8^2$ — Square the radius and then multiply by $\pi$.

$= 72.3822\ldots$ — Write down at least 4 figures of the calculator display.

Area $= 72.4\ \text{m}^2$ — Round the area to 3 significant figures. The units are $\text{m}^2$

## Exercise 22C

If your calculator does not have a $\pi$ button, take the value of $\pi$ to be 3.142
Give answers correct to 3 significant figures.

**1** Work out the areas of circles with these radii.

   **a** 7.2 cm    **b** 14 m    **c** 1.5 cm    **d** 3.7 m    **e** 2.43 cm

**2** Work out the areas of circles with these diameters.

   **a** 3.8 cm    **b** 5.9 cm    **c** 18 m    **d** 0.47 m    **e** 7.42 cm

**3** The radius of a dartboard is 22.86 cm.
Work out its area.

**4** The diameter of Avebury stone circle is 365 m.
Work out the area enclosed by the circle.

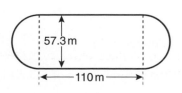

**5** The radius of a semicircle is 2.7 m.
Work out its area.

**6** The diameter of a semicircle is 8.2 cm.
Work out its area.

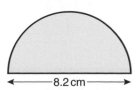

**7** The diagram shows a running track.
The ends are semicircles of diameter 57.3 m and the straights are 110 m long.
Work out the area enclosed by the track.

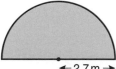

8 The diagram shows a circle of diameter 6 cm inside a square of side 10 cm.

   a Work out the area of the square.

   b Work out the area of the circle.

   c By subtraction work out the area of the shaded part of the diagram.

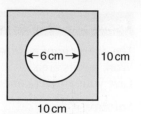

9 The diagram shows a circle of radius 7 cm inside a circle of radius 9 cm.
   Work out the area of the shaded part of the diagram.

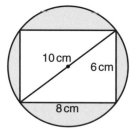

10 The diagram shows a 8 cm by 6 cm rectangle inside a circle of diameter 10 cm.
   Work out the area of the shaded part of the diagram.

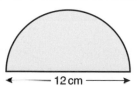

## 22.4 Circumferences and areas in terms of $\pi$

Answers to questions involving the circumference or area of a circle are sometimes given in terms of $\pi$, which is exact, and not as a number, which is approximate.

### Example 5

The diameter of a circle is 8 cm.
Find the circumference of the circle.
Give your answer as a multiple of $\pi$.

*Solution 5*

$\pi \times 8$ | Multiply the diameter by $\pi$.

Circumference $= 8\pi$ cm | Write the 8 before the $\pi$. The units are cm.

### Example 6

The radius of a circle is 3 m.
Find the area of the circle.
Give your answer as a multiple of $\pi$.

*Solution 6*

$\pi \times 3^2 = \pi \times 9$ | Square the radius and then multiply by $\pi$.

Area $= 9\pi$ m$^2$ | Write the 9 before the $\pi$. The units are m$^2$.

### Example 7

The diameter of a semicircle is 12 cm.
Find the perimeter of the semicircle.
Give your answer in terms of $\pi$.

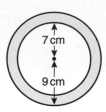

*Solution 7*

The perimeter is the sum of the diameter and half the circumference.

$\dfrac{\pi \times 12}{2} = \dfrac{12\pi}{2} = 6\pi$ | Find the circumference of a circle with a diameter of 12 cm and then divide it by 2

Perimeter $= 6\pi + 12$ cm | To find the perimeter add the diameter and the arc length. The units are cm.

If the circumference of a circle is given as a multiple of $\pi$ its diameter can be found.

### Example 8

The circumference of a circle is $30\pi$ m. Find its radius.

**Solution 8**

$$d = \frac{30\pi}{\pi} = 30$$ | To find the diameter divide the circumference by $\pi$.

$$\frac{30}{2} = 15$$ | To find the radius, divide the diameter by 2

Radius = 15 m | The units are m.

### Exercise 22D

In questions 1–4, give the answers as multiples of $\pi$.

**1** Find the circumference of a circle with a diameter of 7 m.

**2** Find the area of a circle with a radius of 5 cm.

**3** Find the circumference of a circle with a radius of 8 cm.

**4** Find the area of a circle with a diameter of 20 m.

**5** The diameter of a semicircle is 18 cm. Find the perimeter of the semicircle. Give your answer in terms of $\pi$.

**6** The radius of a semicircle is 7 cm. Find the perimeter of the semicircle. Give your answer in terms of $\pi$.

**7** The radius of a semicircle is 10 cm. Find its area. Give your answer as a multiple of $\pi$.

**8** The circumference of a circle is $16\pi$ cm. Find its diameter.

**9** The circumference of a circle is $30\pi$ m. Find its radius.

**10** The circumference of a circle is $14\pi$ cm. Find its area. Give your answer as a multiple of $\pi$.

## 22.5 Arc length and sector area

An **arc** is part of the circumference of a circle.

A **sector** of a circle is formed by an arc and two radii.
The perimeter of a sector is the sum of its arc length and two radii.

90° is $\frac{1}{4}$ of 360° (a full turn) and so for a sector with an angle of 90° at the centre the arc length is $\frac{1}{4}$ of the circumference of the whole circle.

The sector area is $\frac{1}{4}$ of the area of the whole circle.

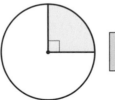

A quarter circle is called a **quadrant**.

For a sector with an angle of 70° at the centre the arc length is $\frac{70}{360}$ of the circumference of the whole circle.

The sector area is $\frac{70}{360}$ of the area of the whole circle.

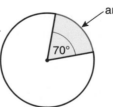

In general, for a sector with an angle of $x°$ at the centre of a circle of radius $r$.

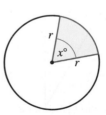

$$\text{arc length} = \frac{x}{360} \times 2\pi r$$

$$\text{sector area} = \frac{x}{360} \times \pi r^2$$

### Example 9

Calculate the perimeter of this sector.
Give your answer correct to 3 significant figures.

80°
7 m

*Solution 9*

$\dfrac{80}{360} \times 2\pi \times 7$

| $\dfrac{80}{360} \times 2\pi \times 7$ gives the arc length. |
|---|

$\qquad + 2 \times 7$

| Add twice the radius. |
|---|

$\qquad\qquad = 9.773 \ldots + 14$

| Write down at least 4 figures of the calculator display. |
|---|

$\qquad\qquad = 23.773 \ldots$

Perimeter $= 23.8$ m

| Round the perimeter to 3 significant figures. The units are m. |
|---|

### Example 10

Calculate the area of this sector.
Give your answer correct to 3 significant figures.

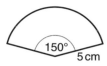

150°
5 cm

*Solution 10*

$\dfrac{150}{360} \times \pi \times 5^2$

| $\dfrac{150}{360}$ is the fraction of a full circle. $\pi \times 5^2$ is the area of the full circle. |
|---|

$\qquad = 32.72 \ldots$

| Write down at least 4 figures of the calculator display. |
|---|

Area $= 32.7$ cm²

| Round the area to 3 significant figures. The units are cm². |
|---|

## 22.6 Segment area

A **segment** of a circle is formed by a chord and an arc.

In the diagram, $AB$ is a chord of a circle, centre $O$.

Segment area = area of **sector** $OAB$ − area of **triangle** $OAB$.

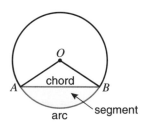

O
chord
A    B
segment
arc

### Exercise 22E

Give answers correct to 3 significant figures unless the question states otherwise.
If your calculator does not have a $\pi$ button, take the value of $\pi$ to be 3.142

**1** Calculate the arc length of each of these sectors.

**a**

30°  4 cm

**b**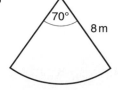

70°
8 m

**c**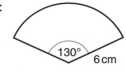

130°  6 cm

**d**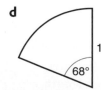

10 cm
68°

**2** Calculate the perimeter of each of these sectors.

**a**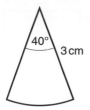
40° 3 cm

**b**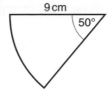
9 cm
50°

**c**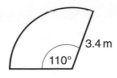
3.4 m
110°

**3** Find the perimeter of each of these sectors. Give each answer in terms of $\pi$.

**a**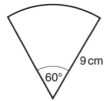
9 cm
60°

**b**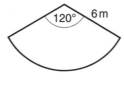
120° 6 m

**4** Calculate the area of each of these sectors.

**a**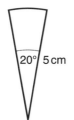
20° 5 cm

**b**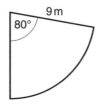
9 m
80°

**c**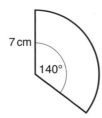
7 cm
140°

**d**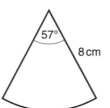
57°
8 cm

**5** Find the area of each of these sectors. Give each answer as a multiple of $\pi$.

**a**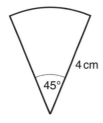
4 cm
45°

**b**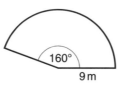
160°
9 m

**6 a** Find the area of the shaded segment.

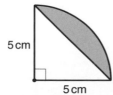
5 cm
5 cm

**b** Find an expression for the area of the shaded segment when the radius of the circle is $r$ cm. Give your answer in terms of $r$ and $\pi$.

## 22.7 Units of area

The diagram shows two congruent squares. The sides of square A are measured in metres and the sides of square B are measured in centimetres.

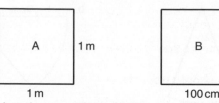

Square A is 1 m by 1 m so that the area of square A is $1 \times 1 \, m^2 = 1 \, m^2$

As 100 cm = 1 m, square B is 100 cm by 100 cm so that the area of square B is $100 \times 100 \, cm^2 = 10\,000 \, cm^2$

The squares have the same area so $1 \, m^2 = 100 \times 100 \, cm^2 = 10\,000 \, cm^2$

There are also similar results for other units.

| Length | Area |
|---|---|
| 1 cm = 10 mm | $1 \, cm^2 = 10 \times 10 = 100 \, mm^2$ |
| 1 m = 100 cm | $1 \, m^2 = 100 \times 100 = 10\,000 \, cm^2$ |
| 1 km = 1000 m | $1 \, km^2 = 1000 \times 1000 = 1\,000\,000 \, m^2$ |

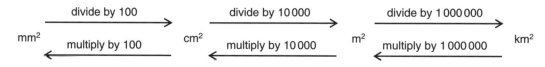

### Example 11

Change $4.6 \, m^2$ to $cm^2$.

*Solution 11*

$4.6 \, m^2 = 4.6 \times 10\,000 \, cm^2$    $1 \, m^2 = 10\,000 \, cm^2$

$\quad\quad\quad = 46\,000 \, cm^2$    Multiply the number of $m^2$ by 10 000

### Exercise 22F

**1** Change to $cm^2$
   **a** $4 \, m^2$      **b** $6.9 \, m^2$      **c** $600 \, mm^2$      **d** $47 \, mm^2$

**2** Change to $m^2$
   **a** $5 \, km^2$      **b** $0.3 \, km^2$      **c** $40\,000 \, cm^2$      **d** $560 \, cm^2$

**3 a** How many mm are there in 1 m?
   **b** How many $mm^2$ are there in $1 \, m^2$?
   **c** Change $8.3 \, m^2$ to $mm^2$.

**4** Find, in $cm^2$, the area of a rectangle
   **a** 3.2 m by 1.4 m      **b** 45 mm by 8 mm

**5** Work out the area of this triangle in
   **a** $cm^2$      **b** $mm^2$

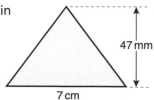

# Chapter summary

> **You should now:**
>
> ★ be able to draw triangles and quadrilaterals accurately using ruler, protractor and compasses
>
> **You should now know:**
>
> ★ how to find the circumference of a circle using
> circumference of a circle = $\pi$ × diameter
> or circumference of a circle = 2 × $\pi$ × radius
>
> or using the formulae $C = \pi d$ and $C = 2\pi r$

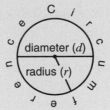

> ★ how to find the diameter (or radius) of a circle if its circumference is known using
>
> diameter = $\dfrac{\text{circumference}}{\pi}$ or using the formula $d = \dfrac{C}{\pi}$
>
> ★ how to find the area of a circle using area of a circle = $\pi$ × radius × radius or
> area of a circle = $\pi$ × (radius)² or using the formula
>
> $A = \pi r^2$
>
> ★ how to solve problems involving the circumference and area of a circle, including compound shapes and shaded areas
>
> ★ how to express answers to questions involving the circumference or area of a circle in terms of $\pi$
>
> ★ how to find arc length using the formula

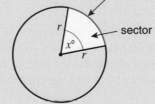

> $$\text{arc length} = \frac{x}{360} \times 2\pi r$$
>
> ★ how to find sector area using the formula
>
> $$\text{sector area} = \frac{x}{360} \times \pi r^2$$
>
> ★ how to find the perimeter of a sector
>
> ★ how to convert between units of area.

# Chapter 22 review questions

**1** Here is a sketch of a triangle.
Use ruler and compasses to **construct** this triangle accurately.
You must show all your construction lines.

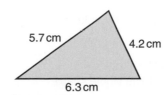

(1387 November 2003)

**2** The diagram shows a sketch of triangle $ABC$.

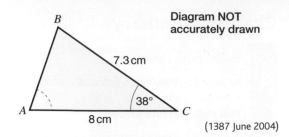

Diagram NOT
accurately drawn

  $BC = 7.3$ cm
  $AC = 8$ cm
  Angle $C = 38°$

  **a** Make an accurate drawing of triangle $ABC$.
  **b** Measure the size of angle $A$ on your diagram.

(1387 June 2004)

**3** A table has a top in the shape of a circle with a
  radius of 45 centimetres.

  **a** Calculate the area of the circular table top.

  The base of the table is also in the shape of a circle.
  The circumference of this circle is 110 centimetres.

  **b** Calculate the diameter of the base of the table.

(1384 November 1996)

**4** A rug is in the shape of a semicircle.
  The diameter of the semicircle is 1.5 m.
  Calculate the perimeter of the rug.
  Give your answer *in cm* correct to the nearest cm.

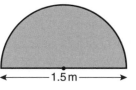

Diagram NOT
accurately drawn

**5** The diagram shows a shape.
  $AB$ is an arc of a circle, centre $O$.
  Angle $AOB = 90°$
  $OA = OB = 6$ cm.
  Calculate the perimeter of the shape.
  Give your answer correct to 3 significant figures.

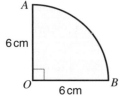

Diagram NOT
accurately drawn

(4400 May 2004)

**6** The diagram shows a circle of diameter 70 cm inside
  a square of side 70 cm.
  Work out the area of the shaded part of the diagram.
  Give your answer correct to 3 significant figures.

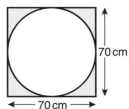

(1384 November 1997)

**7** The diagram shows a right-angled triangle $ABC$ and a circle.
  $A$, $B$ and $C$ are points on the circumference of the circle.
  $AC$ is a diameter of the circle.
  The radius of the circle is 10 cm.
  $AB = 16$ cm and $BC = 12$ cm.
  Work out the area of the shaded part of the circle.
  Give your answer correct to the nearest cm$^2$.

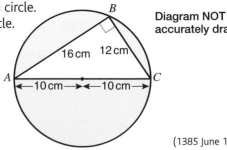

Diagram NOT
accurately drawn

(1385 June 1999)

**8** The diagram shows a shape made from a trapezium
  $ABCD$ and a semicircle with diameter $AB$.
  $AB = 18$ m
  $CD = 10$ m
  The total height of the shape is 21 m.
  Calculate the area of the whole shape.

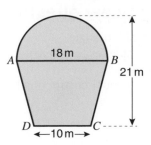

Diagram NOT
accurately drawn

**9** There is an infra-red sensor in a security system.
The sensor can detect movement inside a sector of a circle.
The radius of the circle is 15 m.
The sector angle is 110°.
Calculate the area of the sector.

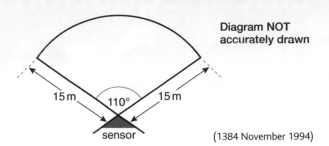

Diagram NOT accurately drawn

15 m   110°   15 m

sensor

(1384 November 1994)

**10** The diagram shows the sector of a circle, centre $O$.
The radius of the circle is 9 cm.
The angle at the centre of the circle is 40°.
Find the perimeter of the sector.
Leave your answer in terms of $\pi$.

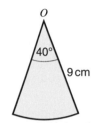

Diagram NOT accurately drawn

(1387 June 2003)

**11** The diagram shows a sector of a circle with a radius of $x$ cm and centre $O$.
$PQ$ is an arc of the circle.
Angle $POQ = 120°$.
Write down an expression in terms of $\pi$ and $x$ for

**a** the area of this sector

**b** the arc length of this sector.

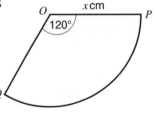

Diagram NOT accurately drawn

(1387 November 2004)

**12 a** Change 7 m² to cm².

(1387 June 2005)

**b** Change 50 000 mm² to cm².

(1387 November 2005)

**13** $B$ is 5 km North of $A$.
$C$ is 4 km from $B$.
$C$ is 7 km from $A$.

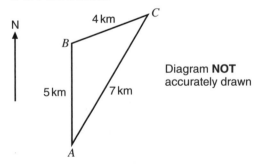

Diagram **NOT** accurately drawn

**a** Make an accurate scale drawing of triangle $ABC$.

Use a scale of 1 cm to 1 km.

**b** From your accurate scale drawing, measure the bearing of $C$ from $A$.

**c** Find the bearing of $A$ from $C$.

(1385 November 2000)

# Linear equations

Here are four examples of **equations**.

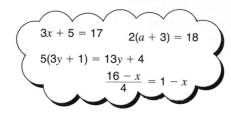

$$3x + 5 = 17 \qquad 2(a + 3) = 18$$
$$5(3y + 1) = 13y + 4$$
$$\frac{16 - x}{4} = 1 - x$$

An equation must always have an = sign because one side of an equation is always equal to the other side.

The expression before the equals sign is called the **left-hand side** (LHS) of the equation.

The expression after the equals sign is called the **right-hand side** (RHS) of the equation.

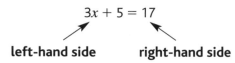

$$3x + 5 = 17$$

**left-hand side**          **right-hand side**

The equation $3x + 5 = 17$ can also be written as $17 = 3x + 5$

The **solution** of an equation is the value of the letter (the 'unknown') that makes the equation true.

## 23.1 The balance method for solving equations

Some equations can be solved mentally. To solve more complicated equations the **balance method** is used.

To keep the balance, whatever you do to the left-hand side you must also do to the right-hand side of the equation.

Using the equation $3x + 5 = 17$ as an example.

| **Balance** | **Explanation** | **Step** |
|---|---|---|
| 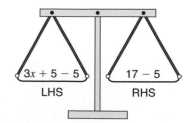 | Start<br>The scales balance because both sides are equal. | $3x + 5 = 17$ |
| 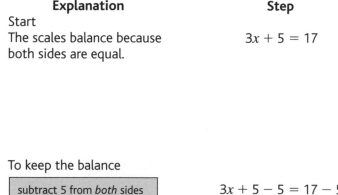 | To keep the balance<br><br>subtract 5 from *both* sides | $3x + 5 - 5 = 17 - 5$ |

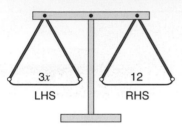

The scales still balance and
show that $3x = 12$

$3x = 12$

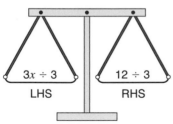

To keep the balance

| divide *both* sides by 3 |

$3x \div 3 = 12 \div 3$

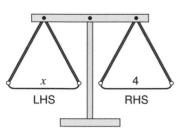

Solution

$x = 4$

The solution of the equation $3x + 5 = 17$ is $x = 4$
To check this, substitute $x = 4$ into the equation.

$$3 \times 4 + 5 = 17$$
$$12 + 5 = 17 \quad \checkmark$$

## Example 1

Use the balance method to solve the equation $4x + 3 = 31$

**Solution 1**

$4x + 3 = 31$

$4x + 3 - 3 = 31 - 3$    Subtract 3 from *both* sides.

$4x = 28$

$4x \div 4 = 28 \div 4$    Divide *both* sides by 4

$x = 7$

Check: substitute $x = 7$,
$$4 \times 7 + 3 = 31$$
$$28 + 3 = 31 \quad \checkmark$$

## Example 2

Use the balance method to solve the equation $27 = 6p - 9$

**Solution 2**

$27 = 6p - 9$

$27 + 9 = 6p - 9 + 9$    Add 9 to *both* sides.

$36 = 6p$

$36 \div 6 = 6p \div 6$    Divide *both* sides by 6

$6 = p$ or $p = 6$    $6 = p$ is usually written as $p = 6$

Check: substitute $p = 6$,
$$27 = 6 \times 6 - 9$$
$$27 = 36 - 9 \quad \checkmark$$

## Example 3

Solve the equation $6 - 7x = 11$

**Solution 3**

$$6 - 7x = 11$$
$$6 - 7x + 7x = 11 + 7x$$ Add $7x$ to both sides so that the sign of the term in $x$ is positive.
$$6 = 11 + 7x$$
$$6 - 11 = 11 + 7x - 11$$ Subtract 11 from *both* sides.
$$-5 = 7x$$
$$\frac{-5}{7} = \frac{7x}{7}$$ Divide *both* sides by 7 then cancel.
$$-\frac{5}{7} = x$$
$$x = -\frac{5}{7}$$

To solve equations which include brackets, one method is to expand the brackets and then use the balance method.

## Example 4

Solve the equation $5(a + 3) = 18$

**Solution 4**

$$5(a + 3) = 18$$
$$5a + 15 = 18$$ Expand the brackets.
$$5a + 15 - 15 = 18 - 15$$ Subtract 15 from both sides.
$$5a = 3$$
$$\frac{5a}{5} = \frac{3}{5}$$ Divide both sides by 5 then cancel.
$$a = \frac{3}{5}$$ This solution can also be given as the exact decimal 0.6

To solve equations which have letters on both sides, the first step is to rearrange the terms so that the letter appears only on one side of the equation. It is helpful if the sign of the letter term is positive.

## Example 5

Solve the equation $4 - d = 5d + 1$

**Solution 5**

$$4 - d = 5d + 1$$ Add $d$ to both sides so that when like terms are collected, $d$ only appears on one side of the equation and the sign of the term in $d$ is positive.
$$4 - d + d = 5d + 1 + d$$
$$4 = 6d + 1$$
$$3 = 6d$$ Subtract 1 from both sides.
$$\frac{3}{6} = \frac{6d}{6}$$ Divide both sides by 6 then cancel.
$$\frac{1}{2} = d$$ Simplify $\frac{3}{6}$
$$d = \frac{1}{2}$$

## Example 6

Solve the equation $5(3y + 2) = 13y + 4$

**Solution 6**

$5(3y + 2) = 13y + 4$

$15y + 10 = 13y + 4$    | Expand the brackets. |

$2y + 10 = 4$    | Subtract $13y$ from both sides. |

$2y = 4 - 10$    | Subtract 10 from both sides. |

$2y = -6$

$y = -3$    | Divide both sides by 2 |

## Exercise 23A

Solve these equations.

**1** $2d + 1 = 19$

**2** $3e - 4 = 23$

**3** $7g + 4 = 4$

**4** $15 + 2r = 36$

**5** $13 = 5 - 4a$

**6** $3(b + 2) = 15$

**7** $4(c + 3) = 30$

**8** $4(e - 3) = 20$

**9** $5(g - 6) = 2$

**10** $2(2h + 1) = 22$

**11** $2(3k + 1) = 26$

**12** $3(4m - 3) = 27$

**13** $2(n + 3) = 11$

**14** $4(p + 3) = 18$

**15** $2(2q + 3) = 16$

**16** $3(r + 2) = 0$

**17** $3(1 - t) = 6$

**18** $5(1 - 2u) = 35$

**19** $12 = 2(4w + 1)$

**20** $20 = 11 + 4(x + 1)$

**21** $6x = 2x + 1$

**22** $5y = 2y + 9$

**23** $3p = p + 6$

**24** $2x = 8 - 2x$

**25** $y = 8 - y$

**26** $4 - 2x = 3x$

**27** $3x + 1 = 2x + 4$

**28** $2a + 4 = a - 7$

**29** $8x + 4 = 3x + 8$

**30** $4(x + 1) = x + 10$

**31** $5(c + 3) = 3c + 1$

**32** $6(x - 1) = 2x + 8$

**33** $4(3y + 2) = 16 + 10y$

**34** $6(x + 5) = 23 - x$

**35** $9y + 7 = 2(1 + 6y)$

**36** $3(2d - 1) = 4d - 7$

**37** $4(x - 1) = 3(x + 1)$

**38** $5(3x - 1) = 7(x + 1)$

**39** $2(2g - 5) = 4(1 - g)$

**40** $5(2x + 3) = 2(3 - x) + 3$

## 23.2 Setting up equations

Equations can be used to solve problems.

### Example 7

Kevin thinks of a number.
He doubles the number then adds 7
His answer is 19
Work out the number that Kevin thinks of.

**Solution 7**
Let $n$ be the number Kevin thinks of.

| | |
|---|---|
| $2n$ | Double the number $n$ giving $2n$ |
| $2n + 7$ | Add 7 |
| $2n + 7 = 19$ | Set up the equation by writing $2n + 7$ equal to Kevin's answer of 19 |
| $2n = 12$ | Solve the equation by subtracting 7 from both sides and dividing both sides by 2 |
| $n = 6$ | |

The number Kevin thinks of is 6

### Example 8

All the angles are measured in degrees.

**a** Write down an expression, in terms of $x$, for the sum of the angles of this triangle.

**b** By setting up an equation, work out
   **i** the value of $x$
   **ii** the size of the largest angle of this triangle.

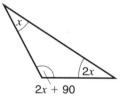

**Solution 8**

**a** Sum of the angles $= x + 2x + 90 + 2x$
                   $= 5x + 90$

> Add the three angles and collect like terms.

**b** **i**    $5x + 90 = 180$

> The sum of the angles of a triangle $= 180°$.

             $5x = 90$

> Subtract 90 from both sides.

              $x = 18$

> Divide both sides by 5.

  **ii** Substitute $x = 18$, $2x + 90 = 2 \times 18 + 90$
       Size of largest angle $= 126°$

> The largest angle is $2x + 90$

### Example 9

Georgina is $x$ years old.
Jessica is 4 years younger than Georgina.

**a** Write down, in terms of $x$, an expression for Jessica's age.

Angela is 3 times as old as Jessica.

**b** Work out the ages of Georgina and Jessica if Angela is 27 years old.

**Solution 9**

**a** Jessica's age is $x - 4$

**b** Angela's age is $3(x - 4)$     | Multiply Jessica's age by 3 |

$3(x - 4) = 27$     | Form the equation using 'Angela is 27 years old'. |

     | Solve the equation. |

$3x - 12 = 27$     | Expand the brackets. |

$3x = 39$     | Add 12 to both sides. |

$x = 13$     | Divide both sides by 3 |

Georgina's age is 13     | Georgina is 13 years old and Jessica is 9 years old
Jessica's age is $x - 4 = 13 - 4 = 9$     | since she is 4 years younger than Georgina. |

---

## Example 10

The diagram shows a square and a rectangle.
The lengths of the edges are given in centimetres.
The perimeter of the square is equal to the perimeter
of the rectangle.

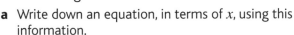

**a** Write down an equation, in terms of $x$, using this
information.

**b** By solving your equation, work out the area of the square.

**Solution 10**

**a** Perimeter of the square $= 4 \times 3x = 12x$

Perimeter of the rectangle
$$= 2 \times (2x + 5) + 2 \times (x + 1)$$
$$= 4x + 10 + 2x + 2$$
$$= 6x + 12$$
$12x = 6x + 12$     | The perimeters are equal. |

**b** $12x = 6x + 12$     | Solve the equation. |

$6x = 12$     | Subtract $6x$ from both sides. |

$x = 2$     | Divide both sides by 6 |

The length of each side of the square is $3x$ cm $= 6$ cm     | Substitute $x = 2$ into $3x$. |
The area of the square $= 6$ cm $\times 6$ cm $= 36$ cm$^2$.

---

### Exercise 23B

**1** Cath thinks of a number. She multiplies the number by 4 then subtracts 5. Her answer is 27
Work out the number that Cath is thinking of.

**2** Bill pays 96 pence for a pen and a pencil. The pen costs five times as much as the pencil.
The pencil costs $x$ pence.

    **a** Write down an expression, in terms of $x$, for the cost of the pen.

    **b** By setting up an equation, work out   **i** the value of $x$   **ii** the cost of the pen.

**3 a** Write down an expression, in terms of $x$, for the sum of the angles of this quadrilateral. Simplify your answer.

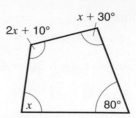

The sum of the angles in a quadrilateral is 360°.

  **b** By setting up an equation, work out
    **i** the value of $x$
    **ii** the size of the largest angle of this quadrilateral.

**4** Eggs are sold in cartons, each containing $y$ eggs.
Viv buys 3 of these cartons of eggs.
One carton has 4 broken eggs in it.
Viv now has just 23 good eggs.
How many eggs does each carton contain?

**5** The lengths, in centimetres, of the sides of a triangle are $2s + 1$, $3s$ and $5s - 3$
The perimeter of the triangle is 38 cm.
Find the length of the smallest side of this triangle.

**6** Mansoor is $p$ years of age.
Mansoor's father is three times as old as Mansoor.

  **a** Write down an expression, in terms of $p$, for the age of Mansoor's father.

In 5 years time Mansoor's father will be 47 years of age.

  **b** Write down an equation in $p$ to show this information.

  **c** Solve your equation to find Mansoor's present age.

**7** A bag contains $b$ yellow balls, $3b + 2$ red balls and $b + 8$ blue balls.

  **a** Write an expression, in terms of $b$, for the total number of balls in the bag.

  **b** The total number of balls in the bag is 45
    **i** By forming an equation find the value of $b$.
    **ii** How many blue balls are in the bag?
    **iii** How many more red balls are there than blue balls in the bag?

**8** A bag contains red sweets, yellow sweets and green sweets only.
There are $2t$ red sweets, $4t - 5$ yellow sweets and $14 - 3t$ green sweets.
There are a total of 21 sweets in the bag.

  **a** By setting up an equation, work out the number of red sweets in the bag.

  **b** Work out the number of yellow sweets in the bag.

**9** Samantha works for $x$ hours each week for 3 weeks.
In the fourth week she works for an extra 10 hours.
She works a total of 150 hours during these 4 weeks.
By setting up an equation, work out

  **a** the value of $x$

  **b** the number of hours that Samantha works in the fourth week.

**10** Four of the angles in a pentagon are each $(x + 70)°$.
The fifth angle is $(x + 100)°$.

  **a** Write down an equation, in terms of $x$, using this information.
    Give your answer in its simplest form.

  **b** By solving your equation in part **a**, work out the value of $x$.

**11** The width of a rectangle is $x$ cm. The length of the rectangle is 6 cm longer than its width.
  **a** Write down an expression, in terms of $x$, for the length of the rectangle.
  The perimeter of the rectangle is 40 cm.
  **b** By setting up an equation, work out
     **i** the value of $x$     **ii** the length of the rectangle.

**12** The diagram shows an L-shape.
  The lengths of the edges are given in centimetres.
  The area of the shape is 44 cm².
  **a** Write down an equation, in terms of $x$, using this information.
  **b** By solving your answer to part **a** work out
     **i** the value of $x$     **ii** the perimeter of the shape.

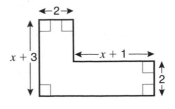

**13** The diagram shows an equilateral triangle and a square.
  The lengths of the edges are given in centimetres.
  The perimeter of the equilateral triangle is equal to the
  perimeter of the square.
  **a** Write down an equation, in terms of $k$, using this
  information.
  **b** By solving your equation work out the area of the square.

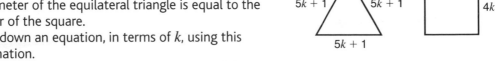

**14** Brian, Clare and Daniel share a sum of money.
  Brian's share is £4$w$
  Clare's share is £$(2w - 1)$
  Daniel's share is £$(8w - 8)$
  Daniel's share is the same as the sum of Brian's share and Clare's share.
  **a** Write down an equation, in terms of $w$, using this information.
  **b** Work out, in £, the total sum of money shared by Brian, Clare and Daniel.

**15** The diagram shows a rectangle.
  Work out the value of $x$.

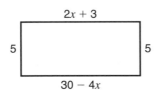

## 23.3 Solving equations with fractional terms

In algebra, the expressions $(x + 3) \div 4$ and $\frac{1}{2}y$ are usually written as $\dfrac{x + 3}{4}$ and $\dfrac{y}{2}$

### Example 11

Solve the equation $\dfrac{4}{p} = 8$

*Solution 11*

$\dfrac{4}{p} = 8$ so $\dfrac{4}{\not{p}} \times \not{p} = 8 \times p$     | Multiply both sides by $p$ then cancel. |

$4 = 8p$

$\dfrac{4}{8} = \dfrac{\not{8}p}{\not{8}}$     | Divide both sides by 8 |

$p = \dfrac{1}{2}$

## Example 12

Solve the equation $\dfrac{3(q+5)}{2} = 6$

### Solution 12

$$\frac{3(q+5)}{2} = 6$$

$$\frac{3(q+5)}{\cancel{2}} \times \cancel{2} = 6 \times 2 \qquad \boxed{\text{Multiply both sides by 2 then cancel.}}$$

$$3(q+5) = 12$$

$$3q + 15 = 12 \qquad \boxed{\text{Expand the brackets.}}$$

$$3q = -3 \qquad \boxed{\text{Subtract 15 from both sides.}}$$

$$q = -1 \qquad \boxed{\text{Divide both sides by 3}}$$

## Example 13

Solve the equation $\dfrac{16-x}{4} = 1 - x$

### Solution 13

$$\frac{16-x}{4} = 1 - x$$

$$4 \times \left(\frac{16-x}{4}\right) = 4 \times (1-x) \qquad \boxed{\begin{array}{l}\text{Multiply both sides} \\ \text{by 4, remember to} \\ \text{put brackets around} \\ \text{the } 1-x.\end{array}}$$

$$\cancel{4} \times \left(\frac{16-x}{\cancel{4}}\right) = 4 \times (1-x) \qquad \boxed{\text{Cancel the 4s on the LHS.}}$$

$$16 - x = 4 - 4x \qquad \boxed{\text{Expand the brackets.}}$$

$$4x - x = 4 - 16 \qquad \boxed{\begin{array}{l}\text{Add } 4x \text{ to both sides} \\ \text{and subtract 16 from} \\ \text{both sides.}\end{array}}$$

$$3x = -12 \qquad \boxed{\text{Simplify.}}$$

$$x = -4 \qquad \boxed{\text{Divide both sides by 3}}$$

## Example 14

Solve the equation $\dfrac{2x-1}{2} - \dfrac{x-5}{3} = \dfrac{5}{4}$

### Solution 14

$$\frac{2x-1}{2} - \frac{x-5}{3} = \frac{5}{4}$$

$$12 \times \left(\frac{2x-1}{2}\right) - 12 \times \left(\frac{x-5}{3}\right) = 12 \times \frac{5}{4} \qquad \boxed{\text{Multiply both sides by 12 since 12 is the lowest common multiple (LCM) of 2, 3 and 4}}$$

$$\overset{6}{\cancel{12}} \times \left(\frac{2x-1}{\cancel{2}}\right) - \overset{4}{\cancel{12}} \times \left(\frac{x-5}{\cancel{3}}\right) = \overset{3}{\cancel{12}} \times \frac{5}{\cancel{4}} \qquad \boxed{\text{Cancel.}}$$

$$6(2x-1) - 4(x-5) = 15$$

$$12x - 6 - 4x + 20 = 15 \qquad \boxed{\text{Expand the brackets.}}$$

$$8x + 14 = 15 \qquad \boxed{\text{Simplify the LHS.}}$$

$$8x = 1 \qquad \boxed{\text{Subtract 14 from both sides.}}$$

$$x = \frac{1}{8} \qquad \boxed{\text{Divide both sides by 8}}$$

**Exercise 23C**

**1** Solve these equations.

**a** $\dfrac{x}{3} = 4$    **b** $\dfrac{1}{x} = 2$    **c** $\dfrac{12}{n} = 2$    **d** $\dfrac{x}{3} + 2 = 5$

**e** $5 + \dfrac{x}{4} = 1$    **f** $\dfrac{x+1}{3} = 2$    **g** $\dfrac{y-2}{4} = 1$    **h** $2 = \dfrac{p-1}{6}$

**i** $5 = \dfrac{2q-1}{3}$    **j** $\dfrac{1-y}{4} = 1$    **k** $\dfrac{4p+9}{2} = 1$    **l** $\dfrac{x}{2} = x - 3$

**m** $\dfrac{5y}{2} = y - 6$    **n** $4 + \dfrac{3x}{5} = x$    **o** $\dfrac{7-x}{3} = 1 - x$    **p** $\dfrac{9-x}{5} = 3 - x$

**q** $2 + 3x = \dfrac{3x}{5}$    **r** $\dfrac{a}{2} + \dfrac{a}{3} = 1$    **s** $\dfrac{3y+6}{10} + \dfrac{5-2y}{5} = 6$

**t** $\dfrac{x+1}{2} - \dfrac{4x-1}{3} = \dfrac{5}{12}$    **u** $\dfrac{3x-4}{2} + \dfrac{2x+1}{5} = \dfrac{8-x}{3}$

**2** Saika thinks of a number. She halves the number then adds 9   Her answer is twice the number she is thinking of.

   **a** If $n$ is the number that Saika is thinking of, write down an equation in terms of $n$.

   **b** Solve the equation to find $n$.

**3** Six multiplied by the reciprocal of a number is equal to $\frac{2}{3}$
   Find the number.

**4** $ABC$ is an isosceles triangle with $AB = AC$.
   Angle $ABC = \frac{2}{3}(p + 20)°$ and angle $ACB = (100 - p)°$.

   **a** Write down an equation in terms of $p$.

   **b** Find the value of $p$.

   **c** Find the size of angle $BAC$.

**5** Terry cycles for $\frac{1}{4}$ hour at an average speed of $x$ kilometres per hour.
   He then cycles for $\frac{1}{3}$ hour at an average speed of $(x + 10)$ kilometres per hour.
   He cycles a total distance of 15 kilometres. Work out the value of $x$.

## 23.4 Simultaneous linear equations

The diagram shows red and green cylinders.
The height of each red cylinder is $y$ cm.
The height of each green cylinder is $x$ cm.

The total height of 3 green cylinders and 1 red cylinder is 16 cm.
The total height of 1 red cylinder and 1 green cylinder is 10 cm.

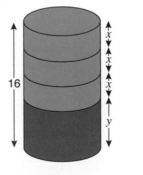
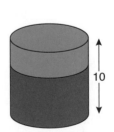

This information can be written as

$$3x + y = 16$$
$$x + y = 10$$

These are known as **simultaneous equations**.
It is possible to use algebra to find the value of $x$ and the value of $y$ which satisfy both of these equations simultaneously (at the same time). This process is called solving simultaneous equations.

## Example 15

Solve the simultaneous equations

$$3x + y = 16$$
$$x + y = 10$$

### Solution 15

$3x + y = 16$    (1)
$x + y = 10$    (2)

| Label the equations as (1) and (2). |

$(1) - (2)$

| Since $+y$ and $+y$ are exactly the same, $y$ can be eliminated by subtracting equation (2) from equation (1). |

$3x + y = 16$    (1)
$\underline{x + y = 10}$    (2)
$2x + 0 = 6$

| $3x - x = 2x$ |
| $+y - +y = 0$ |
| $16 - 10 = 6$ |

$2x = 6$ so $x = 3$

| Divide both sides by 2 |

When $x = 3, 3 + y = 10$

| Substitute $x = 3$ into equation (2). |

So $y = 7$

| Subtract 3 from both sides. |

$x = 3, y = 7$

| Always give the values for both letters. |

This shows that the height of each green cylinder is 3 cm and the height of each red cylinder is 7 cm.

## Example 16

Solve the simultaneous equations

$$3x + 2y = 7$$
$$5x - 4y = 8$$

### Solution 16

$3x + 2y = 7$    (1)
$5x - 4y = 8$    (2)

| Label the equations as (1) and (2). |

$(1) \times 2$ gives $6x + 4y = 14$
                    $5x - 4y = 8$

| Neither the $x$ terms nor the $y$ terms are the same. To get the same number of $y$s in each equation multiply equation (1) by 2 |

$6x + 4y = 14$
$\underline{5x - 4y = 8}$
$11x + 0 = 22$

| Eliminate the $y$ terms by adding the two equations. |
| $6x + 5x = 11x$ |
| $+4y + -4y = 0$ |
| $14 + 8 = 22$ |

$11x = 22$ so $x = 2$

| Divide both sides by 11 |

When $x = 2, 6 + 2y = 7$

| Substitute $x = 2$ into equation (1). |

$2y = 1$ so $y = \frac{1}{2}$

$x = 2, y = \frac{1}{2}$

| Always give the values for both letters. |

## Example 17

Solve the simultaneous equations

$$4y - 3x = 29$$
$$y = 2x + 11$$

### Solution 17

$$4y - 3x = 29 \qquad (1)$$
$$y = 2x + 11 \qquad (2)$$

> Label the equations as (1) and (2).

$$4(2x + 11) - 3x = 29$$

> Substitute $2x + 11$ for $y$ in equation (1).

$$8x + 44 - 3x = 29$$

> Expand the brackets.

$$5x + 44 = 29$$

> Simplify the LHS.

$$5x = -15$$

> Subtract 44 from both sides.

$$x = -3$$

> Divide both sides by 5

When $x = -3$, $y = -6 + 11$

> Substitute $x = -3$ into equation (2).

So $y = 5$

$x = -3$, $y = 5$

---

Sometimes both equations have to be multiplied by numbers before one of the letters can be eliminated.

## Example 18

Solve the simultaneous equations

$$2x - 3y = 7$$
$$3x - 4y = 10$$

### Solution 18

$$2x - 3y = 7 \qquad (1)$$
$$3x - 4y = 10 \qquad (2)$$

> Label the equations as (1) and (2).

$(1) \times 3$ gives $6x - 9y = 21 \qquad (3)$
$(2) \times 2$ gives $6x - 8y = 20 \qquad (4)$

> Neither the $x$ terms nor the $y$ terms are the same. To get the $x$ terms equal multiply equation (1) by 3 and equation (2) by 2. Label the new equations as (3) and (4).

$$6x - 8y = 20 \qquad (4)$$
$$6x - 9y = 21 \qquad (3)$$
$$\overline{\phantom{6x}0 + \phantom{0}y = -1}$$

> $x$ can be eliminated by subtracting equation (3) from equation (4).
> $$6x - 6x = 0$$
> $$-8y - (-9y) = -8y + 9y = +y$$
> $$20 - 21 = -1$$

so $y = -1$

When $y = -1$, $2x - 3 \times -1 = 7$

> Substitute $y = -1$ into equation (1).

$$2x + 3 = 7$$

$$2x = 4 \text{ so } x = 2$$

$x = 2$, $y = -1$

### Exercise 23D

Solve these simultaneous equations.

**1** $x + y = 5$
$x - y = 1$

**2** $2x + y = 5$
$x + y = 3$

**3** $5x + 2y = 3$
$3x - 2y = 5$

**4** $x + y = 8$
$x - y = 1$

**5** $x + y = 2$
$x - y = 3$

**6** $x + y = 5$
$y = x + 1$

**7** $x + y = 4$
$y = x + 2$

**8** $3x + y = 10$
$y = 2 - x$

**9** $3x + 5y = 11$
$2x - y = 3$

**10** $2x + 7y = 5$
$x + 3y = 3$

**11** $5x + 3y = 4$
$x - y = 4$

**12** $7x + y = 5$
$6x - 2y = 10$

**13** $3x + 2y = 8$
$5x + 6y = 16$

**14** $3x + 2y = 7$
$4x - 3y = -2$

**15** $2x + 5y = 16$
$5x - 2y = 11$

**16** $2x + 3y = 1$
$3x - 2y = 8$

**17** $4x + 3y = 3$
$6x + 5y = 7$

**18** $6x + 7y = 9$
$8x + 9y = 12$

**19** $5x - 3y = 7$
$3x + 5y = -6$

**20** $5x - 3y = 14$
$7x - 5y = 20$

## 23.5 Setting up simultaneous linear equations

Simultaneous equations can be used to solve problems.

### Example 19

6 pencils and 2 crayons cost a total of £2.70
5 pencils and 3 crayons cost a total of £2.45
Work out the cost of a pencil and the cost of a crayon.

*Solution 19*

Let the cost of 1 pencil be $a$ pence.
Let the cost of 1 crayon be $b$ pence.

> To avoid working with decimals, change the costs to pence.

$6a + 2b = 270$   (1)
$5a + 3b = 245$   (2)

> Write 6 pencils and 2 crayons cost a total of 270 pence as equation (1).
> Write 5 pencils and 3 crayons cost a total of 245 pence as equation (2).
> Solve the simultaneous equations.

(1) × 3 gives $18a + 6b = 810$   (3)
(2) × 2 gives $10a + 6b = 490$   (4)

> Neither the $a$ terms nor the $b$ terms are the same. To get the $b$ terms equal multiply equation (1) by 3 and equation (2) by 2

$18a + 6b = 810$   (3)
$\underline{10a + 6b = 490}$   (4)
$\ \ 8a + 0 \ = 320$

> Eliminate the $b$ terms by subtracting equation (4) from equation (3).

$8a = 320$ so $a = 40$

When $a = 40$, $240 + 2b = 270$

> Substitute $a = 40$ into equation (1).

$2b = 30$ so $b = 15$

The cost of a pencil is 40p.
The cost of a crayon is 15p.

**Exercise 23E**

1  In my pocket there are $x$ one-pence coins and $y$ two-pence coins.
   There are 8 coins altogether and their total value is 11 pence. Find $x$ and $y$.

2  The sum of two numbers is 15   The difference of the two numbers is 8   Find the larger number.

3  The diagram shows a rectangle.
   All sides are measured in centimetres.
   **a** Show that $5a + 2b = 18$
   **b** Write down another equation in terms of $a$ and $b$.
   **c** Solve the two equations simultaneously to find $a$ and $b$.
   **d** Hence find the area of the rectangle.

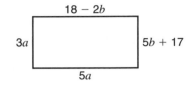

4  Tony and June answered the same thirty questions in a maths quiz.
   The table shows some information.

|  | Number of correct answers | Number of wrong answers | Total points scored |
|---|---|---|---|
| Tony | 20 | 10 | 90 |
| June | 15 | 15 | 60 |

   Each correct answer scored $x$ points.
   Each wrong answer scored $y$ points.
   **a** Use the information in the table for Tony to show that $2x + y = 9$
   **b** Use the information in the table for June to find another equation in $x$ and $y$.
   **c** Solve the two equations simultaneously to find $x$ and $y$.
   **d** James gave only 5 correct answers in the quiz. Find the total number of points he scored.

5  The lengths of the sides of an equilateral triangle are $(3x + 2)$ cm, $(2y - x)$ cm and $(y + 3)$ cm.
   **a** Find $x$ and $y$.          **b** Find the perimeter of the triangle.

6  Each large filing cabinet contains 100 files and has 4 drawers.
   Each small filing cabinet contains 80 files and has 3 drawers.

   In an office, there are $x$ large filing cabinets and $y$ small filing cabinets.
   The total number of files in these filing cabinets is 1560 and the total number of drawers is 61

   Work out the number of large filing cabinets and the number of small filing cabinets in this office.

## Chapter summary

**You should now be able to:**

★  solve an **equation** to find the value of the unknown letter, for example $x = 2$ is the **solution** of the equation $x + 3 = 5$

★  solve equations by the **balance method** in which whatever is done to the right-hand side of the equation must also be done to the left-hand side of the equation

★  solve equations which have **brackets** by expanding the brackets, collecting any like terms and then using the balance method

★  solve a problem by setting up an equation and solving it algebraically

★  set up and solve equations with fractional terms by multiplying both sides of the equation by the lowest common multiple of the denominators

★  set up and solve **simultaneous linear equations**.

# Chapter 23 review questions

**1  a** Solve $3y + 2 = 17$        **b** Solve $\dfrac{x}{3} = 5$

**2** Solve

**a** $2x + 9 = 4x + 6$        **b** $5(x - 2) = 20$        **c** $21 = 3(2x + 11)$

**3  a** Solve $7p + 2 = 5p + 8$        **b** Solve $7r + 2 = 5(r - 4)$        (1387 June 2003)

**4  a** Solve $7x + 18 = 74$        **b** Solve $4(2y - 5) = 32$        **c** Solve $5p + 7 = 3(4 - p)$

(1387 November 2003)

**5** Solve $4(2x + 1) = 2(3 - x)$                                (1387 November 2004)

**6** Solve $5(x - 2) = 8 - 7x$

**7** Fred is $x$ years old.
His sister, Mary, is 4 years older than Fred.
**a  i** Write down an expression, in terms of $x$, for Mary's age.

Sarfraz is twice as old as Fred.
**ii** Write down an expression, in terms of $x$, for the total of Fred's age, Mary's age and Sarfraz's age.

The total of Fred's age, Mary's age and Sarfraz's age is 64 years.
**b** Form an equation and solve it to find Fred's age.

**8** These four blocks $A$, $B$, $C$ and $D$ have a total mass of 62 grams.
The mass, in grams, of each block is shown on the diagram.

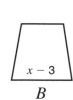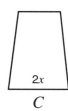

Diagram **NOT** accurately drawn

|   |   |   |   |
|---|---|---|---|
| $x$ | $x - 3$ | $2x$ | 5 |
| $A$ | $B$ | $C$ | $D$ |

**a** Express this information as an equation in terms of $x$.
**b** Solve your equation and write down the masses of blocks $A$, $B$ and $C$.        (1385 November 2001)

**9** $ABCD$ is a quadrilateral.
Work out the size of the largest angle in the quadrilateral.

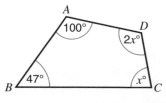

(1387 June 2004)

**10** The lengths, in cm, of the sides of the triangle are $3(x - 3)$, $4x - 1$ and $2x + 5$
**a** Write down, in terms of $x$, an expression for the perimeter of the triangle.

The perimeter of the triangle is 49 cm.
**b** Work out the value of $x$.

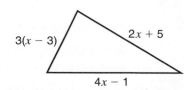

(1388 June 2004)

**11** **a** Solve $6(2x + 3) = 2(4x + 7)$     **b** Solve $\dfrac{x}{4} + 2 = 7$     **c** Solve $\dfrac{3x}{5} = x - 3$

**12** Solve $2x + 1 = \dfrac{5x}{3}$     (1388 March 2004)

**13** **a** Solve $20y - 16 = 18y - 9$     **b** Solve $\dfrac{40 - x}{3} = 4 + x$     (1387 June 2004)

**14** Solve $7(x + 2) = \dfrac{5x + 1}{2}$     (1388 January 2004)

**15** Solve $5(x + 8) = \dfrac{7x - 4}{2}$     (1388 March 2005)

**16** Solve $\dfrac{x}{3} - 5 = 3(x - 2)$     (1388 March 2003)

**17** $ABC$ is an isosceles triangle.
$AB = AC$
Angle $A = x°$

   **a** Find an expression, in terms of $x$, for the size of angle $B$.

   **b** Solve the simultaneous equations

$$3p + q = 11$$
$$p + q = 3$$

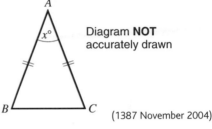

Diagram **NOT** accurately drawn

(1387 November 2004)

**18** Solve the simultaneous equations

$$2x + y = 4$$
$$5x - y = 17$$

(1388 March 2003)

**19** Solve     $x + 2y = 4$
$$3x - 4y = 7$$

(1387 June 2005)

**20** Solve the simultaneous equations

$$6x - 2y = 33$$
$$4x + 3y = 9$$

(1387 June 2004)

**21** Solve the simultaneous equations

$$3x - 4y = 11$$
$$5x + 6y = 12$$

(1388 January 2003)

**22** Solve the simultaneous equations

$$5p + 4q = -4$$
$$2p + 3q = 0.5$$

**23** A company makes compact discs (CDs).
The total cost, $P$ pounds, of making $n$ compact discs is given by the formula

$$P = a + bn$$

where $a$ and $b$ are constants.
The cost of making 1000 compact discs is £58 000
The cost of making 2000 compact discs is £64 000

   **a** Calculate the values of $a$ and $b$.

The company sells the compact discs at £10 each.
The company does not want to make a loss.

   **b** Work out the minimum number of compact discs the company must sell.     (1385 June 1998)

# Percentages

## 24.1 Percentages

also assessed in
Module 3

'…thank you for your contribution this year. Your salary will increase by 6% …'

The pictures show examples of the use of **percentages**.

**Per cent** means 'out of 100'.

35 per cent means 35 out of 100 or as a fraction $\frac{35}{100}$

35 per cent is written as 35%.

Learn these percentages with their fraction and decimal equivalents.

| Percentage | Fraction | Decimal |
|---|---|---|
| 1% | $\frac{1}{100}$ | $\frac{1}{100} = 0.01$ |
| 10% | $\frac{10}{100} = \frac{1}{10}$ | $\frac{10}{100} = 0.1$ |
| 20% | $\frac{20}{100} = \frac{1}{5}$ | $\frac{20}{100} = 0.2$ |
| 25% | $\frac{25}{100} = \frac{1}{4}$ | $\frac{25}{100} = 0.25$ |
| 50% | $\frac{50}{100} = \frac{1}{2}$ | $\frac{50}{100} = 0.5$ |
| 75% | $\frac{75}{100} = \frac{3}{4}$ | $\frac{75}{100} = 0.75$ |
| 100% | $\frac{100}{100} = 1$ | $\frac{100}{100} = 1$ |

### Example 1

Write each of these percentages as    **i** a fraction in its simplest form    **ii** a decimal

**a** 45%          **b** 3%          **c** $17\frac{1}{2}$%

### Solution 1

**a** **i** $45\% = \frac{45}{100}$

The numerator is 45 and the denominator is 100

$$\frac{45}{100} \overset{\div 5}{\underset{\div 5}{=}} \frac{9}{20}$$

5 is the highest common factor of 45 and 100
Divide both 45 and 100 by 5 to get $\frac{9}{20}$
9 and 20 do not have a common factor and so $\frac{9}{20}$ is in its simplest form.

**ii** $45\% = \frac{45}{100} = 0.45$

Change the fraction into a decimal.

**b** **i** $3\% = \frac{3}{100}$

$\frac{3}{100}$ cannot be simplified as there is no number that divides exactly into both 3 and 100

**ii** $3\% = \frac{3}{100} = 0.03$

Write the 3 in the hundredths column and a zero in the tenths column.

**c  i** $17\frac{1}{2}\% = \dfrac{17\frac{1}{2}}{100}$

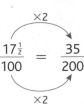

$\dfrac{17\frac{1}{2}}{100} = \dfrac{35}{200}$

> $17\frac{1}{2}\%$ means $17\frac{1}{2}$ out of 100

> Multiply both $17\frac{1}{2}$ and 100 by 2 to change the $17\frac{1}{2}$ to a whole number.

$\dfrac{35}{200} = \dfrac{7}{40}$

> 5 is the highest common factor of 35 and 200
> Divide both 35 and 200 by 5 to get $\frac{7}{40}$
> 7 and 40 do not have a common factor and so $\frac{7}{40}$ is in its simplest form.

**ii** $17\frac{1}{2}\% = 17.5\% = \dfrac{17.5}{100}$

> $\frac{1}{2}$ is the same as 0.5

**Method 1**

$\dfrac{17.5}{100} = 17.5 \div 100$

$17\frac{1}{2}\% = 0.175$

> To work out $17.5 \div 100$ without a calculator move each figure of the 17.5 two places to the right as described in Section 7.5.

**Method 2**

$\dfrac{17.5}{100} = \dfrac{175}{1000}$

$17\frac{1}{2}\% = 0.175$

> Alternatively multiply both 17.5 and 100 by 10 and then write $\frac{175}{1000}$ as a decimal.

**Method 3**

$\dfrac{17.5}{100} = 0.175$

> With a calculator work this out by keying in
>

To find a percentage of a quantity the percentage should first be written as a fraction or a decimal.

## Example 2

**a** Find 25% of 60

**b** Find 16% of 75

*Solution 2*

**a** If a percentage can be written as a simple fraction then use this fraction to work out a percentage of a quantity.

$25\% = \frac{1}{4}$

> This fact should be known.

$60 \div 4 = 15$

> To find $\frac{1}{4}$ of 60, divide 60 by 4

25% of 60 = 15

$\frac{1}{4}$ of 60 is 15 and $\frac{1}{4} \times 60$ is also 15

**b**

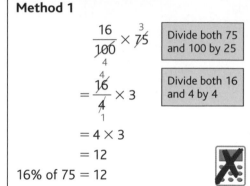

| Method 1 | Method 2 |
|---|---|
| $\dfrac{16}{100} \times \overset{3}{\cancel{75}}$ ・ Divide both 75 and 100 by 25 $=\dfrac{\overset{4}{\cancel{16}}}{\cancel{4}_1} \times 3$ ・ Divide both 16 and 4 by 4 $=4 \times 3$ $=12$ 16% of 75 = 12 | 16% = 0.16 0.16 × 75 = 12 Key in ⓪ ⓪ ① ⑥ ⓧ ⑦ ⑤ ⓿ 16% of 75 = 12 |

## Example 3

Colin invests £1850
The interest rate is 5.2% per year.
How much interest will Colin receive after 1 year?

*Solution 3*

5.2% of 1850

$= \dfrac{5.2}{100} \times 1850$

$= 96.2$

> Key in
> ⑤ ⓪ ② ⓓ ① ⓪ ⓪ ⓧ ① ⑧ ⑤ ⓪ ⓿
> or change 5.2% to a decimal and work out 0.052 × 1850

Colin receives £96.20 interest after 1 year.

> As the answer is an amount of money, there must be two figures after the decimal point.

### Exercise 24A

**1** Write each percentage as a fraction.

    **a** 79%      **b** 9%      **c** 57%      **d** 1%      **e** 7%

**2** Write each percentage as a fraction in its simplest form.

    **a** 50%      **b** 25%      **c** 75%      **d** 30%      **e** 2%

    **f** 84%      **g** 95%      **h** $12\frac{1}{2}$%      **i** $7\frac{1}{4}$%      **j** 4.2%

**3** Write each percentage as a decimal.

    **a** 63%      **b** 98%      **c** 7%      **d** 25%      **e** 60%      **f** 23.5%

**4** 28% of children walk to school. What fraction of children walk to school?
Give your fraction in its simplest form.

**5** 92% of students in a class have a mobile telephone.
Work out the fraction of students that do *not* have a mobile telephone.
Give your fraction in its simplest form.

**6** Work out

    **a** 50% of 300      **b** 25% of 40      **c** 50% of 184      **d** 10% of 60      **e** 25% of 12

    **f** 20% of 30      **g** 50% of 18      **h** 10% of 15      **i** 20% of 50      **j** 75% of 40

**7** Work out
   **a** 20% of £300     **b** 25% of 60 g     **c** 20% of 80 cm     **d** 75% of 400 m
   **e** 80% of £30      **f** 4% of 700 g     **g** 12% of 300 kg    **h** 65% of 300 km

**8** Tony earns £400. He gets a bonus of 15%. Work out Tony's bonus.

**9** There are 150 shop assistants in a large store. 8% of the shop assistants are male.
   How many of the shop assistants are male?

**10** Danya invests £250   The interest rate is 4% per year.
   How much interest will she receive after 1 year?

**11** In Year 11 there are 154 students. 84 of these students are girls. 50% of the girls and 10% of the
   boys attend Spanish lessons. What fraction of these Year 11 students attend Spanish lessons?
   Give your fraction in its simplest form.

**12** Work out
   **a** 34% of 50 m       **b** 12% of £36      **c** 5% of £32.40      **d** 8% of 62 kg
   **e** 95% of £23 000    **f** 14% of 90 m     **g** 4.2% of 60 km     **h** 3.2% of £14 000
   **i** $17\frac{1}{2}$% of £300   **j** $6\frac{1}{4}$% of 40 cm

**13** Anna scored 65% in a test. The test was out of 40 marks. How many marks did Anna score?

**14** There are 2400 students in a school. 17% of the students in the school wear glasses.
   How many of the students wear glasses?

**15** A shop has 4600 DVDs. 23% of the DVDs are thrillers.
   How many of the DVDs in the shop are thrillers?

**16** The price of a washing machine is £520   Ahmid pays a deposit of 35% of the price.
   Work out his deposit.

**17** Martin invests £3500   The interest rate is 4.3% per year.
   How much interest will he receive at the end of 1 year?

## 24.2 Increases and decreases

A **multiplier** is a single number that an amount is multiplied by in order to increase or decrease that
amount.

---

### Example 4

Write down the multiplier that can be used to

**a** increase an amount by 9%              **b** decrease an amount by 12%

*Solution 4*

**a** 100% + 9% = 109%

$$109\% = \frac{109}{100} = 1.09$$

> Think of the original amount as 100%, add 9% to the original amount.
> Write the new percentage as a fraction.
> Divide by 100 to write the fraction as a decimal.

The multiplier that can be used to increase an amount by 9% is 1.09

**b** 100% − 12% = 88%

$$88\% = \frac{88}{100} = 0.88$$

> The original amount is 100%, subtract 12% from the original amount.
> Write the new percentage as a fraction.
> Divide by 100 to write the fraction as a decimal.

The multiplier that can be used to decrease an amount by 12% is 0.88

---

These examples show how to increase and decrease amounts by a given percentage, including the use of multipliers.

### Example 5

Hugh's salary is £25 000 a year. His salary is increased by 4%
Work out his new salary.

**Solution 5**

**Method 1**
4% of £25 000

| This is the increase in Hugh's salary. |

$$= \frac{4}{100} \times 25\,000$$

$$= \frac{4}{\overset{}{\underset{1}{100}}} \times 25\,\overset{250}{\cancel{000}}$$

$$= 4 \times 250$$

$$= 1000$$

| His increase in salary is £1000 |

$25\,000 + 1000 = 26\,000$

| Add the increase to his original salary. |

Hugh's new salary is £26 000

**Method 2**
$100\% + 4\% = 104\%$

| His new salary is 104% of £25 000 |

$$104\% = \frac{104}{100} = 1.04$$

| 1.04 is the multiplier. |

$1.04 \times 25\,000$

| Find 104% of 25 000
(This increases 25 000 by 4%) |

$$= 26\,000$$

Hugh's new salary is £26 000

### Example 6

The value of a car depreciates by 15% each year. The value of a car when new is £14 000
Work out the value of the car after 1 year.

**Solution 6**

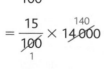

**Method 1**
15% of £14 000

| **Depreciates** means that the value of the car decreases. This is the decrease in the value of the car. |

$$= \frac{15}{100} \times 14\,000$$

| Write the percentage as a fraction. |

$$= \frac{15}{\overset{}{\underset{1}{100}}} \times 14\,\overset{140}{\cancel{000}}$$

| Work out the multiplication. |

$$= \frac{15}{1} \times 140$$

$$= 2100$$

| The depreciation in 1 year is £2100 |

$14\,000 - 2100 = 11\,900$

| Subtract to work out the new value. |

Value after 1 year = £11 900

## Method 2

| | |
|---|---|
| $100\% - 15\% = 85\%$ | The final value is 85% of the original value. |
| $85\% = 0.85$ | 0.85 is called the multiplier. |
| $0.85 \times 14\,000$ | Find 85% of 14 000 to reduce 14 000 by 15% |
| $= 11\,900$ | With this method the actual decrease in the value is not found. |

Value after 1 year $= £11\,900$

## Exercise 24B

**1** Write down the multiplier that can be used to work out an increase of

    **a** 64%        **b** 3%        **c** 14%        **d** 40%

    **e** 13.4%     **f** $12\frac{1}{2}\%$    **g** 100%     **h** 450%

**2** Write down the multiplier that can be used to work out a decrease of

    **a** 7%        **b** 20%       **c** 16%       **d** 27%

    **e** 5.6%      **f** $2\frac{1}{2}\%$    **g** $7\frac{1}{4}\%$    **h** 0.8%

**3** Write down the percentage increase represented by each of the following multipliers.

    **a** 1.06      **b** 1.43      **c** 1.24      **d** 1.01

    **e** 1.035     **f** 1.089     **g** 3        **h** 2.5

**4** Write down the percentage decrease represented by each of the following multipliers.

    **a** 0.85      **b** 0.94      **c** 0.4       **d** 0.75

    **e** 0.65      **f** 0.925     **g** 0.757    **h** 0.995

**5** Jeevan earns £200 per week. He gets a wage rise of 10%.
    How much does Jeevan earn per week after his rise?

**6** In a sale all prices are reduced by 15%. Work out the sale price of each of the following

    **a** a television set that normally costs £300
    **b** a CD player that normally costs £40
    **c** a computer that normally costs £1200

**7** The table shows the salaries of three workers.
Each worker receives a 5% salary increase.
Work out the new salary of each worker.

| Helen | £12 000 |
|---|---|
| Tom | £24 000 |
| Sandeep | £32 000 |

**8** Hanni invests £3000   The interest rate is 4% per year.
How much will Hanni have in his bank account at the end of 1 year?

**9** The price of a computer is £450   Its price is reduced by 15% in a sale.
Work out the sale price of the computer.

**10** Jenny puts £600 into a bank account. At the end of 1 year 3.5% interest is added.
How much is in her account at the end of 1 year?

**11** A holiday normally costs £850   It is reduced by 12%. How much will the holiday now cost?

**12** A year ago, the value of Richard's house was £85 000   Its value has now increased by 9%.
Work out the value of Richard's house now.

**13** Ria buys a car for £2300   The value of the car depreciates by 20% each year.
Work out the value of the car at the end of 1 year.

**14** Value Added Tax (VAT) is charged at a rate of $17\frac{1}{2}$%. A builder's bill is £962 before VAT is added.
What will the bill be after VAT has been added?

**15** Pat normally pays £45 for her train fare. All train fares are increased by 9%.
How much will Pat now have to pay?

**16** The total price of a radio is £46.80 plus VAT at 17.5%. Work out the *total* price of the radio.

To write one quantity as a percentage of another quantity:
- write down the first quantity as a fraction of the second quantity
- change the fraction to a percentage

## Example 7

**a** Change 11 out of 20 to a percentage. Do not use a calculator.

**b** Change 23 out of 40 to a percentage. You may use a calculator.

*Solution 7*

**a** $\dfrac{11}{20}$

| Write the first number as a fraction of the second number. |

$$\dfrac{11}{20} \overset{\times 5}{\underset{\times 5}{=}} \dfrac{55}{100}$$

| Change $\frac{11}{20}$ to a fraction with a denominator of 100 |

$$\dfrac{55}{100} = 55\%$$

11 out of 20 = 55%

**b** $\dfrac{23}{40}$

| Write the first number as a fraction of the second number. |

$$\dfrac{23}{40} = 0.575 = 57.5\%$$

| Change the fraction to a decimal and then to a percentage. |

23 out of 40 = 57.5%

In **a** $\dfrac{11}{20} \times 100 = 55$ and in **b** $\dfrac{23}{40} \times 100 = 57.5$

So to change a fraction to a percentage, multiply the fraction by 100

## Example 8

Express 45 cm as a percentage of 2 m.

### Solution 8

There are two different types of units in the question — centimetres and metres.
**Quantities must have the same units before the percentage is found.**
In this case express both lengths in centimetres.

2 m = 200 cm            | Use 1 m = 100 cm |

$\dfrac{45}{200}$            | Write 45 as fraction of 200 |

$\dfrac{45}{200} \times 100 = 22.5\%$            | Multiply $\frac{45}{200}$ by 100 to change it to a percentage. |

45 cm is 22.5% of 2 m.

Percentage problems sometimes involve percentage profit or percentage loss where

$$\text{Percentage profit} \quad = \quad \frac{\text{profit}}{\text{original amount}} \times 100\%$$

$$\text{Percentage loss} \quad = \quad \frac{\text{loss}}{\text{original amount}} \times 100\%$$

## Example 9

Karen bought a car for £1200   One year later she sold it for £840
Work out her percentage loss. (Do not use a calculator.)

### Solution 9

1200 − 840 = 360            | Subtract the selling price from the original price to find her loss. |

$\dfrac{360}{1200}$            | Write down the fraction $\dfrac{\text{loss}}{\text{original price}}$

It is not necessary to simplify the fraction yet. |

$\dfrac{360}{1200} \times \overset{1}{\underset{12}{100}} = 30\%$            | Multiply $\frac{360}{1200}$ by 100 to change it to a percentage. |

Her percentage loss is 30%.

## Example 10

Tony bought a box of 24 oranges for £4   He sold all the oranges for 21p each.
Work out his percentage profit.

### Solution 10

In this question there are two different units, pounds and pence.
Either pounds **or** pence may be used but it must be the same throughout the question.

24 × 21 = 504            | Work out the total amount in pence Tony received from selling all the oranges. |

504 − 400 = 104p profit            | Subtract the original price from the selling price to find his profit in pence. |

$\dfrac{104}{400}$            | Write down the fraction $\dfrac{\text{profit}}{\text{original price}}$ working in pence. |

$\dfrac{104}{400} \times 100 = 26\%$            | Multiply $\frac{104}{400}$ by 100 to change it to a percentage. |

Percentage profit = 26%.

An **index number** is used to give a measure of how a value has changed. The index number is based on using 100 to represent the value in a particular year.

For example in 2004 a man's daily pay was £200  In 2005 it was £208

208 as a percentage of 200 is $\dfrac{208}{200} \times 100 = 104\%$

Taking 2004 as the base year with an index number of 100, the index number in 2005 is 104

## Example 11

In 2000 a holiday cost £650  In 2005 an identical holiday cost £728
**a**  Express the cost of the holiday in 2005 as a percentage of the cost in 2000
**b**  If the index number in the year 2000 is 100 write down the index number of the holiday in 2005

### Solution 11

**a**  $\dfrac{728}{650} \times 100 = 112\%$

The price in 2005 is 112% of the price in the year 2000

**b**  The index number is 112    | To write down the index number just omit the percentage sign from the answer to **a**. |

## Example 12

In 2005 a Deltan computer cost £500   In 2006 the Deltan computer cost £450
Taking 2005 as the base year with an index number of 100 find the index number in 2006

### Solution 12

$\dfrac{450}{500} \times 100 = 90\%$    | Express the cost of the computer in 2006 as a percentage of its cost in 2005 |

The index number in 2006 is 90

## Exercise 24C

**1**  Write
   **a**  £3 as a percentage of £6
   **c**  4p as a percentage of 10p
   **e**  £80 as a percentage of £400
   **g**  7 kg as a percentage of 35 kg
   **i**  90 km as a percentage of 100 km

   **b**  2 kg as a percentage of 8 kg
   **d**  8 cm as a percentage of 40 cm
   **f**  £3 as a percentage of £30
   **h**  3 m as a percentage of 20 m
   **j**  £36 as a percentage of £48

**2**  Write
   **a**  20p as a percentage of £2
   **c**  600 g as a percentage of 1 kg
   **e**  60p as a percentage of £2.40
   **g**  36 minutes as a percentage of 1 hour

   **b**  25 cm as a percentage of 1 m
   **d**  800 m as a percentage of 1 km
   **f**  15 mm as a percentage of 6 cm
   **h**  50 cm as a percentage of 4 m

**3**  Janet scored 36 out of 40 in a German test. Work out her score as a percentage.

**4**  In Year 10 there are 240 students. 150 of these students are boys.
   What percentage of Year 10 students are    **a**  boys    **b**  girls?

**5** Jerry took 60 bottles to a bottle bank. 27 of the bottles were green.
What percentage of the bottles were green?

**6** There are 80 pages in a book. 30 of the pages have pictures on them.
What percentage of the pages in the book have pictures on them?

**7** Mr Potter buys a house for £200 000   Five years later the value of the house has increased to
£256 000   Work out the percentage increase in the value of the house.

**8** A shopworker's wage was £240 per week. After a pay rise her wage is £254.40
Work out the percentage increase.

**9** Trevor bought a new car for £16 000   He sold it after two years for £11 200
Work out his percentage loss.

**10** In a sale the price of a clock is reduced from £32 to £27.20   Work out the percentage reduction.

**11** In a survey 1440 out of 2500 people surveyed said that they were going on holiday in the summer.
**a** What percentage were going on holiday?
**b** What percentage were *not* going on holiday?

**12** Rob bought a crate of 40 melons for £30   He sold all the melons for £1.05 each.
Work out his percentage profit.

**13** A box of cereal weighs 750 g. It contains 210 g of dried fruit.
What percentage of the cereal is dried fruit?

**14** A 40 g serving of cereal contains 8 g of protein, 24 g of carbohydrates, 4.5 g of fat and 3.5 g of fibre.
What percentage of the serving is
**a** protein          **b** carbohydrates
**c** fat               **d** fibre?

**15** In 2000 the value of a house was £85 000   In 2005 its value was £102 000
**a** Work out the value of the house in 2005 as a percentage of the value of the house in 2000
**b** If the index number in 2000 is 100 write down the index number in 2005

**16** In 2000 a train journey cost £85   In 2003 the same train journey cost £98.60
In 2005 the same train journey cost £107.95
**a** Work out the cost of the train journey in 2003 as a percentage of the cost of the train journey
in 2000
**b** Write down the index number in 2003 based on an index of 100 in 2000
**c** Work out the cost of the train journey in 2005 as a percentage of the cost of the train journey
in 2000
**d** Write down the index number in 2005 based on an index of 100 in 2000

**17** In 2002 the price of an 'Olympic' camera was £250   In 2003, the price of the camera was £225
and in 2004 the price of the camera was £220
**a** Work out the price of the camera in 2003 as a percentage of its price in 2002
**b** Based on an index of 100 in 2002
   **i** write down the index number in 2003
   **ii** find the index number in 2004

## 24.3 Use of multipliers

Banks and building societies pay **compound interest**.
At the end of the first year, interest is paid on the money in an account. This interest is then added to the account. At the end of the second year interest is paid on the *total amount in the account*, that is, the original amount of money plus the interest earned in the first year.

At the end of each year, interest is paid on the *total* amount in the account at the start of that year.

If £200 is invested in a bank account for one year and interest is paid at a rate of 5% then, using the multiplier 1.05

- after 1 year there will be a total of £(200 × 1.05) in the account
- after 2 years there will be a total of £((200 × 1.05) × 1.05) in the account
- after 3 years there will be a total of £(((200 × 1.05) × 1.05) × 1.05) in the account.

To find the amount in the account after **3** years the original £200 is multiplied by 1.05 × 1.05 × 1.05 which is equivalent to $1.05^3$

$1.05^3 = 1.157\,625$ is the single number that £200 is multiplied by to find the amount in the bank account after **3** years.

In general to work out the amount in a savings account after $n$ years if interest is paid at $r$ % per annum, multiply the original amount by $\left(1 + \dfrac{r}{100}\right)^n$

### Example 13

£4000 is invested for 2 years at 5% per annum **compound interest**.
Work out the **total interest** earned over the 2 years.

*Solution 13*

**Method 1** – using multipliers

100% + 5% = 105%

105% = 1.05          | Work out the multiplier for an increase of 5%. |

$4000 \times 1.05^2 = 4410$          | Multiply the original amount by $1.05^2$ to find the amount in the account after **2** years. |

4410 − 4000 = 410          | Subtract the original amount to find the interest. |

The total interest earned over the 2 years is £410

**Method 2** – repeated increase

$\dfrac{5}{100} \times 4000 = 200$          | Work out the interest in the first year. |

4000 + 200 = 4200          | Add the interest to the original amount. |

$\dfrac{5}{100} \times 4200 = 210$          | Work out the interest in the second year. |

200 + 210 = 410          | Find the total interest. |

The total interest earned over the 2 years is £410

## Example 14

**a** Each year the value of a car depreciates by 30%. Find the single number as a decimal that the value of the car can be multiplied by to find its value at the end of 4 years.

**b** The value of a house increases by 16% of its value at the beginning of the year. The next year its value decreases by 3% of its value at the start of the second year. Find the single number as a decimal that the original value of the house can be multiplied by to find its value at the end of the 2 years.

### Solution 14

**a** $100\% - 30\% = 70\%$

| | |
|---|---|
| | Find the multiplier that represents a decrease of 30%. |

$70\% = \dfrac{70}{100} = 0.7$

$0.7 \times 0.7 \times 0.7 \times 0.7 = 0.7^4$

| | |
|---|---|
| | The depreciation is over **4** years so the single multiplier is 0.7 raised to the power of **4** |

$0.7^4 = 0.2401$

0.2401 is the single number

**b** $100\% + 16\% = 116\%$

| | |
|---|---|
| | Find the multiplier for an increase of 16%. |

$116\% = \dfrac{116}{100} = 1.16$

$100\% - 3\% = 97\%$

| | |
|---|---|
| | Find the multiplier for a decrease of 3%. |

$97\% = \dfrac{97}{100} = 0.97$

$1.16 \times 0.97 = 1.1252$

1.1252 is the single number

| | |
|---|---|
| | The value increases and then decreases so find the product of the two multipliers. |

## Example 15

The value of a machine when new is £8000   The value of the machine depreciates by 10% each year. Work out its value after 3 years.

### Solution 15

**Method 1** – using multipliers

$100\% - 10\% = 90\%$

| | |
|---|---|
| | Work out the multiplier for a decrease of 90%. |

$90\% = 0.9$

$8000 \times 0.9^3 = £5832$

| | |
|---|---|
| | Multiply the value when new by $0.9^3$ to find the value after **3** years. |

The value of the machine after 3 years is £5832

**Method 2** – repeated decrease

$\dfrac{10}{100} \times 8000 = 800$

| | |
|---|---|
| | Work out the depreciation in the first year. |

$8000 - 800 = 7200$

| | |
|---|---|
| | Work out the value of the machine at the end of the first year. |

$\dfrac{10}{100} \times 7200 = 720$

| | |
|---|---|
| | Work out the depreciation in the second year. |

$7200 - 720 = 6480$

| | |
|---|---|
| | Work out the value of the machine at the end of the second year. |

$\dfrac{10}{100} \times 6480 = 648$

| | |
|---|---|
| | Work out the depreciation in the third year. |

$6480 - 648 = 5832$

| | |
|---|---|
| | Work out the value of the machine at the end of the third year. |

The value of the machine after 3 years is £5832

**Example 16**

The population of an island is 56 000
The population is increasing at a rate of 8% per year.
After how many years will the population first exceed 100 000?

*Solution 16*

$100\% + 8\% = 108\%$

$\dfrac{108}{100} = 1.08$

> Find the multiplier that represents an increase of 8%.
> Choose a number of years to use in a first trial, say 5

$56\,000 \times 1.08^5 = 82\,282.37\,...$

> Work out the population after 5 years.
> The population is less than 100 000

$56\,000 \times 1.08^7 = 95\,974.15\,...$

> Work out the population after 7 years.
> The population is still less than 100 000

$56\,000 \times 1.08^8 = 103\,652.09\,...$

> Work out the population after 8 years.
> The population is now greater than 100 000

The population first exceeds 100 000 after 8 years.

**Exercise 24D**

1  Work out the multiplier as a single decimal number that represents
   **a**  an increase of 20% for 3 years
   **b**  a decrease of 10% for 4 years
   **c**  an increase of 6% for 2 years
   **d**  a decrease of 15% for 3 years
   **e**  an increase of 20% followed by a decrease of 8%
   **f**  a decrease of 8% followed by a decrease of 20%
   **g**  an increase of 4% followed by an increase of 2%
   **h**  a decrease of 35% followed by a decrease of 20%

2  Ben says that an increase of 40% followed by an increase of 20% is the same as an increase of 60%. Is Ben correct? You must give a reason for your answer.

3  £1000 is invested for 2 years at 5% per annum **compound interest**.
   Work out the total amount in the account after 2 years.

4  £3000 is invested for 4 years at 7% per annum **compound interest**.
   Work out the **total interest** earned over the 4 years.

5  A motorbike is worth £6500   Each year the value of the motorbike depreciates by 35%.
   Work out the value of the motorbike at the end of three years.

6  A house is worth £175 000   Its value increases by 6% each year.
   Work out the value of the house after
   **a**  3 years              **b**  10 years              **c**  25 years.
   Give your answers to the nearest pound.

7  The population of a town is 60 000   The population is increasing at a rate of 13% per year.
   Work out the population of the town after 4 years.

**8** Mrs Bell buys a house for £60 000   In the first year the value of the house increases by 16%. In the second year the value of the house decreases by 4% of its value at the beginning of that year.

   **a** Write down the single number as a decimal that the original value of the house can be multiplied by to find its value after 2 years.

   **b** Work out the value of the house after the 2 years.

**9** Jeremy deposits £3000 in a bank account. Compound interest is paid at a rate of 4% per annum. Jeremy wants to leave the money in the account until there is at least £4000 in the account. Calculate the least number of years Jeremy must leave his money in the bank account.

**10** £500 is invested in a savings account. Compound interest is paid at a rate of 5.5% per annum. Calculate the least number of years it will take for the original investment to double in value.

## 24.4  Reverse percentages

Multipliers can be used to find the original quantity if the final value after a percentage increase or decrease is known.

### Example 17

In a sale all the jackets are reduced by 20%.
The sale price of a jacket is £33.60
Work out the original price of the jacket.

*Solution 17*
**Method 1** – using multipliers

$100\% - 20\% = 80\%$

| | |
|---|---|
| $\dfrac{80}{100} = 0.8$ | Find the multiplier for a **decrease** of 20%. |

Let the original price be $x$

| | |
|---|---|
| $x \times 0.8 = 33.60$ | The original price was multiplied by 0.8 to give £33.60   Write this as an equation. |
| $x = \dfrac{33.60}{0.8}$ | Solve the equation. |
| $= 42$ | |

The original price of the jacket was £42

(Check: $42 \times 0.8 = 33.6$)

**Method 2** – using percentages

$100\% - 20\% = 80\%$

| | |
|---|---|
| $\dfrac{33.60}{80}$ | £33.60 represents 80% of the original price. |
| | Divide 33.60 by 80 to find the value of 1%. |
| $\dfrac{33.60}{80} \times 100 = 42$ | The original price is 100% so multiply the amount that represents 1% by 100 |

The original price of the jacket was £42

(Check: $42 \times 0.8 = 33.6$)

## Example 18

The price of a new washing machine is £376
This price includes Value Added Tax (VAT) at $17\frac{1}{2}$%.
Work out the cost of the washing machine *before* VAT was added.

### Solution 18
**Method 1** – using multipliers
$100\% + 17.5\% = 117.5\%$

| The original cost was increased by 17.5% so find the multiplier for an **increase** of 17.5%. |

$$\frac{117.5}{100} = 1.175$$

Let the original price be $x$

| The original cost was multiplied by 1.175 to give £376 Write this as an equation. |

$$x \times 1.175 = 376$$

$$x = \frac{376}{1.175}$$

| Solve the equation. |

$$= 320$$

The cost of the washing machine before VAT was added was £320

| (Check: 320 × 1.175 = 376) |

**Method 2** – using percentages
$100\% + 17.5\% = 117.5\%$

| £376 represents 117.5% of the original cost. |

$$\frac{376}{117.5}$$

| Divide 376 by 117.5 to find the value of 1%. |

$$\frac{376}{117.5} \times 100 = 320$$

| The original cost is 100% so multiply the amount that represents 1% by 100 |

The cost of the washing machine before VAT was added was £320

| (Check: 320 × 1.175 = 376) |

## Exercise 24E

1   In a sale all the prices are reduced by 25%. The sale price of a dress is £30
    Work out the normal price of the dress.

2   Employees at a firm receive a pay increase of 4%. After the pay increase Linda earns £24 960
    How much did Linda earn before the pay increase?

3   The price of a new television set is £329   This price includes Value Added Tax (VAT) at $17\frac{1}{2}$%.
    Work out the cost of the television set *before* VAT was added.

4   A company bought a new lorry. Each year the value of the lorry depreciates by 20%.
    After 1 year, the lorry was worth £26 000   Work out the original price of the lorry.

5   A holiday is advertised at a price of £403   This represents a 35% saving on the brochure price.
    Work out the brochure price of the holiday.

6   Kunal pays tax at a rate of 22%. After he has paid tax Kunal received £140.40 per week.
    How much does Kunal earn per week before he pays tax?

7   A large firm hires 3% more workers which brings its total number of workers to 12 772
    How many workers did the firm have before the increase?

**8** In 1 year the population of an island increased by 3.2% to 434 472
Work out the population of the island before the increase.

**9** Tasha invests some money in a bank account. Interest is paid at a rate of 8% per annum. After 1 year there is £291.60 in the account. How much money did Tasha invest?

**10** Javed invests some money in a bank account. Compound interest is paid at a rate of 4% per annum. After 2 years there is £3028.48 in the account. How much did Javed invest?

# Chapter summary

**You should now know that:**

★ **per cent** means 'out of 100'
- 35 per cent means 35 out of 100 or as a fraction, $\frac{35}{100}$
- 35 per cent is written as 35%

★

| Percentage | Fraction | Decimal |
|---|---|---|
| 1% | $\frac{1}{100}$ | $\frac{1}{100} = 0.01$ |
| 10% | $\frac{10}{100} = \frac{1}{10}$ | $\frac{10}{100} = 0.1$ |
| 20% | $\frac{20}{100} = \frac{1}{5}$ | $\frac{20}{100} = 0.2$ |
| 25% | $\frac{25}{100} = \frac{1}{4}$ | $\frac{25}{100} = 0.25$ |
| 50% | $\frac{50}{100} = \frac{1}{2}$ | $\frac{50}{100} = 0.5$ |
| 75% | $\frac{75}{100} = \frac{3}{4}$ | $\frac{75}{100} = 0.75$ |
| 100% | $\frac{100}{100} = 1$ | $\frac{100}{100} = 1$ |

★ an **index number** is used to give a measure of how a value has changed. The index number is based on using 100 to represent the value in a particular year.

★ a multiplier is a single number that a quantity can be multiplied by in order to increase or decrease the quantity.

**You should also be able to:**

★ work out the percentage of a quantity by changing the percentage into a fraction and multiplying the quantity by this fraction

★ change a fraction into a percentage by multiplying the fraction by 100%

★ work out the percentage profit or percentage loss

$$\text{Percentage profit} = \frac{\text{profit}}{\text{original amount}} \times 100\%$$

$$\text{Percentage loss} = \frac{\text{loss}}{\text{original amount}} \times 100\%$$

★ work out the amount in a savings account if interest is paid at $r\%$ for $n$ years by multiplying the original amount by $\left(1 + \frac{r}{100}\right)^n$

★ find the original quantity, given the final value after a percentage increase or decrease by dividing by the original multiplier.

# Chapter 24 review questions

**1 a** Write 0.45 as a percentage.

  **b** Write $\frac{3}{4}$ as a percentage.

  **c** Write 30% as a fraction in its simplest form.        (1388 January 2003)

**2** Work out 20% of 6500

**3 a** Write 45 cm as a percentage of 1 m     **b** Write 300 g as a percentage of 2 kg.

**4 a** Change $\frac{7}{8}$ to a decimal.

  **b** Use your answer to part **a** to write $\frac{7}{8}$ as a percentage.

**5** There are 800 students at Prestfield School. 144 of these students were absent from school on Wednesday.

  **a** Work out how many students were *not* absent on Wednesday.

Trudy says that more than 25% of the 800 students were absent on Wednesday.

  **b** Is Trudy correct? Explain your answer.

45% of these 800 students are girls.

  **c** Work out 45% of 800

There are 176 students in Year 10

  **d** Write 176 out of 800 as a percentage.        (1387 June 2004)

**6** Ben bought a car for £12 000
Each year the value of the car depreciated by 10%.
Work out the value of the car 2 years after he bought it.

(1387 June 2003)

**7** Work out 34% of 1500

**8** Jo got 36 out of 80 in an English test.

  **a** Work out 36 out of 80 as a percentage.

Jo got 65% of the total number of marks in a French test.
Jo got 39 marks.

  **b** Work out the total number of marks for the French test.     (1385 June 2000)

**9** Jane buys a box of 25 pens for £8.00
She sells all the pens at 36p each.
Work out her percentage profit.

**10** Annie bought 240 cans of drink for 25 pence each.
At a fete she sold half of these cans of drink for 45 pence each.
At a disco she sold 80 of these cans of drink for 30 pence each.
She gave the rest of the cans of drink away.
Work out the percentage profit that Annie made on the 240 cans of drink.

**11** A can of drink costs 32p.
The cost of the can of drink is increased to 38p.
Jenny calculates that this is a percentage increase of 19%.
Is Jenny's percentage correct?
You must show how you reached your decision.     (1388 January 2003)

**12** In a sale normal prices are reduced by 12%.
The normal price of a camera is £79
Work out the sale price of the camera. (1385 May 2002)

**13** The depreciation of a car is 20% each year.
The value of the car is £8500
Work out the value of the car at the end of 3 years.

**14** £200 is invested for 3 years at 5% per annum **compound interest**.
Work out the **total interest** earned over the 3 years.

(1385 November 2001)

**15** £5000 is invested for 3 years at 4% per annum **compound interest**.
Work out the **total interest** earned over the 3 years. (1385 June 2001)

**16** The price of a new television is £423
This price includes Value Added Tax (VAT) at $17\frac{1}{2}$%.

**a** Work out the cost of the television *before* VAT was added.

At the end of each year the value of a television has fallen by 12% of its value at the start of the year.
The value of a television was £423 at the start of the first year.

**b** Work out the value of the television at the end of the *third* year.
Give your answer to the nearest penny. (1385 June 2000)

**17** In a sale all prices are reduced by 15%.
The normal price of a jacket is £42
Syreeta buys the jacket in the sale.

**a** Work out the sale price of the jacket.

In the same sale, Winston pays £15.64 for a shirt.

**b** Calculate the normal price of the shirt.

(1385 June 2001)

**18** A company bought a van that had a value of £12 000
Each year the value of the van depreciates by 25%.

**a** Work out the value of the van at the end of 3 years.

The company bought a new truck.
Each year the value of the truck depreciates by 20%.
The value of the truck can be multiplied by a single number to find its value at the end of 4 years.

**b** Find this single number as a decimal. (1387 June 2004)

**19** Bill invests £500 on 1st January 2004 at a compound interest rate of $R$% per annum.
The value, £$V$, of this investment after $n$ years is given by the formula

$$V = 500 \times (1.045)^n$$

**a** Write down the value of $R$.

**b** Use your calculator to find the value of Bill's investment after 20 years. (1387 June 2005)

# Graphs (2)

## 25.1 Real life graphs

### Distance–time graphs

Yakub cycled from his home to his cousin's house.
On his way he waited for his sister before continuing his journey.
He and his sister stayed at his cousin's and then they returned home.
Here is a distance–time graph showing Yakub's complete journey.
The point, $A$, shows that Yakub left home at 12:30

The *straight line AB* shows that Yakub cycles, at a **constant speed**. He cycles 10 km in half an hour. At this speed he would cycle 20 km in 1 hour.

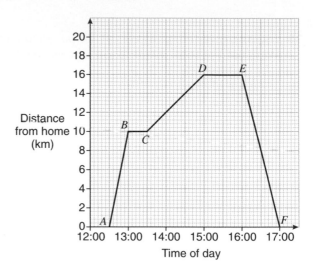

This is the same as saying that Yakub cycles at a constant speed of 20 kilometres per hour for the first half hour of his journey.

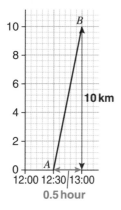

| Average speed = | $\dfrac{\textbf{total distance travelled}}{\textbf{total time taken}}$ |
|---|---|

See Section 9.1

The first part of the graph shows

Yakub's constant speed $= \dfrac{10 \text{ km}}{0.5 \text{ hour}} = 20$ km/h

On the distance–time graph this constant speed could be found by working

out $\dfrac{\text{vertical measurement } (= 10 \text{ km})}{\text{horizontal measurement } (= 0.5 \text{ hour})}$

which is a measure of the steepness of the line $AB$.

The *horizontal line BC* shows that, for the second half hour of his journey, (from 13:00 to 13:30) Yakub is *not moving*. He is still 10 km from home, waiting for his sister.

The *line CD* shows Yakub continues his journey to his cousin's house. His cousin's house is 16 km from home. He cycles the remaining 6 km (16 − 10) in 1.5 hours (from 13:30 to 15:00). During this part of the journey, Yakub cycles, at a constant speed. His speed is 4 kilometres per hour (6 ÷ 1.5 = 4).

The *horizontal line DE* shows that, for one hour, (from 15:00 to 16:00) Yakub is not moving and is still 16 km from home, at his cousin's house.

The *line EF* shows the return journey home. He arrives back home at 17:00 having cycled 16 km in 1 hour at a constant speed. His speed on this final part of his journey is 16 kilometres per hour.

**Example 1**

Peter and Jane go to the seaside, a distance of 30 kilometres from their home.
Peter leaves home on his bicycle at 12:30 pm.
Jane leaves later on her scooter.

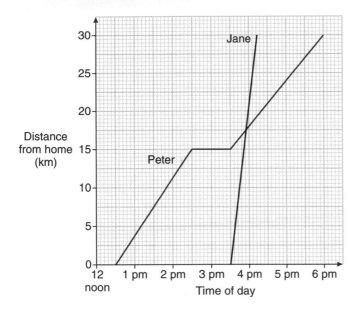

The travel graphs show some information about their journeys.

**a** Peter takes a break on his journey.
  **i**  For how long does Peter take a break?
  **ii** How far from the seaside is he when he takes his break?
**b** Peter cycles more quickly before the break than after it. Explain how the graph shows this.
**c** At what time does Jane leave home?
**d** Estimate the time at which Jane passes Peter?
**e** How many minutes before Peter does Jane arrive at the seaside?
**f** Estimate Jane's speed in km/h.

**Solution 1**

**a  i** 1 hour
   **ii** 30 − 15 = 15 kilometres

> Horizontal line on the graph shows Peter's break from 2:30 pm to 3:30 pm.

**b** The line before the break is steeper than the line after the break

**c** 3:30 pm

**d** about 3:55 pm

> Find where the two graphs cross.

**e** 1 hour 45 minutes = 105 minutes

> From 4:15 pm to 6 pm.

**f** Jane travels 20 km in $\frac{1}{2}$ hour
   Her speed is 40 km/h

> Find distance travelled in the half hour from 3:30 pm to 4 pm.

## Velocity–time graphs

A velocity–time graph shows how velocity changes with time.
When the velocity of a car increases steadily it is said to accelerate with a constant acceleration.

In general,

$$\text{constant acceleration} = \frac{\text{increase in velocity}}{\text{time taken}}$$

### Example 2

A tram travels between two stations.
The diagram represents the velocity–time graph of the tram.

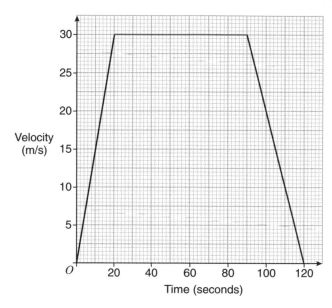

**a**  Write down the maximum velocity of the tram.
**b**  Find the constant acceleration of the tram during the first 20 seconds of the journey.
**c**  For how many seconds did the tram have zero acceleration?
**d**  Describe the journey of the tram for the final 30 seconds.

### Solution 2

**a**  30 m/s

> Highest value of velocity taken from the graph.

**b**  $\dfrac{30 - 0}{20 - 0} = 1.5$

> Use constant acceleration $= \dfrac{\text{increase in velocity}}{\text{time taken}}$

Constant acceleration $= 1.5$ m/s$^2$

> The units are m/s $\div$ s = m/s$^2$

**c**  $90 - 20 = 70$ seconds

> The tram has zero acceleration when the velocity doesn't *change*. The green line represents this part of the journey. From 20 seconds to 90 seconds the velocity is a constant value of 30 m/s.

**d**  The tram slows down steadily (deceleration) from a speed of 30 m/s and then finally stops (velocity = 0 m/s).

> As the red line is straight, the velocity decreases at a constant rate.

## Example 3

The diagram shows a rectangular tank filling up with water.

The diagonal of the surface of the water is of length $d$ when the height of the water is $h$.

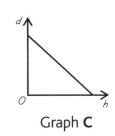

**a** Which of these three graphs describes the relationship between $d$ and $h$?

The water then leaks, at a constant rate, from a hole in the bottom of the tank.

**b** Sketch a graph which describes the relationship between the height of the water $h$ and time $t$.

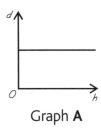 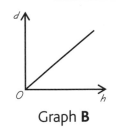 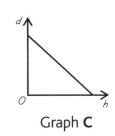

Graph **A**     Graph **B**     Graph **C**

### Solution 3

**a** Graph **A**    $d$ does not change as the height, $h$, of the water increases.

**b**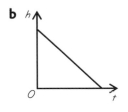

Since water leaks out at a constant rate, after each second the height decreases by the same amount.

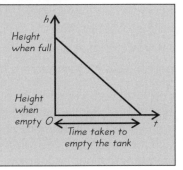

---

## Exercise 25A

**1** The travel graph shows some information about the flight of an aeroplane from London to Rome and back again.

  **a** At what time did the aeroplane arrive in Rome?

  **b** For how long did the aeroplane remain in Rome?

  **c** How many hours did the flight back take?

  **d** Work out the average speed, in kilometres per hour, of the aeroplane from London to Rome.

  **e** Estimate the distance of the aeroplane from Rome
    **i** at 12:30   **ii** at 15:12

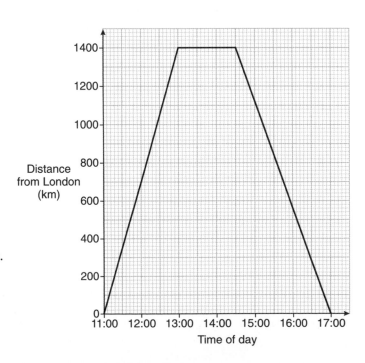

**2** Sangita cycles from Bury to the airport, a distance of 24 miles. The distance–time graph shows some information about her journey.

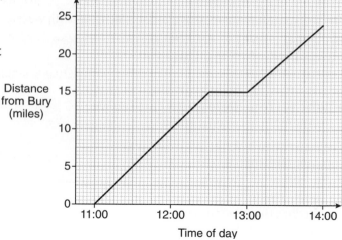

  **a** Sangita stops for lunch. For how many minutes does she stop?

  **b** Explain how the graph shows that Sangita cycles more slowly after lunch?

  **c** Work out Sangita's speed, in miles per hour, for the part of her journey before lunch.

  **d** Simon leaves the airport at 13:00 and travels at a steady speed to Bury. He arrives in Bury at 13:45.

   **i** On the resource sheet draw a distance–time graph for Simon's journey.

   **ii** Use your graph to work out Simon's speed.

   **iii** Use your graph to estimate the time at which Simon and Sangita are at the same distance from the airport.

**3** Mr Jacobs leaves home at 10:15 am and drives 90 kilometres to see a customer. He drives at 90 km/h. Mr Jacobs stays with the customer for 45 minutes and then travels back home at 60 km/h.

  **a** Show this information on a distance–time graph.

  **b** At what time did Mr Jacobs arrive back home?

**4** Ken sets off from home to go to his office. On his way he remembers that he has left his keys in the front door. He returns home to collect them and sets off for work again. The travel graph shows information about his journey.

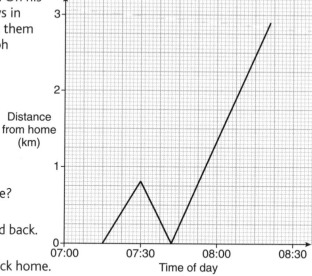

  **a** At what time did Ken first set off?

  **b** How far is Ken's office from his home?

  **c** How far had Ken gone before he turned back?

  **d** At what time did he arrive back home to get his keys?

  **e** How long did the complete journey take?

  **f** Ken travelled more quickly after he picked up his keys than before he turned back. Explain how the graph shows this.

  **g** Work out the speed of Ken's journey back home. Give your answer in km/h.

**5** The diagram represents the velocity–time graph of a van for a journey of 12 seconds. Describe fully this journey.

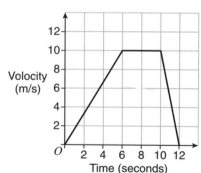

**6** A train leaves a station and steadily accelerates to reach a speed of 30 m/s in 6 seconds.

   **a** Work out the constant acceleration during these 6 seconds.

   **b** The train continues at a constant speed of 30 m/s for 10 seconds. It then makes an emergency stop coming to rest after a further 4 seconds. For this 20 second period, draw the velocity–time graph for the train.

   **c** Find the velocity of the train 2 seconds before it stops.

**7** The diagrams show three containers filling up with water.

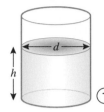 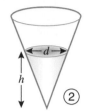  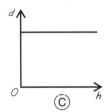

$d$ is the diameter of the surface of the water when the height of the water is $h$.
These graphs each show a relationship between $d$ and $h$.

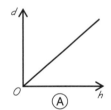 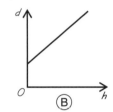 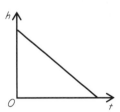

   **a** Match the graphs with the containers.

Water then leaks, at a constant rate, from a hole in the bottom of *one* of the containers. The graph shows the relationship between height $h$ and the time $t$.

   **b** Which container is leaking?

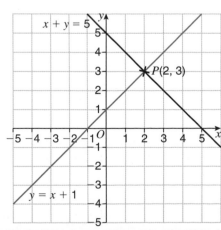

## 25.2 Solving simultaneous equations graphically

In Section 23.4 simultaneous equations are solved algebraically.
Simultaneous equations can also be solved by using graphs.

The diagram shows the graphs of the lines

    $x + y = 5$ and $y = x + 1$

The graphs cross at the point $P(2, 3)$.

$P$ is the only point that lies on both lines.
So the coordinates of $P(2, 3)$ satisfy both equations.

The $x$-coordinate of $P$ is 2 and the
$y$-coordinate of $P$ is 3

The solution of the simultaneous equations

    $x + y = 5$
    $y = x + 1$

is $x = 2, y = 3$

Check: $2 + 3 = 5$ ✓
       $3 = 2 + 1$ ✓

> You solved these simultaneous equations algebraically in Exercise 23D, question 6.

## Example 4

Find graphically the solution of the simultaneous equations

$$x + y = -2$$
$$y = x - 4$$

### Solution 4

For $x + y = -2$:   when $x = 0$, $y = -2$
when $y = 0$, $x = -2$

> Find and plot the points where the line $x + y = -2$ crosses the axes.

For $y = x - 4$:   when $x = 0$, $y = -4$
when $x = 1$, $y = -3$
when $x = 2$, $y = -2$

> Find and plot any three points on the line $y = x - 4$

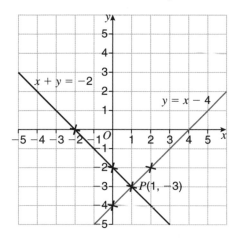

$x = 1$, $y = -3$

> Find the coordinates of the point where the lines cross
> Check: $1 + -3 = -2$ ✓
> $-3 = 1 - 4$ ✓

---

## Exercise 25B

**1** The diagrams show six lines labelled $L_1$, $L_2$, $L_3$, $L_4$, $L_5$ and $L_6$.

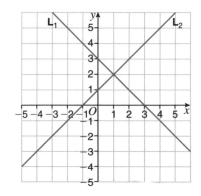 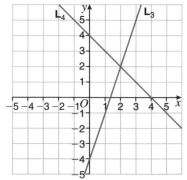 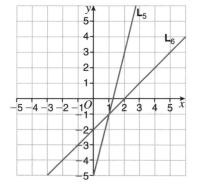

**a** Match three lines to these equations.
  **i** $y = x + 1$     **ii** $x + y = 4$     **iii** $y = 4x - 5$

**b** Use the diagrams to solve these simultaneous equations.
  **i** $x + y = 3$           **ii** $x + y = 4$           **iii** $y = x - 2$
    $y = x + 1$              $y = 3x - 4$            $y = 4x - 5$

For each of **Questions 2–6** draw two linear graphs on the same grid (each axis scaled from −5 to 5) to solve the simultaneous equations.

**2** $x + y = 0$
   $y = x + 2$

**3** $y = x - 2$
   $y = 2x$

**4** $x + y = 1$
   $y = -2x - 1$

**5** $y = x + 1$
   $y = 2x + 4$

**6** $y = 2x - 5$
   $x + y = 4$

## 25.3  The equation $y = mx + c$

The diagram shows the lines with equations

   $y = x$
   $y = x + 1$
   $y = x + 3$
   $y = x - 2$

The four lines are parallel. They cross the y-axis at different points (called **y-intercepts**).

The y-intercept of a line is the value of $y$ when $x = 0$
$y = x$ (or $y = x + 0$) has y-intercept **0**
$y = x + 1$ has y-intercept **1**
$y = x + 3$ has y-intercept **3**
$y = x - 2$ has y-intercept **−2**

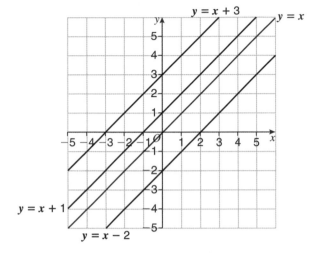

**In general,**
   the line $y = mx + c$ has y-intercept $c$, since it crosses the y-axis at $(0, c)$.

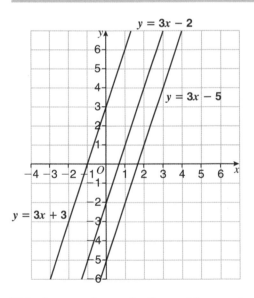

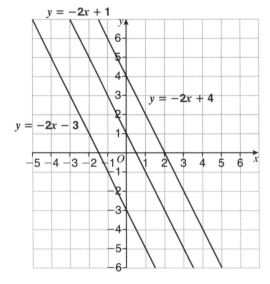

This diagram shows the lines with equations
   $y = 3x + 3$
   $y = 3x - 2$
and   $y = 3x - 5$

The intercepts on the y-axis are
   3, −2 and −5

The three lines are **parallel**.
In the equation of each line,
the $x$ term includes the number **3**

This diagram shows the lines with equations
   $y = -2x - 3$
   $y = -2x + 1$
and   $y = -2x + 4$

The intercepts on the y-axis are
   −3, +1 and +4

The three lines are **parallel**.
In the equation of each line,
the $x$ term includes the number **−2**

For different numbers in the $x$ term, the slopes of the lines are different.
For example the line $y = 3x$ is steeper than the line $y = 2x$.

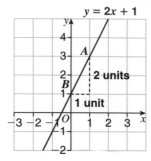

$y = 2x + 1$

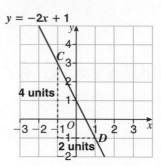

$y = -2x + 1$

This diagram shows the line $y = 2x + 1$
$A$ and $B$ are two points on the line.
The steepness of the slope of this line can be measured by dividing the vertical distance by the horizontal distance.

$$\frac{\text{vertical distance}}{\text{horizontal distance}} = \frac{2 \text{ units}}{1 \text{ unit}} = 2$$

This diagram shows the line $y = -2x + 1$
$C$ and $D$ are two points on the line.
The steepness of the slope of this line can be measured by dividing the vertical distance by the horizontal distance.

$$\frac{\text{vertical distance}}{\text{horizontal distance}} = \frac{4 \text{ units}}{2 \text{ units}} = 2$$

The steepness of the two lines appears to be the same ($= 2$) but the slopes of the two lines clearly look different; from the bottom point on the line, the first line slopes to the right and the second line slopes to the left.

A line that slopes to the right, /, is said to have positive slope.
A line that slopes to the left, \ , is said to have negative slope.

So,      the slope of the line $y = 2x + 1$ is **2**
and      the slope of the line $y = -2x + 1$ is $-2$

The slope of a line is called the **gradient** of the line.
The gradient of the line $y = 2x + 1$ is **2**
The gradient of the line $y = -2x + 1$ is $-2$

> **In general,**
> **for a line with the equation $y = mx + c$, the gradient of the line is the value of $m$.**

## Example 5

**a** Find the equation of the line which has gradient 6 and crosses the $y$-axis at the point $(0, -4)$.
**b** Find the coordinates of the point where the line crosses the $x$-axis.

*Solution 5*
**a** Gradient, $m = 6$
$y$-intercept, $c = -4$

> $y = mx + c$ is the equation of a line with gradient $m$ and $y$-intercept $c$

Equation of line is $y = 6x + (-4)$
$\qquad\qquad\qquad\quad y = 6x - 4$

> Substitute $m = 6$ and $c = -4$ in $y = mx + c$

**b** When $y = 0$, $0 = 6x - 4$

> Points on the $x$-axis have $y$-coordinate $= 0$
> Substitute $y = 0$ in $y = 6x - 4$

$6x = 4$   so $x = \dfrac{\cancel{4}^{2}}{\cancel{6}_{3}} = \dfrac{2}{3}$

> Add 4 to both sides of the equation, then divide by 6

The line $y = 6x - 4$ crosses the $x$-axis at the point $(\frac{2}{3}, 0)$

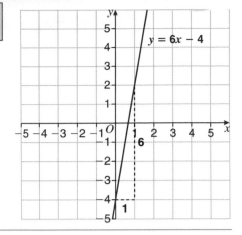

## Example 6

**a** Find the gradient and the $y$-intercept of the line with equation $2y - 6x = 3$
**b** Use your answer to part **a** to draw the graph of $2y - 6x = 3$

*Solution 6*

**a** $2y - 6x = 3$

> Rearrange the equation $2y - 6x = 3$ in the form $y = mx + c$

$2y = 6x + 3$
$y = 3x + 1.5$

> Add $6x$ to both sides
> Divide both sides by 2

$m = 3$ and $c = +1.5$
Gradient $= 3$
$y$-intercept $= 1.5$

> Compare $y = 3x + 1.5$ with $y = mx + c$

**b**

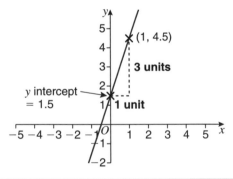

> A gradient of 3 means for every 1 unit to the right move up 3 units.
>
> Starting at the marked point $(0, 1.5)$ move 1 unit to the right then 3 units up then mark the point $(1, 4.5)$
>
> Draw a line through these two points.

## Exercise 25C

In **Questions 1–5** for each of the lines
**a** write down **i** the gradient **ii** the $y$-intercept and **b** sketch the line.

**1** $y = 2x + 3$     **2** $y = x - 5$     **3** $y = -3x$     **4** $y = \dfrac{2}{3}x + 1$     **5** $y = 8 - 3x$

In **Questions 6-10** find **a** the gradient and **b** the $y$-intercept of lines with these equations.

**6** $2y = 6x + 1$          **7** $x + y = 12$          **8** $y = 2(3x + 2)$

**9** $3y + 2x = 12$          **10** $2x - y + 4 = 0$

**11** A line passes through the point (0, 5) and has gradient 7. Find the equation of the line.

**12** Find the equations of the lines **A**, **B**, **C** and **D**.

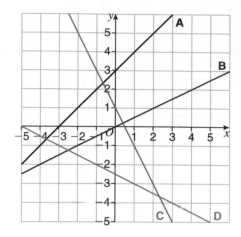

# 25.4 Further uses of $y = mx + c$

## *Parallel lines*

Parallel lines have equal gradients.
For example,

the three lines $y = 3x + 3$, $y = 3x - 2$ and $y = 3x - 5$ are parallel since $m = 3$ in each case.

## *Perpendicular lines*

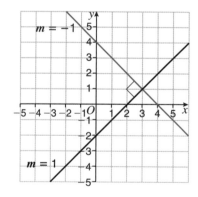
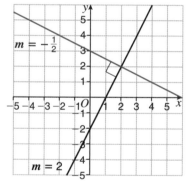
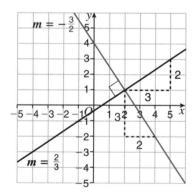

The diagrams show three pairs of perpendicular lines (lines which cross at 90°).

The blue lines have positive gradients.

The red lines have negative gradients.

Multiplying the gradients for each pair of perpendicular lines gives $-1$

$$-1 \times 1 = -1$$

$$-\frac{1}{2} \times 2 = -1$$

$$-\frac{3}{2} \times \frac{2}{3} = -1$$

> **Two lines are perpendicular if the product of their gradients is $-1$**
>
> **If a line has gradient $m$, all lines perpendicular to it have gradient $-\dfrac{1}{m}$**

## Example 7

The equations of three lines are $y = 2x + 1$, $y = -2x + 3$, $2y + x = 4$
Two of these lines are perpendicular.
**a** Find the equations of the two perpendicular lines.
**b** Are any of these lines parallel? Give a reason for your answer.

*Solution 7*
**a**

| To find the gradient of each line write them all in the form $y = mx + c$ |

$2y + x = 4 \quad$ so $\quad 2y = 4 - x$

| Subtract $x$ from both sides. |

$$\text{so } y = \frac{4}{2} - \frac{x}{2}$$

| Divide both sides by 2 |

$$y = -\frac{1}{2}x + 2$$

| Write the RHS in form $mx + c$ |

The gradient of $y = 2x + 1$ is **2**
The gradient of $y = -2x + 3$ is $-2$
The gradient of $2y + x = 4$

| Compare each equation with $y = mx + c$ and read off the value of $m$ for the gradient. |

or $\quad y = -\frac{1}{2}x + 2$ is $\quad -\frac{1}{2}$

Since $-\frac{1}{2} \times 2 = -1$, the two perpendicular lines are

| Lines are perpendicular if the product of their gradients is $-1$ |

$2y + x = 4$ and $y = 2x + 1$

**b** Since the gradients $\left(2, -2 \text{ and } -\frac{1}{2}\right)$ of the lines are

| Parallel lines have equal gradients. |

all different, the lines are not parallel.

## Example 8

Find the gradient of all lines which are **a** parallel **b** perpendicular to the line with equation $4y + 8x = 3$

*Solution 8*
**a**

| Write $4y + 8x = 3$ in the form $y = mx + c$ |

Write $4y + 8x = 3$ so $4y = 3 - 8x$

| Subtract $8x$ from both sides. |

$$\text{so } y = \frac{3}{4} - \frac{8x}{4}$$

| Divide both sides by 4 |

$$y = -2x + \frac{3}{4}$$

| Write the RHS in form $mx + c$ |

The gradient of $4y + 8x = 3$ or
$y = -2x + \frac{3}{4}$ is $-2$

| Compare with $y = mx + c$ and read off the value of $m$ for the gradient. |

Lines parallel to $4y + 8x = 3$ have
gradient $= -2$

| Parallel lines have equal gradients. |

**b** Lines perpendicular to $4y + 8x = 3$
have gradient $= -\frac{1}{-2} = \frac{1}{2}$

| If a line has gradient $m$, lines perpendicular to it have gradient $-\frac{1}{m}$ |

## Example 9

Find *an* equation of the line which passes through the point $(-2, 1)$ and is parallel to the line with equation $y = -2x + 5$

> Since $y = 2x$ and $2y = 4x$ are examples of two equations for the same line, sometimes exam questions ask you to find 'an equation' rather than 'the equation'.

### Solution 9

$y = -2x + 5$ has gradient $= -2$

> Compare with $y = mx + c$ and read off the value of $m$

The gradient of any parallel line is $-2$

> Parallel lines have equal gradients.

and has an equation in the form $y = -2x + c$

> Use $y = mx + c$

$(-2, 1)$ is a point on $y = -2x + c$
so $\quad 1 = -2(-2) + c$
$\quad\quad 1 = 4 + c$

> A line passes through a point if the coordinates of the point satisfy the equation of the line.

so $\quad c = -3$

An equation of the line is
$y = -2x - 3$

> Substitute $c = -3$ in $y = -2x + c$

You could write this equation in many other forms.
For example, $y + 2x + 3 = 0$ or $2y + 4x = -6$

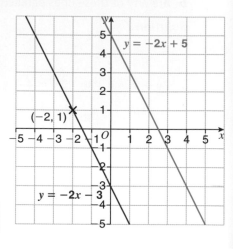

## Example 10

Find the equation of the line which passes through the point $(-3, 2)$ and is perpendicular to the line with equation $2y = 3x - 4$

### Solution 10

$2y = 3x - 4$

so $\quad y = \dfrac{3x - 4}{2}$ or $y = \dfrac{3}{2}x - 2$

> Write $2y = 3x - 4$ in the form $y = mx + c$
> Divide both sides by 2

The gradient of $y = \dfrac{3}{2}x - 2$ is $\dfrac{3}{2}$

> Compare with $y = mx + c$ and read off the value of $m$ for the gradient.

The gradient of the perpendicular
is $-\dfrac{1}{\frac{3}{2}} = -\dfrac{2}{3}$

> If a line has gradient $m$, lines perpendicular to it have gradient $-\dfrac{1}{m}$

and has an equation in the form

$y = -\dfrac{2}{3}x + c$

> Use $y = mx + c$

$(-3, 2)$ is a point on $y = -\dfrac{2}{3}x + c$

> A line passes through a point if the coordinates of the point satisfy the equation of the line.

so $2 = -\dfrac{2}{3}(-3) + c$

$\quad 2 = 2 + c$ means $c = 0$

The equation of the line is

$y = -\dfrac{2}{3}x$

> Substitute $c = 0$ in $y = -\dfrac{2}{3}x + c$

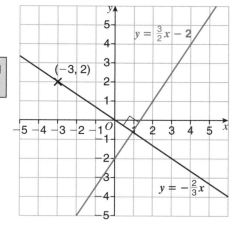

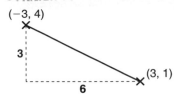

## Example 11

Find the equation of the line which joins the points $(-3, 4)$ and $(3, 1)$.

*Solution 11*

$(-3, 4)$

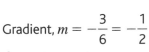

**3**

$(3, 1)$

**6**

| Find the gradient, $m$, of the line joining $(-3, 4)$ and $(3, 1)$ |
|---|

| The line slopes \ so has negative gradient. |
|---|

| Vertical distance $= 4 - 1 = 3$ Horizontal distance $= 3 - (-3) = 6$ |
|---|

Gradient, $m = -\dfrac{3}{6} = -\dfrac{1}{2}$

$y = -\dfrac{1}{2}x + c$

| Use $y = mx + c$ |
|---|

$(3, 1)$ is a point on $y = -\dfrac{1}{2}x + c$

| Substitute $x = 3$, $y = 1$ into the equation $y = -\dfrac{1}{2}x + c$ to find $c$ |
|---|

so $\quad 1 = -1.5 + c$

$\quad\quad 2.5 = c$

$y = -\dfrac{1}{2}x + 2.5$

| Substitute $c = 2.5$ in $y = -\dfrac{1}{2}x + c$ |
|---|

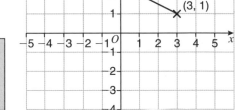

The equation of the line joining $(-3, 4)$ and $(3, 1)$ is

$$y = -\frac{1}{2}x + 2.5$$

---

### Exercise 25D

In **Questions 1–3** find the equation of the line which passes through the point $A$ and has the given gradient.

**1** $A$ $(0, 1)$ gradient 3      **2** $A$ $(2, 3)$ gradient $-2$      **3** $A$ $(-4, 2)$ gradient $\dfrac{1}{2}$

In **Questions 4–6** find an equation of the line which passes through the point $P$ and is parallel to the line **L**.

**4** $P$ $(0, 2)$ **L** is $y = 3x$      **5** $P$ $(3, 2)$ **L** is $y = 3 - x$      **6** $P$ $(4, -3)$ **L** is $2y = 3x + 4$

In **Questions 7–9** find the equation of the line which passes through the point $Q$ and is perpendicular to the line **M**.

**7** $Q$ $(1, 0)$ **M** is $y = -\dfrac{1}{2}x + 3$      **8** $Q$ $(-3, 2)$ **M** is $y = x + 1$

**9** $Q$ $(2, -1)$ **M** is $4y = 8x - 1$

In **Questions 10–12** find the equation of the line which joins the given points.

**10** $(1, 2)$ and $(3, 4)$      **11** $(-3, 5)$ and $(0, -4)$      **12** $(-4, 1)$ and $(2, 4)$

**13** $A$ is the point $(1, 3)$, $B$ is the point $(4, -3)$ and $C$ is the point $(-1, -2)$.
Find the equation of the line which passes through the point $C$ and is parallel to the line joining
the points $A$ and $B$.

**14**

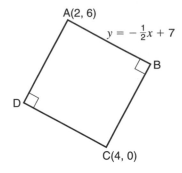

A(2, 6)
$y = -\frac{1}{2}x + 7$
B
D
C(4, 0)

$ABCD$ is a square. $A$ is the point $(2, 6)$. $C$ is the point $(4, 0)$.
The equation of the straight line through

$A$ and $B$ is $y = -\dfrac{1}{2}x + 7$

**a** Find the equation of the straight line through $D$ and $C$

**b** Find the equation of the straight line through $B$ and $C$

**c** Find the equation of the diagonal $AC$.

## Chapter summary

### You should now be able to use the fact:

★ that the coordinates of the point where two linear graphs cross gives the solution to that pair of simultaneous equations

★ that the equation of a straight line written in the form $y = mx + c$, where $m$ and $c$ are numbers, has

- $y$-intercept $c$, since the line crosses the $y$-axis at $(0, c)$

- and gradient $m$

★ that parallel lines have equal gradients

★ that two lines are perpendicular if the product of their gradients is $-1$ .

★ that, for a line with gradient $m$,

- all lines parallel to it have gradient $m$

- all lines perpendicular to it have gradient $-\dfrac{1}{m}$

### You should also be able to:

★ draw and interpret **distance–time graphs**, recognising that:

- straight lines represent constant speed

- horizontal lines represent no movement

★ draw and interpret **velocity–time graphs**, recognising that:

- a velocity–time graph shows how velocity changes with time

- straight lines represent constant acceleration

- **constant acceleration** $= \dfrac{\textbf{increase in velocity}}{\textbf{time taken}}$

- horizontal lines represent constant velocity

# Chapter 25 review questions

**1** Anil cycled from his home to the park. Anil waited in the park.
Then he cycled back home.
Here is a distance–time graph for Anil's complete journey.

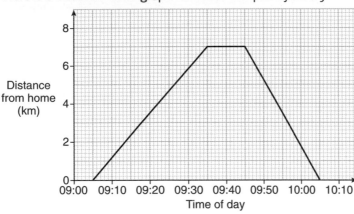

**a** At what time did Anil leave home?

**b** What is the distance from Anil's home to the park?

**c** How many minutes did Anil wait in the park?

**d** Work out Anil's average speed on his journey home. Give your answer in kilometres per hour.

(1387 June 2004)

**2** A car is travelling at a constant velocity of 15 m/s. It steadily accelerates at 5 m/s$^2$ for 4 seconds.
It then continues at a constant velocity for a further 12 seconds. The car then brakes and comes
to rest with a steady deceleration of 7 m/s$^2$.

**a** Work out the constant velocity during this period.

**b** Draw the velocity–time graph for the car.

**c** Find the time taken from the car starting to accelerate to it coming to rest.

**3 a** On the same axes, draw the graphs of **i** $x + y = 10$ **ii** $y = 3x$

**b** Hence solve the simultaneous equations $x + y = 10$
$$y = 3x$$

**4** The diagram shows 4 straight lines, labelled **P, Q, R** and **S**.
The equations of the straight lines are

   **A:** $y = 2x$
   **B:** $y = 3 - 2x$
   **C:** $y = 2x + 3$
   **D:** $y = 3$

Match each straight line, **P, Q, R** and **S** to its equation.

(1388 March 2004)

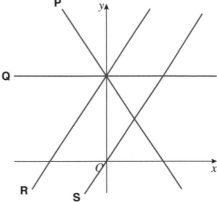

**5** A straight line has equation $y = 2(3 - 4x)$
Find the gradient of the straight line.

(1388 June 2004)

**6** A straight line, **L**, has equation $3y = 5x - 6$
Find **i** the gradient of **L**,
**ii** the $y$-coordinate of the point where **L** cuts the $y$-axis.

**7** Find the gradient of the straight line with equation $5y = 3 - 2x$. (1388 March 2003)

**8** A straight line has equation $2y - 6x = 5$

   **a** Find the gradient of the line.

   The point $(k, 6)$ lies on the line.

   **b** Find the value of $k$. (1388 January 2004)

**9** The straight line $L_1$ has equation $y = 2x + 3$
   The straight line $L_2$ is parallel to the straight line $L_1$.
   The straight line $L_2$ passes through the point $(3, 2)$.
   Find an equation of the straight line $L_2$. (1387 November 2004)

**10** A straight line, **L**, passes through the point with coordinates $(4, 7)$ and is perpendicular to the line
   with equation $y = 2x + 3$
   Find an equation of the straight line **L**. (1387 November 2003)

**11** A straight line passes through the points $(0, 5)$ and $(3, 17)$.
   Find the equation of the straight line. (1388 January 2005)

**12** $ABCD$ is a rectangle. $A$ is the point $(0, 1)$. $C$ is the point $(0, 6)$.
   The equation of the straight line through $A$ and $B$ is $y = 2x + 1$

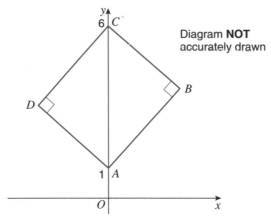

Diagram **NOT** accurately drawn

   **a** Find the equation of the straight line through $D$ and $C$.
   **b** Find the equation of the straight line through $B$ and $C$. (1388 June 2004)

**13** The diagram shows two straight lines intersecting at point $A$.
   The equations of the lines are $y = 4x - 8$ and $y = 2x + 3$
   Work out the coordinates of $A$.

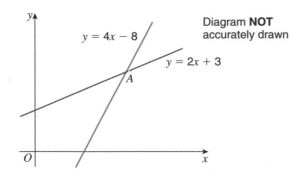

Diagram **NOT** accurately drawn

(1388 March 2005)

**14** The diagram shows three points $A\,(-1, 5)$, $B\,(2, -1)$ and $C\,(0, 5)$.
The line **L** is parallel to $AB$ and passes through $C$.

**a** Find the equation of the line **L**.

The line **M** is perpendicular to $AB$ and passes through $(0, 0)$.

**b** Find the equation of the line **M**.

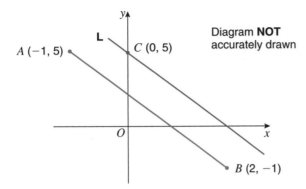

Diagram **NOT**
accurately drawn

(1388 June 2005)

# Transformations

## 26.1 Introduction

A **transformation** changes the position or size of a shape.

Using the 'Nudge' instruction or the arrow keys, a computer can move a shape to a new position.

For example

This is a **translation**

A computer can also rotate, reflect (flip) and enlarge a shape.
For example

A **rotation**          A **reflection**          An **enlargement**

The starting shape is called the **object.** In each case, the object is the yellow triangle.
The final shape, called the **image**, is the red triangle.
The object **maps onto** the image.

Translation, rotation, reflection and enlargement are **transformations**.

## 26.2 Translations

In this diagram, shape **A** has been mapped onto shape **B** by a **translation**.
In a translation, all the points of the shape move the same distance
and in the same direction. All the points of shape **A** move 3 squares
to the right and 5 squares up.

A translation is a transformation in which all points of an object move
the same distance in the same direction.

In a translation:
- the lengths of the sides of the shape do not change
- the angles of the shape do not change
- the shape does not turn.

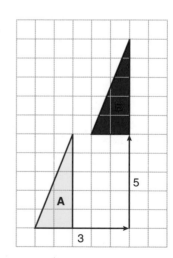

**Example 1**

Describe the translation that maps triangle **P** onto triangle **Q**.

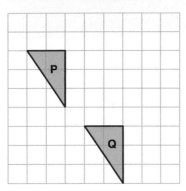

**Solution 1**

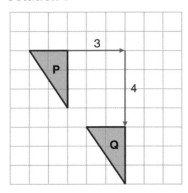

Choose one corner of triangle **P**.

Count the number of squares to the right and the number of squares down from this corner on triangle **P** to the same corner on triangle **Q**.

The translation from triangle **P** to triangle **Q** is 3 squares to the right and 4 squares down.

The translation of 3 squares to the right and 4 squares down can be written as $\begin{pmatrix} 3 \\ -4 \end{pmatrix}$.

This is a **vector**. Vectors can be used to describe translations.

The top number shows the number of squares moved to the right or left.

The bottom number shows the number of squares moved up or down.

The rules for which directions are positive and which are negative are the same as for coordinates.

To the right and up are positive.

To the left and down are negative.

Some translations and their vectors are shown on the grid.

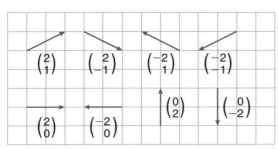

## Example 2

**a** Describe the translation that maps shape **A** onto **i** shape **B**, **ii** shape **C**.

**b** Translate shape **A** by the vector $\begin{pmatrix} -3 \\ -5 \end{pmatrix}$.

Label this new shape **D**.

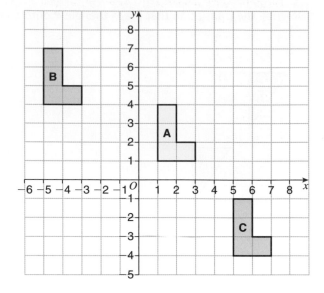

## Solution 2

**a  i** From **A** to **B** is the translation with vector $\begin{pmatrix} -6 \\ 3 \end{pmatrix}$

Count the number of squares moved to the left (negative) and up (positive) from any corner in **A** to the same corner in **B**

**ii** From **A** to **C** is the translation with vector $\begin{pmatrix} 4 \\ -5 \end{pmatrix}$

Count the number of squares moved to the right (positive) and down (negative) from any corner in **A** to the same corner in **C**

**b**

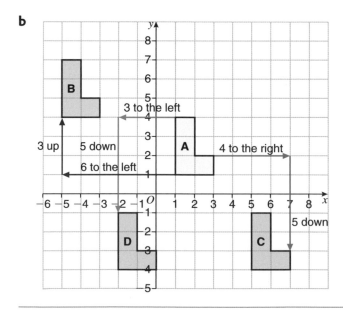

$\begin{pmatrix} -3 \\ -5 \end{pmatrix}$ means 3 to the left and 5 down

Choose one corner of shape **A**. Count from this corner 3 squares to the left and then count 5 squares down to find where this corner has moved to. The new shape is the same as shape **A**. Draw the new shape and label it **D**.

## Exercise 26A

**1** Describe, with a vector, the translation that maps triangle **A** onto

  **a** triangle **B**

  **b** triangle **C**

  **c** triangle **D**

  **d** triangle **E**

  **e** triangle **F**

  **f** triangle **G**.

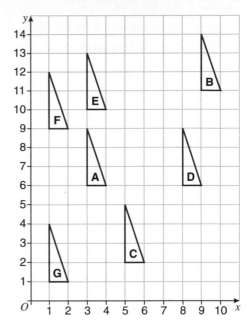

**2** Use the resource sheet. Translate triangle **A**

  **a** 5 to the right and 4 up. Label your new triangle **B**.

  **b** 4 to the right and 6 down. Label your new triangle **C**.

  **c** 7 to the left. Label your new triangle **D**.

  **d** by the vector $\begin{pmatrix} 3 \\ 2 \end{pmatrix}$. Label your new triangle **E**.

  **e** by the vector $\begin{pmatrix} -6 \\ -4 \end{pmatrix}$. Label your new triangle **F**.

  **f** by the vector $\begin{pmatrix} 0 \\ 2 \end{pmatrix}$. Label your new triangle **G**.

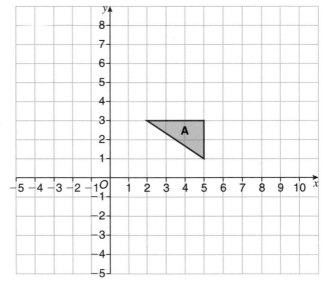

**3** The coordinates of the point $A$ of the kite are $(-2, 1)$. The kite is translated so that the point $A$ is mapped onto the point $(3, 4)$.

  **a** Use the resource sheet and draw the image of the kite after this translation.

  **b** Describe this translation with a vector.

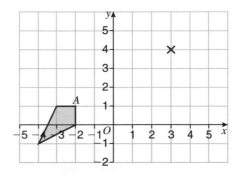

4   Use the resource sheet and

a   translate the kite **A** by the vector $\begin{pmatrix} 4 \\ 7 \end{pmatrix}$.

Label this new kite **B**

b   translate kite **B** by the vector $\begin{pmatrix} -6 \\ -3 \end{pmatrix}$.

Label this new kite **C**.

c   Describe, with a vector, the translation that maps kite **A** onto kite **C**.

d   Describe, with a vector, the translation that maps kite **C** onto kite **A**.

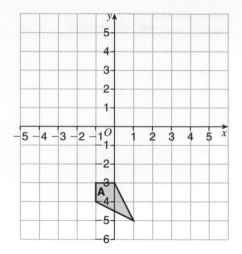

## 26.3 Rotations

To rotate means to turn. A cycle wheel, the hands of a clock and the drum of a washing machine all turn or rotate.

This face on a stick has rotated 60° **clockwise** ( ⌣ ) about the point $O$.
The point $O$ is the **centre of rotation.**
The size of the face has not changed.

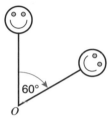

To describe a rotation give

- the angle of turn
- the direction of turn (clockwise or anticlockwise)
- the point the shape turns about (the centre of rotation).

In a rotation

- the lengths of the sides of the shape do not change
- the angles of the shape do not change
- the shape turns
- the centre of rotation does not move.

### Example 3

Rotate the triangle a quarter turn clockwise about the point $A$.

✕ $A$

✕ $A$

### Solution 3
Tracing paper can be used to rotate the shape.

Trace the triangle and mark the point $A$.

Fix the point $A$ with a pencil or a compass point so that the point $A$ does not move.

Turn the tracing paper about $A$, clockwise through a quarter turn (90°).

Now the position of the image of the triangle can be seen.

## Example 4

Describe the transformation that maps triangle **A** onto triangle **B**.

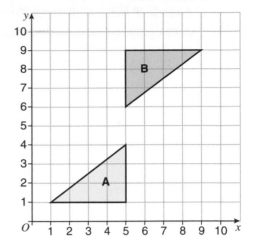

### Solution 4

Triangle **A** is mapped onto triangle **B** by a rotation of 180° (a half turn) about the point (5, 5).

> Use tracing paper to check that the transformation is a rotation of 180°. Then check that the centre of rotation is the point (5, 5).

## Exercise 26B

**1** Use the resource sheet and

  **a** rotate trapezium **A** a half turn about the origin $O$. Label the new trapezium **B**.

  **b** rotate trapezium **A** a quarter turn clockwise about the origin $O$. Label the new trapezium **C**.

  **c** rotate trapezium **A** a quarter turn anticlockwise about the origin $O$. Label the new trapezium **D**.

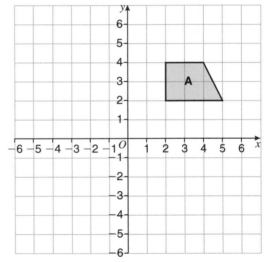

**2** The resource sheet contains 3 copies of the diagram showing trapezium **P**.

  **a** On copy 1 of the diagram, rotate trapezium **P** 180° about the point (2, 0). Label the new trapezium **Q**.

  **b** On copy 2 of the diagram, rotate trapezium **P** 90° clockwise about the point (−2, 2). Label the new trapezium **R**.

  **c** On copy 3 of the diagram, rotate trapezium **P** 90° anticlockwise about the point (−1, −1). Label the new trapezium **S**.

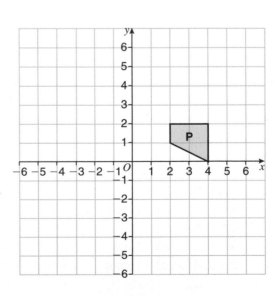

**3 a** Describe fully the rotation that maps shape **A** onto
   **i** shape **B**, **ii** shape **C**, **iii** shape **D**.

   **b** Describe fully the rotation that maps shape **B** onto shape **A**.

   **c** Describe fully the rotation that maps shape **B** onto shape **D**.

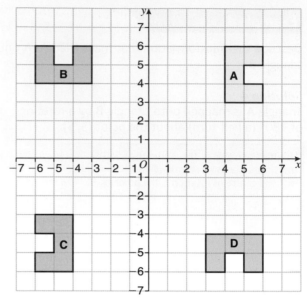

**4 a** Describe fully the rotation that maps triangle **A** onto
   **i** triangle **B**, **ii** triangle **C**, **iii** triangle **D**, **iv** triangle **E**, **v** triangle **F**.

   **b** Describe the transformation that maps triangle **B** onto triangle **E**.

   **c** Describe the transformation that maps **i** triangle **D** onto triangle **B**, **ii** triangle **F** onto triangle **E**.

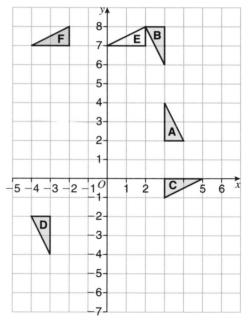

# 26.4 Reflections

Look in a mirror. You see a reflection.
In the mirror line, the reflection of point $A$ is point $A'$.

- Point $A'$ is the same distance behind the mirror line as point $A$ is in front.
- The line joining points $A$ and $A'$ is perpendicular to the mirror line.

Reflecting each corner of triangle **P** in the mirror line gives the corners of triangle **Q**.

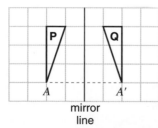

mirror line

Triangle **Q** is the **reflection** of triangle **P** in the mirror line.

Also, triangle **P** is the reflection of triangle **Q** in the mirror line.

(In mathematics, mirror lines are two-way mirrors.)

The image is as far behind the mirror line as the object is in front.

$A$ and $A'$ are called **corresponding** points, because the point $A'$ is the image of the point $A$. In this case, each point is the corner with the smallest angle.

To describe a reflection
● give the mirror line.

In a reflection
● the lengths of the sides of the shape do not change
● the angles of the shape do not change
● the image is as far behind the mirror line as the object is in front.

## Example 5

Reflect trapezium **T** in the mirror line. Label the new trapezium **U**.

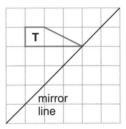

### Solution 5
**Method 1**
Reflect each corner in the mirror line so that its image is the same distance behind the mirror line as the corner is in front.

Notice that
● the line joining each corner to its image is perpendicular to the mirror line
● the image of the corner which is on the mirror line is also on the mirror line

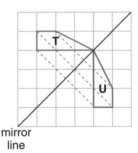

**Method 2**
● Put the edge of a sheet of tracing paper on the mirror line and make a tracing of the trapezium.
● Turn the tracing paper over and put the edge of the tracing paper back on the mirror line.
● Mark the images of the corners with a pencil or compass point.

Method 2 is particularly useful when the shape is not a polygon or not drawn on a grid.

## Example 6

The diagram shows a triangle **S** and its image, triangle **T**, after a reflection.

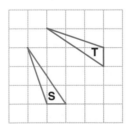

Draw the mirror line of the reflection.

### Solution 6

Join each corner of triangle **S** to its image on triangle **T**. The mirror line passes through the midpoints (marked with crosses) of these lines.

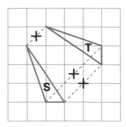

Draw the mirror line by joining the crosses.

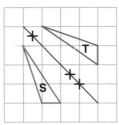

## Example 7

Describe fully the transformation which maps triangle **P** onto triangle **Q**.

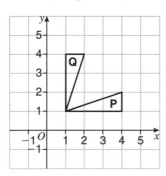

### Solution 7

The transformation is a reflection in the line with equation $y = x$

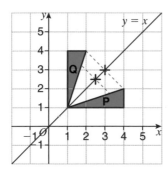

See Section 13.1

## Exercise 26C

1 The resource sheet contains each of the diagrams below. Reflect each shape in the mirror line, shown in red.

a
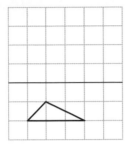

b
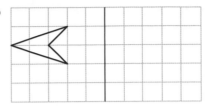

c
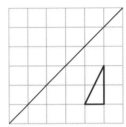

d
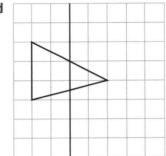

2 The resource sheet contains each of the diagrams shown below. On the resource sheet, draw in the mirror line.

a
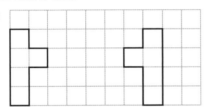

b
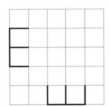

c
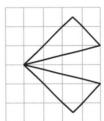

3 On the resource sheet

   a reflect triangle **P** in the line $x = 1$.
     Label this new triangle **Q**.

   b reflect triangle **P** in the line $y = 2$.
     Label this new triangle **R**.

   c Describe the reflection that maps triangle **Q** onto triangle **T**.

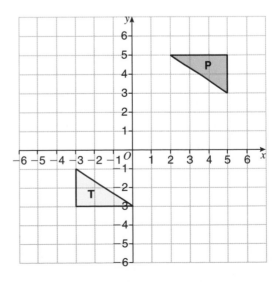

**4** On the resource sheet

   **a** reflect triangle **A** in the line $y = x$.
Label this new triangle **B**.

   **b** reflect triangle **A** in the line $y = -x$.
Label this new triangle **C**.

   **c** Describe fully the transformation that maps triangle **B** onto triangle **A**.

   **d** Describe fully the transformation that maps triangle **B** onto triangle **C**.

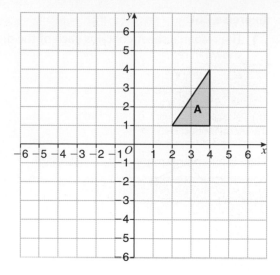

**5** The diagram shows 8 triangles.

   **a** Triangle **A** maps onto *four* of the unshaded triangles by a reflection.
Name these four triangles and describe the reflection in each case.

   **b** Triangle **A** maps onto *three* of the unshaded triangles by a rotation.
Name these three triangles and describe the rotation in each case.

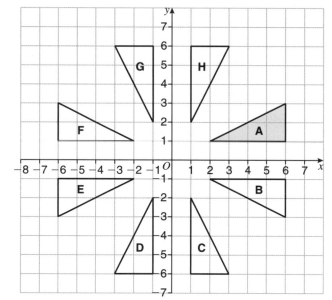

**6** Use the resource sheet.

   **a** Draw the mirror line of the reflection that maps
     **i** shape **P** onto shape **Q**,
     **ii** shape **P** onto shape **R**.

   **b** Describe fully the transformation that maps shape **Q** onto shape **P**.

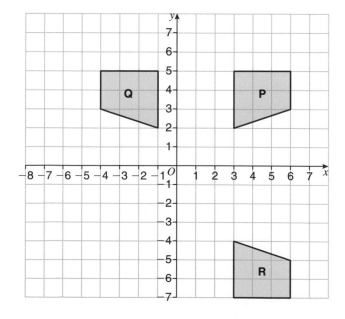

## 26.5 Enlargements

Here is a photograph of a windmill.

2 cm

3 cm

Here is an **enlargement** of the photograph.

8 cm

12 cm

The shapes in the two photographs are the same but each length in the enlargement is 4 times the corresponding length in the original photograph.

For example, the length of a sail of the windmill in the enlargement is 4 times the length of the sail in the original photograph.

The **scale factor** of an enlargement is the number by which lengths have been multiplied.
The width of the enlargement is 12 cm and the width of the original photograph is 3 cm

$$\text{so the scale factor of the enlargement} = \frac{12}{3} = 4$$

Similarly, the height of the enlargement is 8 cm and the height of the original photograph is 2 cm

$$\text{so the scale factor of the enlargement} = \frac{8}{2} = 4$$

The larger photograph is an enlargement with **scale factor 4** of the smaller photograph.

In general,

$$\text{scale factor of enlargement} = \frac{\text{length of side in image}}{\text{length of corresponding side in object}}$$

Notice that each angle in the original photograph is the same as the corresponding angle in the enlargement.

In an enlargement

- the lengths of the sides of the shape change
- the angles of the shape do not change.

**Example 8**

Shape **B** is an enlargement of shape **A**.

**a** Find the scale factor of the enlargement that maps shape **A** onto shape **B**.

**b** Find the scale factor of the enlargement that maps shape **B** onto shape **A**.

> Notice that pairs of corresponding sides are parallel.

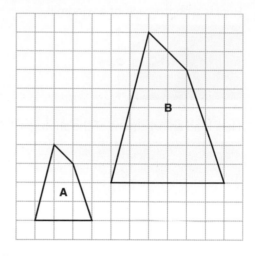

**Solution 8**

**a** Scale factor of the enlargement which maps shape **A** onto shape **B**

$$= \frac{6}{3}$$

Scale factor = 2

> The base of shape **A** is 3 squares wide.
> The base of shape **B** is 6 squares wide.

> The lengths of the sides of shape **B** are twice the lengths of the sides of shape **A**.

**b** Scale factor of the enlargement which maps shape **B** onto shape **A**

$$= \frac{3}{6}$$

Scale factor = $\frac{1}{2}$

> In mathematics, the word 'enlargement' is also used when a shape gets smaller.
> This answer means that the lengths of the sides of shape **A** are $\frac{1}{2}$ the lengths of the sides of shape **B**.

---

**Exercise 26D**

**1** This rectangle has a length of 4 cm and a width of 3 cm.
The rectangle is enlarged with a scale factor of 2.
Find the length and the width of the new rectangle.

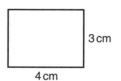

4 cm

3 cm

**2** Use the resource sheet and draw an enlargement, scale factor 2, of the shaded shape.

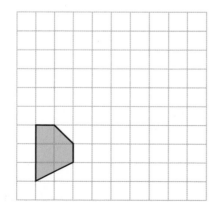

**3** The right-angled triangle has a base of 8 cm and a height of 6 cm.

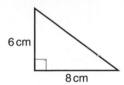

The triangle is enlarged with a scale factor of $\frac{1}{2}$

Find the base and the height of the new triangle.

**4** Use the resource sheet to draw an enlargement, scale factor $\frac{1}{3}$ of the shaded shape.

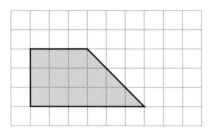

**5** On the resource sheet draw

    **a** an enlargement of shape **A** with scale factor 3.
       Label this enlargement shape **B**.

    **b** an enlargement of shape **A** on the same diagram
       with scale factor $\frac{1}{2}$
       Label this enlargement shape **C**.

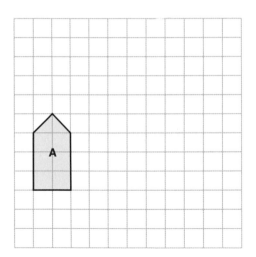

**6** The diagram shows 4 triangles drawn on a grid.

    **a** Triangle **B** is an enlargement of triangle **A**.
       Work out the scale factor.

    **b** Triangle **C** is an enlargement of triangle **D**.
       Work out the scale factor.

    **c** Triangle **C** is an enlargement of triangle **A**.
       Work out the scale factor.

    **d** Triangle **D** is an enlargement of triangle **A**.
       Work out the scale factor.

    **e** Triangle **D** is an enlargement of triangle **B**.
       Work out the scale factor.

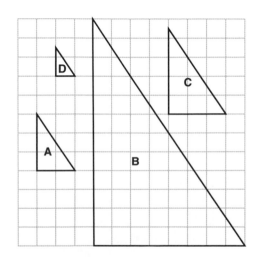

**7** Rectangle **P** has a base of 4 cm and a height of 2 cm.
Rectangle **Q** is an enlargement of rectangle **P** with a scale factor of 2
Rectangle **R** is an enlargement of rectangle **P** with a scale factor of 3

**a** On the resource sheet, draw rectangles **Q** and **R**.

**b** Find the perimeters of **i** rectangle **P**, **ii** rectangle **Q**, **iii** rectangle **R**.

**c** Find the areas of **i** rectangle **P**, **ii** rectangle **Q**, **iii** rectangle **R**.

**d** Work out the values of **i** $\dfrac{\text{Perimeter of }\mathbf{Q}}{\text{Perimeter of }\mathbf{P}}$, **ii** $\dfrac{\text{Perimeter of }\mathbf{R}}{\text{Perimeter of }\mathbf{P}}$.
Write down anything that you notice about these values.

**e** Work out the values of **i** $\dfrac{\text{Area of }\mathbf{Q}}{\text{Area of }\mathbf{P}}$, **ii** $\dfrac{\text{Area of }\mathbf{R}}{\text{Area of }\mathbf{P}}$.
Write down anything that you notice about these values.

**f** Rectangle **S** is an enlargement of rectangle **P** with a scale factor of 8. What is the perimeter of rectangle **S**?

# 26.6  Centre of enlargement

In the diagram, triangle **P** has been enlarged by a scale factor of 2 to give triangle **Q**.

The corner $A$ of triangle **P** is mapped onto the corner $A'$ of triangle **Q**. A line has been drawn joining $A$ and $A'$. Lines have also been drawn joining the other pairs of corresponding points of triangles **P** and **Q**.

The lines meet at a point $C$, called the **centre of enlargement**.

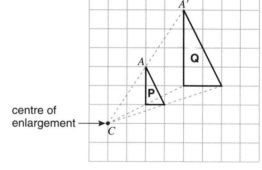

$C$ to $A$ is 2 squares across and 3 squares up.
$C$ to $A'$ is 4 squares across and 6 squares up.

So $\dfrac{CA'}{CA} = 2$, the scale factor of the enlargement.

To describe an enlargement give

● the scale factor

● the centre of enlargement.

In general,

for an enlargement with centre $C$ and scale factor $k$,
$CA' = k \times CA$, for any point $A$ of the object and the corresponding point $A'$ of the image.

## Example 9

Enlarge the triangle with scale factor 2 and centre $O$.

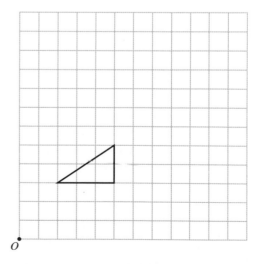

**Solution 9**

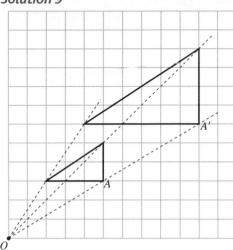

Choose any point $A$ of the triangle. Join $O$, the centre of enlargement, to $A$ and extend this line.

$O$ to $A$ is 5 to the right and 3 up.

The scale factor is **2**

$\mathbf{2} \times 5 = 10$ and $\mathbf{2} \times 3 = 6$

so for $A'$, the image of $A$, $O$ to $A'$ is 10 to the right and 6 up.

Mark $A'$ on the diagram.

Either repeat the process to find the other two corners of the enlarged triangle or draw it using the fact that its sides are twice the length of those of the original triangle.
Always measure from the centre of the enlargement.

## Negative scale factors

When a shape is enlarged by a negative scale factor, the image is on the opposite side of the centre of enlargement.

### Example 10

Enlarge triangle $PQR$ by a scale factor of $-2$ with the point $C(3, 2)$ as the centre of enlargement.

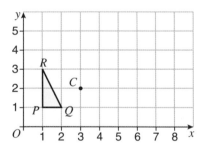

**Solution 10**

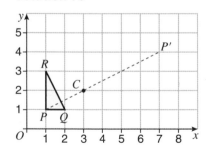

From $P$ to $C$ is 2 squares to the right and 1 square up. So to find $P'$, the image of $P$, *start at $C$* and move 4 squares to the right and 2 squares up.

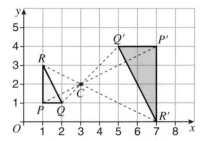

Repeat the process to find $Q'$ and $R'$, the images of $Q$ and $R$.

Complete the enlargement.

The lengths of the sides of the transformed triangle are twice those of the original triangle, which has also been rotated through 180°.

*Example 11*

Describe fully the transformation which maps
**a** trapezium **P** onto trapezium **Q**
**b** trapezium **Q** onto trapezium **P**.

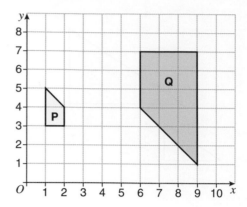

*Solution 11*

**a**

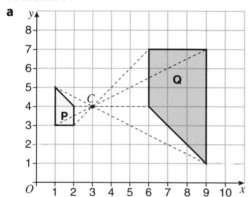

The lengths of the sides of trapezium **Q** are 3 times those of trapezium **P**, which has also been rotated through 180°.

So the transformation is an enlargement with scale factor −3.

To find the centre of enlargement, join each vertex of trapezium **P** to the corresponding vertex of trapezium **Q**.

The centre of enlargement $C$ is the point where the red lines cross.

The transformation is an enlargement with scale factor −3, centre (3, 4).

**b** The transformation is an enlargement with scale factor $-\frac{1}{3}$ centre (3, 4).

The lengths of the sides of trapezium **P** are $\frac{1}{3}$ of those of trapezium **Q**, which has also been rotated through 180°.

So the transformation is an enlargement with scale factor $-\frac{1}{3}$
The centre of enlargement has already been found in part **a**.

---

## Exercise 26E

**1** On the resource sheet draw the enlargement of the shapes with the given scale factor and centre of enlargement marked with a dot (•).

**a**

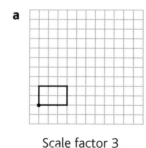

Scale factor 3

**b**

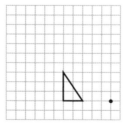

Scale factor 2

**c**

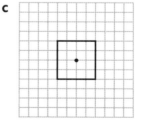

**i** Scale factor 2
**ii** Scale factor 0.5

Draw both enlargements on the same diagram.

**d**

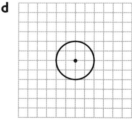

**i** Scale factor 3
**ii** Scale factor 2
**iii** Scale factor $\frac{1}{2}$
Draw all three enlargements on the same diagram.

**2**

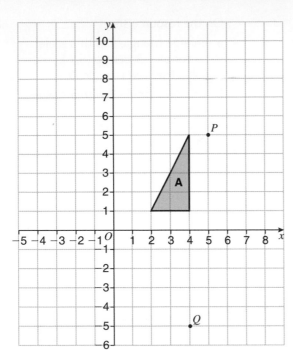

On the resource sheet

**a** enlarge triangle **A** by a scale factor of 2 and centre $O$ (0, 0). Label the new triangle **B**.

**b** enlarge triangle **A** by a scale factor of 3 and centre $P$ (5, 5). Label the new triangle **C**.

**c** enlarge triangle **A** by a scale factor of $\frac{1}{2}$ and centre $Q$ (4, $-5$). Label the new triangle **D**.

**d** Triangle **C** is an enlargement of triangle **D**.
  **i** Find the scale factor of this enlargement.
  **ii** By drawing lines on your diagram, find the centre of this enlargement. Mark this centre with a cross (✗) and the letter $R$.

**3** Use the resource sheet.

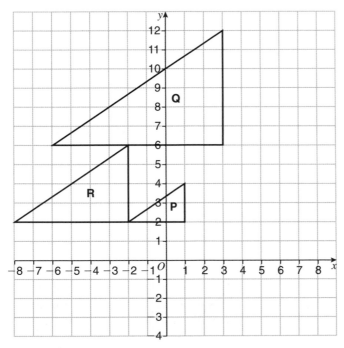

**a** Describe fully the transformation that maps triangle **P** onto triangle **Q**.

**b** Describe fully the transformation that maps triangle **P** onto triangle **R**.

**c** Describe fully the transformation that maps triangle **R** onto triangle **P**.

**4**

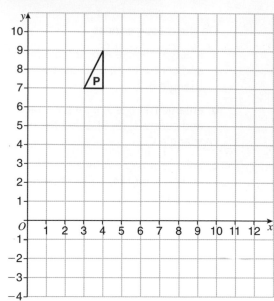

On the resource sheet

**a** enlarge triangle **P** with a scale factor of −3, centre (5, 6). Label the new triangle **Q**.

**b** enlarge triangle **P** with a scale factor of −2, centre (3, 7). Label the new triangle **R**.

**5**

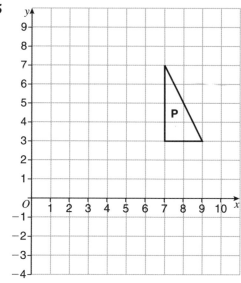

On the resource sheet

**a** enlarge triangle **P** with a scale factor of $-\frac{1}{2}$, centre (5, 5). Label the new triangle **Q**.

**b** enlarge triangle **P** with a scale factor of $-\frac{1}{2}$, centre (7, 1). Label the new triangle **R**.

**6 a** Describe the transformation which maps trapezium **P** onto trapezium **Q**.

**b** Describe the transformation which maps trapezium **Q** onto trapezium **P**.

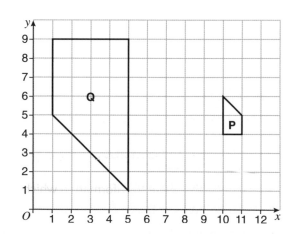

## 26.7 Combinations of transformations

It is sometimes possible to find a single transformation which has the same effect as a combination of transformations.

---

### Example 12

a  Reflect triangle **P** in the line $x = 3$. Label the new triangle **Q**.
b  Reflect triangle **Q** in the line $y = 5$. Label the new image **R**.
c  Describe fully the *single* transformation which maps triangle **P** onto triangle **R**.

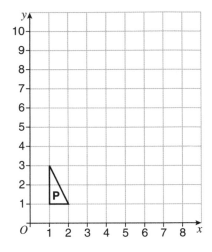

### Solution 12

**a, b**

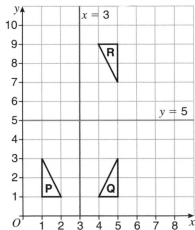

c  The single transformation which maps triangle **P** onto triangle **R** is a rotation of 180° about the point (3, 5).

---

### Example 13

a  Rotate triangle **P** 90° anticlockwise about (3, 5). Label the new triangle **Q**.
b  Rotate triangle **Q** 90° clockwise about (6, 2). Label the new triangle **R**.
c  Describe fully the *single* transformation which maps triangle **P** onto triangle **R**.

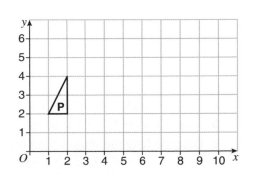

*Solution 13*

**a, b**

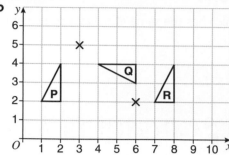

**c** The single transformation which maps triangle **P** onto triangle **R** is a translation with vector $\begin{pmatrix} 6 \\ 0 \end{pmatrix}$.

---

### Exercise 26F

**1** On the resource sheet

    **a** reflect triangle **P** in the $x$-axis. Label the new triangle **Q**.

    **b** reflect triangle **Q** in the $y$-axis. Label the new triangle **R**.

    **c** Describe fully the single transformation which maps triangle **P** onto triangle **R**.

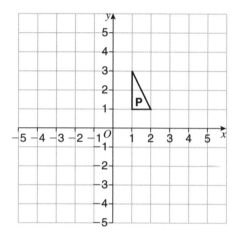

**2** On the resource sheet

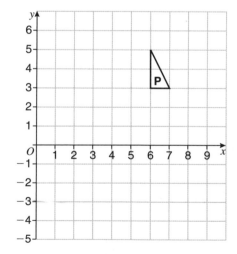

    **a** reflect triangle **P** in the line $y = 1$. Label the new triangle **Q**.

    **b** rotate triangle **Q** $180°$ about $(4, 1)$. Label the new triangle **R**.

    **c** Describe fully the single transformation which maps triangle **P** onto triangle **R**.

**3** On the resource sheet

  **a** reflect trapezium **T** in the line $x = 3$.
Label the new trapezium **U**.

  **b** reflect trapezium **U** in the line $x = 7$.
Label the new trapezium **V**.

  **c** Describe fully the single transformation which maps trapezium **T** onto trapezium **V**.

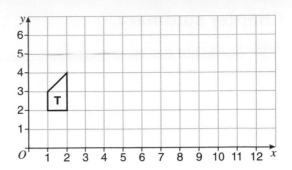

**4** On the resource sheet

  **a** reflect triangle **A** in the line $y = x$.
Label the new triangle **B**.

  **b** rotate triangle **B** 90° anticlockwise about $(4, 4)$.
Label the new triangle **C**.

  **c** Describe fully the single transformation which maps triangle **A** onto triangle **C**.

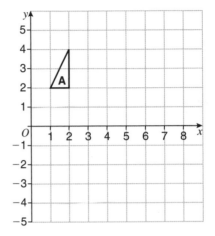

**5** On the resource sheet

  **a** translate triangle **P** by the vector $\begin{pmatrix} 3 \\ 1 \end{pmatrix}$.
Label the new triangle **Q**.

  **b** rotate triangle **Q** 90° clockwise about $(6, 3)$.
Label the new triangle **R**.

  **c** Describe fully the single transformation which maps triangle **P** onto triangle **R**.

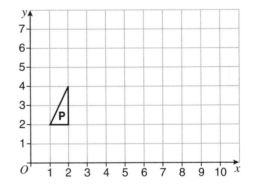

**6** On the resource sheet

  **a** reflect triangle **P** in the line $y = x$.
Label the new triangle **Q**.

  **b** rotate triangle **Q** 180° about $(0, 0)$.
Label the new triangle **R**.

  **c** Describe fully the single transformation which maps triangle **P** onto triangle **R**.

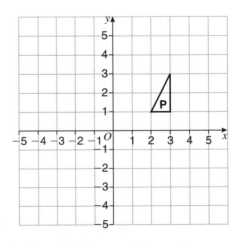

## Chapter summary

**You should now know and be able to use these facts:**

★ A transformation of a shape is a change in position or size of the shape

★ The starting shape is the **object** and the final shape is the **image**

★ In a **translation** all points of the object move the same distance in the same direction

★ A translation can also be described by a vector so that, for example, $\begin{pmatrix} 9 \\ -6 \end{pmatrix}$ means 9 to the right and 6 down.

★ A **rotation** is described by giving the angle of turn, the direction of turn and the **centre of rotation**

★ In a **reflection** the image is as far behind the **mirror line** as the object is in front

★ A reflection is described by giving the position or equation of the mirror line

★ The **scale factor** of an **enlargement** is the number by which lengths have been multiplied

★ Scale factor of enlargement $= \dfrac{\text{length of side in image}}{\text{length of corresponding side in object}}$

★ In an enlargement all lengths are multiplied by the scale factor

★ An enlargement is described by giving the scale factor and the **centre of enlargement**

★ If the scale factor is **negative**, the object and the image are on opposite sides of the centre of enlargement

★ Translations, rotations, reflections have no effect on lengths and angles

★ Enlargement changes the size of a shape but has no effect on angles

★ It is sometimes possible to find a single transformation which has the same effect as a combination of transformations.

## Chapter 26 review questions

1 On the resource sheet rotate the triangle through 180° about centre $A$, and draw the new position of the triangle.

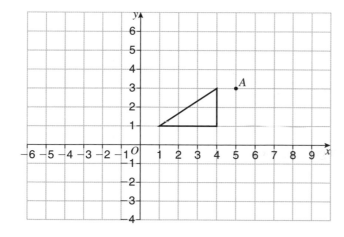

(1384 November 1994)

**2** On the resource sheet

    **a** rotate triangle **A** 180° about $O$.
Label your new triangle **B**.

    **b** enlarge triangle **A** by scale factor $\frac{1}{2}$ centre $O$.
Label your new triangle **C**.

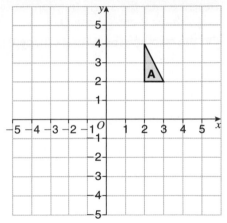

(1387 June 2003)

**3** Describe fully the single transformation that maps shape **P** onto shape **Q**.

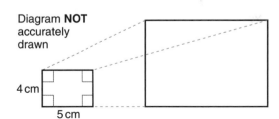

(1388 January 2003)

**4** The diagram represents two photographs.

    **a** Work out the area of the small photograph.

The photograph is to be enlarged by scale factor 3

    **b** Write down the measurements of the enlarged photograph.

    **c** How many times bigger is the area of the enlarged photograph than the area of the small photograph?

Diagram **NOT** accurately drawn

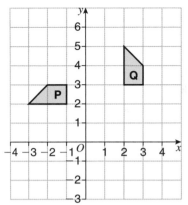

4 cm

5 cm

(1385 June 1998)

**5 a** Triangle **P** has a line of symmetry.
Write down the equation of the line of symmetry.

    **b** On the resource sheet reflect triangle **P** in the $y$-axis.
Label your new triangle **Q**.

    **c** On the same diagram rotate triangle **P** 90° **clockwise** about $(0, 0)$.
Label your new triangle **R**.

    **d** On the same diagram translate triangle **P** by the vector $\begin{pmatrix} -5 \\ -4 \end{pmatrix}$.
Label your new triangle **S**.

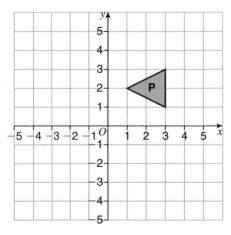

**6 a** Describe fully the single transformation that maps shape **P** onto shape **Q**.

**b** On the resource sheet rotate shape **P** 90° anticlockwise about the point $A$ (1, 1).

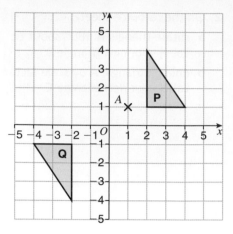

(1385 June 2000)

**7 a** On the resource sheet enlarge triangle **A** by the scale factor $\frac{1}{3}$ with centre the point $P$ $(-7, 7)$.

**b** Describe fully the single transformation which maps triangle **A** onto triangle **C**.

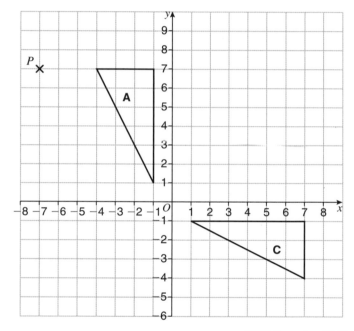

(1385 November 2000)

**8** The shape **P** has been drawn on the grid.

**a** On the resource sheet reflect the shape **P** in the $y$-axis. Label the image **Q**.

**b** On the same diagram rotate the shape **Q** through 180° about (0, 0). Label this image **R**.

**c** Describe fully the single transformation which maps the shape **P** to the shape **R**.

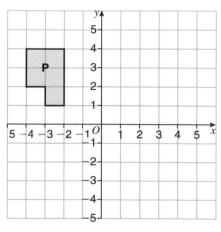

(1385 November 1998)

**9** **a** On the resource sheet reflect the shaded shape **S** in the y-axis. Label the new shape **T**.

   **b** On the same diagram rotate the new shape **T** through an angle of 90° anticlockwise using (0, 0) as the centre of rotation. Label the new shape **U**.

   **c** Describe the single transformation which maps shape **S** onto shape **U**.

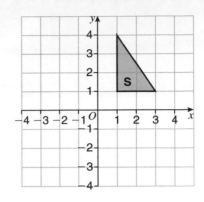

**10** **a** On the resource sheet reflect the shaded triangle **S** in the line $y = 2$ Label the new triangle **T**.

   **b** On the resource sheet translate triangle **T** by the vector $\begin{pmatrix} 0 \\ -4 \end{pmatrix}$. Label the new triangle **U**.

   **c** Describe the single transformation which maps triangle **T** onto triangle **U**.

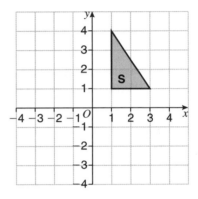

# Inequalities

In this chapter the **inequality signs**, $<$, $>$, $\leqslant$ and $\geqslant$ are used.

> $<$ **means less than, for example** $3 < 4$
> $>$ **means greater than , for example** $5 > 4$
> $\leqslant$ **means less than or equal to**
> $\geqslant$ **means greater than or equal to.**

## 27.1 Inequalities on a number line

The inequality $x > 1$ is shown on a number line.
The *open* circle shows that $x = 1$ is *not included*.

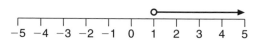

The inequality $x \leqslant 4$ is shown on a number line.
The *filled* circle shows that $x = 4$ is *included*.

This number line shows the values of $x$ which satisfy
both $x > 1$ and $x \leqslant 4$
We write this as $1 < x \leqslant 4$

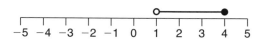

---

### Example 1

Write down the inequality shown on the number line.

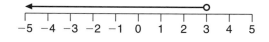

*Solution 1*

$x < 3$

> The open circle indicates that 3 is not included.

---

### Example 2

Show the inequality $x \geqslant -2$ on a number line.

*Solution 2*

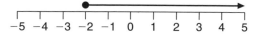

> The filled circle indicates that $-2$ is included.

---

### Exercise 27A

**1** Write down the inequality shown on the number line:

**a**

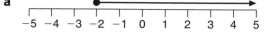

**b**

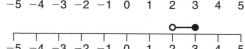

**c**

**d**

**e**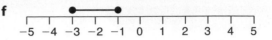

**f**

**g**

Answer questions **2** to **4** on the resource sheet.

2  Show these inequalities on a number line.

   **a** $x > 2$        **b** $x < 5$        **c** $x \geq 0$        **d** $x \leq -1$

3  Show on a number line the values of $x$ which satisfy **both**

   **a** $x > 1$ and $x < 3$        **b** $x > -1$ and $x \leq 0$        **c** $x \geq -4$ and $x \leq -1$

4  Show these inequalities on a number line.

   **a** $-1 < x \leq 3$        **b** $-4 \leq x < 0$        **c** $-5 < x \leq -2$

## 27.2 Solving inequalities

Inequalities can be solved in a similar way to equations but the inequality sign must be kept throughout and the solution is an inequality with the letter on its own.

### Example 3

Solve the inequality $2x - 1 < 7$

*Solution 3*

$$2x - 1 < 7$$

$$2x - 1 + 1 < 7 + 1$$   Add 1 to both sides.

$$2x < 8$$   Divide both sides by 2.

$$x < 4$$   The solution is an inequality with the letter on its own.

### Example 4

**a** Solve the inequality $5x \leq x + 10$

**b** Show your solution on a number line.

*Solution 4*

**a**     $5x \leq x + 10$

$$5x - x \leq x + 10 - x$$   Subtract $x$ from both sides.

$$4x \leq 10x$$

$$x \leq 2.5$$   Divide both sides by 4

**b**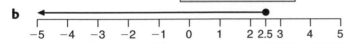

The inequality $2 > 1$ is true.

Multiplying both sides by $-3$ gives $-6 > -3$ which is *not* true but $-6 < -3$ is true.

In general, multiplying or dividing both sides of an inequality by a negative number changes the direction of the inequality.

To avoid this problem rearrange the terms so that the final term in $x$ is positive as in the next example.

## Example 5

Solve the inequality $2\left(\dfrac{x}{3} + 1\right) < 1$

### Solution 5

$2\left(\dfrac{x}{3} + 1\right) < 1$

$\dfrac{2x}{3} + 2 < 1$  | Expand the bracket. |

$\dfrac{2x}{3} < -1$  | Subtract 2 from both sides. |

$2x < -3$  | Multiply both sides by 3 |

$x < -1.5$  | Divide both sides by 2 |

## Example 6

Solve the inequality $-2x > 6$

### Solution 6

> When solving $-2x > 6$ the common error is to divide both sides by $-2$ to get $x > -3$
> This is incorrect since $x = 1$ satisfies $x > -3$ but when $x = 1$, $-2x = -2$ which is *not* greater than 6

$-2x > 6$  | Rearrange the terms so that the $x$ term is positive. |

$-2x + 2x > 6 + 2x$  | Add $2x$ to both sides. |

$0 > 6 + 2x$

$-6 > 2x$  | Subtract 6 from both sides. |

$-3 > x$  | Divide both sides by 2 |

$x < -3$  | $-3$ is greater than $x$ means the same as $x$ is less than $-3$ |

## Example 7

Solve the inequality $2(x - 1) < 7(x + 2)$

### Solution 7

$2(x - 1) < 7(x + 2)$

$2x - 2 < 7x + 14$  | Expand the brackets. |

$-2 - 14 < 7x - 2x$  | Rearrange the terms so that the final term in $x$ is positive. |

$-16 < 5x$

$-3.2 < x$  | Divide both sides by 5 |

$x > -3.2$

## Exercise 27B

**1** Solve these inequalities.

    **a** $x + 1 < 5$      **b** $x - 2 > 7$      **c** $2x < 8$      **d** $3x \geqslant 6$      **e** $4x \leqslant 3$

**2** Solve these inequalities.

    **a** $2x + 1 < 9$      **b** $3x - 1 > 8$      **c** $4x + 3 < 15$      **d** $5x + 12 \geqslant 2$      **e** $6x - 3 \leqslant 0$

**3 a** Solve the inequality $10x + 1 > 16$

    **b** Show your solution on a number line.

**4** Solve these inequalities.

    **a** $3x < 2x + 10$      **b** $5x > x + 12$      **c** $4x \leqslant x - 6$      **d** $7x \geqslant 2x + 35$      **e** $9x \geqslant 5x - 18$

**5 a** Solve the inequality $3x \leqslant 2 - x$     **b** Show your solution on a number line.

**6** Solve the inequality $5x + 7 \leqslant 19$

**7** Solve these inequalities.

  **a** $\dfrac{x}{5} > 4$     **b** $\dfrac{x-1}{3} < 4$     **c** $\dfrac{5x}{8} \geqslant 2$     **d** $2 \leqslant \dfrac{2x-1}{3}$     **e** $x < \dfrac{3x}{4} + 2$

**8** Solve these inequalities.

  **a** $4(x + 1) < 7$     **b** $5x < 2(x + 6)$     **c** $9(x - 2) \leqslant 5(x + 2)$
  **d** $1 + 7(x + 2) \geqslant 5x + 3$     **e** $5(x + 3) - 2 < 3(x + 4)$

**9** Solve these inequalities.

  **a** $-4x > 8$     **b** $2 - 5x < 8 - 2x$     **c** $2(4 - 5x) \leqslant 19$
  **d** $1 + 2(3x + 2) \geqslant 5(2x - 3)$     **e** $7 - 4(x - 3) < 2x$

**10** Solve these inequalities.

  **a** $3x \leqslant 2\left(\dfrac{4x}{5} + 7\right)$     **b** $1 - \dfrac{x}{2} < 2(x - 1)$     **c** $4 + \dfrac{x}{2} > 3(x + 1)$

  **d** $\dfrac{x-2}{3} < \dfrac{x+1}{2}$     **e** $\dfrac{2-5x}{4} \geqslant \dfrac{x}{2} - 3$

## 27.3 Integer solutions to inequalities

The inequality $-4 < x \leqslant 2$ is shown on the number line.

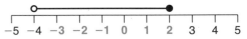

Integers are whole numbers. They can be positive or negative or zero.

So the integer values which satisfy the inequality $-4 < x \leqslant 2$ are $-3, -2, -1, 0, 1, 2$

---

### Example 8

$-4 < 2n \leqslant 3$
$n$ is an integer. Find all the possible values of $n$.

*Solution 8*
$-4 < 2n \leqslant 3$

$-4 < 2n$     AND     $2n \leqslant 3$     | Split $-4 < 2n \leqslant 3$ |

$-2 < n$     AND     $n \leqslant 1.5$     | Solve both inequalities. |

$-2 < n \leqslant 1.5$     | Combine the two inequalities. |

The possible values of $n$ are $-1, 0, 1$     | Write down the integer values satisfying $-2 < n \leqslant 1.5$ |

---

### Example 9

$-4 \leqslant 2p + 1 < 12$
$p$ is a *positive* integer. Find all the possible values of $p$.

*Solution 9*

$-4 \leqslant 2p + 1 < 12$

| | | |
|---|---|---|
| $-4 \leqslant 2p + 1$ | AND | $2p + 1 < 12$ |

Split $-4 \leqslant 2p + 1 < 12$

| | | |
|---|---|---|
| $-5 \leqslant 2p$ | AND | $2p < 11$ |
| $-2.5 \leqslant p$ | AND | $p < 5.5$ |

Solve both inequalities.

$-2.5 \leqslant p < 5.5$

Combine the two inequalities.

The possible values of $p$ are 1, 2, 3, 4, 5

Write down the *positive* integer values satisfying $-2.5 \leqslant p < 5.5$
Note 0 is not a positive integer.

## 27.4 Problems involving inequalities

### Example 10

A rectangular field has length $(3x + 5)$ metres and width $(2x - 3)$ metres.
**a** Explain why $x > 1.5$
**b** The perimeter of the field is no more than 60 metres. Find the greatest possible length of the field.

*Solution 10*

**a** $2x - 3 > 0$ since the width must be positive $2x > 3$ so $x > 1.5$

**b** Perimeter $= 2(3x + 5) + 2(2x - 3)$

Perimeter $= 2l + 2w$

$2(3x + 5) + 2(2x - 3) \leqslant 60$

Perimeter $\leqslant 60$ (no more than 60 so it can be equal to 60).

$6x + 10 + 4x - 6 \leqslant 60$
$10x + 4 \leqslant 60$
$10x \leqslant 56$
$x \leqslant 5.6$

Solve the inequality.

$3x + 5 \leqslant 3 \times 5.6 + 5$
Length $\leqslant 21.8$

$x \leqslant 5.6$ so $3x \leqslant 3 \times 5.6$ and $3x + 5 \leqslant 3 \times 5.6 + 5$

The greatest possible length is 21.8 metres

### Exercise 27C

**1 a** Show the inequality $2 \leqslant y < 4$ on a number line.
   **b** If $y$ is an integer, use your number line to write down all the possible values of $y$.

**2 a** Show the inequality $-3 \leqslant p < 2$ on a number line.
   **b** If $p$ is an integer, use your number line to write down all the possible values of $p$.

**3** $-8 < y < -5$
   $y$ is an integer.
   Write down all the possible values of $y$.

**4 a** Solve $-6 < 2x \leqslant 4$
   **b** Given that $x$ is an integer and $-6 < 2x \leqslant 4$, write down all the possible values of $x$.

**5 a** Solve $-3 < 4w \leqslant 13$
   **b** Given that $w$ is an integer and $-3 < 4w \leqslant 13$, write down all the possible values of $w$.

**6 a** Solve $-3 < x - 1 \leqslant 5$

**b** Given that $x$ is an integer and $-3 < x - 1 \leqslant 5$, write down all the possible values of $x$.

**7** $2p \leqslant 9$
$p$ is a *positive* integer.
Find all the possible values of $p$.

**8** $5q - 1 < 19$
$q$ is a *positive* integer.
Find all the possible values of $q$.

**9** $2x + 10 > 1$
$x$ is a *negative* integer.
Find all the possible values of $x$.

**10 a** Solve $-7 \leqslant 2p + 3 < 9$

**b** $p$ is a *positive* integer and $-7 \leqslant 2p + 3 < 9$
Find all the possible values of $p$.

**11 a** Solve $-1 < \dfrac{2x}{3} \leqslant 2$     **b** $4 < 5(x + 1) < 10$     **c** $-3 \leqslant 2(1 - 2x) \leqslant 6$

**12** A rectangular field has length $(2x + 8)$ metres and width $(4x - 3)$ metres.
The perimeter of the field is no more than 112 metres. Find the greatest possible area of the field.

**13** The Smiths have four daughters. The table shows some information about their ages in years.

| | Ann | Betty | Cath | Debra |
|---|---|---|---|---|
| **Ages in years** | $x$ | $2x + 3$ | 12 | 20 |

Betty is older than Cath but younger than Debra.
Work out all Ann's possible ages.

## 27.5 Solving inequalities graphically

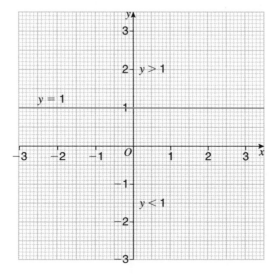

The diagram shows the line with equation $y = 1$
The coordinates of all points **on** this line satisfy the equation $y = 1$
All points **above** this line have a $y$-coordinate greater than 1 and so satisfy the inequality $y > 1$
All points **below** this line have a $y$-coordinate less than 1 and so satisfy the inequality $y < 1$

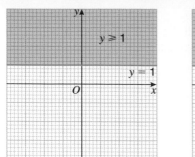     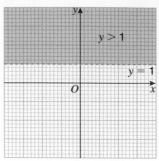

A solid line as a boundary is considered to be part of the shaded region.
A dashed boundary is not part of the shaded region.

Each diagram above shows a region shaded above the line $y = 1$

When the line is solid, the coordinates of all points in the shaded region satisfy the inequality $y \geqslant 1$

When the line is dashed, the coordinates of all points in the shaded region satisfy the inequality $y > 1$

The diagram shows a shaded region bounded by the lines $x = -1$ and $x = 2$

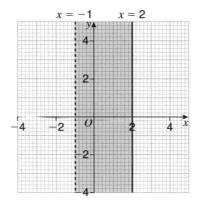

The coordinates of all points in this region satisfy *both* the inequalities

$x > -1$ (since this boundary is dashed)
$x \leqslant 2$ (since this boundary is solid)

This is written as $-1 < x \leqslant 2$

The diagram shows the shaded region bounded by the lines $x = -1$, $x = 2$, $y = 1$ and $y = 4$

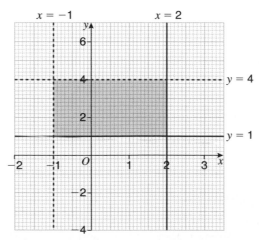

The coordinates of all points in this region satisfy both $-1 < x \leqslant 2$ and $1 \leqslant y < 4$

## Example 11

On a grid, shade the region of points whose coordinates satisfy the inequalities
$-3 \leqslant x < 0$ and $-2 \leqslant y < 1$

### Solution 11

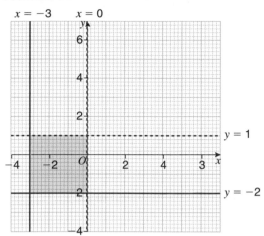

A dashed line is drawn very close to the $y$-axis to show that points on this boundary are not included.

The diagram shows the line with equation $x + y = 3$

The coordinates of all points **on** this line satisfy the equation $x + y = 3$

The coordinates of all points **above** this line satisfy the inequality $x + y > 3$

The coordinates of all points **below** this line satisfy the inequality $x + y < 3$

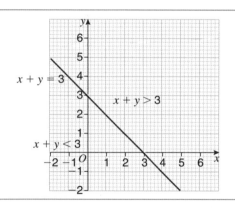

For example, the point (4, 2) lies above the line $x + y = 3$ and since $4 + 2 = 6$, its coordinates satisfy the inequality $x + y > 3$

## Example 12

**a** Write down the three inequalities which must be satisfied by the coordinates of all points in the shaded region.

**b** $x$ and $y$ are integers. Mark with a cross (**✗**) the three points which satisfy the three inequalities.

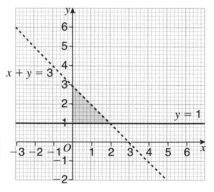

### Solution 14

**a**    $y \geqslant 1$    The shaded region is *above* the *solid* line $y = 1$

       $x \geqslant 0$    The shaded region is to the *right* of the *solid* $y$-axis ($x = 0$)

    $x + y < 3$    The shaded region is *below* the *dashed* line $x + y = 3$

**b**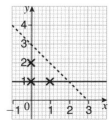

The three points must be in the shaded region which includes the lines $x = 0$ and $y = 1$
The coordinates of the three points are (0, 1), (0, 2) and (1, 1).

**Example 13**

**a**  **i** On a grid scaled from 0 to +8 on each axis, draw the line with equation $x + y = 5$

 **ii** On the same grid draw the line with equation $2y = x + 4$

**b** On the grid, shade the region of points whose coordinates satisfy the three inequalities $x + y \geqslant 5$, $2y \leqslant x + 4$, $2x < 7$

**c** From your graph, write down the solution of the simultaneous equations

$$x + y = 5$$
$$2y = x + 4$$

**d** Write down the least value of $x$ in the shaded region.

*Solution 13*

**a b**

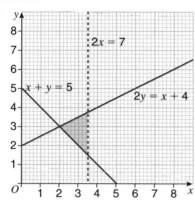

**c** $x = 2, y = 3$

**d** Least value of $x$ is 2

**a  i** $x + y = 5$ passes through (0, 5) and (5, 0)

 **ii** To draw the line $2y = x + 4$ plot and join some of the points

(0, 2)   (2, 3)   (4, 4)   (6, 5)

Or write the equation in the form $y = 0.5x + 2$ and draw the line with gradient 0.5 and intercept 2 on the $y$-axis.

**b** Draw the dashed vertical line $2x = 7$ or $x = 3.5$ for the inequality $2x < 7$.
The shaded region will be:

● to the left of the line $2x = 7$ (since $2x < 7$)

● above the solid line $x + y = 5$ (since $x + y \geqslant 5$)

● below the solid line $2y = x + 4$ (since $2y \leqslant x + 4$)

Find the coordinates of the point where the line $x + y = 5$ and the line $2y = x + 4$ intersect.

Find the $x$-coordinate of the point which is furthest to the left in the shaded region.

**Exercise 27D**

**1** For each of these questions use the resource sheet with a grid scaled from −5 to +5 on each axis to shade the region of points whose coordinates satisfy

**a** $x \leqslant 3$  **b** $y > -1$  **c** $0 \leqslant x < 2$  **d** $-4 \leqslant y < 4$

**e** $-1 \leqslant x \leqslant 3$ and $-2 \leqslant y \leqslant 0$  **f** $1 \leqslant x < 5$ and $0 \leqslant y \leqslant 3$

**g** $-2 \leqslant x < 3$ and $-4 \leqslant y < 1$  **h** $-3 < x < 1$ and $-3 \leqslant y < -1$

**i** $0 < x < 3$ and $-2 < y < 0$  **j** $y \geqslant x$  **k** $y < -x$

**l** $y > x - 1$  **m** $x + y \leqslant 4$

**n** $x + y \geqslant 1$ and $x + y < 3$  **o** $2x + y \leqslant 4$, $y < x + 1$ and $y > 0$

**2** The diagram shows a shaded region bounded by three lines.

**a** Write down the equation of each of the three lines.

**b** Write down the three inequalities satisfied by the coordinates of the points in the shaded region.

**c** Write down the greatest value of $x$ in the shaded region.

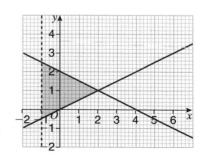

**3** The diagram shows the line with equation $y = 2x - 3$

   **a** On a copy of the grid, draw the line $2x + y = 6$

   **b** Shade the region for which $y \geqslant 2x - 3$, $2x + y \leqslant 6$ and $x > 1$

   **c** $x$ and $y$ are integers. Write down the coordinates of the two points which satisfy the three inequalities.

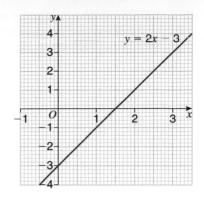

**4 a** Show that the points $(0, 4)$ and $(6, 0)$ lie on the line with equation $3y + 2x = 12$

   **b** The diagram shows a shaded region.

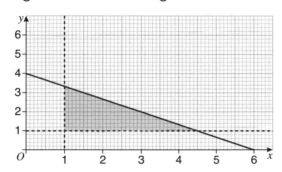

Write down the *three* inequalities which must be satisfied by the coordinates of all points in the shaded region.

   **c** $x$ and $y$ are integers. Write down the coordinates of the two points which satisfy the three inequalities.

**5** The diagram shows the lines with equations $y = 2x$ and $y = x + 3$

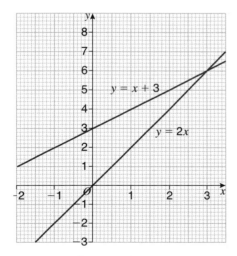

   **a** From the graph, write down the solution of the simultaneous equations
$y = 2x$
$y = x + 3$

   **b** Shade the region for which $x + y > -1$, $y \leqslant x + 3$, $y \geqslant 2x$ and $2y < 5$

   **c** $x$ and $y$ are integers. Write down the coordinates of the set of points which satisfy *all* four inequalities, $x + y > -1$, $y \leqslant x + 3$, $y \geqslant 2x$ and $2y < 5$

**6 a** $6x + 5y < 30$     $x$ and $y$ are both integers.
Write down two possible pairs of values, $(x = ..., y = ...)$,
that satisfy this inequality.

**b** $6x + 5y < 30, 2x < y, y > 1, x > 0$
$x$ and $y$ are both integers.
On the resource sheet, mark with a cross (**X**), each
of the *three* points which satisfy *all* these four inequalities.

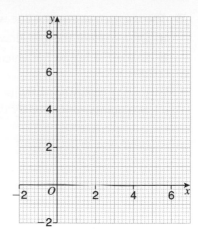

## Chapter summary

**You should now know that:**

★  the **signs** used for **inequalities** are $<$, $>$, $\leqslant$ and $\geqslant$

> $<$ **means less than**
> $>$ **means greater than**
> $\leqslant$ **means less than or equal to**
> $\geqslant$ **means greater than or equal to**

★  to show an inequality on a number line, you use a $\circ$ together with a line for $<$ and $>$ and use a $\bullet$ together with a line for $\leqslant$ and $\geqslant$

★  **solving** inequalities is similar to solving equations but the inequality sign must be kept throughout and the solution is an inequality with a letter of its own

★  multiplying or dividing both sides of an inequality by a negative number will change the direction of an inequality

★  to find **integer** solutions to inequalities, a number line can be used, for example, the integer solutions to the inequality $-2 < p \leqslant 1$ are $-1, 0$ and $1$

★  to solve inequalities graphically, draw regions bounded by straight lines

★  a solid line as a boundary is considered to be part of the shaded region, a dashed boundary is not part of the shaded region.

## Chapter 27 review questions

**1**  $-6 < y < -3$
$y$ is an integer.
Write down all the possible values of $y$.

**2**  Solve $5x + 3 > 19$                                                                         (1388 March 2005)

**3 a**  Solve the equation $5 - 3x = 2(x + 1)$
  **b**  $-3 \leqslant y < 3$
$y$ is an integer.
Write down all the possible values of $y$.                                          (1388 June 2005)

**4** $5p \leqslant 17$
$p$ is a positive integer.
Write down all the possible values of $p$.

**5** **a** Solve the inequality $3x \leqslant 7 - x$
　　**b** Show your answer to part **a** on a number line.

**6** $-5 < 2m \leqslant 9$
$m$ is an integer.
Write down all the possible values of $m$.

**7** $3x + 20 > 3$
$x$ is a negative integer.
Write down all the possible values of $x$.

**8** $2 < 6y + 19$
$y$ is a negative integer.
Write down all the possible values of $y$.

**9** Solve the inequality $3x + 2 > -7$ 　　　　　　　　　　　　　　(1388 June 2003)

**10** Solve $3x - 4 < -16$ 　　　　　　　　　　　　　　　　　　　(1388 Mock)

**11** **a** Solve the inequality $5x + 12 > 2$ 　　**b** Expand and simplify $(x - 6)(x + 4)$. 　(1388 January 2003)

**12** Solve $4 < x - 2 \leqslant 7$ 　　　　　　　　　　　　　　　　　　(1388 March 2003)

**13** $n$ is a whole number such that $7 \leqslant 3n < 15$
List all the possible values of $n$. 　　　　　　　　　　　　　　(1388 January 2004)

**14** **a** Solve the equation $4x + 3 = 2(x - 3)$ 　　**b** Solve the inequality $2x + 3 \leqslant 8$ 　　(1385 June 1999)

**15** $n$ is a whole number such that $6 < 2n < 13$
List all the possible values of $n$. 　　　　　　　　　　　　　　(1385 November 2000)

**16** **a** Solve the equation $7(x - 1) = 2x - 1$
　　**b** **i** Solve the inequality $4y + 3 \geqslant 1$
　　　　**ii** Write down the smallest **integer** value of $y$ which satisfies the inequality $4y + 3 \geqslant 1$
　　　　　　　　　　　　　　　　　　　　　　　　　　　　　　　(1385 June 2002)

**17** $n$ is an integer such that $-5 < 2n \leqslant 6$
　　**a** List all the possible values of $n$. 　　**b** Solve the inequality $5 + x > 5x - 11$ 　(1387 Mock)

**18** **a** Solve the inequality $4x - 3 < 7$

An inequality is shown on the number line.

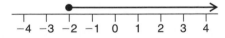

　　**b** Write down the inequality. 　　　　　　　　　　　　　　(1388 March 2004)

**19 a** Solve the inequality $7x - 3 > 17$

    $x$ is a whole number such that $7x - 3 > 17$

  **b** Write down the smallest value of $x$.                     (1388 November 2005)

**20 a** $4x + 3y < 12$

    $x$ and $y$ are both integers.

    Write down two possible pairs of values, $(x = ..., y = ...)$, that satisfy this inequality.

  **b** $4x + 3y < 12,$     $y < 3x,$     $y > 0,$     $x > 0$

    $x$ and $y$ are both integers.

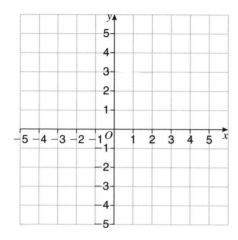

On the resource sheet, mark with a cross (**X**) each of the *three* points which satisfy *all* these four inequalities.                   (1387 November 2005)

**21 a** On the resource sheet draw the line with equation $x + 2y = 6$

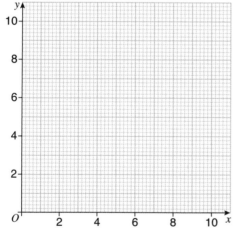

  **b** Shade the region for which $x + 2y \leqslant 6, 0 \leqslant x \leqslant 4$ and $y \geqslant 0$.     (1385 November 1997)

**22** The perimeter of this rectangle has to be more than 11 cm and less than 20 cm.

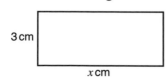

3 cm                                Diagram **NOT** accurately drawn

$x$ cm

  **a** Show that $5 < 2x < 14$

  **b** $x$ is an integer. List all the possible values of $x$.          (1385 November 2002)

**23 a** $-2 < x \leqslant 1$, $x$ is an integer.
   Write down all the possible values of $x$.

   **b** $-2 < x \leqslant 1$, $y > -2$, $y < x + 1$,
   $x$ and $y$ are integers.

   On the resource sheet, mark with a cross (**X**) each
   of the six points which satisfies *all* these three
   inequalities.                                   (1387 June 2003)

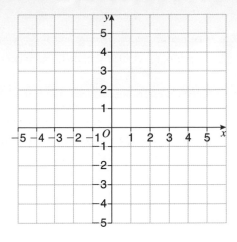

**24 a** On the resource sheet, draw straight lines and use
   shading to show the region $R$ that satisfies the
   inequalities $x > 1$, $y > x + 1$ and $x + y \leqslant 7$

   **b** Write down the coordinates of all the points of
   $R$ whose coordinates are both integers.

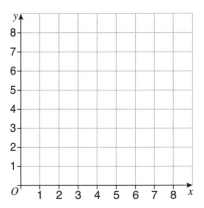

# Formulae

A **formula** is a way of describing a fact or a rule. A formula can be written using algebraic expressions.
A formula must have an = sign.
In Section 12.5 the area ($A$) of a trapezium was found using

$$A = \tfrac{1}{2}(a + b)h$$

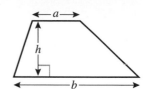

This is called an **algebraic formula**.
$A$ appears just once and only on the left-hand side of the formula,
so $A$ is called the **subject of the formula**.
The letters $a$ and $b$ represent the lengths of the parallel sides of the trapezium and $h$ represents the height of the trapezium.

## 28.1 Using an algebraic formula

The formula $A = \tfrac{1}{2}(a + b)h$ can be used to work out the area of a trapezium.
The value of $A$ is calculated by **substituting** values of $a$, $b$ and $h$ into this formula.

### Example 1

The area, $A$ cm², of this hexagon is given by the formula

$$A = h(a + b)$$

Find the area of the hexagon when $a = 5$ cm, $b = 9$ cm and $h = 4$ cm.

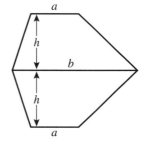

**Solution 1**
$A = h(a + b)$

$\quad = 4(5 + 9)$     Substitute $a = 5$, $b = 9$ and $h = 4$

$\quad = 4 \times 14 = 56$

Area of hexagon = 56 cm².

### Example 2

Use the formula $F = 3n^2 - 5n$ to work out the value of $F$ when $n = -2$

**Solution 2**
$F = 3n^2 - 5n$

$\quad = 3 \times (-2)^2 - 5 \times (-2)$    Substitute $n = -2$

$\quad = 3 \times 4 + 10$    $(-2)^2 = +4$   and   $-5 \times (-2) = +10$, negative $\times$ negative = positive.

$F = 22$

**Example 3**

This formula can be used to change temperatures in degrees Fahrenheit ($F$) to temperatures in degrees Celsius ($C$).

$$C = \tfrac{5}{9}(F - 32)$$

Change $-4$ degrees Fahrenheit to degrees Celsius.

*Solution 3*

$C = \tfrac{5}{9}(F - 32)$

$\quad = \tfrac{5}{9}(-4 - 32)$    Substitute $F = -4$

$\quad = \tfrac{5}{9} \times -36$    Work out the brackets first.

$\quad = -20$

$-4°F = -20°C$

THE LIBRARY
NORTH WEST KENT COLLEG
DERING WAY, GRAVESEND

**Exercise 28A**

**1** Sunita's pay is worked out using the formula

$$P = rh$$

$P$ = pay, $r$ = rate of pay for each hour worked and $h$ = number of hours worked.
Use this formula to work out Sunita's pay when she works for 5 hours at a rate of £8 an hour.

**2** $T = c + 3d$
Use this formula to work out the value of $T$, when
   **a** $c = 12, d = 9$                  **b** $c = 10, d = -6$

**3** The formula $m = \dfrac{y - c}{x}$ can be used to work out the gradient of a straight line.
Use this formula to work out the value of $m$ when
   **a** $x = 2, y = 11$ and $c = 4$        **b** $x = 3, y = 1$ and $c = 10$

**4** The formula $v = u + at$ is used to work out velocity, $v$.
Use this formula to work out the value of $v$ when
   **a** $u = 5, a = 10$ and $t = 3$      **b** $u = 4, a = -5$ and $t = 2$
   **c** $u = -20, a = -2$ and $t = 8$

**5** This formula can be used to change temperatures in degrees Celsius ($C$) to temperatures in degrees Fahrenheit ($F$).

$$F = \tfrac{9}{5}C + 32$$

Use this formula to find the value of $F$ when
   **a** $C = 30$         **b** $C = 0$                **c** $C = -40$

**6** Use the formula

$$P = \sqrt{3x}$$

to work out the value of $P$ when
   **a** $x = 3$          **b** $x = 12$

**7**  $W = 5t^2 + t$
Find the value of $W$ when

**a**  $t = 3$                          **b**  $t = -4$

**8**  $P = kr^2 - 2rs$
Work out the value of $P$ when

**a**  $k = 10, r = 4$ and $s = 5$         **b**  $k = 3, r = -5$ and $s = 10$

**9**  The cooking time, $T$ (minutes), to cook a joint of meat is given by the formula

> $T = 20w + 30$

where $w$ (kg) is the weight of the joint of meat.
Use this formula to work out the cooking time when

**a**  $w = 4$                          **b**  $w = 2.5$

**10**  This formula is used to work out the distance ($d$) travelled, when the time taken is $t$ and the average speed is $s$.

> $d = st$

Use this formula to work out the value of $d$

**a**  when $s = 30$ and $t = 2$   **b**  when $s = 5.5$ and $t = 6$   **c**  when $s = 20$ and $t = 3.4$

**11**  $T = 2\pi\sqrt{\dfrac{l}{g}}$
Work out the value of $T$ when $\pi = 3.14$, $l = 12$ and $g = 9.8$

**12**  $s = ut + \frac{1}{2}at^2$
Find the value of $s$ when $u = 24, t = 6$ and $a = -2.4$

## 28.2 Writing an algebraic formula

The diagram shows a rectangle.

>
> $a$ = the length of the rectangle.
> $b$ = the width of the rectangle.

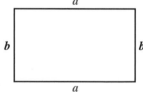

The perimeter of a shape is the total distance around the edges of the shape.
The perimeter of this rectangle $= a + b + a + b$.
Collecting the like terms together gives the perimeter $= 2a + 2b$.
Let $P$ represent the perimeter of the rectangle.

This can then be written as the algebraic formula

> $P = 2(a + b)$

When writing an algebraic formula it is important to define what each letter stands for.

### Example 4

In some football matches, 3 points are awarded for a win, 1 point is awarded for a draw and no points are awarded for a loss.
Write down an algebraic formula that can be used to work out the total points awarded to a football team. You must define the letters used.

*Solution 4*

Let $P$ = the total number of points awarded to each team.

Let $w$ = the number of matches won.

Let $d$ = the number of matches drawn.

The number of points awarded for wins = 3 × the number of matches won = $3 \times w = 3w$.

The number of points awarded for draws =1 × the number of matches drawn =$1 \times d = d$.

The formula for the total number of points awarded to each team is $P = 3w + d$.

## Exercise 28B

**1** A florist sells $t$ tulips at 40p each and $d$ daffodils at 30p each. If $C$ pence is the total cost of the flowers sold, write down, in terms of $t$ and $d$, a formula for $C$.

**2** A shop sells eating apples and baking apples.
Write down an algebraic formula that can be used to work out the total number of apples that the shop sells. You must define the letters used.

**3** Dan hires a car for his holidays. He pays a fixed charge of £75 plus £30 for each day that he has the car.

  **a** Write down a formula that Dan can use to work out the total cost of hiring the car.
  You must define the letters used.

  **b** Use your formula to work out the cost of hiring this car for 7 days.

**4 a** Write down a formula that can be used to work out the perimeter of an equilateral triangle.
  You must define the letters used.

  **b** Use your formula to work out the perimeter of an equilateral triangle of side 5 m.

**5** Andrew earns $p$ pounds per hour. He works for $t$ hours. He earns a bonus of $b$ pounds.

  **a** Write down a formula for the total amount he earns, $T$ pounds.

  **b** $p = 5, t = 38$ and $b = 20$

  Use your formula to work out the value of $T$.

**6** A box of nails costs £1.50    A box of screws costs £2
Bob buys $n$ boxes of nails and $s$ boxes of screws. The total cost is $C$ pounds.

  **a** Write a formula for $C$ in terms of $n$ and $s$.

  **b** Work out the value of $C$ when $n = 4$ and $s = 3$

**7** Jean sits a mathematics exam, an English exam and a science exam. Her mean mark is $x$%.
Write down a formula that can be used to work out the value of $x$. You must define the letters used.

**8** A bus can seat $p$ passengers and 8 passengers are allowed to stand.

  **a** Write a formula that can be used to work out the total number, $T$, of passengers that can travel on $b$ buses.

  **b** **i** Work out the value of $T$ when $b = 6$ and $p = 60$

   **ii** What does your value of $T$ tell you?

## 28.3 Changing the subject of a formula

The subject of the formula appears just once and only on the left-hand side of the formula.

$p$ is the subject of the formula

$$p = 2q + 3$$

The formula can be rearranged, using the balance method, as follows

$p - 3 = 2q + 3 - 3$ | Subtract 3 from both sides.

$p - 3 = 2q$

$\dfrac{p - 3}{2} = \dfrac{2q}{2}$ | Divide both sides by 2

$\dfrac{p - 3}{2} = q$

This is written as $q = \dfrac{p - 3}{2}$

The subject of the formula $p = 2q + 3$ has now been changed to make $q$ the subject of the formula.

### Example 5

Make $R$ the subject of the formula $V = IR$.

*Solution 5*

$V = IR$

$\dfrac{V}{I} = \dfrac{\cancel{I}R}{\cancel{I}}$ | Divide both sides by $I$ and then cancel.

$\dfrac{V}{I} = R$

$R = \dfrac{V}{I}$ | Write the subject $(R)$ on the left-hand side.

### Example 6

$P = 2(a + b)$ can be used to work out the perimeter, $P$, of a rectangle of length $a$ and width $b$. Make $b$ the subject of the formula.

*Solution 6*

$P = 2(a + b)$

$P = 2a + 2b$ | Expand the brackets.

$P - 2a = 2b$ | Subtract $2a$ from both sides to get only the $b$ term on one side.

$b = \dfrac{P - 2a}{2}$ | Divide both sides by 2 and write $b$ on the left-hand side.

## Example 7

Make $x$ the subject of the formula $y = \dfrac{3 - 4x}{5}$.

**Solution 7**

$y = \dfrac{3 - 4x}{5}$

$y \times 5 = \dfrac{3 - 4x}{\cancel{5}} \times \cancel{5}$    | Multiply both sides by 5, then cancel. |

$5y = 3 - 4x$

| Now rearrange this to get a positive $x$ term on one side. |

$5y + 4x = 3 - 4x + 4x$    | Add $4x$ to both sides. |

$5y + 4x = 3$

$5y + 4x - 5y = 3 - 5y$    | Subtract $5y$ from both sides to get $4x$ on its own. |

$4x = 3 - 5y$

$\dfrac{\cancel{4}x}{\cancel{4}} = \dfrac{3 - 5y}{4}$    | Divide both sides by 4, then cancel. |

$x = \dfrac{3 - 5y}{4}$

## Example 8

**a** Given that $t$ is positive, rearrange the formula $W = u + at^2$ to make $t$ the subject.

**b** Find the value of $t$ when $W = 120$, $u = 80$ and $a = 10$

**Solution 8**

**a** $W = u + at^2$    | Rearrange to get only the $t$ term on one side. |

$W - u = u + at^2 - u$    | Subtract $u$ from both sides... |

$W - u = at^2$    | ... to get the $t$ term on one side. |

$\dfrac{W - u}{a} = \dfrac{\cancel{a}t^2}{\cancel{a}}$    | Divide both sides by $a$, then cancel. |

$\sqrt{\dfrac{W - u}{a}} = \sqrt{t^2}$    | Take the square root of both sides to find $t$. |

$t = \sqrt{\dfrac{W - u}{a}}$

**b** $t = \sqrt{\dfrac{120 - 80}{10}}$    | Substitute $W = 120$, $u = 80$ and $a = 10$ into the formula $t = \sqrt{\dfrac{W - u}{a}}$. |

$t = \sqrt{4}$

$t = 2$    | $t$ is positive so ignore the $-2$ solution. |

## Exercise 28C

**1** Make $x$ the subject of the formula

     **a** $y = x + 7$        **b** $y = 4x$        **c** $y = 2x + 1$

     **d** $y = 3x - 5$       **e** $y = 1 - x$       **f** $y = 6 - 2x$

**2** Make $w$ the subject of the formula $A = lw$.

**3** Make $t$ the subject of the formula $s = ut + u^2$.

**4** Make $m$ the subject of the formula $W = 3n - 2m$.

**5** A formula to find velocity is given by $v = u + at$.
Rearrange this formula to make

    **a** $u$ the subject of the formula                 **b** $a$ the subject of the formula

    **c** $t$ the subject of the formula.

**6** Make $P$ the subject of the formula

    **a** $D = 3(P + Q)$         **b** $D = 4(2P - 3Q)$         **c** $D = 2(5Q - P)$

**7** Make $a$ the subject of the formula $c = \dfrac{a - b}{3}$

**8** Make $f$ the subject of the formula $g = \dfrac{e + 3f}{2}$

**9** Make $n$ the subject of the formula $a = \dfrac{5(b - 3n)}{2}$

**10** The area, $A$, of a circle is given by the formula $A = \pi r^2$, where $r$ is the radius.
Make $r$ the subject of the formula.

**11**  **a**  Make $u$ the subject of the formula $v^2 = u^2 + 2as$.

    **b**  Find the positive value of $u$ when $v = 20$, $a = 10$ and $s = 12.8$

## 28.4 Expressions, identities, equations and formulae

We know from Section 8.1 that $3x - 6$ is called an **expression**.

The expression $3x - 6$ can be factorised to give $3(x - 2)$.
$3x - 6 = 3(x - 2)$ is called an **identity** because the left-hand side, $3x - 6$, says the same as the right-hand side, $3(x - 2)$. Another example of an identity is $5x = x + 4x$.

From Section 23.1, we know that $3x - 6 = 0$ is called an **equation**, which can be solved to find the value of $x$.
$P = 3x - 6$ is called a **formula**. The value of $P$ can be worked out if the value of $x$ is known.
A formula has at least two letters.

---

### Example 9

Here is a mixture of some terms and some signs.

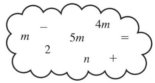

Using some of the above, write down an example of

**a**  an expression     **b**  an identity     **c**  an equation     **d**  a formula.

*Solution 9*

**a**  $4m - n$      | An expression does not have an = sign. |

**b**  $4m = 5m - m$    | An identity is true for all values of the letter. |

**c**  $5m = 2 + m$     | An equation can be solved. In this case the solution is $m = \tfrac{1}{2}$ |

**d**  $n = 5m + 2$     | A formula must have at least two letters and an = sign. Here $n$ is the subject of the formula. |

---

## Exercise 28D

**1** Write down whether each of the following is an expression or an identity or an equation or a formula.

**a** $4x = x + x + x + x$

**b** $y = Ax$

**c** $3m + 2m = 5m$

**d** $y - 3 = 2$

**e** $5s + t$

**f** $A = \dfrac{bh}{2}$

**g** $x^2 + 4x = 3$

**h** $p^3 + q^2$

**i** $C = 3r - 2s$

**j** $6(y + 4) = 6y + 24$

**k** $\dfrac{x}{3} = 4$

**l** $y = mx + c$

**m** $c + c + d + d + d = 2c + 3d$

**n** $q^2 + q - 3 = 0$

**o** $7r^2 + 3s$

**2** Here is a mixture of some terms and some signs.

Using some of the above, write down an example of

**a** an expression      **b** an identity      **c** an equation      **d** a formula.

## 28.5 Further changing the subject of a formula

To change the subject of some formulae, further steps including factorisation and squaring of algebraic expressions are required.

### Example 10

Make $t$ the subject of the formula $v = ut + 2nt$.

**Solution 10**

$$v = ut + 2nt$$

> All the $t$ terms are on the same side.

$$v = t(u + 2n)$$

> There is more than one term involving $t$ so factorise the right-hand side.

$$\frac{v}{u + 2n} = \frac{t(\cancel{u + 2n})}{\cancel{u + 2n}}$$

> Divide both sides by $(u + 2n)$, then cancel.

$$t = \frac{v}{u + 2n}$$

> Write the new subject, $t$, on the left-hand side.

## Example 11

Make $A$ the subject of the formula $r = \dfrac{1}{2}\sqrt{\dfrac{A}{\pi}}$

**Solution 11**

$r = \dfrac{1}{2}\sqrt{\dfrac{A}{\pi}}$ | Rearrange so that only $\sqrt{\dfrac{A}{\pi}}$ is on one side.

$2 \times r = 2 \times \dfrac{1}{2}\sqrt{\dfrac{A}{\pi}}$ | Multiply both sides by 2

$2r = \sqrt{\dfrac{A}{\pi}}$

$(2r)^2 = \dfrac{A}{\pi}$ | Remove the square root sign by squaring both sides.

$4r^2 = \dfrac{A}{\pi}$

$\pi \times 4r^2 = A$ | Multiply both sides by $\pi$ and then cancel.

$A = 4\pi r^2$ | Write the new subject, $A$, on the left-hand side.

## Example 12

Make $c$ the subject of $a + 3c = 4(bc + 2d)$

**Solution 12**

$a + 3c = 4(bc + 2d)$ | There are $c$ terms on both sides.

$a + 3c = 4bc + 8d$ | Expand the brackets.

$a + 3c - 4bc = 8d$ | Subtract $4bc$ from both sides to get all $c$ terms on one side.

$3c - 4bc = 8d - a$ | Subtract $a$ from both sides.

$c(3 - 4b) = 8d - a$ | There is more than one term involving $c$. Factorise the left-hand side.

$c = \dfrac{8d - a}{3 - 4b}$ | Divide both sides by $(3 - 4b)$.

## Example 13

The radius of a circle is $r$ cm.
The length of a rectangle is $(x + 2)$ cm and its width is $x$ cm.
The area of the circle and the area of the rectangle are equal.
Express $r$ in terms of $x$ and $\pi$.

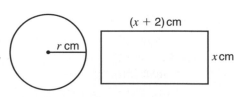

**Solution 13**

Area of circle $= \pi r^2$    Area of rectangle $= x(x + 2)$

$\pi r^2 = x(x + 2)$ | Area of the circle = area of the rectangle.

$r^2 = \dfrac{x(x + 2)}{\pi}$ | Divide both sides by $\pi$.

$r = \sqrt{\dfrac{x(x + 2)}{\pi}}$ | Take the square root of both sides.

**Exercise 28E**

**1** Make $x$ the subject of $y + 2x = 1 - 3x$.

**2** Rearrange $\dfrac{y + x}{3} = 2(x - 1)$ to make $x$ the subject.

**3** The diagram shows a right-angled triangle and a rectangle.
The area of the triangle is equal to the area of the rectangle.
Express $h$ in terms of $b$ and $x$.

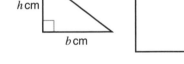

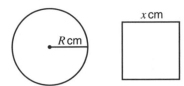

**4** Make $m$ the subject of the formula $W = a(3 - m) + 5m$

**5** Make $c$ the subject of $a + bc = 5 - 3c$

**6** Rearrange $2p + 3q = r(q - 4)$ to make $q$ the subject.

**7** The radius of a circle is $R$ cm.
The length of a side of a square is $x$ cm.
The area of the circle and the area of the square are equal.
   **a** Express $R$ in terms of $x$ and $\pi$.
   **b** Hence find the radius of the circle when the perimeter of the square is 40 cm.
   Give your answer in terms of $\pi$.

**8** Rearrange $5(x - 3) = y(7 + 2x)$ to make $x$ the subject.

**9** Make $w$ the subject of the formula $y = \dfrac{3 + 2w}{w}$

**10** Make $x$ the subject of $\dfrac{1 - x}{1 + x} = y$

**11** Make $x$ the subject of $\dfrac{x}{a} = 1 + bx$

**12** Make $A$ the subject of the formula
$P = \sqrt{\dfrac{\pi A - 5}{A}}$

**13** Make **a** $l$ **b** $g$
the subject of the formula $T = 2\pi\sqrt{\dfrac{l}{g}}$

## Chapter summary

**You should now know:**

★ that a **formula** can be written using algebraic expressions to describe a rule or a relationship. For example, the area of a rectangle can be written as an **algebraic formula** $A = lw$, where $A$ = area, $l$ = length and $w$ = width.

★ that in a formula, for example, $A = lw$, $A$ is called the **subject of the formula**.

★ the meanings of **expression**, **identity**, **equation** and **formula**.

**You should also be able to:**

★ find the value of the subject of a formula by **substituting** given values for the letters on the right-hand side.

★ **change the subject** of a simple formula by rearranging the terms in the formula.

★ change the subject of a formula which may involve square roots, or the required subject occurring twice.

## Chapter 28 review questions

**1** $D = 3s - 7t$
$s = -4, t = 2$
Work out the value of $D$.

**2** $M = \dfrac{Q(P + 2)}{6}$
$P = -4, Q = 30$
Work out the value of $M$.

**3** $P = Q^2 - 2Q$
Find the value of $P$ when $Q = -3$                              (1388 March 2004)

**4** The number of diagonals, $D$, of a polygon with $n$ sides is given by the formula

$$D = \frac{n^2 - 3n}{2}$$

A polygon has 20 sides.
Work out the number of diagonals of this polygon.               (1388 March 2005)

**5** A bicycle has 2 wheels.
A tricycle has 3 wheels.
In a shop there are $x$ bicycles and $y$ tricycles.
The total number of wheels on the bicycles and the tricycles in the shop is given by $W$.
Write down a formula for $W$, in terms of $x$ and $y$.

**6** The cost, in pounds, of hiring a car can be worked out using this rule.

> Add 3 to the number of days' hire
> Multiply your answer by 10

The cost of hiring a car for $n$ days is $C$ pounds.
Write down a formula for $C$ in terms of $n$.                    (1387 June 2005)

**7** A ruler costs 45 pence. A pen costs 30 pence.
Louisa buys $x$ rulers and $y$ pens. The total cost is $C$ pence.

**a** Write down a formula for $C$ in terms of $x$ and $y$.

$C = 240, x = 2$

**b** Work out the value of $y$.

**8** Pat plays a game with red cards and green cards.
Red cards are worth 5 points each.
Green cards are worth 3 points each.
Pat has $r$ red cards and $g$ green cards.
His total number of points is $N$.
Write down, in terms of $r$ and $g$, a formula for $N$.             (1388 March 2003)

**9** Which of the following is an identity?
$4a + 3 = 8a - 4$          $C = 2\pi r$          $2(c + 1) = 2c + 2$          $4m + 3$

**10** Copy the table and write either **expression** or **equation** or **identity** or **formula** in each space. The first has been done for you.

*(1388 March 2005)*

| | formula |
|---|---|
| $A = \frac{1}{2}bh$ | |
| $x^2 + 5 = 86$ | |
| $7m + 5n$ | |
| $3p + 2q = p + p + p + q + q$ | |

**11** Make $k$ the subject of the formula $p = 9k + 20$

**12** Jo uses the formula $F = \frac{9}{5}C + 32$ to change degrees Celsius $(C)$ to degrees Fahrenheit $(F)$.

   **a** Use this formula to find
     **i** the value of $F$ when $C = 40$      **ii** the value of $C$ when $F = 50$

   **b** Rearrange the formula to make $C$ the subject of the formula.

**13** Make $p$ the subject of the formula $m = 3n + 2p$          *(1388 March 2005)*

**14 a** Solve $4(x + 3) = 6$             **15** Make $a$ the subject of the

   **b** Make $t$ the subject of the
     formula $v = u + 5t$   *(1387 June 2005)*     formula $s = \dfrac{a}{4} + 8u$   *(1388 March 2005)*

**16** Use a calculator for this question.

$$P = \pi r + 2r + 2a$$
$$P = 84 \quad r = 6.7$$

   **a** Work out the value of $a$.
     Give your answer correct to three significant figures.

   **b** Make $r$ the subject of the formula $P = \pi r + 2r + 2a$        *(1387 June 2005)*

**17** Make $m$ the subject of the formula $2(2p + m) = 3 - 5m$      *(1388 January 2003)*

**18** Make $a$ the subject of the formula $2(3a - c) = 5c + 1$      *(1388 January 2004)*

**19** The diagram shows a shape made up of a right-angled triangle and a semicircle. The area of the triangle is equal to the area of the semicircle.

   **a** Express $h$ in terms of $\pi$ and $d$.

   **b** Express $d$ in terms of $\pi$ and $h$.

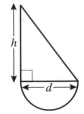

*(1388 January 2004)*

**20** Make $y$ the subject of the formula $T = \dfrac{w(1 - y)}{y}$

**21** Make $x$ the subject of the formula $y = \dfrac{x}{a - x}$      *(1385 June 2002)*

**22 a** Simplify $(3xy^3)^4$.    **b** Rearrange $\sqrt{\dfrac{x - 4}{5}} = 2y$ to give $x$ in terms of $y$.   *(1385 June 2001)*

**23** The fraction, $p$, of an adult's dose of medicine which should be given to a child who weighs $w$ kg is given by the formula     $p = \dfrac{3w + 20}{200}$

   **a** Use the formula to find the weight of a child whose dose is the same as an adult's dose.

   **b** Make $w$ the subject of the formula.

   **c** Express $A$ in terms of $w$ in the following formula.

$$\frac{3w + 20}{200} = \frac{A}{A + 12}$$

*(1387 November 2003)*

# Pythagoras' theorem and trigonometry (1)

## 29.1 Pythagoras' theorem

Pythagoras was a famous mathematician in Ancient Greece.
The theorem which is named after him is an important result about **right-angled triangles**.

Here is a right-angled triangle $ABC$.

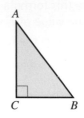

The angle at $C$ is the right angle.
The side, $AB$, opposite the right angle is called the **hypotenuse**.
It is the longest side in the triangle.

The right-angled triangle in the diagram on
the right has sides of length 3 cm, 4 cm and 5 cm.
Squares have been drawn on each side of the triangle and
each square has been divided up into squares of side 1 cm.

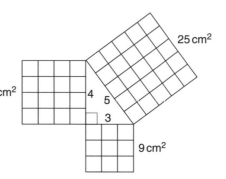

The area of the square on the side of length 3 cm is 9 cm².
The area of the square on the side of length 4 cm is 16 cm².
The area of the square on the side of length 5 cm
(the hypotenuse) is 25 cm².

Notice that 25 cm² = 9 cm² + 16 cm²

that is, $5^2 = 3^2 + 4^2$

In other words $5^2$ (the area of the square on the hypotenuse) is equal to the sum of $3^2$ and $4^2$
(the areas of the squares on the other two sides added together).

This is an example of Pythagoras' theorem. It is only true for right-angled triangles.

**Pythagoras' theorem states:**

> **In a right-angled triangle, the area of the square on the hypotenuse is equal to the sum of areas of the squares on the other two sides.**

Area of square R = area of square P + area of square Q

Pythagoras' theorem can be used to find
the length of the third side of a right-angled triangle
when the lengths of the other two sides are known.
For this, the theorem is usually stated in terms of the
lengths of the sides of the triangle.

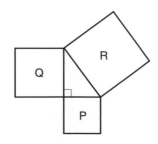

That is  $c^2 = a^2 + b^2$

Pythagoras' theorem can also be written

$$DE^2 = EF^2 + DF^2$$

($DE^2$ means that the length of the side $DE$ is squared.)

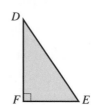

## 29.2  Finding lengths

Pythagoras' theorem can be used to work out the length of the hypotenuse of a right-angled triangle when the lengths of the two shorter sides are given.

---

### Example 1

Work out the length of the hypotenuse in this triangle.

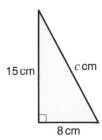

**Solution 1**

$c^2 = a^2 + b^2$ | State Pythagoras' theorem.

$c^2 = 8^2 + 15^2$ | Substitute the given lengths.

$c^2 = 64 + 225$ | Work out $8^2$ and $15^2$ and add the results.

$c^2 = 289$

$c = \sqrt{289} = 17$ | Find $\sqrt{289}$

Length of hypotenuse $= 17$ cm | The answer is sensible because the hypotenuse is longer than the other two sides.

---

It is important to be able to apply Pythagoras' theorem when the triangle is in a different position.

---

### Example 2

In triangle $XYZ$, angle $X = 90°$, $XY = 8.6$ cm and $XZ = 13.9$ cm.
Work out the length of $YZ$.
Give your answer correct to 3 significant figures.

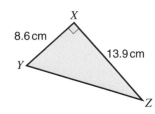

**Solution 2**

| The hypotenuse is the side opposite the right angle. Angle $X$ is the right angle so the hypotenuse is $YZ$.

$YZ^2 = XY^2 + XZ^2$ | State Pythagoras' theorem.

$YZ^2 = 8.6^2 + 13.9^2$ | Substitute the given lengths.

$YZ^2 = 73.96 + 193.21$ | Work out $8.6^2$ and $13.9^2$ and add the results.

$YZ^2 = 267.17$

$YZ = \sqrt{267.17} = 16.34...$ | Find $\sqrt{267.17}$  Write down at least 4 figures.

$YZ = 16.3$ cm (to 3 s.f.) | Give the final answer correct to 3 significant figures.

Pythagoras' theorem can also be used to work out the length of one of the shorter sides in a right-angled triangle when the lengths of the other two sides are known.

## Example 3

In triangle $ABC$, angle $A = 90°$, $BC = 17.4$ cm and $AC = 5.8$ cm.
Work out the length of $AB$. Give your answer correct to 3 significant figures.

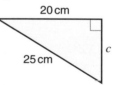

*Solution 3*

| Angle $A$ is the right angle so the hypotenuse is $BC$. |

$$BC^2 = AC^2 + AB^2$$

| State Pythagoras' theorem. |

$$17.4^2 = 5.8^2 + AB^2$$

| Substitute the given lengths. |

$$302.76 = 33.64 + AB^2$$

| Work out $17.4^2$ and $5.8^2$ |

$$302.76 - 33.64 = AB^2$$

| Subtract 33.64 from both sides. |

$$269.12 = AB^2$$

$$AB = \sqrt{269.12} = 16.40...$$

| Find $\sqrt{269.12}$  Write down at least 4 figures. |

$$AB = 16.4 \text{ cm (to 3 s.f.)}$$

| Give the final answer correct to 3 significant figures. |

## Exercise 29A

**1** Work out the length of the sides marked with letters in these triangles.

**a**     12 cm, $a$, 5 cm

**b**     40 cm, $b$, 9 cm

**c**     6 cm, 8 cm, $c$

**2** Work out the length of the sides marked with letters in these triangles.

**a**     $a$, 25 cm, 24 cm

**b**     12 cm, 37 cm, $b$

**c**     20 cm, 25 cm, $c$

**3** Work out the length of the sides marked with letters in these triangles.
Give each answer correct to 3 significant figures.

**a**     7.9 cm, $a$, 7.4 cm

**b**     4.8 cm, $b$, 9.1 cm

**c**     $c$, 6.2 cm, 8.3 cm

**d**     $d$, 10.6 cm, 4.8 cm

**4** Work out the length of the sides marked with letters in these triangles.
Give each answer correct to 3 significant figures.

**a**

10.7 cm
4.8 cm
*a*

**b**

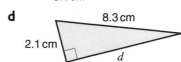

11.3 cm
*b*
8.1 cm

**c**

*c*
1.8 cm
12.4 cm

**d**

8.3 cm
2.1 cm
*d*

**5 a** In triangle $ABC$
angle $A = 90°$, $AB = 3.4$ cm and $AC = 12.1$ cm.
Work out the length of $BC$.
Give your answer correct to 3 significant figures.

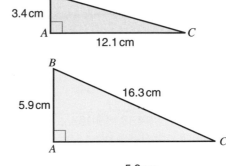

B
3.4 cm
A
12.1 cm
C

**b** In triangle $ABC$
angle $A = 90°$, $AB = 5.9$ cm and $BC = 16.3$ cm.
Work out the length of $AC$.
Give your answer correct to 3 significant figures.

B
16.3 cm
5.9 cm
A
C

**c** In triangle $PQR$
angle $R = 90°$, $PR = 5.9$ cm and $QR = 13.1$ cm.
Work out the length of $PQ$.
Give your answer correct to 3 significant figures.

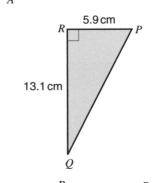

5.9 cm
R
P
13.1 cm
Q

**d** In triangle $PQR$
angle $R = 90°$, $PQ = 11.2$ cm and $QR = 9.6$ cm.
Work out the length of $RP$.
Give your answer correct to 3 significant figures.

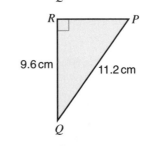

R
P
9.6 cm
11.2 cm
Q

**e** In triangle $XYZ$
angle $X = 90°$, $XY = 12.6$ cm and $XZ = 16.5$ cm.
Work out the length of $YZ$.
Give your answer correct to 3 significant figures.

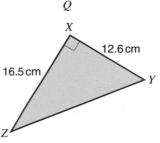

X
12.6 cm
16.5 cm
Y
Z

**f** In triangle $DEF$
angle $E = 90°$, $DF = 10.1$ cm and $EF = 7.8$ cm.
   **i** Draw a sketch of the right-angled triangle $DEF$ and label sides $DF$ and $EF$ with their lengths.
   **ii** Work out the length of $DE$.
      Give your answer correct to 3 significant figures.

## 29.3 Applying Pythagoras' theorem

Pythagoras' theorem can be used to solve problems.

### Example 4

A boat travels due north for 5.7 km. The boat then turns and travels due east for 7.2 km. Work out the distance between the boat's finishing point and its starting point. Give your answer correct to 3 significant figures.

### Solution 4

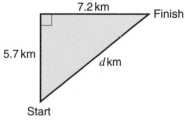

| Draw a sketch of the boat's journey |
|---|

| Remember that the points of the compass are | 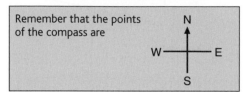 |
|---|---|

$$d^2 = 5.7^2 + 7.2^2$$
$$d^2 = 32.49 + 51.84$$
$$d^2 = 84.33$$
$$d = \sqrt{84.33} = 9.183...$$
Distance = 9.18 km (to 3 s.f.)

| The sketch is of a right-angled triangle so that Pythagoras' theorem can be used. |
|---|

| The distance between the starting point and the finishing point is the length of the hypotenuse of the triangle, marked $d$ km in the sketch. |
|---|

Isosceles triangles can be split into two right-angled triangles and Pythagoras' theorem can then be used.

### Example 5

The diagram shows an isosceles triangle $ABC$. The midpoint of $BC$ is the point $M$.
In the triangle, $AB = AC = 8$ cm and $BC = 6$ cm.
**a** Work out the height, $AM$, of the triangle.
   Give your answer correct to 3 significant figures.
**b** Work out the area of triangle $ABC$.
   Give your answer correct to 3 significant figures.

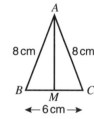

### Solution 5

Pythagoras' theorem cannot be used in triangle $ABC$ as this triangle is not right-angled.

**a**

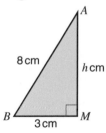

| As $M$ is the midpoint of the base of the isosceles triangle, the line $AM$ is the line of symmetry of triangle $ABC$. So $AM$ is perpendicular to the base and angle $AMB = 90°$. |
|---|

| $BM = 3$ cm as $M$ is the midpoint of $BC$. |
|---|

| Draw a sketch of triangle $ABM$. |
|---|

By Pythagoras

$$AB^2 = AM^2 + BM^2$$
$$8^2 = h^2 + 3^2$$
$$64 = h^2 + 9$$
$$64 - 9 = h^2$$
$$55 = h^2$$
$$h = \sqrt{55} = 7.416...$$
$$h = 7.42$$
Height of triangle = 7.42 cm (to 3 s.f.)

| Triangle $ABM$ is right-angled with hypotenuse $AB$. The height, $AM$, of the triangle is marked $h$ cm on the sketch. |
|---|

**b**          Area $= \frac{1}{2} \times 6 \times 7.416...$

             Area $= 22.24...$

             Area $= 22.2$ cm² (to 3 s.f.)

> area of a triangle $= \frac{1}{2} \times$ base $\times$ height
>
> For triangle $ABC$ base $= 6$ cm and height $= 7.416...$ cm.

---

## Exercise 29B

**1** The diagram shows a ladder leaning against a vertical wall.
The foot of the ladder is on horizontal ground.
The length of the ladder is 5 m.
The foot of the ladder is 3.6 m from the wall.
Work out how far up the wall the ladder reaches.
Give your answer correct to 3 significant figures.

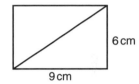

**2** The diagram shows a rectangle of length 9 cm and width 6 cm.
Work out the length of a diagonal of the rectangle.
Give your answer correct to 3 significant figures.

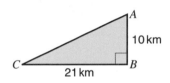

**3** Aiton ($A$), Beeville ($B$) and Ceaborough ($C$) are three
towns as shown in this diagram.
Beeville is 10 km due south of Aiton and 21 km due
east of Ceaborough.
Work out the distance between Aiton and Ceaborough.
Give your answer correct to the nearest km.

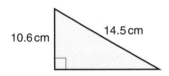

**4** Work out the area of the triangle.
Give your answer correct to 3 significant figures.

**5** Work out the perimeter of the triangle.
Give your answer correct to 3 significant figures.

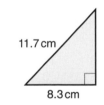

**6** The diagram represents the end view of a tent,
triangle $ABC$, two guy-ropes, $AP$ and $AQ$, and a
vertical tent pole, $AN$. The tent is on horizontal ground
so that $PBNCQ$ is a straight horizontal line.
Triangles $ABC$ and $APQ$ are both isosceles triangles.

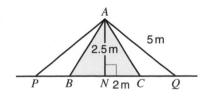

$BN = NC = 2$ m, $AN = 2.5$ m and $AP = AQ = 5$ m

   **a** Work out the length of the side $AC$ of the tent.
      Give your answer correct to 3 significant figures.

   **b** Work out the length of
      **i** $NQ$             **ii** $CQ$.
      Give your answers correct to 3 significant figures.

   There is a tent peg at $P$ and a tent peg at $Q$.

   **c** Work out the distance between the two tent pegs at $P$ and $Q$.
      Give your answer correct to 3 significant figures.

**7** The diagram shows two right-angled triangles.

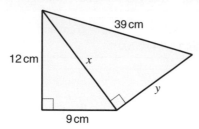

**a** Work out the length of the side marked $x$.
**b** Hence work out the length of the side marked $y$.

**8** The diagram shows two right-angled triangles

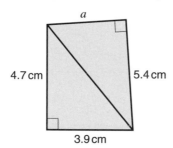

Work out the length of the side marked $a$.
Give your answer correct to 3 significant figures.

**9** The lengths of the sides of a triangle are 8 cm, 15 cm and 17 cm.
**a** Find the value of    **i** $8^2 + 15^2$    **ii** $17^2$
**b** What do you notice about the two answers in **a**?
**c** What information does this give about the triangle?

**10** Here are the lengths of sides of six triangles.
   Triangle 1    5 cm, 12 cm and 13 cm    Triangle 2    9 cm, 40 cm and 41 cm
   Triangle 3    10 cm, 17 cm and 18 cm    Triangle 4    20 cm, 21 cm and 29 cm
   Triangle 5    8 cm, 17 cm and 20 cm    Triangle 6    33 cm, 56 cm and 65 cm
   Find by calculation which of these triangles are right-angled triangles.

# 29.4 Line segments and Pythagoras' theorem
## Length of a line segment

The diagram shows the points $A(1, 1)$ and $B(9, 5)$.
The right-angled triangle $ABC$ has been drawn so
that $AC = 8$ and $BC = 4$

Pythagoras' theorem can be used
to find the length of $AB$.
$AB^2 = 8^2 + 4^2$
$AB^2 = 64 + 16 = 80$
$AB = \sqrt{80} = 8.94$ (to 3 s.f.)

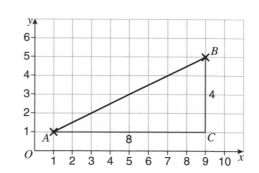

## Example 6

Find the length of the line joining
**a** $A(3, 2)$ and $B(15, 7)$　　　　　　**b** $P(-9, 4)$ and $Q(7, -5)$

*Solution 6*

**a**

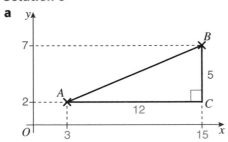

Draw a sketch showing $A$ and $B$ and complete the right-angled triangle $ABC$.

$AC = 15 - 3 = 12$

$BC = 7 - 2 = 5$

$AB^2 = 12^2 + 5^2$

$AB^2 = 144 + 25 = 169$

$AB = \sqrt{169} = 13$

Use Pythagoras' theorem to find the length of $AB$.

**b**

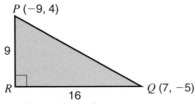

Draw a sketch showing $P$ and $Q$ and complete the right-angled triangle $PQR$.

$QR = 7 - {-9} = 7 + 9 = 16$

$PR = 4 - {-5} = 4 + 5 = 9$

$PQ^2 = 16^2 + 9^2$

$PQ^2 = 256 + 81 = 337$

$PQ = \sqrt{337} = 18.4$ (to 3 s.f.)

Use Pythagoras' theorem to find the length of $PQ$.

### Exercise 29C

**1** Work out the length of the line joining each of these pairs of points.
   **a** $(3, 1)$ and $(11, 7)$　　**b** $(2, 5)$ and $(12, 29)$　　**c** $(-6, 9)$ and $(8, 13)$
   **d** $(-4, -6)$ and $(6, 12)$　　**e** $(9, -15)$ and $(-11, 6)$　　**f** $(0, -5)$ and $(9, -11)$

**2** The point $A$ has coordinates $(5, 2)$, the point $B$ has coordinates $(8, 6)$ and the point $C$ has coordinates $(1, 5)$.
   **a** Work out the length of
     **i** $AB$　　　　　　**ii** $BC$　　　　　　**iii** $AC$.
   **b** What does your answer to part **a** tell you about triangle $ABC$?

**3** The point $P$ has coordinates $(5, 3)$, the point $Q$ has coordinates $(6, 6)$ and the point $R$ has coordinates $(-6, 10)$.
   **a** Work out the length of each side of triangle $PQR$.
   **b** Use your answers to part **a** to show that triangle $PQR$ is a right-angled triangle.
   **c** Work out the area of triangle $PQR$.

**4** The points $A(2, 6)$ and $B(18, 36)$ are the ends of a diameter of a circle.
   **a** Find the coordinates of the centre of the circle.
   **b** Work out the
     **i** diameter of the circle　　　　　　**ii** radius of the circle.

**5** A circle has centre $O(4, 2)$. The point $A(9, 14)$ lies on the circle.

  **a** Work out the radius of the circle.

  **b** Determine by calculation which of the following points also lie on the circle.

    **i** $B(16, 7)$        **ii** $C(-1, -10)$        **iii** $D(7, 16)$        **iv** $E(4, 15)$.

**6** The point $A$ has coordinates $(-3, -8)$ and the point $B$ has coordinates $(8, 9)$.

  **a** Find the coordinates of the midpoint of the line segment $AB$.

  **b** Work out the length of the line segment $AB$. Give your answer correct to 3 significant figures.

**7** The points $A\ (5, 1)$, $B\ (29, 8)$, $C\ (9, 23)$ and $D\ (-15, 16)$ are the vertices of quadrilateral $ABCD$.

  **a** Work out the length of

    **i** $AB$        **ii** $BC$        **iii** $CD$        **iv** $DA$.

  **b** Explain what your answers to **a** tell you about the quadrilateral $ABCD$.

**8** The point $A$ has coordinates $(a, b)$ and the point $B$ has coordinates $(p, q)$.

Show that the length of the line segment $AB$ is $\sqrt{(p - a)^2 + (q - b)^2}$

# 29.5 Trigonometry – introduction

Trigonometry means 'triangle measure'.
It is used to work out lengths and angles in triangles and in shapes that can be divided up into triangles.
Trigonometry is important in bridge building and tunnel building where it is important to know accurate distances and accurate angle sizes.
It is also used in many other areas of surveying, engineering and architecture.

## *Trigonometric ratios*

Here are two right-angled triangles.
The triangle with hypotenuse 2 is an enlargement with scale factor 2 of the triangle with hypotenuse 1, that is, its sides are twice as long.

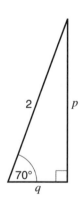

    $p = 2s$ and $q = 2c$

So if $s$ and $c$ are known for the right-angled triangle with hypotenuse 1, $p$ and $q$ can be calculated.
If the length of its hypotenuse is known, the lengths of the sides of any right-angled triangle which is an enlargement of these triangles can be calculated.

The values of $s$ and $c$ are known accurately and can be found on any standard scientific calculator.
The length $s$ is called the *sine of 70°* written *sin 70°*. Not all calculators are the same but the following key sequence to find sin 70° applies to many calculators.

Make sure that the angle mode of your calculator is degrees, usually shown by $\boxed{\text{D}}$ on the calculator screen.

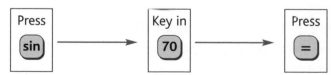

The number $0.93969262$ should appear on your calculator display.

So correct to 4 decimal places sin 70° = 0.9397

The length $c$ is called *the cosine of 70°* and is written *cos 70°*. As above but using the button $\boxed{\text{cos}}$

correct to 4 decimal places cos 70° = 0.3420

Using the triangles opposite and writing $s$ as sin 70° and $c$ as cos 70°

$$p = 2\sin 70° \quad \text{and} \quad q = 2\cos 70°$$

So for any right-angled triangle

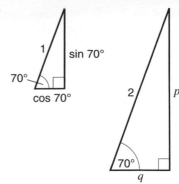

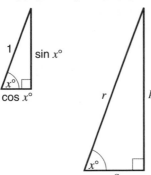

$$p = r \sin x° \text{ and } q = r \cos x°$$

or $\quad \sin x° = \dfrac{p}{r}$ and $\cos x° = \dfrac{q}{r}$

The **hypotenuse** (hyp) of a right-angled triangle is the side opposite the right angle and is the longest side of the triangle. The sides of the triangle are named according to their position relative to the angle given or the angle to be found. If this angle is $x°$ then the side opposite this angle is called the **opposite side** (opp). The side next to this angle is called the **adjacent side** (adj).

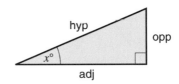

The results above become

$$\text{opp} = \text{hyp} \times \sin x° \quad \text{and} \quad \text{adj} = \text{hyp} \times \cos x°$$

or $\quad \sin x° = \dfrac{\text{opp}}{\text{hyp}}$ and $\cos x° = \dfrac{\text{adj}}{\text{hyp}}$

$\sin x°$ is used when opposite and hypotenuse are involved.
$\cos x°$ is used when adjacent and hypotenuse are involved.

When opposite and adjacent are involved, a third result called $\tan x°$, is needed where

$$\tan x° = \dfrac{\text{opp}}{\text{adj}} \quad \text{or} \quad \text{opp} = \text{adj} \times \tan x°$$

$\tan x°$ means the tangent of $x°$.

*SOHCAHTOA* might help you to remember these results.

**Sin Opp Hyp Cos Adj Hyp** Tan Opp Adj

# 29.6 Finding lengths using trigonometry

## Example 7

Work out the length of each of the marked sides. Give each answer correct to 3 significant figures.

**a**

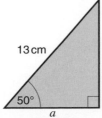

13 cm
50°
$a$

**b**

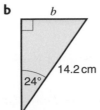

$b$
14.2 cm
24°

**c**

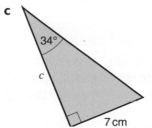

34°
$c$
7 cm

**Solution 7**

**a**

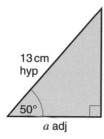

13 cm is the hypotenuse
$a$ is adjacent to the 50° angle.

adj and hyp are involved so use cos

$$\cos = \frac{\text{adj}}{\text{hyp}} \quad \text{or} \quad \text{adj} = \text{hyp} \times \cos$$

$a = 13 \times \cos 50° = 13\cos 50°$
$a = 13 \times 0.6427...$
$a = 8.356...$
$a = 8.36$ cm

Give your answer correct to 3 significant figures.

**b**

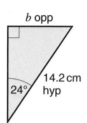

14.2 cm is the hypotenuse
$b$ is opposite to the 24° angle.

$$\sin = \frac{\text{opp}}{\text{hyp}} \quad \text{or} \quad \text{opp} = \text{hyp} \times \sin$$

$b = 14.2 \times \sin 24° = 14.2\sin 24°$
$b = 14.2 \times 0.4067...$
$b = 5.7756...$
$b = 5.78$ cm

Give your answer correct to 3 significant figures.

**c**

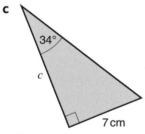

Opposite and adjacent are involved so use

$$\tan = \frac{\text{opp}}{\text{adj}} \quad \text{or} \quad \text{opp} = \text{adj} \times \tan$$

When using tan to find a length it is easier to find the opposite side. Relative to the angle of 34°, 7 cm is the opposite side and $c$ is the adjacent side.

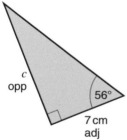

The third angle in the triangle is $(180 - 90 - 34)° = 56°$

Relative to the angle of 56°, c is the opposite side and 7 cm is the adjacent side.

$c = 7 \times \tan 56° = 7 \tan 56°$
$c = 7 \times 1.4825...$
$c = 10.3779...$
$c = 10.4$ cm

Give your answer correct to 3 significant figures.

---

## Example 8

In triangle $ABC$, angle $CAB = 90°$, angle $ABC = 37°$ and $AB = 8.4$ cm.
Calculate the length of $BC$.
Give your answer correct to 3 significant figures.

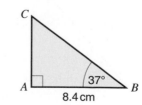

## Solution 8

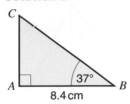

In triangle $ABC$, $BC$ is the hypotenuse and 8.4 cm is adjacent to the 37° angle so use

$$\cos = \frac{\text{adj}}{\text{hyp}} \text{ or } \text{adj} = \text{hyp} \times \cos$$

Substitute the known values in adj = hyp × cos

$$8.4 = BC \times \cos 37°$$

$$BC = \frac{8.4}{\cos 37°} = 10.5179...$$

Make $BC$ the subject.

$$BC = 10.5 \text{ cm}$$

Give your answer correct to 3 significant figures.

As $BC$ is the hypotenuse its length must be greater than 8.4 cm so this is a sensible answer.

## Exercise 29D

**1** Use a calculator to find the value of each of the following.
Give each answer correct to 4 decimal places, where necessary.
   **a** sin 20°    **b** sin 72°    **c** cos 60°    **d** tan 86°    **e** tan 45°
   **f** cos 18.9°    **g** tan 4°    **h** sin 14.7°    **i** cos 75.3°

**2** Work out the lengths of the sides marked with letters. Give each answer correct to 3 significant figures.

**a**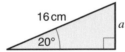

**b**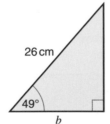

**c**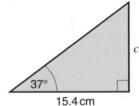

**d**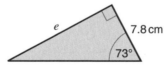

**e**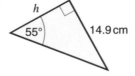

**f**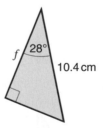

**g**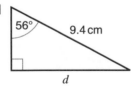

**h**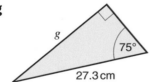

**i**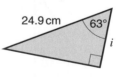

**3** Triangle $PQR$ is right-angled at $Q$. In each part calculate the length of $QR$.
Give each answer correct to 3 significant figures.
   **a** $PQ = 7.3$ cm, angle $QPR = 68°$
   **b** $PR = 17.2$ m, angle $QRP = 39°$
   **c** $PR = 12.6$ cm, angle $QPR = 59°$

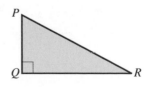

**4** In triangle $ABD$ the point $C$ lies on $AD$ so that $BC$ and $AD$ are perpendicular.

  **a** Using triangle $ABC$, work out the length of

   **i** $BC$                    **ii** $AC$

   Give each answer correct to 3 significant figures.

  **b** Work out the length of $CD$ correct to 3 significant figures.

  **c** Hence calculate the length of $AD$ correct to 3 significant figures.

  **d** Calculate the area of triangle $ABD$. Give your answer correct to the nearest cm².

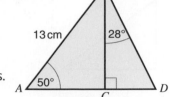

**5** Calculate the length of $BC$ in these triangles. Give each answer correct to 3 significant figures.

  **a**

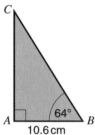

  **b**

# 29.7 Finding angles using trigonometry

Work out the size of each of the angles marked with letters.
Give each answer correct to 1 decimal place.

**a**

**b**

**c**

*Solution 9*

**a**

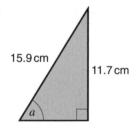

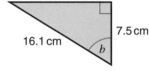

15.9 cm is the hypotenuse
11.7 cm is opposite angle $a$.

$\sin = \dfrac{\text{opp}}{\text{hyp}}$

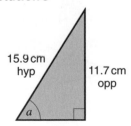

$\sin a = \dfrac{11.7}{15.9} = 0.7358...$

$a = 47.379...°$

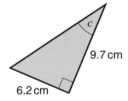

Use your calculator to find $\sin^{-1} 0.7358...$ which is $47.379...°$

$a = 47.4°$

Give your answer correct to 1 decimal place.

**b**

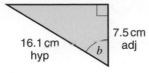

16.1 cm hyp

7.5 cm adj

*b*

16.1 cm is the hypotenuse
7.5 cm is adjacent to angle *b*.

$$\cos = \frac{adj}{hyp}$$

$$\cos b = \frac{7.5}{16.1} = 0.4658...$$

$$b = 62.235...°$$

Use your calculator to find $\cos^{-1} 0.4658...$ which is $62.235...°$

$$b = 62.2°$$

Give your answer correct to 1 decimal place.

**c**

*c*

9.7 cm adj

6.2 cm opp

6.2 cm is opposite angle *c*
9.7 cm is adjacent to angle *c*.

$$\tan = \frac{opp}{adj}$$

$$\tan c = \frac{6.2}{9.7} = 0.6391...$$

$$c = 32.585...°$$

Use your calculator to find $\tan^{-1} 0.6391...$ which is $32.585...°$

$$c = 32.6°$$

Give your answer correct to 1 decimal place.

## Exercise 29E

**1** Use a calculator to find the value of *x* in each of the following.
Give answers correct to 1 decimal place where necessary.

**a** $\cos x° = 0.5$     **b** $\sin x° = 0.43$     **c** $\cos x° = 0.6$     **d** $\tan x° = 0.96$

**e** $\sin x° = 0.8516$   **f** $\tan x° = 2.03$     **g** $\sin x° = 0.047$   **h** $\tan x° = \frac{2}{7}$

**2** Work out the size of each of the marked angles. Give each answer correct to 1 decimal place.

**a**

18 cm    11 cm

*a*

**b**

13 cm

*b*

9 cm

**c**

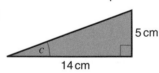

5 cm

*c*

14 cm

**d**

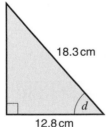

18.3 cm

*d*

12.8 cm

**e**

17 cm

15.8 cm

*e*

**f**

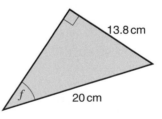

13.8 cm

*f*

20 cm

**3** Triangle *ABC* is right-angled at *B*. Give each answer correct to 0.1°.

**a** $AB = 8.9$ cm and $BC = 12.1$ cm. Calculate the size of angle *ACB*.

**b** $BC = 15.5$ cm, $AC = 24.7$ cm. Calculate the size of angle *BAC*.

**c** $AB = 6.3$ cm, $AC = 11.8$ cm. Calculate the size of angle *ACB*.

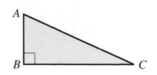

*A*

*B*          *C*

4 In triangle $ACD$ the point $B$ lies on $AD$ so that $CB$ and $AD$ are perpendicular.

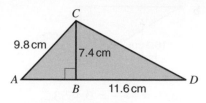

a Using triangle $ABC$ calculate the size of angle $ACB$. Give your answer correct to 1 decimal place.

b Using triangle $BCD$ calculate the size of angle $BCD$. Give your answer correct to 1 decimal place.

c Hence calculate the size of angle $ACD$. Give your answer to the nearest degree.

## 29.8 Trigonometry problems

Trigonometry can be used to solve problems. Sometimes Pythagoras' theorem is needed as well. Some questions involve bearings (see Section 7.5).

---

### Example 10

Two towns, Aytown and Beeville, are 40 km apart. The bearing of Beeville from Aytown is 067°.

a Calculate how far north and how far east Beeville is from Aytown. Give your answers correct to 3 significant figures.

Ceeham is 60 km east of Beeville.

b Calculate the distance between Aytown and Ceeham. Give your answer to the nearest km.

c Calculate the bearing of Ceeham from Aytown. Give your answer to the nearest degree.

*Solution 10*

a

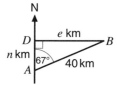

| Draw a diagram showing the positions of Ayton ($A$) and Beeville ($B$). |

| From $B$ draw a line west to meet at D the 'north' line from $A$. |

| In the right-angled triangle $ABD$, the length of $AD$ is the distance that $B$ is north of $A$ ($n$ km). The length of $DB$ is the distance that $B$ is east of $A$ ($e$ km). |

$e = 40 \sin 67° = 36.82...$

| In triangle $ABD$ the 40 km is the hypotenuse and $e$ km is opposite the 67° angle<br><br>$\sin = \dfrac{\text{opp}}{\text{hyp}}$ or opp $=$ hyp $\times$ sin |

Distance east $= 36.8$ km.

| Give your answer correct to 3 significant figures. |

$n = 40 \cos 67° = 15.629...$

| $n$ km is adjacent to the 67° angle<br><br>$\cos = \dfrac{\text{adj}}{\text{hyp}}$ or adj $=$ hyp $\times$ cos |

Distance north $= 15.6$ km.

| Give your answer correct to 3 significant figures. |

b

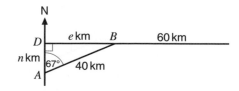

| Mark the point $C$ (for Ceeham) on the diagram 60 km east of $B$. |

| Ceeham is 15.6 km north of Ayton. |

| Ceeham is $60 + 36.8 = 96.8$ km east of Ayton. |

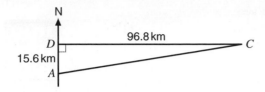

Draw triangle $ADC$.

The distance between Aytown and Ceeham is the length of $AC$.

$AC^2 = AD^2 + DC^2 = 15.6^2 + 96.8^2$

Find the length of $AC$ using Pythagoras' theorem.

$AC^2 = 9613.6$

$AC = 98.04...$

Distance between Aytown and Ceeham is 98 km

Give your answer to the nearest km.

To find the bearing of $C$ from $A$ calculate the size of angle $DAC$.

$\tan DAC = \dfrac{96.8}{15.6} = 6.205...$

$\tan = \dfrac{\text{opp}}{\text{adj}}$

$DAC = 80.8...°$

$DAC = 81°$

Give your answer to the nearest degree.

Bearing of Ceeham from Aytown is 081°.

A bearing must have 3 figures.

## Exercise 29F

Where necessary give lengths correct to 3 significant figures and angles correct to 1 decimal place.

**1** **a** Calculate the length of the line marked $x$ cm.

   **b** Work out the size of the angle marked $y°$.

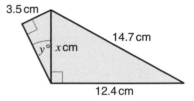

**2** A ladder is 5 m long. The ladder rests against a vertical wall, with the foot of the ladder resting on horizontal ground. The ladder reaches up the wall a distance of 4.8 m.

   **a** Work out how far the foot of the ladder is from the bottom of the wall.

   **b** Work out the angle that the ladder makes with the ground.

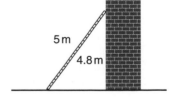

**3** The diagram shows the plans for the sails of a boat. Work out the length of the side marked

   **a** $a$

   **b** $b$

   **c** $c$

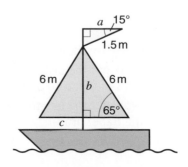

**4** The diagram shows a vertical building standing on
horizontal ground. The points $A$, $B$ and $C$ are in a straight
line on the ground. The point $T$ is at the
top of the building so that $TC$ is vertical. The angle
of elevation of $T$ from $A$ is 40° as shown in the diagram.

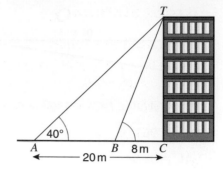

   **a** Work out the height, $TC$, of the building.

   **b** Work out the size of the angle of elevation
   of $T$ from $B$.

**5** The points $P$ and $Q$ are marked on a horizontal field. The distance from $P$ to $Q$ is 100 m.
The bearing of $Q$ from $P$ is 062°. Work out how far

   **a** $Q$ is north of $P$          **b** $Q$ is east of $P$.

**6** The diagram shows a circle centre $O$.
The line $ABC$ is the tangent to the circle at $B$.

   **a** Work out the radius of the circle.

   **b** Work out the size of angle $OCB$.

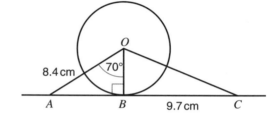

**7** $A$, $B$ and $C$ are three buoys marking the course
of a yacht race.

   **a**   Calculate how far $B$ is
      **i** north of $A$          **ii** east of $A$.

   **b** Calculate how far $C$ is
      **i** north of $B$       **ii** east of $B$.

   **c** Hence calculate how far $C$ is
      **i** north of $A$          **ii** east of $A$.

   **d** Calculate the distance and bearing of $C$ from $A$.

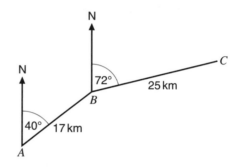

**8** A rock, $R$, is 40 km from a harbour, $H$, on a bearing of 040°. A port, $P$, is 30 km from $R$ on a
bearing of 130°.

   **a** Draw a sketch showing the points $H$, $R$ and $P$ and work out the size of angle $HRP$.

   **b** Work out the distance $HP$.

   **c** Work out the bearing of $P$ from $H$.

**9** The diagram shows an isosceles triangle.
Calculate the area of the triangle.
Give your answer to the nearest cm².

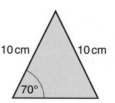

**10** The diagram shows an isosceles trapezium.

   **a** Work out the distance, $h$ cm, between the two parallel
   sides of the trapezium.

   **b** Work out the length of the longer parallel side of
   the trapezium.

   **c** Calculate the area of the trapezium. Give your answer to the nearest cm²

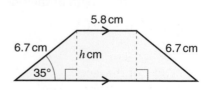

# Chapter summary

> **You should now know:**
>
> ★ that in a right-angled triangle the side opposite the right angle is called the hypotenuse. It is the longest side in the triangle
>
> ★ Pythagoras' theorem for right-angled triangles
>
>  $$c^2 = a^2 + b^2$$
>
> ★ trigonometric ratios for right-angled triangles
>
>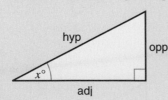
>
> $$\sin x° = \frac{opp}{hyp}$$
> $$\cos x° = \frac{adj}{hyp}$$
> $$\tan x° = \frac{opp}{adj}$$
>
> opp = hyp $\times$ sin $x°$
> adj = hyp $\times$ cos $x°$
> opp = adj $\times$ tan $x°$
>
> **You should now be able to:**
>
> ★ use Pythagoras' theorem in right-angled triangles
>   - to find the length of the hypotenuse when the lengths of the other two sides are known
>   - to find the length of one of the shorter sides of the triangle when the lengths of the other two sides are known
>
> ★ use Pythagoras' theorem to find the length of a line segment, given the coordinates of the end points of the line segment
>
> ★ use trigonometry in right-angled triangles to find the length of an unknown side and to find the size of an unknown angle
>
> ★ apply Pythagoras' theorem and trigonometry to right-angled triangle problems, including bearings.

# Chapter 29 review questions

**1** $ABC$ is a right-angled triangle.
$AB = 8$ cm, $BC = 11$ cm.
Calculate the length of $AC$.
Give your answer correct to
3 significant figures.

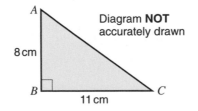

Diagram **NOT** accurately drawn

(1388 March 2003)

**2** Angle $MLN = 90°$
$LM = 3.7$ m
$MN = 6.3$ m
Work out the length of $LN$.
Give your answer correct to
3 significant figures.

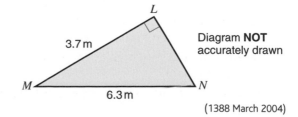

Diagram **NOT** accurately drawn

(1388 March 2004)

**3** Work out the length in centimetres of *AM*.
Give your answer correct to
2 decimal places.

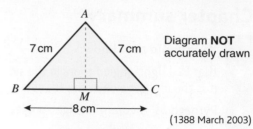

Diagram **NOT** accurately drawn

(1388 March 2003)

**4** Ballymena is due west of Larne.
Woodburn is 15 km due south of Larne.
Ballymena is 32 km from Woodburn.

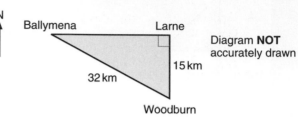

Diagram **NOT** accurately drawn

**a** Calculate the distance of Larne from Ballymena.
Give your answer in kilometres, correct to 1 decimal place.

**b** Calculate the bearing of Ballymena from Woodburn.

(1385 June 1998)

**5** Angle *ABC* = 90°
Angle *ACB* = 24°
*AC* = 6.2 cm
Calculate the length of *BC*.
Give your answer correct to 3 significant figures.

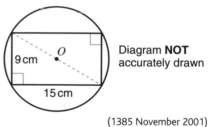

Diagram **NOT** accurately drawn

(1388 March 2004)

**6** The diagram shows a rectangle drawn inside a circle.
The centre of the circle is at *O*.
The rectangle is 15 cm long and 9 cm wide.
Calculate the circumference of the circle.
Give your answer correct to 3 significant figures.

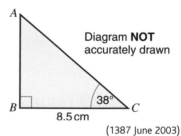

Diagram **NOT** accurately drawn

(1385 November 2001)

**7** The diagram shows triangle *ABC*.
*BC* = 8.5 cm
Angle *ABC* = 90°
Angle *ACB* = 38°
Work out the length of *AB*.
Give your answer correct to 3 significant figures.

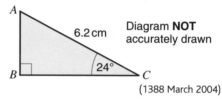

Diagram **NOT** accurately drawn

(1387 June 2003)

**8** *ABD* and *DBC* are two right-angled triangles.
*AB* = 9 m
Angle *ABD* = 35°
Angle *DBC* = 50°
Calculate the length of *DC*.
Give your answer correct to 3 significant figures.

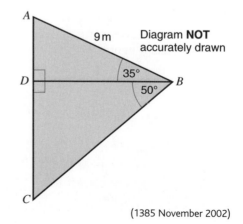

Diagram **NOT** accurately drawn

(1385 November 2002)

**9** The diagram shows the positions of three telephone masts $A$, $B$ and $C$.
Mast $C$ is 5 kilometres due east of Mast $B$.
Mast $A$ is due north of Mast $B$ and 8 kilometres from Mast $C$.

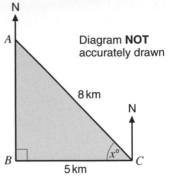

Diagram **NOT** accurately drawn

**a** Calculate the distance of $A$ from $B$.
Give your answer in kilometres, correct to 3 significant figures.

**b**  **i** Calculate the size of the angle marked $x°$.
Give your angle correct to 1 decimal place.

 **ii** Calculate the bearing of $A$ from $C$.
Give your answer correct to 1 decimal place.

 **iii** Calculate the bearing of $C$ from $A$.
Give your bearing correct to 1 decimal place.

(1385 June 1999)

**10** $A$ and $B$ are points on a centimetre grid.
$A$ is the point $(3, 2)$
$B$ is the point $(7, 8)$

Calculate the distance $AB$.
Give your answer correct to 3 significant figures.

**11** $ABCD$ is a trapezium.
$AD$ is parallel to BC.
Angle $A$ = angle $B$ = 90°
$AD = 2.1$ m   $AB = 1.9$ m   $CD = 3.2$ m
Work out the length of $BC$.
Give your answer correct to
3 significant figures.

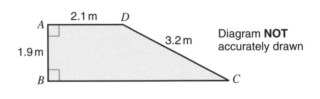

Diagram **NOT** accurately drawn

(1388 January 2003)

**12** $ABC$ is a right-angled triangle.
$D$ is the point on $AB$ such that $AD = 3DB$.
$AC = 2DB$ and angle $A$ = 90°.

Show that $\sin C = \dfrac{k}{\sqrt{20}}$ where $k$ is an integer.

Write down the value of $k$.

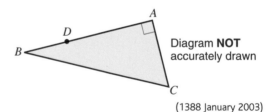

Diagram **NOT** accurately drawn

(1388 January 2003)

**13**

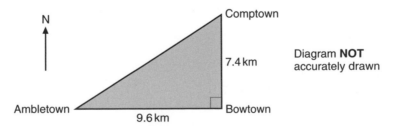

Diagram **NOT** accurately drawn

Ambletown, Bowtown and Comptown are three towns.
Ambletown is 9.6 km due west of Bowtown.
Bowtown is 7.4 km due south of Comptown.
Calculate the bearing of Ambletown from Comptown.
Give your answer correct to 1 decimal place.

(1388 January 2003)

**14**

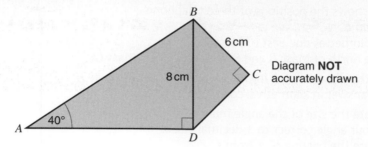

$ABCD$ is a quadrilateral.

Angle $BDA = 90°$,      angle $BCD = 90°$,      angle $BAD = 40°$

$BC = 6$ cm, $BD = 8$ cm

**a** Calculate the length of $DC$. Give your answer correct to 3 significant figures.

**b** Calculate the size of angle $DBC$. Give your answer correct to 3 significant figures.

**c** Calculate the length of $AB$. Give your answer correct to 3 significant figures.

(1385 November 2000)

# Ratio and proportion

## 30.1 Introduction to ratio

Scale 1 : 72

Fertiliser
**50 kg**
1 : 2 : 5

The pictures show examples of **ratio**.

Ratios are used to **compare** quantities.
Ratios are written using a colon (:)

For example if there are 7 boys and 9 girls in a class the ratio of the number of boys to the number of girls is 7 : 9
The order of the numbers is important.
The ratio 9 : 7 is the ratio of the number of girls to the number of boys.

Ratios can be simplified like fractions.
The ratio 7 : 9 cannot be simplified because 7 and 9 have no common factors.
If there are 6 boys and 9 girls the ratio is 6 : 9
This ratio 6 : 9 can be simplified because 3 divides exactly into both 6 and 9
Dividing both 6 and 9 by 3 gives the ratio 2 : 3
This means that for every 2 boys in the class there are 3 girls.
2 : 3 cannot be simplified and is called the **simplest form** of the ratio.
The simplest form of a ratio has whole numbers with no common factor.

### Example 1

In a box there are 24 red, 16 blue and 40 black pencils.

**a** Write down the ratio of the number of red pencils to the number of blue pencils to the number of black pencils.
Give your ratio in its simplest form.

**b** What fraction of the pencils are blue? Give your fraction in its simplest form.

*Solution 1*

**a** 24 : 16 : 40
  = 3 : 2 : 5
  3 : 2 : 5 is the simplest form

> Write down the numbers in the same order as the colours appear in the question.
> 8 is the highest common factor of 24, 16 and 40
> Divide all three numbers by 8

**b** 24 + 16 + 40 = 80
  $\frac{16}{80} = \frac{2}{10} = \frac{1}{5}$
  $\frac{1}{5}$ of the pencils are blue

> Calculate the total number of pencils.
> 16 out of every 80 pencils are blue.
> (Alternatively use the answer to part **a**
>
> Fraction $= \dfrac{2}{3 + 2 + 5} = \dfrac{2}{10} = \dfrac{1}{5}$)

## Example 2

Write each ratio in its simplest form.

**a**  $0.8:1.2$                    **b**  $\frac{2}{5}:\frac{1}{3}$

### Solution 2

**a**  $0.8:1.2$
   $=8:12$
   $=2:3$
   $2:3$ is the simplest form

> Change both numbers in the ratio to whole numbers by multiplying both numbers by 10
> Divide both numbers by 4

**b**  $\frac{2}{5}:\frac{1}{3}$

   $\frac{2}{\cancel{5}_1}\times\cancel{15}^{3}:\frac{1}{\cancel{3}_1}\times\cancel{15}^{5}$

   $=6:5$
   $6:5$ is the simplest form

> Change both fractions in the ratio to whole numbers by multiplying them by 15, the LCM of 5 and 3

If different units are used for the quantities in a ratio, start by writing the quantities in the same units.

## Example 3

Write 8 m to 60 cm as a ratio in its simplest form.

### Solution 3

$8\,\text{m} = 8\times100\,\text{cm}$
   $= 800\,\text{cm}$

$800:60$
   $= 40:3$
$40:3$ is the simplest form

> It is easier to change the larger units into the smaller units. So change metres to centimetres and write down the ratio.

> Divide both numbers by 20

Ratios are sometimes given in the form $1:n$ where $n$ is a number.

## Example 4

In school there are 480 boys and 360 girls.
**a**  Write down the ratio of the number of boys to the number of girls.
   Give your ratio in its simplest form.
**b**  Give your ratio in the form $1:n$.

### Solution 4

**a**  $480:360$
   $= 48:36$
   $= 4:3$
   $4:3$ is the simplest form

> Write down the ratio. The number of boys goes first.
> Divide both numbers by 10
> Divide both numbers by 12

**b**  $4:3 = 1:\frac{3}{4}$
   (or $1:0.75$)

> Divide both numbers by 4 to give the ratio in the form $1:n$

**Exercise 30A**

**1** Write each ratio in its simplest form.

    **a** 15 : 5                 **b** 6 : 9                 **c** 18 : 24

    **d** 45 : 30              **e** 48 : 16              **f** 250 : 350

    **g** 56 : 64              **h** 120 : 600

**2** Write each ratio in its simplest form.

    **a** 30p : £1            **b** 2 m : 50 cm        **c** 20 mm : 5 cm

    **d** 600 m : 2 km      **e** £2 : 80p            **f** 4 cm : 35 mm

    **g** £2.50 : £4          **h** 3.5 kg : 500 g

**3** Write each ratio in its simplest form.

    **a** 0.3 : 1.2           **b** 2.5 : 1.5           **c** 1.4 : 3.5

    **d** 0.6 : 2             **e** 3.2 : 5             **f** 0.25 : 0.3

    **g** 0.2 : 0.24 : 3     **h** 0.08 : 0.4 : 1

**4** Write each ratio in its simplest form.

    **a** $\frac{1}{2} : \frac{1}{3}$          **b** $\frac{1}{5} : \frac{1}{10}$        **c** $\frac{3}{4} : \frac{1}{2}$        **d** $\frac{2}{3} : \frac{1}{4}$

    **e** $\frac{4}{5} : \frac{3}{4}$          **f** $\frac{2}{9} : 1$            **g** $\frac{2}{3} : \frac{1}{2} : 1$      **h** $\frac{3}{4} : 1 : 1\frac{1}{2}$

**5** Write each ratio in the form 1 : $n$.

    **a** 4 : 12            **b** 6 : 24           **c** 2 : 7           **d** 2 : 1

    **e** 10 : 3          **f** 5 : 3            **g** $\frac{1}{4} : 2$        **h** $\frac{3}{4} : \frac{2}{5}$

**6** A model car has a length of 6 cm. The real car has a length of 3 m.
Write down the ratio of the length of the model car to the length of the real car.
Give your ratio in its simplest form.

**7** The weight of a small bag of flour is 500 g. The weight of a large bag of flour is 5 kg.

    **a** Find in its simplest form the ratio of the weight of the small bag of flour to the weight of the
large bag of flour.

    **b** Write down the simplest form of the ratio of the weight of the large bag of flour to the weight
of the small bag of flour.

**8** In a school there are 120 computers. There are 720 students in the school. Write down the ratio
of the number of computers to the number of students. Give your ratio in the form 1 : $n$.

**9** In a school assembly there are 160 girls and 200 boys.

    **a** What fraction of the students are girls?

    **b** Write down the ratio of the number of girls to the number of boys.
Give your ratio in its simplest form.

    **c** Write your answer to **b** in the form 1 : $n$.

**10** The length of a model aeroplane is 16 cm. The length of the real aeroplane is 60 m.
Work out the ratio of the length of the model aeroplane to the length of the real aeroplane.
Write your answer in the form 1 : $n$.

## 30.2 Problems

If the ratio of two quantities is given and one of the quantities is known, then the other quantity can be found.

### Example 5

To make concrete, 1 part of cement is used to every 5 parts of sand.

**a** Write down the ratio of cement to sand.

**b** 2 buckets of cement are used. How many buckets of sand will be needed?

**c** 20 buckets of sand are used. How many buckets of cement will be needed?

*Solution 5*

**a** $1:5$

**b** $1:5$

$\quad = 2:10$

10 buckets of sand will be needed

> For every 1 bucket of cement, 5 buckets of sand will be needed. The amount of cement has been multiplied by 2 so multiply the amount of sand by 2 as well.

**c** $20 \div 5 = 4$

$\quad 1:5$

$\quad = 4:20$

4 buckets of cement will be needed

> The amount of sand has been multiplied by 4 so multiply the amount of cement by 4 as well.

### Exercise 30B

**1** The ratio of the number of red beads to the number of green beads in a bag is $1:3$
Work out the number of green beads if there are

   **a** 2 red beads          **b** 6 red beads          **c** 15 red beads

**2** In a recipe for pastry the ratio of the weight of flour to the weight of fat is $2:1$.
Work out the weight of fat needed for

   **a** 40 g of flour          **b** 120 g of flour          **c** 1 kg of flour

**3** Brass is made from copper and zinc in the ratio $5:3$ by weight.

   **a** If there are 6 kg of zinc work out the weight of copper.

   **b** If there are 25 kg of copper work out the weight of zinc.

**4** On a map 1 cm represents 2 km. What distance on the map will represent a real distance of

   **a** 8 km          **b** 13 km          **c** 2.6 km?

**5** On a map 1 cm represents 5 km. Work out the real distance between two towns if their distance apart on the map is

   **a** 6 cm          **b** 4.2 cm          **c** 7.5 cm

**6** Damian makes a scale model of a yacht. He uses a scale of $1:12$
The length of the yacht is 18 m. Work out the length of his scale model.

**7** The scale of a map is $1:10\,000$   Work out the distance on the map between two towns if the real distance between the towns is

   **a** 700 m          **b** 2 km

**8** The scale of a map is $1:20\,000$   On the map, the distance between two towns is 9 cm.
Work out the real distance between the towns. Give your answer in kilometres.

**9** George and Henry share some money in the ratio $7:9$
If George receives £875 work out how much money Henry receives.

10 The ratio of the widths of two pictures is 6 : 7
If the width of the first picture is 1.02 m calculate the width of the second picture.

11 The ratio of the volumes of two bottles is 3 : 7
If the volume of the second bottle is 875 ml, work out the volume of the first bottle.

12 A model boat is built using the scale 1 : 40
   **a** The model is 14.5 cm long. How long is the real boat?
   **b** The real boat is 2.2 m wide. How wide is the model?

13 A map is drawn using a scale of 1 : 500 000
On the map the distance between two towns is 16.7 cm.
Work out the real distance between the towns. Give your answer in kilometres.

14 In a school the ratio of the number of students to the number of computers is $1 : \frac{2}{15}$
If there are 100 computers in the school, work out the number of students in the school.

15 A recipe says that $\frac{1}{4}$ of a teaspoon of baking powder should be used for every 125 g of flour.
Sarah uses 2 kg of flour. Work out how many teaspoons of baking powder she uses.

## 30.3 Sharing a quantity in a given ratio

There are two methods for sharing a quantity in a given ratio.

### Example 6

Pavinder and Salid share £35 in the ratio 4 : 3
Work out how much each boy gets.

*Solution 6*
**Method 1**

$4 + 3 = 7$ — Add 4 and 3 to get the total number of shares.

$35 \div 7 = 5$ — Divide 35 by 7 to work out what each share is worth.

$5 \times 4 = 20$ — Pavinder gets 4 shares so multiply 5 by 4

$5 \times 3 = 15$ — Salid gets 3 shares so multiply 5 by 3

Pavinder gets £20
Salid gets £15 — Check that the sum of the two amounts is £35 (£20 + £15 = £35)

**Method 2**

$4 + 3 = 7$ — Add 4 and 3 to get the total number of shares.

Pavinder $\frac{4}{7}$
Salid $\frac{3}{7}$ — Work out the fraction of £35 each person receives.

$\frac{4}{7} \times 35 = 20$
$\frac{3}{7} \times 35 = 15$ — Work out $\frac{4}{7}$ of 35 and $\frac{3}{7}$ of 35

Pavinder gets £20
Salid gets £15 — Check that the sum of the two amounts is £35 (£20 + £15 = £35)

## Example 7

Anna, Faye and Harriet share 24 sweets in the ratio 1 : 2 : 3
How many sweets does each girl get?

### Solution 7

$1 + 2 + 3 = 6$     | Add 1, 2 and 3 to get the total number of shares. |

$24 \div 6 = 4$     | Divide 24 by 6 to work out what each share is worth. |

Anna gets 4 sweets     | Anna gets 1 share |

Faye gets $4 \times 2 = 8$ sweets     | Faye gets 2 shares so multiply 4 by 2 |

Harriet gets $4 \times 3 = 12$ sweets     | Harriet gets 3 shares so multiply 4 by 3 (Check: $4 + 8 + 12 = 24$) |

## Exercise 30C

**1 a** Share £40 in the ratio 1 : 3     **b** Share £15 in the ratio 2 : 3
   **c** Share £60 in the ratio 5 : 1     **d** Share £100 in the ratio 3 : 2

**2** A bag contains only red beads and blue beads.
The ratio of the number of red beads to the number of blue beads is 4 : 1
There are 35 beads in the bag. How many blue beads are there in the bag?

**3** In a class the ratio of the number of girls to the number of boys is 3 : 5
There are 32 students in the class. Work out the number of girls in the class.

**4** Ben and Harry share 50 toy cars in the ratio 2 : 3    Work out how many toy cars Ben gets.

**5** Sally is 9 years old. Alex is 11 years old. They share £120 in the ratio of their ages.
How much money does each girl get?

**6** A box contains plain, white and milk chocolates in the ratio 1 : 2 : 3
There are 36 chocolates in the box. How many of each type of chocolate are there?

**7** Andy, Gary and Alan share £80 in the ratio 2 : 3 : 5    How much money does each boy receive?

**8** A recipe for crumble topping uses sugar, fat and flour in the ratio 1 : 2 : 3
How much of each ingredient is needed to make 900 g of crumble?

**9** Share £6 in the ratio 2 : 3 : 5.

**10** Ahmed, Ken and Conrad share £71.10 in the ratio 2 : 3 : 4    Work out how much money Ken gets.

**11** Three boys washed some cars. They earned a total of £87.60    They share the money in the ratio of the amount of time that each of them worked. James worked for 5 hours. Sam worked for $3\frac{1}{2}$ hours and Jack also worked for $3\frac{1}{2}$ hours. Calculate the amount of money James gets.

**12** Three waitresses received £28.70 in tips. They share this money in the ratio of the amount of time that each of them worked. Sasha worked for $2\frac{1}{2}$ hours. Eloise worked for 3 hours and Sarah worked for $4\frac{3}{4}$ hours. Calculate the amount of money Sarah gets.

## 30.4 Direct proportion

If 1 pencil costs 15p, then   2 pencils cost 30p (2 × 15p)
3 pencils cost 45p (3 × 15p)
4 pencils cost 60p (4 × 15p) and so on.

The cost depends on the number of pencils. As the number of pencils increases, the cost increases. The cost is said to increase in the same **proportion** as the number of pencils and so the cost is **proportional** to the number of pencils. For example 4 pencils cost twice as much as 2 pencils and 18 pencils cost 6 times as much as 3 pencils.

The two quantities of cost and price are said to be in **direct proportion**.

There are many examples of direct proportion in everyday life. The amount of money a motorist pays for petrol is proportional to how much petrol they put in their tank. The number of dollars a tourist gets in exchange for their pounds is proportional to the number of pounds.

### Example 8

5 buns cost £1.50   Work out the cost of 7 of these buns.

#### Solution 8

This is called the **unitary** method because it finds the cost of **one** bun first.

$$\begin{array}{r} 30 \\ 5\overline{)150} \end{array}$$

| Work out the cost of 1 bun. Divide the cost of five buns by 5 |

$30 \times 7 = 210$

| Work out the cost of 7 buns. Multiply the cost of one bun by 7 |

7 buns cost £2.10

| As the answer is more than £1 give the answer in pounds. |

### Example 9

The weight of card is directly proportional to its area.
A piece of card has an area of 36 cm² and a weight of 15 grams.
A larger piece of the same card has an area of 48 cm².
Calculate the weight of the larger piece of card.

#### Solution 9

$\frac{15}{36} = \frac{5}{12}$

| Work out the weight of 1 cm² of the card. The answer will be less than 1 so write this as a fraction in its simplest form. |

$\frac{5}{\cancel{12}_1} \times \cancel{48}^{4}$

$= 20$

| Work out the weight of a 48 cm² piece of card. Multiply the weight of 1 cm² by 48 |

The weight of the larger piece of card is 20 grams

### Example 10

Here is a list of the ingredients needed to make carrot soup for 4 people.

200 g carrots          2 onions
40 g butter            300 ml stock

Work out the amount of each ingredient needed to make carrot soup for **12** people.

**Solution 10**

$12 \div 4 = 3$

> Carrot soup for 12 people needs 3 times as much of each ingredient as carrot soup for 4 people.

$200 \times 3 = 600$

> So multiply each amount by 3

$2 \times 3 = 6$

$40 \times 3 = 120$

$300 \times 3 = 900$

The amount of each ingredient is

|  |  |
|---|---|
| 600 g carrots | 6 onions |
| 120 g butter | 900 ml stock |

## Example 11

**a** Janet went on holiday to France.
She changed £200 into euros.
The exchange rate was £1 = 1.62 euros.
Work out the number of euros Janet got.

**b** Janet came home.
She had 62 euros left.
She changed her 62 euros to pounds.
The new exchange rate was £1 = 1.55 euros.
Work out how much Janet got in pounds for 62 euros.

**Solution 11**

**a** $200 \times 1.62 = 324$

Janet got 324 euros

> Janet got 1.62 euros for every £1 so **multiply** the number of pounds by 1.62

**b** $62 \div 1.55 = 40$

Janet got £40

> Janet got £1 for every 1.55 euros so **divide** the number of euros by 1.55

## Exercise 30D

**1** A car travels at a steady speed of 60 miles each hour.
Work out the number of hours it takes to travel

   **a** 120 miles        **b** 300 miles

**2** 3 pencils cost 96p. Work out the cost of 5 of these pencils.

**3** Four 1 litre tins of paint cost a total of £36.60  Work out the cost of seven of the 1 litre tins of paint.

**4** Karen is paid £48 for 6 hours' work. How much is she paid for 4 hours' work?

**5** The cost of 5 pens is £1.20  Work out the cost of 7 of these pens.

**6** The height of a pile of 6 identical books is 15 cm. Work out the height of a pile of 8 of these books.

**7** This is a recipe for 10 cookies.

|  |  |
|---|---|
| 80 g butter | 80 g sugar |
| 2 eggs | 100 g flour |

   **a** Work out the amount of butter needed to make 20 cookies.

   **b** Work out the amount of flour needed to make 15 cookies.

**8** This is a recipe for chickpea curry for 4 people.

| | |
|---|---|
| 2 onions | 200 g of sweet potato |
| 400 g of chickpeas | 1 teaspoon curry powder |

Work out the amounts needed to make chickpea curry for 10 people.

**9** Joe is paid £54 for 8 hours' work in a supermarket. How much is he paid for 5 hours' work?

**10** Prateek buys 15 cakes for £9.75   Work out the cost of 9 of these cakes.

**11** The cost of ribbon is directly proportional to its length. A 3.5 m piece of ribbon costs £1.89
Work out the cost of 5 m of this ribbon.

**12** The length of the shadow of an object at noon is directly proportional to the height of the
object. A lamppost of height 5.4 m has a shadow of length 2.1 m at noon.
Work out the length of the shadow at noon of a man of height 1.8 m.

**13** In a recipe 180 g of flour is needed to make 24 small cakes. How much flour will be needed to
make 60 small cakes?

**14** A machine fills 720 packets of crisps in 1 hour. How long will the machine take to fill 1680
packets of crisps? Give your answer in hours and minutes.

**15** The exchange rate is £1 = $1.80
    **a** Convert £200 to dollars            **b** Convert $270 to pounds

**16** Eric went on holiday to America. He changed £450 into dollars. The exchange rate was
£1 = $1.85   Work out how many dollars Eric got.

**17** Susan bought a coat for €116 in France. The exchange rate was £1 = €1.45
Work out the cost of the coat in pounds.

**18** Angela buys a pair of jeans in England for £45   She then goes on holiday to America and sees an
identical pair of jeans for $55   The exchange rate is £1 = $1.75   In which country are the jeans
cheaper and by how much?

## 30.5 Inverse proportion

A car travelling at a steady speed of 50 km/hr travels 200 km in 4 hours.
A car travelling at a steady speed of 100 km/hr travels 200 km in 2 hours.

As the speed increases, the time decreases.
More specifically as the speed is multiplied by 2, the time is divided by 2

Two quantities are said to be in **inverse proportion** if one quantity increases at the same rate as the
other quantity decreases.

The product of the speed and the time gives the same number (the distance).

$$50 \times 4 = 200$$
$$100 \times 2 = 200$$

When two quantities are **inversely proportional** their product is constant.

## Example 12

It takes 5 cleaners 6 hours to clean a school.
Work out how long it would take 15 cleaners to clean the school.

### Solution 12

5 cleaners take 6 hours

> If there are more cleaners the job will get done more quickly which means a smaller number of hours.

$15 \div 5 = 3$

$6 \div 3 = 2$
15 cleaners would take 2 hours.

> Divide the new number of cleaners by the original number of cleaners. The original number of cleaners has been **multiplied** by 3 ($5 \times 3 = 15$)
> **Divide** the number of hours by 3
> Check: $5 \times 6 = 30$ and $15 \times 2 = 30$

## Example 13

It takes 3 men 4 days to build a wall.
Work out how long it will take 2 men to build the wall.

### Solution 13

$3 \times 4 = 12$

> 1 man would take 12 days to build the wall.

$12 \div 2 = 6$
2 men will take 6 days.

> There are now 2 men so divide by 2
> Check: $3 \times 4 = 12$ and $2 \times 6 = 12$

### Exercise 30E

**1**  It takes 4 men 5 days to cut a hedge. Work out how long it will take to cut the hedge if there are
   **a**  2 men              **b**  10 men.

**2**  One examiner takes a total of 36 hours to mark some exam papers.
   How long would it take 4 examiners to mark the same number of papers?

**3**  A quantity of hay is enough to feed 5 horses for 8 days.
   Work out the number of days that the same quantity of hay will feed 4 horses.

**4**  6 computers process a certain amount of information in 6 hours.
   Work out how long it will take 9 computers to process the same amount of information.

**5**  A factory uses 3 machines to complete a job in 15 hours.
   If 2 extra machines are used, how long will the job take?

**6**  It takes 6 technicians 8 hours to enter data into a computer.
   How long would it take 4 technicians to enter the data?

**7**  It takes 3 machines 6 days to harvest a crop. How long would it take 2 machines?

**8**  6 men take a total of 1 hour to blow up a number of balloons.
   How long will 8 men take to blow up the same number of balloons?

**9**  A large ball of wool is used to knit a scarf. The scarf is 40 stitches wide and 120 cm long. If the same size ball of wool is used to knit a scarf 25 stitches wide, work out the length of the new scarf.

**10**  50 small boxes each hold 20 cans. A large box holds 25 cans.
   Work out how many large boxes will be needed to hold all the cans.

**11** A quantity of soup will fill 24 bowls if 3 ladles of soup are put in each bowl. How many extra bowls would be filled from the same quantity of soup if 2 ladles of soup are put in each bowl?

**12** A document will fit onto exactly 32 pages if there are 500 words on a page. If the number of words on each page is reduced to 400, how many *more* pages will there be in the document?

## Chapter summary

**You should now know that:**

★ a ratio *compares* quantities. An example of a ratio is 4 : 5

★ the order of numbers in a ratio matters. For example if there are 6 red and 7 blue cars then
   ● the ratio of the number of red cars to the number of blue cars is 6 : 7
   ● the ratio of the number of blue cars to the number of red cars is 7 : 6

★ if the units of the quantities in a ratio are different, put the quantities into the same units before starting a calculation

★ to *simplify a ratio* divide each number in the **ratio** by the same number

★ the simplest form of a ratio has whole numbers with no common factor

★ two quantities are in **direct proportion** if their ratio stays the same as the quantities increase or decrease

★ two quantities are in **inverse proportion** if one quantity increases at the same rate as the other quantity decreases. The product of the two quantities is constant

**You should know how to:**

★ find the other quantity if the ratio of two quantities is given and one of the quantities is known

★ share a quantity in a given ratio

★ solve problems involving direct proportion and inverse proportion.

## Chapter 30 review questions

**1** 4 identical notebooks cost 84p. Work out the cost of 9 of these notebooks.

**2** On a map 1 cm represents 2 km. What distance on the map will represent a real distance of
   **a** 8 km        **b** 20 km        **c** 9 km?

**3** On a map 1 cm represents 4 km. Work out the real distance between two towns if their distance apart on the map is
   **a** 3 cm        **b** 1.2 cm        **c** 12.8 cm.

**4** John and Adam share 30 sweets in the ratio 2 : 3    Work out how many sweets Adam gets.

**5** The ratio of girls to boys in a school is 2 : 3
   **a** What fraction of these students are boys?
   In Year 8 the ratio of girls to boys is 1 : 3    There are 300 students in Year 8.
   **b** Work out the number of girls in Year 8.            (1388 January 2004)

6   It takes 6 men 4 days to build a wall. How long would it take 2 men?

7   A greengrocer has a quantity of potatoes. If he puts 3 kg of potatoes in a bag, he fills 30 bags.
    How many bags will he fill if he puts 5 kg of potatoes in each bag?

8   The weight of a piece of pipe is directly proportional to its length. The weight of a 12 cm length
    of pipe is 18 g. Work out the weight of a 20 cm length of pipe.

9   This is a recipe for making Tuna Bake for **4** people.

    > *Tuna Bake*
    > Ingredients for 4 people
    >
    > 400 g of tuna
    > 400 g of mushroom soup
    > 100 g of grated cheddar cheese
    > 4 spring onions
    > 250 g of breadcrumbs

    Work out the amounts needed to make a Tuna Bake for **10** people.        (1385 June 2001)

10  Here are the ingredients needed to make 500 ml of custard.
    a  Work out the amount of sugar needed to make 2000 ml
       of custard.
    b  Work out the amount of milk needed to make 750 ml of
       custard.

    > *Custard*
    > Makes 500 ml
    >
    > 400 ml of milk
    > 3 large egg yolks
    > 50 g sugar
    > 2 teaspoons of cornflour

    (1387 June 2005)

11  Angela and Michelle shared some money in the ratio 4 : 9
    *Then* Angela gave Daniel a half of her share.
    Michelle gave Daniel a third of her share.
    Daniel was given a total of £20
    Work out how much money was shared originally by Angela and Michelle.

12  Michael buys 3 files.
    The total cost of these 3 files is £5.40
    Work out the cost of 7 of these files.                                    (1387 June 2005)

13  Kelly bought 4 identical computer disks for £3.60
    Work out the cost of 9 of these computer disks.                           (1385 November 2002)

14  The weight of a piece of wire is directly proportional to its length.
    A piece of wire is 25 cm long and has a weight of 6 grams.
    Another piece of the same wire is 30 cm long.
    Calculate the weight of the 30 cm piece of wire.                          (1388 January 2004)

15  Margaret goes on holiday to Switzerland.
    The exchange rate is £1 = 2.10 francs.
    She changed £450 into francs.
    How many francs should she get?                                           (1387 June 2005)

16  In Portugal, a suitcase costs 23 euros.
    In England, an identical suitcase costs £15
    The exchange rate is £1 = 1.45 euros.
    In which country is the suitcase cheaper and by how much?

**17** Here is a list of ingredients needed to make leek soup for **6** people.

> *Leek soup*
> Ingredients for 6 people
>
> 600 g leeks
> 30 g butter
> 3 tablespoons of wholemeal flour
> 450 ml milk
> 600 ml water

Work out the amount of each ingredient needed to make leek soup for **4** people.

**18** Bill gave his three daughters a total of £32.40
The money was shared in the ratios 4 : 3 : 2
Jane had the largest share.
Work out how much money Bill gave to Jane.                                    (1385 November 2001)

**19** Stephen and Joanne share £210 in the ratio 6 : 1
How much more money does Stephen get than Joanne?                             (1385 June 2001)

**20** Prendeep bought a necklace in the United States of America.
Prendeep paid 108 dollars ($).
Arthur bought an identical necklace in Germany.
Arthur paid 117 Euros (€).

> £1 = $1.44
> £1 = 1.6€

Calculate in pounds the difference between the prices paid for the two necklaces.
Show how you worked out your answer.                                          (1387 November 2003)

# Three-dimensional shapes (2)

All the photographs show examples of **three-dimensional shapes**.
Three-dimensional shapes have length, breadth and height. They are not flat, like squares and circles.

Three-dimensional is often written as **3-D**.

## 31.1 Planes of symmetry

When a mirror is placed on a line of symmetry of a two-dimensional shape and looked at from either side, the shape looks the same. In other words, each half of the shape is a mirror image of the other half.

In a similar way, when a plane cuts a 3-D shape in two so that each half is a mirror image of the other half, the plane is called a **plane of symmetry**.

The diagram shows one of the planes of symmetry of a cuboid.
A cuboid is a prism with a rectangle as its cross-section.
A prism is a 3-D shape which is the same all along its length.

A cuboid has two more planes of symmetry.

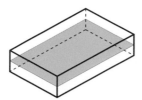
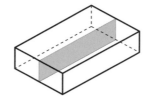
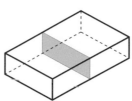

## Example 1

The diagram shows a prism with an isosceles trapezium as its cross-section.
The prism has two planes of symmetry.
On the diagrams below, draw the planes of symmetry.

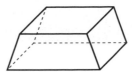

*Solution 1*

## Exercise 31A

Answer this exercise on the resource sheet.

**1** Draw *one* plane of symmetry on each of the 3-D shapes.

**a** 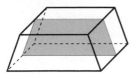   **b** 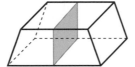   **c**

**d**    **e** 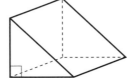   **f**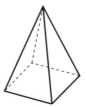

**2** The diagram shows a prism with an equilateral triangle as its cross-section.

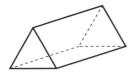

It has four planes of symmetry.
On four separate diagrams, draw a different plane of symmetry.

**3** The diagram shows a prism.

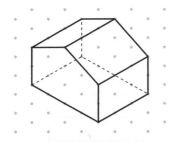

**a** On the diagram, draw a plane of symmetry.
**b** How many planes of symmetry has the prism?

## 31.2 Plans and elevations

The diagram shows a prism with an isosceles triangle as its cross-section.

The view from above is called the **plan**

The view from the front is called the **front elevation**

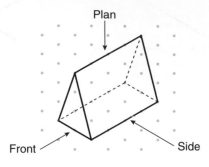

The view from the side is called the **side elevation**

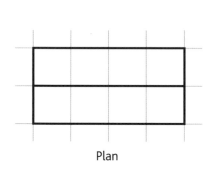

Plan

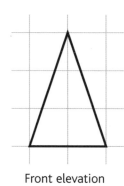

Front elevation

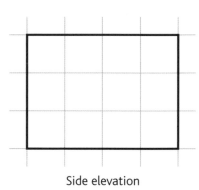

Side elevation

**Example 2**

The diagram shows a prism.

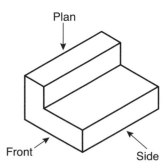

Draw a sketch of the plan, front elevation and side elevation.

**Solution 2**

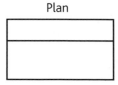

Plan

Front elevation

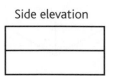

Side elevation

## Exercise 31B

**1** On centimetre squared paper, draw the plan, front elevation and side elevation for this cuboid.

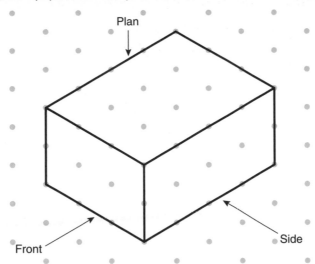

Plan

Front

Side

**2** On centimetre squared paper, draw the plan, front elevation and side elevation for this prism.

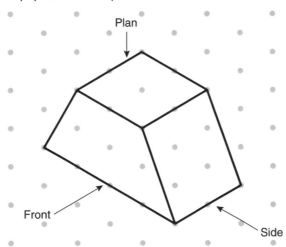

Plan

Front

Side

**3** Sketch the plan, front elevation and side elevation for each of these 3-D shapes.

**a**    **b** 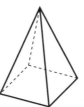   **c**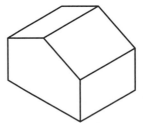

**4** Here are the front elevation and side elevation of a prism.
The front elevation shows the cross-section of the prism.

**a** On squared paper, draw a plan of the prism.

**b** Draw a 3-D sketch of the prism.

Front elevation          Side elevation

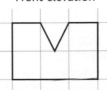   

## 31.3 Volume of three-dimensional shapes

### Volume of a cylinder

A cylinder is a prism with a circle as its cross-section.

The radius of the end of a cylinder is 3 cm.
Its length is 10 cm.

Work out the volume of the cylinder.
Give your answer correct to three significant figures.

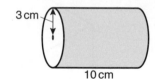

*Solution 3*

Area of cross-section $= \pi \times 3^2$

$\qquad\qquad\qquad = \pi \times 9$

$\qquad\qquad\qquad = 28.27433...\ \text{cm}^2$

> Use $A = \pi r^2$ to work out the area of the circular cross-section.
>
> Use the $\pi$ button on your calculator, if it has one. Otherwise, take the value of $\pi$ to be 3.142
>
> Do not round at this stage. Leave all the figures on your calculator display.

Volume $= 28.27433... \times 10$

$\qquad\quad = 282.7433...\ \text{cm}^3$

> Multiply the area of cross-section by the length to get the volume of the cylinder.

Volume $= 283\ \text{cm}^3$

> Give the volume correct to three significant figures.

The volume, $V$, of a cylinder of radius $r$ and length $h$ is given by the formula

$$V = \pi r^2 h$$

### Exercise 31C

For Questions 1, 2 and 3, use the $\pi$ button on your calculator, if it has one.
Otherwise, take the value of $\pi$ to be 3.142

**1** The radius of the end of a cylinder is 5 cm.
Its length is 15 cm.
Work out the volume of the cylinder.
Give your answer correct to three significant figures.

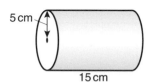

**2** The *diameter* of the end of a cylinder is 8.6 cm.
Its length is 14 cm.
Work out the volume of the cylinder.
Give your answer correct to three significant figures.

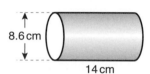

**3** The volume of a cylinder is 500 cm³.
The radius of its end is 7.8 cm.
Work out the length of the cylinder.
Give your answer correct to three significant figures.

## Volume of a pyramid

A pyramid is a 3-D shape in which lines drawn from the edge of the base meet at a point.
A pyramid is usually defined in terms of its base.
If the base is a polygon, all the other faces are triangles.
If the base is a circle, the shape is called a cone.

Triangle-based pyramid

Square-based pyramid

Cone

The diagram shows how a cube of side 12 cm can be split into six congruent square-based pyramids by drawing the four diagonals of the cube.

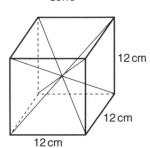

The volume of cube $= 12 \times 12 \times 12 = 1728 \text{ cm}^3$.

The volume of each pyramid $= \dfrac{1728}{6} = 288 \text{ cm}^3$.

In general, the volume of a pyramid is given by the formula

> **volume of a pyramid $= \frac{1}{3} \times$ base area $\times$ height**

For this pyramid,
base area $= 12 \times 12 = 144 \text{ cm}^2$
height $= 6$ cm

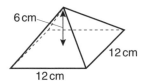

Volume $= \frac{1}{3} \times$ base area $\times$ height
$\quad\quad\; = \frac{1}{3} \times 144 \times 6 = 288 \text{ cm}^3$

For a cone with base radius $r$ and height $h$, the base area $= \pi r^2$ and the volume is given by the formula

> **volume of a cone $= \frac{1}{3} \pi r^2 h$**

### Example 4

The base of a pyramid is an 8 cm by 5 cm rectangle.
The height of the pyramid is 6 cm.
Calculate the volume of the pyramid.

**Solution 4**

Base area $= 8 \times 5$
$\quad\quad\quad\;\; = 40 \text{ cm}^2$

> Calculate the area of the rectangular base.

Volume $= \frac{1}{3} \times 40 \times 6$
$\quad\quad\;\; = 80 \text{ cm}^3$

> Substitute 40 for the base area and 6 for the height in the expression $\frac{1}{3} \times$ base area $\times$ height.
> The units are cm³.

### Example 5

The radius of the base of a cone is 7 m and its height is 10 m.
Calculate the volume of the cone.
Give your answer correct to three significant figures.

**Solution 5**

$\frac{1}{3} \times \pi \times 7^2 \times 10$

$= 513.12...$

| Substitute $r = 7$ and $h = 10$ in the formula $V = \frac{1}{3}\pi r^2 h$ for the volume of a cone. Write down at least four figures of the calculator display. |

Volume $= 513$ m$^3$

| Round the volume to three significant figures. The units are m$^3$. |

---

## Example 6

A cone has a base radius of 2.3 cm and a volume of 43 cm$^3$.

Calculate its height.

Give your answer correct to three significant figures.

**Solution 6**

$43 = \frac{1}{3} \times \pi \times 2.3^2 \times h$

| Substitute $V = 43$ and $r = 2.3$ in the formula $V = \frac{1}{3}\pi r^2 h$ |

$43 = 5.5396... \times h$

| Evaluate $\frac{1}{3} \times \pi \times 2.3^2$. Write down at least four figures of the calculator display. |

$h = \dfrac{43}{5.5396...} = 7.762...$

| Divide 43 by 5.5396... Write down at least four figures of the calculator display. |

Height $= 7.76$ cm

| Round the height to three significant figures. |

---

A **frustum** is the solid remaining when a cone is cut by a plane parallel to its base and the top cone is removed.

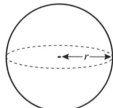

## Volume of a sphere

A sphere is a 3-D shape which, like a ball, is circular in every direction.

The volume of a sphere with radius $r$ is given by the formula

**volume of a sphere $= \frac{4}{3}\pi r^3$**

---

## Example 7

The radius of a sphere is 5.7 cm.

Calculate the volume of the sphere.

Give your answer correct to three significant figures.

**Solution 7**

$\frac{4}{3} \times \pi \times 5.7^3$

$= 775.73...$

| Substitute $r = 5.7$ in the formula $V = \frac{4}{3}\pi r^3$ for the volume of a sphere. Write down at least four figures of the calculator display. |

Volume $= 776$ cm$^3$

| Round the volume to three significant figures. The units are cm$^3$. |

---

### Exercise 31D

Where necessary, give answers correct to three significant figures.

If your calculator does not have a $\pi$ button take the value of $\pi$ to be 3.142, unless the question instructs otherwise.

1   A pyramid has a square base of side 7 cm and a height of 15 cm. Calculate its volume.

2   The base of a pyramid is a 6 m by 3 m rectangle. Its height is 10 m. Calculate its volume.

3   The diagram shows the triangular base of a pyramid.
    The height of the pyramid is 18 cm.
    Calculate its volume.

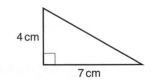

4   The radius of the base of a cone is 3 cm and its height is 8 cm.
    Calculate the volume of the cone.

5   The radius of the base of a cone is 4.9 cm and its height is 8.3 cm.
    Calculate the volume of the cone.

6   The radius of the base of a cone is 4 cm and its height is 9 cm.
    Find the volume of the cone. Give your answer as a multiple of $\pi$.

7   The radius of a sphere is 2.6 cm. Calculate its volume.

8   The diameter of a sphere is 8.2 cm. Calculate its volume.

9   The radius of a sphere is 6 cm. Find its volume. Give your answer as a multiple of $\pi$.

10  The Great Pyramid of Egypt had a square base of side 230 m and a height of 146.5 m.
    Calculate its volume.

11  The radius of a tennis ball is 3.75 cm. Calculate its volume.

12  Calculate the volume of a hemisphere with a diameter of 4.6 cm.

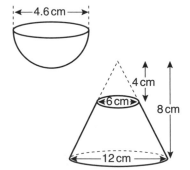

13  A frustum is made by removing a cone with a base of diameter
    6 cm and height 4 cm from a cone with a base of diameter 12 cm
    and a height of 8 cm.
    Calculate the volume of the frustum.

14  A solid is made from a cylinder and a cone.
    Their radius is 2.7 cm.
    The length of the cylinder is 8.9 cm and the height
    of the cone is 6.4 cm.
    Calculate the volume of the solid.

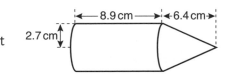

15  A hemispherical wooden bowl has an internal radius
    of 8 cm and an external radius of 9 cm.
    Calculate the volume of wood.

16  A pyramid has a height of 24 cm and a volume of 72 cm³.
    Calculate the area of its base.

17  A square-based pyramid has a height of 15 cm and a volume of 320 cm³. Calculate the length of
    the sides of its base.

18  A cone has a base radius of 7.1 cm and a volume of 650 cm³. Calculate its height.

**19** The height of a cone is 9.7 cm and its volume is 180 cm³. Calculate the radius of its base.

**20** The volume of a sphere is 300 cm³. Calculate its radius.

## Units of volume

The units of volume appearing so far in this chapter have been cm³, m³ and mm³, depending on the units which have been used in a question.

To convert between cm³ and mm³, find the volume of a centimetre cube in both cm³ and in mm³. Remember that 1 cm = 10 mm.

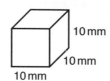

Volume = 1 cm³

Volume = 10 × 10 × 10 mm³
= 1000 mm³

This shows that

**1 cm³ = 1000 mm³**

So, to change cm³ to mm³, multiply by 1000 and to change mm³ to cm³, divide by 1000

To convert between m³ and cm³, find the volume of a metre cube in both m³ and in cm³. Remember that 1 m = 100 cm.

Volume = 1 m³

Volume = 100 × 100 × 100
= 1 000 000 cm³

This shows that

**1 m³ = 1 000 000 cm³**

So, to change m³ to cm³, multiply by 1 000 000 and to change cm³ to m³, divide by 1 000 000

### Example 8

**a** Change 8 m³ to cm³.          **b** Change 700 000 cm³ to m³.

*Solution 8*

**a** 8 × 1 000 000 = 8 000 000 cm³     | Multiply by 1 000 000 |

**b** 700 000 ÷ 1 000 000 = 0.7 m³     | Divide by 1 000 000 |

Another important unit of volume is the **litre**.

Litres are used to measure the **capacity** of a container, which is the amount, often of liquid, that a container can hold.
For example, the capacity of a Ford Fiesta's petrol tank is 42 litres.

**1 litre = 1000 cm³**

## Exercise 31E

**1** Change

   **a** 3 cm³ to mm³        **b** 4500 mm³ to cm³        **c** 0.65 cm³ to mm³

**2** Change

   **a** 2.8 m³ to cm³        **b** 6 000 000 cm³ to m³        **c** 3 200 000 cm³ to m³

**3** The diagram shows a cuboid.
Work out the volume of the cuboid

   **a** in cm³

   **b** in mm³

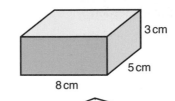

**4** The diagram shows a prism.
Work out the volume of the prism

   **a** in m³

   **b** in cm³

**5** A cylindrical bowl has a radius of 15 cm.
It is filled with water to a depth of 12 cm.
Work out the volume of water in the bowl.
Give your answer in litres, correct to three significant figures.
(Use the $\pi$ button on your calculator, if it has one.
Otherwise, take the value of $\pi$ to be 3.142)

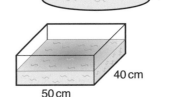

**6** The diagram shows a container which is a cuboid.
15.6 litres of water are poured into the container.
Work out the depth of water.

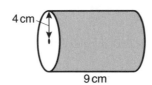

# 31.4 Surface area of three-dimensional shapes

### Example 9

The radius of the end of a solid cylinder is 4 cm.
Its length is 9 cm.

Work out the total surface area of the cylinder.
Give your answer correct to three significant figures.

### Solution 9

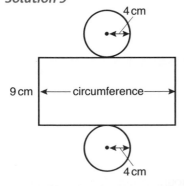

> The net of the curved surface of a cylinder is a rectangle.
> The length of this rectangle is equal to the circumference ($2\pi r$)
> of each circular end of the cylinder.
> The two circular ends complete the net.
>
> The total surface area is the sum of the area of the
> curved surface and the areas of the two circular ends.

Total surface area

   $= 2 \times \pi \times 4 \times 9 + 2 \times \pi \times 4^2$

   $= 226.19 \ldots + 100.53 \ldots$

   $= 327 \text{ cm}^2$

> The area of the curved surface is the area of the rectangle $2 \times \pi \times 4 \times 9$
> The area of each circular end is $\pi \times 4^2$
> Give the area correct to three significant figures.

The total surface area, $S$, of a solid cylinder of radius $r$ and length $h$ is given by the formula

$$S = 2\pi rh + 2\pi r^2$$

## Surface area of a sphere

The surface area of a sphere with radius $r$ is given by the formula

surface area of a sphere $= 4\pi r^2$

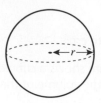

### Example 10

The radius of a sphere is 7.4 cm.
Calculate the surface area of the sphere.
Give your answer correct to three significant figures.

**Solution 10**

$4 \times \pi \times 7.4^2$

$\quad = 688.13...$

| Substitute $r = 7.4$ in the formula $A = 4\pi r^2$ for the surface area of a sphere. Write down at least four figures of the calculator display. |

Surface area $= 688$ cm$^2$

| Round the answer to three significant figures. |

## Surface area of a cone

The net of a hollow cone is a sector of a circle.

The sector can be folded to make the cone.
The radius, $l$, of the sector becomes the *slant height* of the cone and the arc length, $a$, becomes the circumference of the base of the cone.

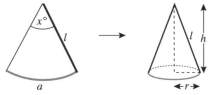

Arc length $a = \dfrac{x}{360} \times 2\pi l$ and circumference of the base $= 2\pi r$

So $\dfrac{x}{360} \times 2\pi l = 2\pi r$

Dividing both sides by $2\pi$ gives $\dfrac{x}{360} \times l = r$

Sector area $= \dfrac{x}{360} \times \pi l^2 = \pi \times \dfrac{x}{360} \times l \times l$

But $\dfrac{x}{360} \times l = r$ and so sector area $= \pi rl$.

The sector area, $\pi rl$, is equal to the area of the curved surface of the cone

curved surface area of a cone $= \pi rl$

For a *solid cone*, the circular base of the cone is added to the net and the total surface area is the sum of the cone's curved surface area, $\pi rl$, and the area of its base $\pi r^2$.

total surface area of a solid cone $= \pi rl + \pi r^2$

### Example 11

A solid cone has a base radius of 3.6 cm and a slant height of 8.5 cm.
Calculate the total surface area of the cone.
Give your answer correct to three significant figures.

**Solution 11**

$(\pi \times 3.6 \times 8.5) + (\pi \times 3.6^2)$

| | |
|---|---|
| $= 96.13... + 40.71...$ | Substitute $r = 3.6$ and $l = 8.5$ in the formula $S = \pi rl + \pi r^2$ for the total surface area of a cone. |
| $= 136.84...$ | Write down at least four figures of the calculator display. |

| | |
|---|---|
| Total surface area $= 137$ cm$^2$ | Round the answer to three significant figures. |

---

## Example 12

The diagram shows the net of a cone.
Calculate the total surface area of the cone.

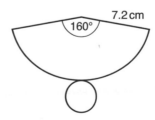

**Solution 12**

| | |
|---|---|
| $\dfrac{160}{360} \times 2 \times \pi \times 7.2$ | Calculate the arc length of the sector. |
| $= 20.106...$ | |

| | |
|---|---|
| $2\pi r = 20.106...$ | The arc length of the sector is equal to the circumference of the base of the cone. Use this fact to calculate the radius $r$ of the cone's base. |
| $r = \dfrac{20.106}{2\pi} = 3.2$ | |

| | |
|---|---|
| $(\pi \times 3.2 \times 7.2) + (\pi \times 3.2^2)$ | Substitute $r = 3.2$ and $l = 7.2$ in the formula $S = \pi rl + \pi r^2$ for the total surface area of a cone. |
| $= 72.38... + 32.16...$ | Write down at least four figures of the calculator display. |
| $= 104.55...$ | |

| | |
|---|---|
| Total surface area $= 105$ cm$^2$ | Round the answer to three significant figures. |

---

## Exercise 31F

Where necessary, give answers correct to three significant figures.
If your calculator does not have a $\pi$ button take the value of $\pi$ to be 3.142, unless the question instructs otherwise.

1 Work out the surface area of each of these 3-D shapes. For each shape, draw a sketch of the net.

 a A solid cylinder of radius 4.7 cm and length 6.3 cm.
 Give your answer correct to three significant figures.

 b A hollow cylinder which is open at one end. Its radius is 5.1 cm and its length is 9.4 cm.
 Give your answer correct to three significant figures.

2 Calculate the surface area of a sphere of radius 6.9 cm.

3 The radius of the base of a cone is 4.3 cm and its slant height is 9.7 cm.
 Calculate its curved surface area.

4 The radius of the base of a cone is 5.3 cm and its slant height is 8.1 cm.
 Calculate its total surface area.

5 The radius of the base of a cone is 3.7 cm and its slant height is 9.4 cm.
 Calculate its total surface area.

**6** The diagrams show the nets of three cones. Calculate the total surface area of each cone.

**a**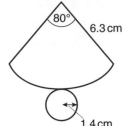

6.3 cm

1.4 cm

**b**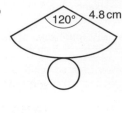

4.8 cm

**c**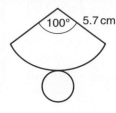

5.7 cm

**7** The radius of a sphere is 3 cm. Find its surface area, giving the answer as a multiple of $\pi$.

**8** The radius of the base of a cone is 4 cm and its slant height is 6 cm.
Find its total surface area, giving the answer as a multiple of $\pi$.

**9** The surface area of a sphere is 60 cm². Calculate its radius.

# Chapter summary

**You should now know:**

★ that **three-dimensional** (3-D) shapes have length, breadth and height

★ that a **plane of symmetry** cuts a 3-D shape in two so that each half is a mirror image of the other half

★ how to draw and interpret **plans** and **elevations** of 3-D shapes

★ how to find the **surface area** of cylinders, spheres and cones, including the use of nets

★ that the volume, $V$, of a cylinder of radius $r$ and length $h$ is given by the formula
$$V = \pi r^2 h$$

★ that the curved surface area, $C$, of a cylinder of radius $r$ and length $h$ is given by the formula
$$C = 2\pi r h$$

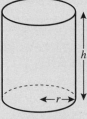

★ that the total surface area, $A$, of a solid cylinder of radius $r$ and length $h$ is given by the formula
$$A = 2\pi r h + 2\pi r^2$$

★ that the volume, $V$, of a cone of base radius $r$ and height $h$ is given by the formula
$$V = \tfrac{1}{3} \pi r^2 h$$

★ that the curved surface area, $A$, of a cone of base radius $r$ and slant height $l$ is given by the formula
$$A = \pi r l$$

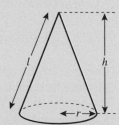

★ that the total surface area, $S$, of a cone of base radius $r$ and slant height $l$ is given by the formula
$$S = \pi r l + \pi r^2$$

★   that the volume, $V$, of a sphere of radius $r$ is given by the formula
$$V = \tfrac{4}{3}\pi r^3$$

★   that the surface area, $A$, of a sphere of radius $r$ is given by the formula
$$A = 4\pi r^2$$

★   that $1\,\text{cm}^3 = 1000\,\text{mm}^3$ and use it to convert between $\text{cm}^3$ and $\text{mm}^3$

★   that $1\,\text{m}^3 = 1\,000\,000\,\text{cm}^3$ and use it to convert between $\text{m}^3$ and $\text{cm}^3$

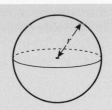

# Chapter 31 review questions

Where necessary, give answers correct to three significant figures.
If your calculator does not have a $\pi$ button take the value of $\pi$ to be 3.142, unless the question instructs otherwise.

**1**   Here are the plan and front elevation of a prism. The front elevation shows the cross-section of the prism.

    **a**   On squared paper, draw a side elevation of the prism.

    **b**   Draw a 3-D sketch of the prism.

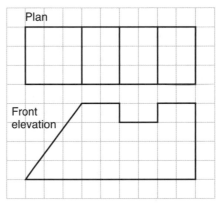

(1387 June 2003)

**2**   A can of drink is in the shape of a cylinder.
The can has a radius of 4 cm and a height of 15 cm.

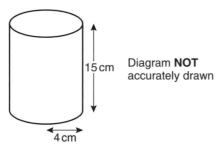

15 cm    Diagram **NOT** accurately drawn

4 cm

Calculate the volume of the cylinder.
Give your answer correct to three significant figures.

(1387 November 2003)

**3**   A child's toy is made out of plastic.
The toy is solid.
The top of the toy is a cone of height 10 cm and base radius 4 cm.
The bottom of the toy is a hemisphere of radius 4 cm.

Calculate the volume of plastic needed to make the toy.

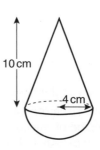

10 cm    4 cm

(1384 November 1997)

**4** A cylindrical can has a radius of 6 centimetres.
The capacity of the can is 2000 cm³.
Calculate the height of the can.
Give your answer correct to one decimal place.

(1384 June 1995)

**5** The diagram represents a large cone of height 30 cm and base diameter 12 cm.

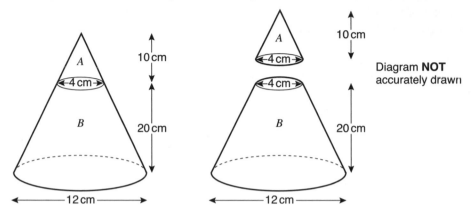

Diagram **NOT**
accurately drawn

The large cone is made by placing a small cone $A$ of height 10 cm and base diameter 4 cm on top of a frustum $B$.
Calculate the volume of frustum $B$.

**6** The diagram shows a sector, $VAB$, of a circle, centre $V$.
The radius of the circle is 12 cm.
Angle $AVB = 72°$.

**a** Calculate the length of the arc $AB$.

$VA$ and $VB$ are joined so that the sector makes the curved surface of a cone.

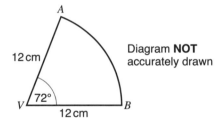

Diagram **NOT**
accurately drawn

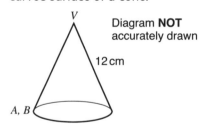

Diagram **NOT**
accurately drawn

**b** Calculate the curved surface area of the cone.

**7**

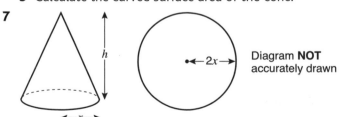

Diagram **NOT**
accurately drawn

The radius of the base of a cone is $x$ cm and its height is $h$ cm.
The radius of a sphere is $2x$ cm.
The volume of the cone and the volume of the sphere are equal.
Express $h$ in terms of $x$.
Give your answer in its simplest form.

(1387 June 2005)

## 32.1 Graphs of quadratic functions

All these expressions contain a letter **squared**.
They are called **quadratic** expressions, or **quadratic functions**.
The highest power in a quadratic expression is 2
All quadratic functions using the variable $x$ can be written
in the form $ax^2 + bx + c$ where $a(\neq 0)$, $b$ and $c$ represent
numbers. $ax^2 + bx + c = 0$ is called a **quadratic equation**.
Graphs of quadratic functions can sometimes be used to
solve quadratic equations.

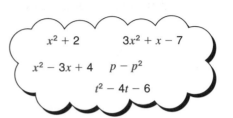

$x^2 + 2 \qquad 3x^2 + x - 7$

$x^2 - 3x + 4 \qquad p - p^2$

$t^2 - 4t - 6$

To draw the graph of a quadratic function a table of values can be used.

---

### Example 1

Draw the graph of $y = x^2$.
Use values of $x$ from $x = -3$ to $x = 3$

*Solution 1*
When $x = 3$, $\qquad y = 3 \times 3 \qquad\qquad = 9$
When $x = 2$, $\qquad y = 2 \times 2 \qquad\qquad = 4$
When $x = 1$, $\qquad y = 1 \times 1 \qquad\qquad = 1$
When $x = 0$, $\qquad y = 0 \times 0 \qquad\qquad = 0$
When $x = -1$, $\quad y = (-1) \times (-1) = 1$
When $x = -2$, $\quad y = (-2) \times (-2) = 4$
When $x = -3$, $\quad y = (-3) \times (-3) = 9$

These results can be shown in a table of values.

| $x$ | $-3$ | $-2$ | $-1$ | 0 | 1 | 2 | 3 |
|---|---|---|---|---|---|---|---|
| $y$ | 9 | 4 | 1 | 0 | 1 | 4 | 9 |

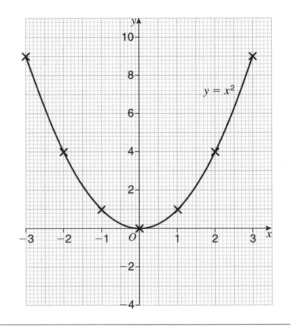

---

The lowest point of this graph is where the graph turns. It is the origin (0, 0).
It is called the **minimum point**.

---

### Example 2

**a** Complete this table of values for $y = 4 - x^2$.

| $x$ | $-3$ | $-2$ | $-1$ | 0 | 1 | 2 | 3 |
|---|---|---|---|---|---|---|---|
| $y$ | $-5$ | | 3 | | 3 | | |

**b** Draw the graph of $y = 4 - x^2$ from $x = -3$ to $x = 3$

**c** Write down the values of $x$ where the graph crosses the $x$-axis.

**Solution 2**

**a**
$$y = 4 - x^2$$

 **i** When $x = 3$,    $y = 4 - 3 \times 3$     $= 4 - 9 = -5$

 **ii** When $x = 2$,    $y = 4 - 2 \times 2$     $= 4 - 4 = 0$

 **iii** When $x = 0$,    $y = 4 - 0 \times 0$     $= 4$

 **iv** When $x = -2$,   $y = 4 - (-2) \times (-2) = 4 - 4 = 0$

| $x$ | −3 | −2 | −1 | 0 | 1 | 2 | 3 |
|---|---|---|---|---|---|---|---|
| $y$ | −5 | 0 | 3 | 4 | 3 | 0 | −5 |

**b**

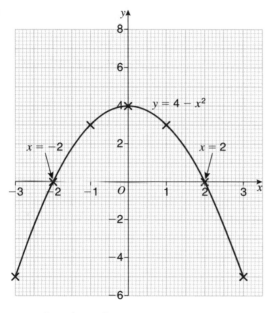

**c** $x = -2$ and $x = 2$

---

The highest point of this graph is where the graph turns.
It is the point (0, 4).
It is called the **maximum point**.

Graphs of quadratic functions always look like ⌢ or ⌣.

---

## Example 3

**a** Complete this table of values for $y = x^2 + x - 6$

| $x$ | −4 | −3 | −2 | −1 | 0 | 1 | 2 | 3 |
|---|---|---|---|---|---|---|---|---|
| $y$ | 6 | 0 | −4 | | −6 | | | 6 |

**b** Draw the graph of $y = x^2 + x - 6$ from $x = -4$ to $x = 3$.

**c** Write down the values of $x$ where the graph crosses the $x$-axis.

**d**  **i** Draw the line of symmetry of your graph.
  **ii** Write down the equation of this line of symmetry.

**e** Use your graph to find an estimate for the minimum value of $y$.

**Solution 3**

**a**
$$y = x^2 + x - 6$$

**i** When $x = 2$, $\quad y = 2 \times 2 + 2 - 6 \qquad = 4 + 2 - 6 = 0$

**ii** When $x = 1$, $\quad y = 1 \times 1 + 1 - 6 \qquad = 1 + 1 - 6 = -4$

**iii** When $x = -1$, $\quad y = (-1) \times (-1) + (-1) - 6 = 1 - 1 - 6 = -6$

| $x$ | $-4$ | $-3$ | $-2$ | $-1$ | 0 | 1 | 2 | 3 |
|---|---|---|---|---|---|---|---|---|
| $y$ | 6 | 0 | $-4$ | $-6$ | $-6$ | $-4$ | 0 | 6 |

**b**

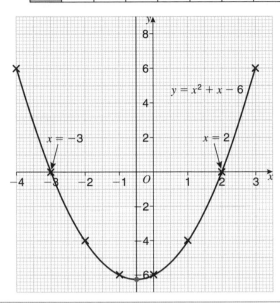

**c** $x = -3$ and $x = 2$

**d** **i** The **blue** line is the line of symmetry of the graph.

   **ii** $x = -0.5$

**e** Minimum value of $y$ is $-6.3$

---

## Exercise 32A

**1** Here is a table of values for $y = x^2 - 1$

| $x$ | $-3$ | $-2$ | $-1$ | 0 | 1 | 2 | 3 |
|---|---|---|---|---|---|---|---|
| $y$ | 8 | 3 | 0 | $-1$ | 0 | 3 | 8 |

   **a** Draw the graph of $y = x^2 - 1$ from $x = -3$ to $x = 3$

   **b** On the same grid draw the graph of $y = x^2 - 5$

**2 a** Copy and complete this table of values for $y = 2 - x^2$.

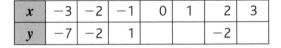

| $x$ | $-3$ | $-2$ | $-1$ | 0 | 1 | 2 | 3 |
|---|---|---|---|---|---|---|---|
| $y$ | $-7$ | $-2$ | 1 | | | $-2$ | |

   **b** Draw the graph of $y = 2 - x^2$ from $x = -3$ to $x = 3$

   **c** Write down the coordinates of the maximum point.

   **d** Estimate the values of $x$ where the graph crosses the $x$-axis.

**3 a** Copy and complete this table of values for $y = 2x^2$.

| $x$ | $-3$ | $-2$ | $-1$ | 0 | 1 | 2 | 3 |
|---|---|---|---|---|---|---|---|
| $y$ | 18 | 8 | | 0 | | 8 | |

   **b** Draw the graph of $y = 2x^2$ from $x = -3$ to $x = 3$

**4 a** Copy and complete this table of values for $y = x^2 - x$.

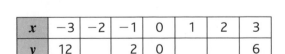

| $x$ | $-3$ | $-2$ | $-1$ | 0 | 1 | 2 | 3 |
|---|---|---|---|---|---|---|---|
| $y$ | 12 | | 2 | 0 | | | 6 |

   **b** Draw the graph of $y = x^2 - x$ from $x = -3$ to $x = 3$

   **c** Write down the values of $x$ where the graph crosses the $x$-axis.

   **d** **i** Draw the line of symmetry of your graph.
      **ii** Write down the equation of this line of symmetry.

   **e** Use your graph to find an estimate for the minimum value of $y$.

**5 a** Copy and complete this table of values for
$y = x^2 + 2x - 3$

| x | −4 | −3 | −2 | −1 | 0 | 1 | 2 |
|---|----|----|----|----|---|---|---|
| y | 5 | 0 | −3 | | −3 | | |

**b** Draw the graph of $y = x^2 + 2x - 3$ from
$x = -4$ to $x = 2$

**c** Write down the values of $x$ where the graph crosses the $x$-axis.

**d** Write down the coordinates of the minimum point.

## 32.2 Using graphs of quadratic functions to solve equations

The grid shows the graph of $y = x^2 + x - 6$
This graph can be used to solve the quadratic equation $x^2 + x - 6 = 0$

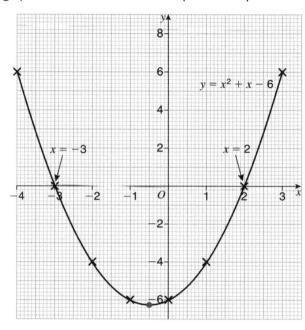

The solutions of the equation $x^2 + x - 6 = 0$ are the values of $x$ where $y = 0$, that is the values of $x$
where the graph crosses the $x$-axis.
From the graph, when $y = 0$, $x = -3$ and $x = 2$

$x = -3$ and $x = 2$ are the solutions of the equation $x^2 + x - 6 = 0$

### Example 4

The diagram shows the graph of $y = 2x^2 - x - 3$

Use the graph to write down the solutions of
the equation $2x^2 - x - 3 = 0$

**Solution 4**
Comparing the equation      $2x^2 - x - 3 = 0$
with the graph          $y = 2x^2 - x - 3$      gives $y = 0$

From the graph, when $y = 0$,
the values of $x$ are $-1$ and $1.5$

The solutions of the equation
$2x^2 - x - 3 = 0$ are $x = -1$ and $x = 1.5$

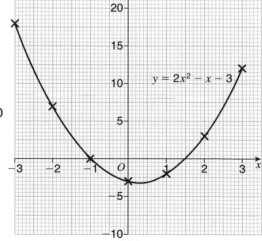

### Example 5

The diagram shows the graph of $y = 6 - x^2$.

**a** Use the graph to find estimates of the solutions to the equation $6 - x^2 = 0$

**b** **i** On the grid draw the graph of $y = 2$
   **ii** Write down the values of the $x$-coordinates of the points where the two graphs cross.

**c** Use the graph of $y = 6 - x^2$ to find estimates of the solutions to the equation $6 - x^2 = -2$

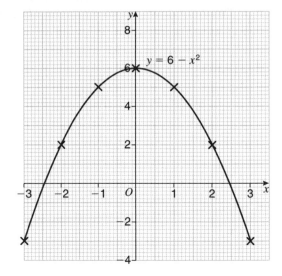

### Solution 5

**a** Comparing the equation $6 - x^2 = 0$
  with the graph $\qquad y = 6 - x^2 \qquad$ gives $y = 0$

From the graph, when $y = 0$, the estimated solutions are $x = -2.4$ and $x = 2.4$

**b** **i** The **red** line shows the graph of $y = 2$
  **ii** $x = -2$ and $x = 2$

**c** Comparing the equation $6 - x^2 = -2$
  with the graph $\qquad y = 6 - x^2 \qquad$ gives $y = -2$

The **green** line shows the graph of $y = -2$

From the graph, when $y = -2$, the estimated solutions are $x = -2.8$ and $x = 2.8$

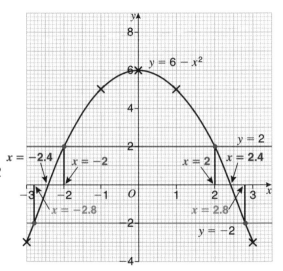

---

### Exercise 32B

**1** The diagram shows the graph of $y = x^2 - x - 2$

Use the graph to write down the solutions of the equation $x^2 - x - 2 = 0$

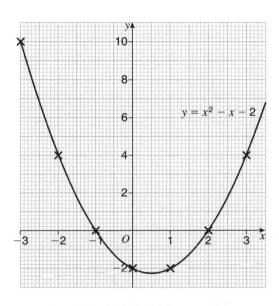

**2 a** Draw the graph of $y = x^2$ from $x = -3$ to $x = 3$

  **b i** On the same axes, draw the graph of $y = 4$
    **ii** Write down the values of the $x$-coordinates of the points where the two graphs cross.

  **c** Use the graph of $y = x^2$ to find estimates of the solutions to the equation $x^2 = 7$

**3** The diagram shows the graph of $y = 2x^2 - 3x$.

  **a** Use the graph to find the solutions of the equation
    $2x^2 - 3x = 0$

  **b** Use the graph to find the solutions to the equation
    $2x^2 - 3x = 5$

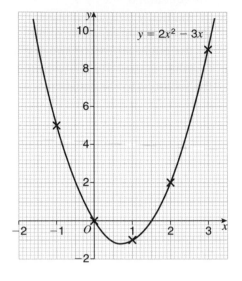

**4** The diagram shows the graph of $y = 8 + 3x - 2x^2$.

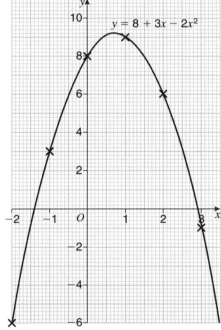

  **a** Use the graph to find estimates of the solutions to the equation $8 + 3x - 2x^2 = 0$

  **b** Use the graph to find the solutions to the equation
    $8 + 3x - 2x^2 = 3$

  **c** Use the graph to find the solutions to the equation
    $8 + 3x - 2x^2 = 6$

**5 a** Draw the graph of $y = 3 + x - x^2$ from $x = -2$ to $x = 3$

  **b** Use your graph to find
    **i** an estimate for the maximum value of $y$
    **ii** estimates of the solutions of the equation $3 + x - x^2 = 0$

# 32.3 Using graphs of quadratic and linear functions to solve quadratic equations

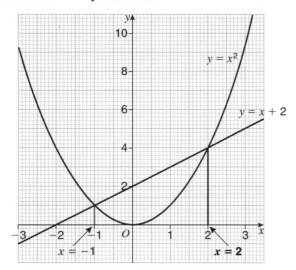

The diagram shows the graphs of $y = x^2$ and $y = x + 2$
The graphs cross at two points.
The coordinates of these two points are $(-1, \mathbf{1})$ and $(\mathbf{2}, \mathbf{4})$.
These coordinates satisfy both equations $y = x^2$ ⠀⠀⠀⠀and⠀$y = x + 2$

simultaneously because

$$1 = (-1)^2 \text{ and } 1 = -1 + 2$$
$$4 = (2)^2 \ \ \text{ and } \ 4 = 2 + 2$$

Eliminating $y$ from $y = x^2$ and $y = x + 2$ gives the equation $x^2 = x + 2$ or $x^2 - x - 2 = 0$
From the graph, the solutions of the quadratic equation $x^2 - x - 2 = 0$ are $x = -1$ and $\boldsymbol{x = 2}$

## Example 6

The diagram shows the graph of $y = x^2$.

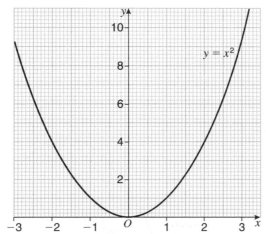

a By drawing suitable straight lines on this graph it is possible to solve the equations
⠀i $x^2 = 2x + 3$ and⠀⠀⠀ii $x^2 - 5x + 6 = 0$

Find the equation of each straight line.

b By drawing a suitable straight line on this graph estimate the solutions of the equation $x^2 + x = 3$

**Solution 6**

**a** **i** Comparing $\qquad x^2 = 2x + 3$
with the graph $\quad y = x^2 \qquad$ gives $y = 2x + 3$

| The equation $x^2 = 2x + 3$ can be compared directly with $y = x^2$. |
| --- |

The equation of the required line is $y = 2x + 3$

**ii** $x^2 - 5x + 6 = 0$ so $x^2 = 5x - 6$

| The equation $x^2 - 5x + 6 = 0$ cannot be directly compared with $y = x^2$ so rearrange. |
| --- |

Comparing $\qquad x^2 = 5x - 6$
with the graph $\quad y = x^2 \qquad$ gives $y = 5x - 6$

| The equation $x^2 = 5x - 6$ can be compared with $y = x^2$. |
| --- |

The equation of the required line is $y = 5x - 6$

**b** $x^2 + x = 3$ so $x^2 = -x + 3$

| The equation $x^2 + x = 3$ cannot be directly compared with $y = x^2$ so rearrange. |
| --- |

Comparing $\qquad x^2 = -x + 3$
with the graph $\qquad y = x^2 \qquad$ gives $y = -x + 3$

| The equation $x^2 = -x + 3$ can be compared with $y = x^2$. |
| --- |

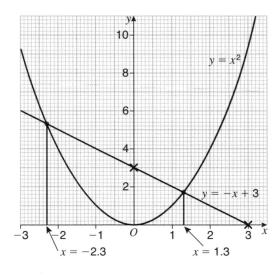

| Draw the graph of $y = -x + 3$ by drawing a line through the points $(0, 3)$ and $(3, 0)$. $x$-coordinates of the points of intersection of the line and curve give solutions of $x^2 + x = 3$ |
| --- |

$x = -2.3$ and $x = 1.3$

| Read off the solutions. |
| --- |

---

## Example 7

The diagram shows the graph of $y = 2x - x^2$.

By drawing a suitable straight line on this graph estimate the solutions of the equation
$x^2 - 4x + 3 = 0$

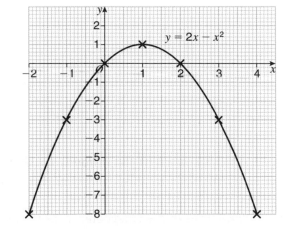

*Solution 7*

$x^2 - 4x + 3 = 0$

| The equation $x^2 - 4x + 3 = 0$ cannot be directly compared with $y = 2x - x^2$ so rearrange. |

$3 = -x^2 + 4x$

| Subtract $x^2$ from both sides and add $4x$ to both sides. |

$3 = -x^2 + 2x + 2x$

| Write $4x$ as $2x + 2x$ since $2x$ is needed to compare with $y = 2x - x^2$. |

$-2x + 3 = 2x - x^2$

| Subtract $2x$ from both sides and then rearrange. |

Comparing           $-2x + 3 = 2x - x^2$
with the graph            $y = 2x - x^2$ gives $y = -2x + 3$

| The equation $2x - x^2 = -2x + 3$ can be compared with $y = 2x - x^2$. |

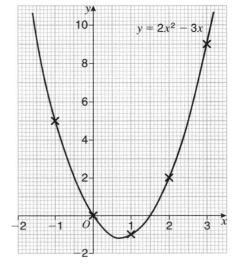

| Draw the graph of $y = -2x + 3$ by drawing a line through the points (0,3) and (1.5, 0). |
| $x$-coordinates of the points of intersection of the line and curve give solutions of $x^2 - 4x + 3 = 0$ |

$x = 1$ and $x = 3$

| Read off the solutions. |

---

## Exercise 32C

**1** The diagram shows the graph of $y = 2x^2 - 3x$.

By drawing suitable straight lines on this graph it is possible to solve the equations

**a** $2x^2 - 3x = x + 5$
**b** $2x^2 - 5x + 1 = 0$

Find the equation of each straight line.

**2 a** Draw the graph of $y = x^2$ from $x = -3$ to $x = 3$
  **b** On the same axes, draw the graph of $y = x + 6$
  **c** **i** Write down the $x$-coordinates of the two points where the graphs cross.
    **ii** Hence find the quadratic equation whose solutions are these values.

3  By drawing suitable straight lines on the graph of $y = x^2$, estimate the solutions of the equation
   a  $x^2 = x + 1$             b  $x^2 + 2x = 1$                    c  $x^2 - 3x + 1 = 0$

4  a  Draw the graph of $y = x^2 - 2x$ from $x = -2$ to $x = 4$
   b  By drawing suitable straight lines on the graph, estimate the solutions of the equation
      i  $x^2 - 2x = 2x - 1$       ii  $x^2 - x = 3$

5  a  Draw the graph of $y = 5 - 2x^2$ from $x = -2$ to $x = 2$
   b  By drawing suitable straight lines on the graph, find the solutions of the equation
      i  $5 - 2x^2 = x + 2$        ii  $4 - 2x - 2x^2 = 0$          iii  $2x^2 - x - 3 = 0$

6  a  Draw the graph of $y = x^2 + x - 2$ from $x = -3$ to $x = 3$.
   b  By drawing suitable straight lines on the graph estimate the solutions of the equation
      i  $x^2 + x - 2 = x + 5$     ii  $x^2 + 2x - 2 = 0$          iii  $x^2 - 2x - 1 = 0$

## Chapter summary

> **You should now know that:**
>
> ★  a **quadratic function** is an expression that can be written in the form $ax^2 + bx + c$ where $a(\neq 0)$, $b$ and $c$ represent numbers. The highest power in a quadratic function is 2
>
> ★  to draw the graph of a quadratic function from a given equation and given values of $x$ you
>
> - draw a table of values for calculated values of $y$
>
> - plot the points from the table of values
>
> - join the points with a smooth curve
>
> ★  the graph of a quadratic function has either a **minimum point** or a **maximum point** and
>
> - the minimum point is the lowest point and is where the graph turns
>
> - the maximum point is the highest point and is where the graph turns
>
> ★  all **quadratic equations** can be written in the form $ax^2 + bx + c = 0$ where $a(\neq 0)$, $b$ and $c$ represent numbers
>
> ★  $x^2 + 5x = 0$ is an example of a quadratic equation, which can be solved from the graph of $y = x^2 + 5x$ by finding the values of $x$ at the points where $y = 0$, that is where the graph crosses the $x$-axis
>
> ★  the points of intersections of the graphs of $y = ax^2 + bx + c$ and the line $y = mx + c$ can be used to solve the equation $ax^2 + bx + c = mx + c$. For example to solve the equation $x^2 - 3x + 2 = 0$ from the graph of $y = x^2$ draw the graph of the line $y = 3x - 2$ and read off the $x$-coordinates of the points where the graphs cross.

## Chapter 32 review questions

1  a  Copy and complete the table of values for $y = x^2 + x$.

| $x$ | −3 | −2 | −1 | 0 | 1 | 2 | 3 |
|---|---|---|---|---|---|---|---|
| $y$ | 6 | 2 |  | 0 |  | 6 |  |

   b  Draw the graph of $y = x^2 + x$ from $x = -3$ to $x = 3$                    (1388 November 2005)

**2 a** Copy and complete this table of values for $y = 6 - x^2$.

| x | −3 | −2 | −1 | 0 | 1 | 2 | 3 |
|---|----|----|----|---|---|---|---|
| y | −3 |    |    | 6 |   | 2 |   |

   **b** Draw the graph of $y = 6 - x^2$ from $x = -3$ to $x = 3$.

   **c** Write down the coordinates of the maximum point.

   **d** Estimate the values of $x$ where the graph crosses the $x$-axis.

**3 a** Copy and complete this table of values for $y = x^2 - 2x$.

| x | −2 | −1 | 0 | 1 | 2 | 3 |
|---|----|----|---|---|---|---|
| y |    | 3  |   |   |   | 3 |

   **b** Draw the graph of $y = x^2 - 2x$ from $x = -2$ to $x = 3$

   **c** Write down the values of $x$ where the graph crosses the $x$-axis.

   **d i** Draw the line of symmetry of your graph.
     **ii** Write down the equation of this line of symmetry.

   **e** Use your graph to find an estimate for the minimum value of $y$.

**4 a** Copy and complete the table of values for $y = x^2 - 3x - 1$

| x | −2 | −1 | 0 | 1 | 2 | 3 | 4 |
|---|----|----|---|---|---|---|---|
| y |    | 3  | −1 | −3 |  |   | 3 |

   **b** Draw the graph of $y = x^2 - 3x - 1$ from $x = -2$ to $x = 4$

   **c** Use your graph to find an estimate for the minimum value of $y$.    (1388 March 2003)

**5** The diagram shows the graph of the equation $y = 2x^2 - 4x - 3$
Use the graph to find the approximate values of $x$ when $2x^2 - 4x - 3 = 0$

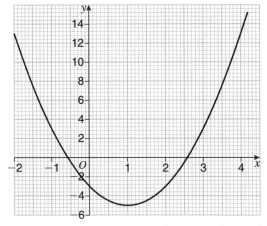

(1388 November 2005)

**6 a** Draw the graph of $y = x^2 - x - 4$ Use values of $x$ between $-2$ and $+3$

   **b** Use your graph to write down an estimate for
     **i** the minimum value of $y$
     **ii** the solutions of the equation $x^2 - x - 4 = 0$    (1385 June 1998)

**7** By drawing the graph of $y = x^2 + x - 6$, for values of $x$ from $-4$ to $+3$
   **a** write down the solutions of the equation $x^2 + x - 6 = 0$
   **b** write down an estimate for the solutions of the equation $x^2 + x - 6 = 1$

**8 a** Copy and complete the table of values for the graph of $y = 4x(11 - 2x)$

| x | 0 | 1 | 2 | 3 | 4 | 5 | 6 |
|---|---|---|---|---|---|---|---|
| y | 0 |   |   | 60 |  |   | −24 |

   **b** Draw the graph of $y = 4x(11 - 2x)$ from $x = 0$ to $x = 6$

   **c** Use your graph to find the maximum value of $y$.    (1388 January 2004)

**9 a** Copy and complete the table of values for
$y = x^2 - 6x + 10$

| $x$ | 0 | 1 | 2 | 3 | 4 | 5 | 6 |
|---|---|---|---|---|---|---|---|
| $y$ | 10 | | | | | | 10 |

**b** Draw the graph of $y = x^2 - 6x + 10$ from
$x = 0$ to $x = 6$

(1388 January 2005)

**10 a** Draw the graph of $y = 5 + 2x - x^2$ for $-2 \leqslant x \leqslant 4$

**b** By drawing a suitable straight line on your graph, find the approximate solutions
to $x + 4 = 5 + 2x - x^2$.

(1385 November 1997)

**11 a** Copy and complete the table for
$y = x^2 - 3x + 1$

| $x$ | $-2$ | $-1$ | 0 | 1 | 2 | 3 | 4 |
|---|---|---|---|---|---|---|---|
| $y$ | 11 | | 1 | $-1$ | | 1 | 5 |

**b** Draw the graph of $y = x^2 - 3x + 1$ from
$x = -2$ to $x = 4$

**c** Use your graph to find an estimate for the minimum value of $y$.

**d** Use a graphical method to find estimates of the solutions to the equation
$x^2 - 3x + 1 = 2x - 4$

(1387 November 2003)

**12** The graph shows part of the curve with equation
$y = 2x^2 - 6x + 3$

**a i** Use the graph to find estimates of the solutions to
the equation $2x^2 - 6x + 3 = 0$

**ii** Use the graph to find estimates of the solutions to
the equation $2x^2 - 6x + 2 = 0$

**b** By drawing a clearly labelled straight line on the
resource sheet, find estimates of the solutions to the
equation $2x^2 - 7x + 5 = 0$

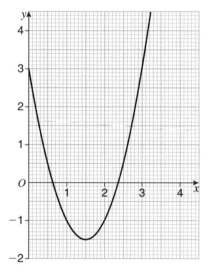

In Chapters 13 and 25, graphs of equations of the form $y = mx + c$ are drawn. These are all straight line graphs.

In Chapter 32, graphs of quadratic functions ($y = ax^2 + bx + c$) are drawn. The shape of these graphs is either $\smile$ (if $a > 0$) or $\frown$ (if $a < 0$).

Graphs of cubic, reciprocal and exponential functions will be drawn in this chapter.

## 33.1 Graphs of cubic, reciprocal and exponential functions

Expressions of the form $ax^3 + bx^2 + cx + d$, where the highest power of $x$ is 3, are called **cubic expressions**. So $2x^3 - x - 4$ and $-2x^3 + x^2$ are two examples of cubic expressions.

The graph of a **cubic function** ($y = ax^3 + bx^2 + cx + d$) can be drawn by completing a table of values, plotting the points and joining the points by a smooth curve.

### Example 1

**a** Complete this table of values for $y = x^3 - 3x^2 + 3$

**b** Draw the graph of $y = x^3 - 3x^2 + 3$ for $-1 \leqslant x \leqslant 3$

| $x$ | −1 | 0 | 1 | 2 | 3 |
|-----|----|----|----|----|----|
| $y$ | | 3 | 1 | | |

### Solution 1

**a** When $x = -1$,   $y = (-1)^3 - 3(-1)^2 + 3 = -1$

When $x = 2$,     $y = (2)^3 - 3(2)^2 + 3 \quad = -1$

When $x = 3$,     $y = (3)^3 - 3(3)^2 + 3 \quad = 3$

| $x$ | −1 | 0 | 1 | 2 | 3 |
|-----|----|----|----|----|----|
| $y$ | −1 | 3 | 1 | −1 | 3 |

> Substitute $x = -1$, $x = 2$ and $x = 3$ into $y = x^3 - 3x^2 + 3$

**b**

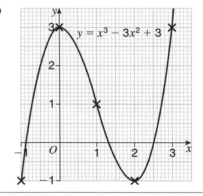

The graph of the cubic function $y = x^3 - 3x^2 + 3$ has a **maximum** point at (0, 3) and a **minimum** point at (2, 1).

The simplest cubic expression is $x^3$.

The graph of $y = x^3$ has no maximum point or minimum point.

In general, the graph of a cubic function has:

● both a maximum point and a minimum point, or

● neither a maximum point nor a minimum point

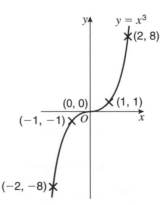

The diagram shows the general shapes of the graphs of cubic functions.

| Cubics | |
|---|---|
| $y = ax^3 + bx^2 + cx + d$ | 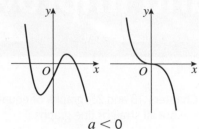 $a > 0$　　　　　　　$a < 0$ |

Notice that the graphs are continuous (no breaks) and that the ends of each graph are going in opposite directions.

### Example 2

Each of the equations $y = x^3 - 6x + 10$ and $y = x^3 - 2x^2 - 4x - 3$ represents one of these graphs.
Give the equation of each graph.

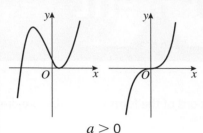

**Solution 2**
Note that graph **A** crosses the positive $y$-axis and graph **B** crosses the negative $y$-axis.

When $x = 0$, $x^3 - 6x + 10 = 10$ and $x^3 - 2x^2 - 4x - 3 = -3$ so

Substitute $x = 0$ into both equations.

　　　graph **A** has equation $y = x^3 - 6x + 10$
and graph **B** has equation $y = x^3 - 2x^2 - 4x - 3$

## Graphs of reciprocal functions

The reciprocal of $x$ is $\dfrac{1}{x}$ (see section 6.5). $\dfrac{1}{x}$ is called a **reciprocal expression**.

The reciprocal of 0 is not defined since division by 0 is not possible.

So the graph of the **reciprocal function** $y = \dfrac{1}{x}$ does not have a point on the $y$-axis where $x = 0$

### Example 3

Draw the graph of $y = \dfrac{1}{x}$, where $x \neq 0$

**Solution 3**

| When $x = 1$, | $y = 1$ | When $x = -1$, | $y = -1$ |
|---|---|---|---|
| When $x = 2$, | $y = \frac{1}{2}$ | When $x = -2$, | $y = -\frac{1}{2}$ |
| When $x = 3$, | $y = \frac{1}{3}$ | When $x = -3$, | $y = -\frac{1}{3}$ |
| When $x = \frac{1}{2}$, | $y = 2$ | When $x = -\frac{1}{2}$, | $y = -2$ |
| When $x = \frac{1}{4}$, | $y = 4$ | When $x = -\frac{1}{4}$, | $y = -4$　and so on |

Substitute values of $x$ into $y = \dfrac{1}{x}$.

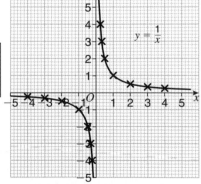

| $x$ | $-3$ | $-2$ | $-1$ | $-\frac{1}{2}$ | $-\frac{1}{4}$ | $\frac{1}{4}$ | $\frac{1}{2}$ | $1$ | $2$ | $3$ |
|---|---|---|---|---|---|---|---|---|---|---|
| $y$ | $-\frac{1}{3}$ | $-\frac{1}{2}$ | $-1$ | $-2$ | $-4$ | $4$ | $2$ | $1$ | $\frac{1}{2}$ | $\frac{1}{3}$ |

The graph of $y = \dfrac{1}{x}$ is in two parts, that is, the graph is discontinuous (breaks) at $x = 0$

The graph of $y = \dfrac{1}{x}$ has no maximum or minimum point and it does not cross or touch either the $x$-axis or the $y$-axis.

In general, the graph of $y = \dfrac{k}{x}$, where $k$ is a positive

number, has the same shape as the graph of $y = \dfrac{1}{x}$.

The diagram shows the general shapes of the graphs of reciprocal functions.

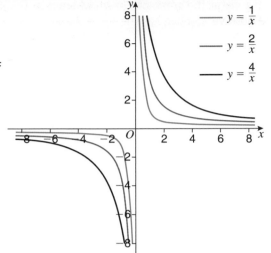

| Reciprocals | | |
|---|---|---|
| $y = \dfrac{k}{x}$ | 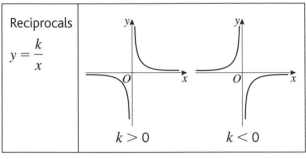 | |
| | $k > 0$ | $k < 0$ |

Notice that the graphs:

- have no maximum or minimum point
- are discontinuous and have two parts, each in diagonally opposite quadrants
- do not cross or touch the $x$-axis or the $y$-axis.

## Graphs of exponential functions

Expressions like $a^x$, where $a$ is a positive number, are called **exponential expressions**.

### Example 4

Draw the graph of $y = 2^x$.

**Solution 4**

When $x = 1$,    $y = 2^1 = 2$

When $x = 2$,    $y = 2^2 = 4$

When $x = 3$,    $y = 2^3 = 8$

When $x = 0$,    $y = 2^0 = 1$

When $x = -1$,    $y = 2^{-1} = \dfrac{1}{2} = 0.5$

When $x = -2$,    $y = 2^{-2} = \dfrac{1}{2^2} = 0.25$

When $x = -3$,    $y = 2^{-3} = \dfrac{1}{2^3} = 0.125$

> Substitute values of $x$ into $y = 2^x$.

> For any non-zero value of $a$, $a^0 = 1$
>
> and $a^{-n} = \dfrac{1}{a^n}$

| $x$ | $-3$ | $-2$ | $-1$ | 0 | 1 | 2 | 3 |
|---|---|---|---|---|---|---|---|
| $y$ | 0.125 | 0.25 | 0.5 | 1 | 2 | 4 | 8 |

The graph of $y = 2^x$ has no maximum or minimum point. It crosses the $y$-axis at the point $(0, 1)$.
The value of $y$ increases rapidly as the value of $x$ increases.
For negative values of $x$, the value of $y$ *approaches* 0.
The graph never crosses the $x$-axis.

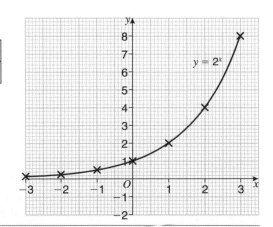

In general, the graph of $y = a^x$, where $a$ is a positive number, has the same shape as the graph of $y = 2^x$.

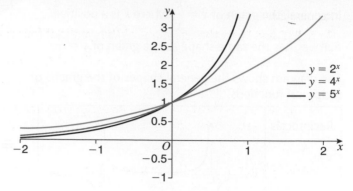

The diagram shows the general shapes of the graphs of exponential functions

| Exponentials $y = a^x, y = a^{-x}$, where $a$ is a positive number | 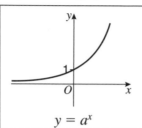 $y = a^x$ | 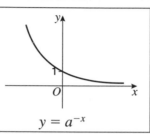 $y = a^{-x}$ |
|---|---|---|

Notice that the graphs:
- have no maximum or minimum point
- are continuous and always above the $x$-axis
- increase very rapidly at one end and approach the $x$-axis (without crossing it) at the other end
- cross the $y$-axis at $(0, 1)$.

## Example 5

This sketch shows part of the graph with equation $y = pq^x$, where $p$ and $q$ are positive constants.
The points with coordinates $(1, 6)$, $(3, 24)$ and $(4, k)$ lie on the graph.

**a** Calculate the values of $p$, $q$ and $k$.

**b** Find the coordinates of the point where the graph crosses the $y$-axis.

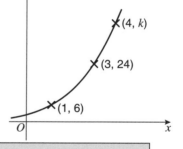

### Solution 5

**a** $6 = p \times q^1$

$24 = p \times q^3$

| | Write $y = pq^x$, as $y = p \times q^x$. Substitute the coordinates of the points $(1, 6)$ and $(3, 24)$ into $y = pq^x$. |
|---|---|

$\dfrac{24}{6} = \dfrac{p \times q^3}{p \times q}$ so $4 = q^2$

| Eliminate $p$ by dividing. $\dfrac{q^3}{q} = q^2$. |
|---|

$q = 2$

| Find the positive square root since $q$ is positive. |
|---|

$2p = 6$ so $p = 3$

| Substitute $q = 2$ in $pq = 6$. |
|---|

$k = 3 \times 2^4 = 3 \times 16 = 48$

| Substitute the coordinates of the point $(4, k)$ into $y = 3 \times 2^x$. |
|---|

$p = 3, q = 2$ and $k = 48$

**b** When $x = 0$, $y = 3 \times 2^0$

$y = 3 \times 1 = 3$

| Substitute $x = 0$ into $y = 3 \times 2^x$. |
|---|

The graph of $y = pq^x$, crosses the $y$-axis at the point $(0, 3)$

## Exercise 33A

1 **A**, **B**, **C** and **D** are the equations of the graphs of four curves.

$$\textbf{A} \; y = x^3 - 2x + 2 \quad \textbf{B} \; y = \frac{8}{x} \quad \textbf{C} \; y = 6^x \quad \textbf{D} \; y = 8 - x^3$$

  **a** Which of these graphs cross the $x$-axis?

  **b**  **i** Which one of these graphs does not cross the $y$-axis?

     **ii** For each of the other three graphs write down the coordinates of the point where it crosses the $y$-axis.

2 **a** Copy and complete this table of values for $y = x^3 - 12x$.

| $x$ | $-4$ | $-3$ | $-2$ | $-1$ | 0 | 1 | 2 | 3 | 4 |
|---|---|---|---|---|---|---|---|---|---|
| $y$ | | 9 | | | 0 | | | $-9$ | 16 |

  **b** Draw the graph of $y = x^3 - 12x$ for $-4 \leqslant x \leqslant 4$

3 Draw the graph of $y = x^3 - 10x^2 + 25x$ for $0 \leqslant x \leqslant 6$

4 On the same axes, sketch the graphs of $y = 3^x$ and $y = 6^x$.
Label each graph clearly with its equation.

5 **a** Copy and complete this table of values for $y = \dfrac{12}{x + 1}$

| $x$ | $-5$ | $-4$ | $-3$ | $-2$ | $-1.8$ | $-0.2$ | 0 | 1 | 2 | 3 |
|---|---|---|---|---|---|---|---|---|---|---|
| $y$ | $-3$ | | | | $-15$ | | | | | |

  **b** Draw the graph of $y = \dfrac{12}{x + 1}$ for values of $x$ between $-5$ and 3

  **c** Write down the value of $x$ at which the graph of $y = \dfrac{12}{x + 1}$ is discontinuous.

6 Draw the graph of $y = -\dfrac{2}{x}$ for values of $x$ between $-5$ and 5

7 The points with coordinates $(0, 0.5)$ and $(3, 2048)$ lie on the graph with equation $y = pq^x$, where $p$ and $q$ are constants.
Calculate the values of $p$ and $q$.

## 33.2 Trial and improvement

The equation $x^3 + 2x = 3$ has an exact solution $x = 1$, because $1^3 + 2 \times 1 = 3$
The equation $x^3 + 2x = 12$ has an exact solution $x = 2$, because $2^3 + 2 \times 2 = 12$
However, not all equations can be solved exactly, for example,
$x^3 + 2x = 7$

The diagram shows that the line $y = 7$ crosses the graph of
$y = x^3 + 2x$ at one point. To find the $x$-coordinate of this
point the equation $x^3 + 2x = 7$ needs to be solved.
From the diagram the value of $x$ is between 1 and 2.

An approximate solution to the equation $x^3 + 2x = 7$ can
be found by first guessing the solution and then using the
method of **trial and improvement** to obtain a more
accurate answer.

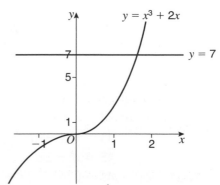

### Example 6

**a** Show that the equation $x^3 + 2x = 7$ has a solution between 1 and 2
**b** Use a trial and improvement method to find this solution. Give your answer correct to one decimal place.

### Solution 6

**a**

| Trial | Left-hand side | Right-hand side |
|-------|----------------|-----------------|
| $x$ | $x^3 + 2x$ | 7 |
| 1 | $1^3 + 2 \times 1 = 3$ | Since $3 < 7, x = 1$ is **too small** |
| 2 | $2^3 + 2 \times 2 = 12$ | Since $12 > 7, x = 2$ is **too big** |

> Solution is between 1 and 2

**b** It is not obvious that the solution is closer to 1 or to 2 so as a first trial take the middle value of the interval, 1.5

| $x$ | $x^3 + 2x$ | 7 |
|-----|-----------|---|
| 1.5 | $1.5^3 + 2 \times 1.5 = 6.375$ | Since $6.375 < 7,$ $x = 1.5$ is **too small** |

> Solution is between 1.5 and 2

So now take a higher value in the interval, 1.6

| 1.6 | $1.6^3 + 2 \times 1.6 = 7.296$ | Since $7.296 > 7,$ $x = 1.6$ is **too big** |
|-----|-----|-----|

> Solution is between 1.5 and 1.6

To decide if the solution is closer to 1.5 or 1.6, try the middle value, 1.55, of the interval.

| $x$ | $x^3 + 2x$ | 7 |
|-----|-----------|---|
| 1.55 | $1.55^3 + 2 \times 1.55 = 6.823\,875$ | Since $6.823\,875 < 7,$ $x = 1.55$ is **too small** |

> Solution is between 1.55 and 1.6

```
        1.5        1.55        1.6
```

All values between 1.55 and 1.6 are closer to 1.6 than 1.5 so, correct to one decimal place, the solution of $x^3 + 2x = 7$ which lies between 1 and 2 is $x = 1.6$

### Example 7

A container in the shape of a cuboid with a square base is to be constructed. The height of the cuboid is to be 2 metres less than the length of a side of its base and the container is to have a volume of 45 cubic metres.

**a** Taking $x$ metres as the length of a side of the base, show that $x$ satisfies the equation $x^3 - 2x^2 - 45 = 0$
**b** Use a trial and improvement method to find the solution of the equation $x^3 - 2x^2 - 45 = 0$ that lies between 4 and 5. Give your answer correct to two decimal places.
**c** Find the height of the container correct to the nearest centimetre.

### Solution 7

**a** The height of the cuboid is $(x - 2)$ metres since it is 2 metres less than the length of a side of its base.

Volume of cuboid $= x \times x \times (x - 2)$

$$45 = x^2(x - 2)$$
$$45 = x^3 - 2x^2$$

So $x^3 - 2x^2 - 45 = 0$

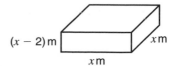

**b**

| $x$ | $x^3 - 2x^2 - 45$ | 0 |
|---|---|---|
| 4 | $4^3 - 2 \times 4^2 - 45 = -13$ | Since $-13 < 0$, $x = 4$ is **too small** |
| 4.5 | $4.5^3 - 2 \times 4.5^2 - 45 = 5.625$ | Since $5.625 > 0$, $x = 4.5$ is **too big** <br><br> Solution is between 4 and 4.5 |
| 4.3 | $4.3^3 - 2 \times 4.3^2 - 45 = -2.473$ | Since $-2.473 < 0$, $x = 4.3$ is **too small** <br><br> Solution is between 4.3 and 4.5 |
| 4.4 | $4.4^3 - 2 \times 4.4^2 - 45 = 1.464$ | Since $1.464 > 0$, $x = 4.4$ is **too big** <br><br> Solution is between 4.3 and 4.4 |
| 4.36 | $4.36^3 - 2 \times 4.36^2 - 45 = -0.137\,344$ | Since $-0.137\,344 < 0$, $x = 4.36$ is **too small** <br><br> Solution is between 4.36 and 4.4 |
| 4.37 | $4.37^3 - 2 \times 4.37^2 - 45 = 0.259\,653$ | Since $0.259\,653 > 0$, $x = 4.37$ is **too big** <br><br> Solution is between 4.36 and 4.37 |

To decide if the solution is closer to 4.36 or 4.37, try the middle value, 4.365, of the interval.

| $x$ | $x^3 - 2x^2 - 45$ | 0 |
|---|---|---|
| 4.365 | $4.365^3 - 2 \times 4.365^2 - 45$ <br> $= 0.060\,877...$ | Since $0.060\,877... > 0$, $x = 4.365$ is **too big** <br><br> Solution is between 4.36 and 4.365 |

```
   |--------------|--------------|
  4.36          4.365          4.37
```

All values between 4.36 and 4.365 are closer to 4.36 than 4.37 so, correct to two decimal places, the solution of $x^3 - 2x^2 - 45 = 0$ which lies between 4 and 5 is $x = 4.36$

**c** The height of the container is $x - 2 = 4.36 - 2 = 2.36$ m to the nearest centimetre.

---

## Example 8

**a** Show that $y = -1$ is a solution of the equation $y^2 = \dfrac{1}{y} + 2$

**b** Use a trial and improvement method to find the solution of $y^2 = \dfrac{1}{y} + 2$ that lies between 1 and 2
Give your answer correct to one decimal place.

*Solution 8*

**a** $y = -1$ is a solution of the equation $y^2 = \dfrac{1}{y} + 2$ if $(-1)^2 = \dfrac{1}{(-1)} + 2$ that is $(+1) = (-1) + 2$
which is true so $y = -1$ is a solution.

**b** $y^2 = \dfrac{1}{y} + 2$

| $y$ | $y^2$ | $\dfrac{1}{y} + 2$ | |
|---|---|---|---|
| 1 | $1^2 = 1$ | $\dfrac{1}{1} + 2 = 3$ | Since $1 < 3$, $y = 1$ is **too small** |
| 1.5 | $1.5^2 = 2.25$ | $\dfrac{1}{1.5} + 2 = 2.666...$ | Since $2.25 < 2.666...$, $y = 1.5$ is **too small** |
| 1.7 | $1.7^2 = 2.89$ | $\dfrac{1}{1.7} + 2 = 2.588...$ | Since $2.89 > 2.588...$, $y = 1.7$ is **too big** <br> Solution is between 1.5 and 1.7 |
| 1.6 | $1.6^2 = 2.56$ | $\dfrac{1}{1.6} + 2 = 2.625$ | Since $2.56 < 2.625$, $y = 1.6$ is **too small** <br> Solution is between 1.6 and 1.7 |

To decide if the solution is closer to 1.6 or 1.7, try the middle value, 1.65, of the interval

| $y$ | $y^2$ | $\dfrac{1}{y} + 2$ | |
|---|---|---|---|
| 1.65 | $1.65^2 = 2.7225$ | $\dfrac{1}{1.65} + 2 = 2.6060...$ | Since $2.7225 > 2.606...$, $x = 1.65$ is **too big** <br> Solution is between 1.6 and 1.65 |

         1.6        1.65        1.7

All values between 1.6 and 1.65 are closer to 1.6 than 1.7 so, correct to one decimal place, the
solution of $y^2 = \dfrac{1}{y} + 2$ which lies between 1 and 2 is $y = 1.6$

**Exercise 33B**

**1** Show that each of these equations has a solution between 0 and 1

  **a** $x^2 + 4x = 2$     **b** $x^3 + 2x = 1$     **c** $y^3 + 3y^2 = 3$     **d** $x^2(x - 3) + 1 = 0$

**2** Use a trial and improvement method to find the solution of each of these equations
between 0 and 1
Give your answers correct to one decimal place.

  **a** $x^2 + 4x = 2$     **b** $x^3 + 2x = 1$     **c** $y^3 + 3y^2 = 3$     **d** $x^2(x - 3) + 1 = 0$

**3** The equation $x^3 + 2x = 9$ has a solution between 1 and 2
Use a trial and improvement method to find this solution.
Give your answer correct to two decimal places.

**4** The equation $x^3 - 4x = 8$ has a solution between 2 and 3
Use a trial and improvement method to find this solution.
Give your answer correct to one decimal place.

**5** **a** Show that the equation $x^3 - 4x = 76$ has a solution between 4 and 5

   **b** Use a trial and improvement method to find this solution.
     Give your answer correct to one decimal place.

**6** A container in the shape of a cuboid with a square base is to be constructed.
The height of the cuboid is to be 2 metres more than the length of a side of its base and the container is to have a capacity of 80 cubic metres.

   **a** Taking $x$ metres as the length of a side of the base, show that $x$ satisfies the equation
     $x^3 + 2x^2 - 80 = 0$

   **b** Use a trial and improvement method to find the solution of the equation $x^3 + 2x^2 - 80 = 0$
     that lies between 3 and 4. Give your answer correct to two decimal places.

   **c** Find the height of the container correct to the nearest centimetre.

**7** The equation $x^3 = 5x + 6$ has a solution between 2 and 3
Use a trial and improvement method to find this solution.
Give your answer correct to one decimal place.

**8** **a** Show that the equation $t^3 = 3(t^2 - 1)$ has a solution between 1 and 2

   **b** Use a trial and improvement method to find this solution.
     Give your answer correct to one decimal place.

**9** Use a trial and improvement method to find the solution of $x^3 + x = 5$
Give your answer correct to one decimal place.

**10** Use a trial and improvement method to find the solution of $x^2 - \dfrac{1}{x} = 4$ that lies between 2 and 3

Give your answer correct to one decimal place.

**11** Use a trial and improvement method to find the solution of $\dfrac{2}{x} = x^2 - 7$ that lies between 2 and 3

Give your answer correct to one decimal place.

## Chapter summary

**You should now know:**

★ that expressions of the form $ax^3 + bx^2 + cx + d$, where the highest power of $x$ is 3, are called **cubic expressions**

★ that expressions of the form $\dfrac{k}{x}$ where $k$ is a number, are called **reciprocal expressions**

★ that expressions of the form $a^x$, where $a$ is a positive number, are called **exponential expressions**

★　the shapes of the graphs of cubic functions, reciprocal functions and exponential functions

### Cubics $y = ax^3 + bx^2 + cx + d$

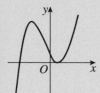

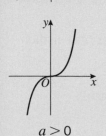

$a > 0$

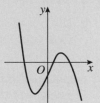

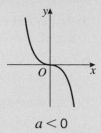

$a < 0$

- Either a minimum point and a maximum point or no maximum/minimum point
- Both ends of the graph are going in different directions
- Continuous curve

### Reciprocals $y = \dfrac{k}{x}$

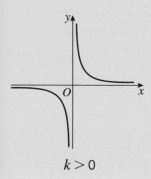

$k > 0$

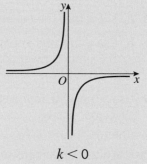

$k < 0$

- No maximum/minimum point
- Curve is discontinuous; it has two parts each in diagonally opposite quadrants
- Graph does not cross or touch the $x$-axis or the $y$-axis

### Exponentials $y = a^x, y = a^{-x}$, where $a > 0$

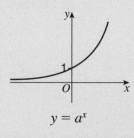

$y = a^x$

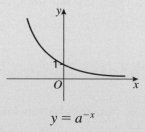

$y = a^{-x}$

- No maximum or minimum point
- Continuous curve and always above the $x$-axis
- Graphs increase very rapidly at one end and approach the $x$-axis (without crossing or touching it) at the other end
- Graphs cross the $y$-axis at $(0,1)$

★　how to find approximate solutions to equations which cannot be solved exactly by using the method of **trial and improvement**.

# Chapter 33 review questions

**1** The equation $x^3 - 2x - 37 = 0$ has a solution between 3 and 4
The method of trial and improvement is used to solve this equation.
The table shows some values.

Write down this solution correct to one decimal place.
Explain your answer.

| $x$ | $x^3 - 2x - 37$ |
|-----|-----------------|
| 3.5 | $-1.125$ |
| 3.6 | 2.456 |
| 3.55 | 0.6388... |

**2** The equation $x^3 - 2x = 67$ has a solution between 4 and 5
Use a trial and improvement method to find this solution.
Give your answer correct to one decimal place. You must show **ALL** your working.    (1387 June 2004)

**3** The equation $x^3 - 4x = 24$ has a solution between 3 and 4
Use a trial and improvement method to find this solution.
Give your answer correct to 1 decimal place. You must show **ALL** your working.    (1387 June 2005)

**4** The equation $x^3 + x = 16$ has a solution between 2 and 3
Use a trial and improvement method to find this solution.
Give your answer correct to 2 decimal places. You must show **ALL** your working.

(1385 November 2002)

**5** The equation $x^3 + x^2 = 220$ has a solution between 5 and 6
Use a trial and improvement method to find this solution.
Give your answer correct to 1 decimal place. You must show **ALL** your working.

**6** Use a trial and improvement method to find the solution of $x^2 = \dfrac{1}{x} + 5$ that lies between 2 and 3

Give your answer correct to 1 decimal place.

**7** A cuboid has a square base of side $x$ cm.
The height of the cuboid is 1 cm more than the length $x$ cm.
The volume of the cuboid is 230 cm$^3$.

**a** Show that $x^3 + x^2 = 230$

The equation $x^3 + x^2 = 230$ has a solution between $x = 5$ and $x = 6$

**b** Use a trial and improvement method to find this solution.
Give your answer correct to one decimal place.
You must show **ALL** your working.

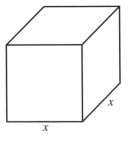

(1387 June 2003)

**8 a** Copy and complete the table of values for $y = x^3 - 3x + 1$

| $x$ | $-2$ | $-1.5$ | $-1$ | $-0.5$ | 0 | 0.5 | 1 | 1.5 | 2 |
|-----|------|--------|------|--------|---|-----|---|-----|---|
| $y$ | $-1$ |        | 3    | 2.375  | 1 | $-0.375$ |   | $-0.125$ | 3 |

**b** Draw the graph of $y = x^3 - 3x + 1$ for $-2 \leqslant x \leqslant 2$

**9 a** Copy and complete this table of values
for $y = x^3 + x - 2$
**b** Draw the graph of $y = x^3 + x - 2$ for $-2 \leqslant x \leqslant 2$

| $x$ | $-2$ | $-1$ | 0 | 1 | 2 |
|-----|------|------|---|---|---|
| $y$ | $-12$ |     |   | 0 | 2 |

**10** *A* 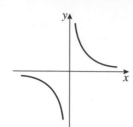    *B* 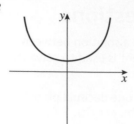    *C*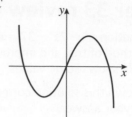

*D* 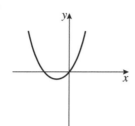    *E* 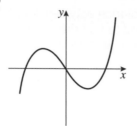    *F*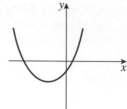

Each of the equations in the table represents one of the graphs *A* to *F*.

Copy and complete the table by writing the letter of each graph in the correct place.

(1385 November 1998)

| Equation | Graph |
|---|---|
| $y = x^2 + 3x$ | |
| $y = x - x^3$ | |
| $y = x^3 - 2x$ | |
| $y = x^2 + 2x - 4$ | |
| $y = \dfrac{4}{x}$ | |
| $y = x^2 + 3$ | |

**11** *A* 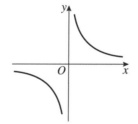    *B* 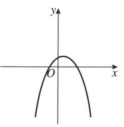    *C*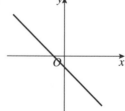

*D* 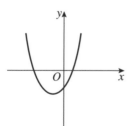    *E* 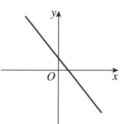    *F*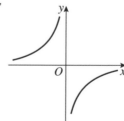

*G* 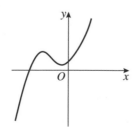    *H* 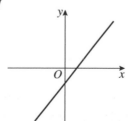    *I*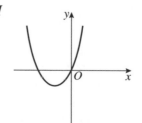

Write down the letter of the graph which could have the equation

**a** $y = 3x - 2$          **b** $y = 2x^2 + 5x - 3$          **c** $y = \dfrac{3}{x}$

**12 a** Simplify
    **i** $pq \times q^2$
    **ii** $3^x \times 3^y$

$2^c\, 8^{2c} = 2^k$

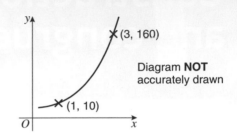

X (3, 160)

Diagram **NOT**
accurately drawn

X (1, 10)

**b** Express $k$ in terms of $c$.

The sketch graph shows a curve with equation
$y = pq^x$, where $q > 0$.

The curve passes through the points (1, 10) and (3, 160)

**c** Calculate the values of $p$ and $q$.                       (1385 November 2001)

**13** The sketch graph shows a curve with equation $y = pq^x$.
The curve passes through the points (1, 5) and (4, 320).
Calculate the value of $p$ and the value of $q$.

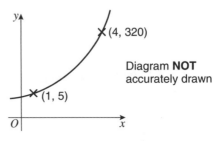

X (4, 320)

Diagram **NOT**
accurately drawn

X (1, 5)

(1387 November 2005)

**14** $y = \dfrac{k}{(x + a)^2}$

    **a** Rearrange the formula to express $x$ in terms of $k$, $y$ and $a$.

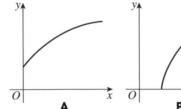 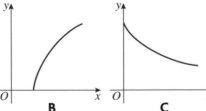 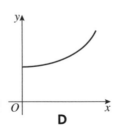 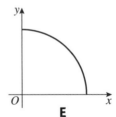

      **A**                 **B**                 **C**                 **D**                 **E**

One of the diagrams above shows a sketch of the graph of $y = \dfrac{k}{(x + a)^2}$ for $x \geqslant 0$

    **b**  **i** Write down the letter of the diagram.
          **ii** Write down the coordinates of any points where the curve meets the axes.     (1385 June 1998)

**15** Mr Patel has a car.
The value of the car on January 1st 2000 was £1600
The value of the car on January 1st 2002 was £400
The sketch graph shows how the value, £$V$,
of the car changes with time.
The equation of the sketch graph is

      $V = pq^t$

where $t$ is the number of years after January 1st 2000.
$p$ and $q$ are positive constants.

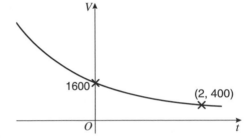

1600 X

(2, 400)
X

    **a** Use the information on the graph to find the value of $p$ and the value of $q$.
    **b** Using your values of $p$ and $q$ in the formula $V = pq^t$, find the value of the car
       on January 1st 1998.                                 (1387 June 2004)

# Constructions, loci and congruence

## 34.1 Constructions

The accurate drawing of shapes in Chapter 12 involved the use of rulers for measuring lengths. There are other constructions in which a ruler is not used. They are sometimes called **straight edge** and **compasses** constructions.

### Example 1

Construct an equilateral triangle with base $AB$.

**Solution 1**

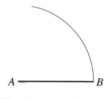

Draw an arc, centre $A$, with radius equal to the length of $AB$.

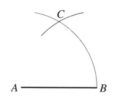

Draw another arc, centre $B$, with the same radius. The two arcs cross at $C$.

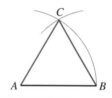

Complete the equilateral triangle $ABC$.

Each angle of an equilateral triangle is 60° and so the method in Example 1 can be used to construct an angle of 60°.

### Example 2

Draw a circle, radius 2 cm, and construct a regular hexagon inside it.

**Solution 2**

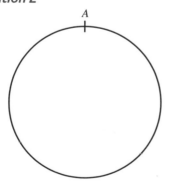

Draw a circle, radius 2 cm, and mark a point $A$ on its circumference.

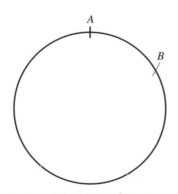

Keeping the compasses set at 2 cm, draw an arc, centre $A$, which cuts the circle at $B$.

$B$ is the centre of the next arc.

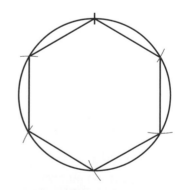

Repeat the process until six points are marked on the circumference.

Join the points to make a regular hexagon.

## Example 3

Construct the perpendicular bisector of the line $AB$.

### Solution 3

| Draw an arc, centre $A$, with radius more than half the length of $AB$. | Draw another arc, centre $B$, with the same radius. The two arcs cross at $C$ and $D$. | Draw a line through $C$ and $D$. The line is the perpendicular bisector of $AB$ as it crosses the line $AB$ at right angles at $M$, the midpoint of $AB$. |

Every point on the perpendicular bisector of the line $AB$ is the same distance from $A$ as it is from $B$. This property will be used later in the chapter.

## Example 4

Construct the bisector of angle $BAC$.

### Solution 4

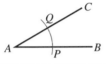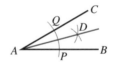

| Draw an arc, centre $A$, to cross $AB$ at $P$ and $AC$ at $Q$. | Draw an arc, centre $P$, and an arc, centre $Q$, with the same radius. The two arcs cross at $D$. | Draw a line from $A$ through $D$. This line is the bisector of angle $BAC$. |

Every point on the bisector of angle $BAC$ is the same distance from the line $AB$ as it is from the line $AC$. This property will also be used later in the chapter.

## Example 5

Construct the perpendicular from the point $C$ to the line $AB$.

### Solution 5

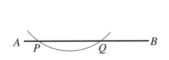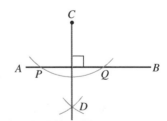

| Draw an arc, centre $C$, to cross $AB$ at $P$ and $Q$. | Draw an arc, centre $P$, with radius more than half the length of $PQ$. Draw another arc, centre $Q$, with the same radius. The two arcs cross at $D$. | Draw a line from $C$ through $D$. This line is the perpendicular from $C$ to the line $AB$. |

## Example 6

Construct the perpendicular to the line $AB$ from a point $C$ on the line $AB$.

### Solution 6

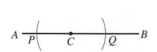

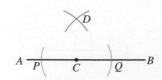

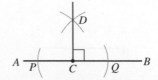

| Draw two arcs, centre $C$, with the same radius to cross $AB$ at $P$ and $Q$. | Draw an arc, centre $P$, with radius more than $PC$. Draw another arc, centre $Q$, with the same radius. The two arcs cross at $D$. | Draw a line from $C$ through $D$. This line is the perpendicular at $C$ to the line $AB$. |

---

Other constructions can be developed from the five basic ones described above. For example, the construction in Example 1 can be extended to an angle of 120° by initially drawing a longer arc and drawing a second arc, centre $C$, with the same radius. An angle of 30° can be constructed by constructing a 60° angle (Example 1) and then bisecting it (Example 4).

### Exercise 34A

Answer Questions **1** and **3–11** on the resource sheet.

**1** Construct an equilateral triangle with $PQ$ as its base.

$P \text{——————} Q$

**2** Draw a circle with radius 3 cm and construct a regular hexagon inside it.

**3** Construct the perpendicular bisector of the line $PQ$.

$P \text{——————} Q$

**4** Construct the bisector of angle $QPR$.

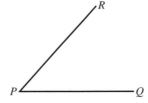

**5** Construct the bisector of angle $QPR$.

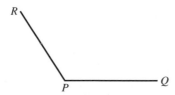

**6** Construct the perpendicular from the point $R$ to the line $PQ$.

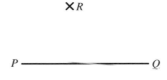

**7** Construct the perpendicular from the point $R$ to the line $PQ$.

**8** Construct the perpendicular at $P$ to the line $PQ$.

**9** Construct the perpendicular bisector of each of the sides of the triangle *PQR*.

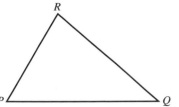

**10** Construct the bisector of each of the angles of the triangle *PQR*.

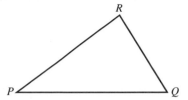

**11** Construct the perpendicular from *R* to the side *PQ* of the triangle.

**12** Construct an angle of 120°.

**13** Construct an angle of 30°.

## 34.2 Loci

The **locus** of a point is its path when it obeys given rules or conditions. **Loci** is the plural of locus.

There are two simple loci, on which many others are based. One is the locus of a point which moves so that it is always the same distance from a fixed point. The other is the locus of a point which moves so that it is always the same distance from a fixed line.

### Example 7

A point moves so that it is always 2 cm from a fixed point *A*. Draw its locus.

*Solution 7*

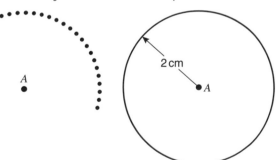

Here are some points which are 2 cm from *A*.

Combining all the points forms the locus, which is a circle, centre *A*, with a radius of 2 cm.

### Example 8

A point moves so that it is always 1 cm from a fixed line *AB*. Draw its locus.

*Solution 8*

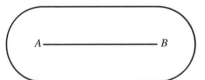

Here are some points which are 1 cm from the line *AB*. They make lines parallel to *AB* and 1 cm away from it.

Here are some points which are 1 cm away from point *A* and some points which are 1 cm away from point *B*. They make semicircles, centres *A* and *B*, with a radius of 1 cm.

Combining the two parallel lines and the two semicircles gives the complete locus.

The locus of a point which moves so that it is always an equal distance from two fixed points is the perpendicular bisector of the line joining the two fixed points (Example 3).

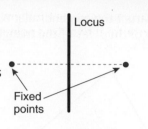

The word **equidistant**, which means 'equal distance', is sometimes used. So this could also be described as the locus of a point which moves so that it is always equidistant from two fixed points.

The locus of a point which moves so that it is always equidistant from each of two fixed lines is the bisector of the angle between the two fixed lines (Example 4).

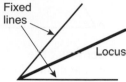

Sometimes a moving point has to satisfy two loci conditions. The position, or positions, of the point are where the loci cross (or intersect).

## Example 9

A point $P$ moves so that it is equidistant from $A$ and $B$. It is also 2 cm from $C$.
On the centimetre grid, find the *two* possible positions of $P$. Mark them with a cross.

### Solution 9

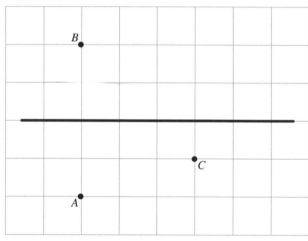

> The locus of a point which moves so that it is equidistant from $A$ and $B$ is the perpendicular bisector of the line $AB$.

> Draw this perpendicular bisector.

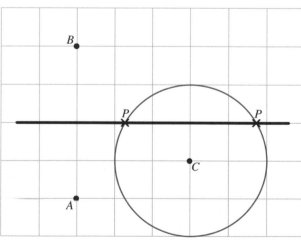

> The locus of a point which moves so that it is 2 cm from $C$ is a circle, centre $C$, with a radius of 2 cm.

> The two possible positions of $P$ are where the line and the circle cross.

## Exercise 34B

Answer this exercise on the resource sheet.

**1** A point moves so that it is always 3 cm away from a fixed point $A$. Draw its locus.

**2** A point moves so that it is always 2 cm away from a fixed line $AB$. Draw its locus.

**3** A point moves so that it is always an equal distance from two fixed points $A$ and $B$. Draw its locus.

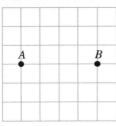

**4** A point moves so that it is always an equal distance from two fixed points $A(2, 1)$ and $B(2, 3)$. Draw its locus.

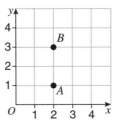

**5** A point moves so that it is always an equal distance from two fixed points $A$ and $B$. Draw its locus.

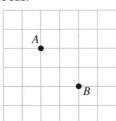

**6** A point moves so that it is always an equal distance from two fixed points $A$ and $B$. Construct its locus.

$A \bullet$

$\bullet B$

**7** A point moves so that it is always the same distance from two fixed lines $AB$ and $AC$. Construct its locus.

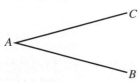

**8** A point moves so that it is always the same distance from two fixed lines $AB$ and $AC$. Draw its locus.

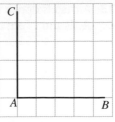

**9** A point moves outside this rectangle so that it is always 2 cm from the edges of the rectangle. Draw its locus.

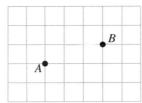

**10** A point $P$ moves so that it is 2 cm from $A$ and 3 cm from $B$. Find the two possible positions of $P$. Mark them with a cross.

**11** A point $P$ moves so that it is equidistant from $A$ and $B$ and 2 cm from $C$. Find the two possible positions of $P$. Mark them with a cross.

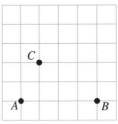

**12** The diagram represents Mr. Khan's garden. It is drawn to a scale of 0.5 cm to 5 m. He plants a tree which is the same distance from the path as it is from the wall. The tree is also 12 m from the hedge. Find the position of the tree. Mark it with a cross and label it $T$.

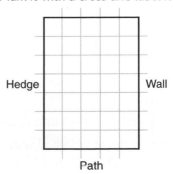

**13** A map is drawn to a scale of 1 cm to 10 km. $A$, $B$ and $C$ are three ports.
A ship is 75 km from $A$. It is also the same distance from $B$ as it is from $C$.
Find the position of the ship. Mark it with a cross and label it $S$.

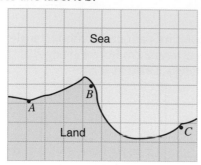

**14** $AB$ and $AC$ are two fixed lines. A point $P$ moves so that it is equidistant from $A$ and $B$ and 1 cm from the line $AC$. Find the two possible positions of $P$.
Mark them with crosses.

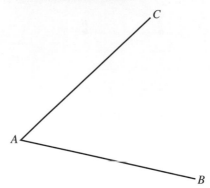

## 34.3 Regions

Sometimes, when a moving point obeys a rule, the point has to lie inside a **region** rather than on a particular path made up of lines or curves.

### Example 10

A point moves so that it is always less than 2 cm from a fixed point $A$.
Show, by shading, the region which satisfies this condition.

*Solution 10*

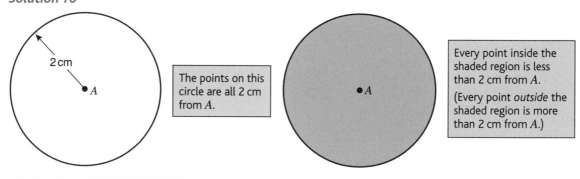

The points on this circle are all 2 cm from $A$.

Every point inside the shaded region is less than 2 cm from $A$.
(Every point *outside* the shaded region is more than 2 cm from $A$.)

### Example 11

$A$ and $B$ are two fixed points.
A point moves so that it is always nearer to $A$ than $B$.
Show, by shading, the region which satisfies this condition.

*Solution 11*

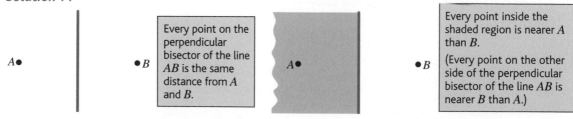

Every point on the perpendicular bisector of the line $AB$ is the same distance from $A$ and $B$.

Every point inside the shaded region is nearer $A$ than $B$.
(Every point on the other side of the perpendicular bisector of the line $AB$ is nearer $B$ than $A$.)

## Example 12

*AB* and *AC* are two fixed lines.
A point moves so that it is always nearer *AC* than *AB*.
Show, by shading, the region which satisfies this condition.

### Solution 12

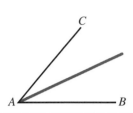

| Every point on the bisector of angle *BAC* is the same distance from *AB* and *AC*. |

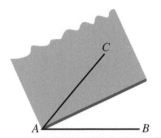

| Every point inside the shaded region is nearer *AC* than *AB*.<br>(Every point on the other side of the bisector of angle *BAC* is nearer *AB* than *AC*.) |

If a moving point has to satisfy two conditions, the region in which both conditions are satisfied may be the overlap (**intersection**) of two regions.

## Example 13

*A*, *B* and *C* are three fixed points.
A point moves so that it is always nearer *B* than *A*
*and* less than 2 cm from *C*.
Show, by shading, the region which satisfies
both these conditions.

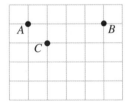

### Solution 13

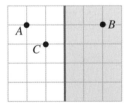

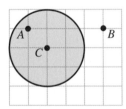

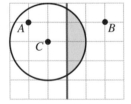

| Every point inside the shaded region is nearer *B* than *A*. | Every point inside the shaded region is less than 2 cm from *C*. | Both conditions are satisfied where the two regions overlap. |

### Exercise 34C

Answer this exercise on the resource sheet.

1 A point moves so that it is always less than 3 cm away from a fixed point *A*.
Show, by shading, the region which satisfies this condition.

2 A point moves so that it is always less than 2 cm away from a fixed line *AB*.
Show, by shading, the region which satisfies this condition.

A ———————— B

3 *A* and *B* are two fixed points. A point moves so that it is always nearer *A* than *B*.
Show, by shading, the region which satisfies this condition.

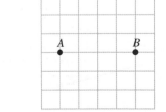

**4** $A(2, 1)$ and $B(2, 3)$ are two fixed points. A point moves so that it is always nearer $B$ than $A$.
Show, by shading, the region which satisfies this condition.

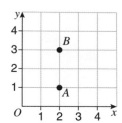

**5** $A$ and $B$ are two fixed points. A point moves so that it is always nearer $A$ than $B$.
Show, by shading, the region which satisfies this condition.

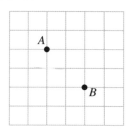

**6** $A$ and $B$ are two fixed points. A point moves so that it is always nearer $B$ than $A$.
Show, by shading, the region which satisfies this condition.

$A \bullet$

$\bullet B$

**7** $AB$ and $AC$ are two fixed lines. A point moves so that it is always nearer $AB$ than $AC$.
Show, by shading, the region which satisfies this condition.

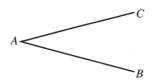

**8** $AB$ and $AC$ are two fixed lines. A point moves so that it is always nearer $AC$ than $AB$.
Show, by shading, the region which satisfies this condition.

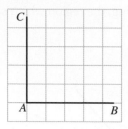

**9** A point moves outside this rectangle so that it is always less than 2 cm from the edges of the rectangle.
Show, by shading, the region which satisfies this condition.

**10** A point moves so that it is less than 2 cm from $A$ and more than 3 cm from $B$.
Show, by shading, the region which satisfies this condition.

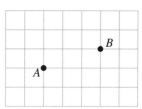

**11** $A$, $B$ and $C$ are three fixed points.
A point moves so that it is always nearer $A$ than $B$ and less than 2 cm from $C$.
Show, by shading, the region which satisfies both these conditions.

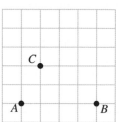

**12** $A(0, 2)$ is a fixed point. A point moves so that its is always less than 2 cm from $A$ and nearer the $x$-axis than the $y$-axis. Show, by shading, the region which satisfies this condition.

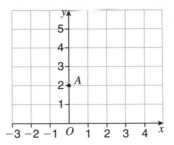

**13** The map shows two ports, $A$ and $B$. The scale of the map is 1 cm to 10 km.
A ship is less than 25 km from $A$ and less than 35 km from $B$. Show, by shading, the region where the ship is.

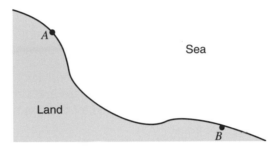

## 34.4 Drawing triangles

Triangles can be drawn using the methods of Section 22.1.

Sometimes it is possible to draw more than one triangle from given measurements.

**Example 14**

Show that there are two possible triangles $ABC$ in which $AB = 6.2$ cm, $BC = 3.7$ cm and angle $A = 31°$.

*Solution 14*

$A$ ——————— $B$

> Draw the line $AB$ with length 6.2 cm.

> Using a protractor, draw an angle of 31° at $A$.

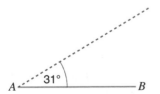

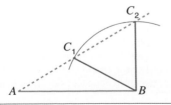

> Draw an arc of radius 3.7 cm, centre $B$, to locate the two possible positions of $C$.
> Triangle $ABC_1$ and triangle $ABC_2$ both have the given measurements.

**Exercise 34D**

**1** Use the given measurements to make an accurate drawing of triangle $ABC$ with $AB$ as base.

   **a** $AB = 6.4$ cm, $AC = 5.6$ cm, angle $A = 43°$.

   **b** $AB = 5.5$ cm, $BC = 5.3$ cm, angle $B = 127°$.

   **c** $AB = 6.2$ cm, $BC = 4.8$ cm, $AC = 5.6$ cm.

   **d** $AB = 7.3$ cm, angle $A = 47°$, angle $B = 62°$.

   **e** $AB = 6.1$ cm, angle $A = 69°$, angle $B = 50°$.

   **f** $AB = 4.9$ cm, angle $A = 90°$, $BC = 5.7$ cm.

   **g** $AB = 5.3$ cm, angle $B = 90°$, $AC = 6.3$ cm.

**2** Make accurate drawings of two triangles with the given measurements.

   **a** Angle $A = 42°$, angle $B = 73°$, angle $C = 65°$.

   **b** Angle $A = 134°$, angle $B = 20°$, angle $C = 26°$.

   **c** $AB = 7.2$ cm, $BC = 5.1$ cm, angle $A = 40°$.

# 34.5 Congruent triangles

Two triangles are **congruent** if they have exactly the same shape and size. To prove that two triangles are congruent, one of these four conditions must be proved for the two triangles:

- two sides and the included angle are equal (SAS)
- two angles and a corresponding side are equal (AAS)
- three sides are equal (SSS)
- a right angle, the hypotenuse and one other side are equal (RHS).

The four conditions relate to given measurements from which only one triangle can be constructed. For example, in Questions **1a** and **1b** of Exercise 34D, two sides and the included angle are given and, in Question **1c**, three sides are given.

Example 15

In a quadrilateral $ABCD$, $AB = AD$, and $AC$ bisects angle $BAD$.

**a** Prove that triangles $ABC$ and $ADC$ are congruent.

**b** Hence prove that $AC$ bisects angle $BCD$.

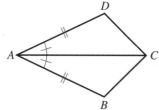

*Solution 15*

**a**

> Mark equal lengths and equal angles on the diagram.

$AB = AD$ (given)

angle $BAC$ = angle $CAD$ (given)

$AC$ is common

> Explain each step in the proof with reasons.

Triangles $\begin{matrix} ABC \\ ADC \end{matrix}$ are congruent (SAS)

> State which of the four conditions for congruency has been satisfied – two sides and the included angle in this case.

**b** In the congruent triangles $ABC$ and $ADC$, angle $ACB$ corresponds to angle $ACD$, and so angle $ACB$ = angle $ACD$ i.e. $AC$ bisects angle $BCD$

> Explain why angle $ACB$ = angle $ACD$. Angles $ACB$ and $ACD$ correspond because they are in the same relative positions in triangles $ABC$ and $ADC$ – between $AC$ and sides of equal length.

## Exercise 34E

**1** In the isosceles triangle $ABC$, $AB = AC$ and $AD$ bisects angle $BAC$.
  **a** Prove that triangles $ABD$ and $ACD$ are congruent.
  **b** Hence prove that angle $B$ = angle $C$.

**2** In the quadrilateral $ABCD$, $BC = CD$ and the angles at $B$ and $D$ are right angles.
  **a** Prove that triangles $ABC$ and $ADC$ are congruent.
  **b** Hence prove that $CA$ bisects angle $BAD$.

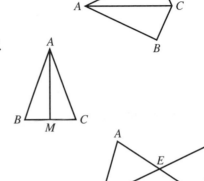

**3** In triangle $ABC$, $AB = AC$ and $M$ is the midpoint of $BC$.
  **a** Prove that triangles $ABM$ and $ACM$ are congruent.
  **b** Hence prove that angle $BAM$ = angle $CAM$.

**4** Two straight lines, $AB$ and $CD$, bisect each other at $E$.
  **a** Prove that triangles $AEC$ and $BED$ are congruent.
  **b** Hence prove that $AC = BD$.

**5** $A$ and $B$ are points on a circle, centre $O$.
  $D$ is the point where the perpendicular from $O$ meets the chord $AB$.
  **a** Prove that triangles $OAD$ and $OBD$ are congruent.
  **b** Hence prove that $D$ is the midpoint of the chord $AB$.

**6** In triangle $ABC$, $AB = AC$.
  $D$ and $E$ are points on $BC$ such that
  angle $BAD$ = angle $CAE$.
  **a** Prove that triangles $BAD$ and $CAE$ are congruent.
  **b** Hence prove that $BD = CE$.

## 34.6 Proofs of standard constructions

Proofs involving congruent triangles can be used to verify the standard straight edge and compass constructions.

### Proof 1  The perpendicular bisector of a line

In triangles $ACD$ and $BCD$
- $AC = BC$ (equal radii)
- $AD = BD$ (equal radii)
- $CD$ is common.

  Triangles $\begin{matrix} ACD \\ BCD \end{matrix}$ are congruent (SSS).

In triangles $ACE$ and $BCE$
- angle $ACE$ = angle $BCE$
- $AC = BC$ (equal radii)
- $CE$ is common.

  Triangles $\begin{matrix} ACE \\ BCE \end{matrix}$ are congruent (SAS).

So $AE = BE$ and angle $AEC$ = angle $BEC$,
but angle $AEC$ and angle $BEC$ are angles on a straight line and their sum is 180°.
As they are equal, angle $AEC$ = angle $BEC$ = 90°.
i.e. $CD$ is the perpendicular bisector of the line $AB$.

## Proof 2  The bisector of an angle

In triangles $BPD$ and $BQD$:

- $BP = BQ$ (equal radii)
- $PD = QD$ (equal radii)
- $BD$ is common.

Triangles $\begin{matrix} BPD \\ BQD \end{matrix}$ are congruent (SSS).

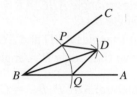

So angle $PBD$ = angle $QBD$. i.e. $BD$ is the bisector of angle $ABC$.

## Proof 3  The perpendicular from a point to a line

In triangles $PCD$ and $QCD$:

- $CP = CQ$ (equal radii)
- $PD = QD$ (equal radii)
- $CD$ is common.

Triangles $\begin{matrix} PCD \\ QCD \end{matrix}$ are congruent (SSS).

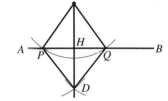

In triangles $PCH$ and $QCH$:

- angle $PCH$ = angle $QCH$
- $CP = CQ$ (equal radii)
- $CH$ is common.

Triangles $\begin{matrix} PCH \\ QCH \end{matrix}$ are congruent (SAS).

So angle $CHP$ = angle $CHQ$
but angle $CHP$ and angle $CHQ$ are angles on a straight line and their sum is 180°.
As they are equal, angle $CHP$ = angle $CHQ$ = 90°.
i.e. $CH$ is the perpendicular from $C$ to the line $AB$.

## Proof 4  The perpendicular to a line from a point on the line

In triangles $DCP$ and $DCQ$:

- $CP = CQ$ (equal radii)
- $DP = DQ$ (equal radii)
- $CD$ is common.

Triangles $\begin{matrix} DCP \\ DCQ \end{matrix}$ are congruent (SSS).

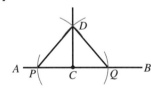

So angle $DCP$ = angle $DCQ$
but angle $DCP$ and angle $DCQ$ are angles on a straight line and their sum is 180°.
As they are equal, angle $DCP$ = angle $DCQ$ = 90°.
i.e. $CD$ is the perpendicular from $C$ to the line $AB$.

## Chapter summary

**You should now know:**

★ how to construct –
- an equilateral triangle with a given base
- a regular hexagon inside a circle
- the perpendicular bisector of a line
- the bisector of an angle
- the perpendicular from a point to a line
- the perpendicular from a point on a line

★ how to draw the **locus** of a point which moves so that it –
  - is a given distance from a fixed point
  - is a given distance from a fixed line
  - is **equidistant** from two fixed points
  - is equidistant from two fixed lines
  - satisfies two of the above conditions

★ how to show **regions** which satisfy given conditions, including overlapping regions

★ that two triangles are **congruent** if they have exactly the same shape and size

**You should also be able to:**

★ prove that two triangles are congruent by proving that they satisfy one of these four conditions –
  - two sides and the included angle are equal (SAS)
  - two angles and a corresponding side are equal (AAS)
  - three sides are equal (SSS)
  - a right angle, the hypotenuse and one other side are equal (RHS)

★ write down the proofs involving congruent triangles used to verify the standard **straight edge and compass constructions.**

# Chapter 34 review questions

Answer Questions **1–8** on the resource sheet.

**1** A point moves so that it is always 1 cm away from a fixed point $A$. Draw its locus.

**2** A point moves so that it is always more than 1 cm away from a fixed point $A$. Show, by shading, the region which satisfies this condition.

**3** $P$ and $Q$ are two points marked on the grid opposite.

Construct accurately the locus of all the points which are equidistant from $P$ and $Q$.

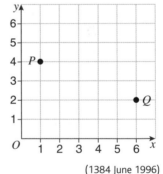

(1384 June 1996)

**4** Draw the locus of all points which are 3 cm away from the line $AB$.

B

A

(1385 November 2002)

**5** $A$, $B$ and $C$ represent three radio masts on a plan. Signals from a mast $A$ can be received 300 km away, from mast $B$ 350 km away and from mast $C$ 200 km away. Show, by shading, the region in which signals can be received from all three masts.

• B

• A

• C

Scale: 1 cm represents 100 km.

(1384 November 1994)

**6** A treasure chest is buried on an island. *P* and *Q* are two trees on this island.

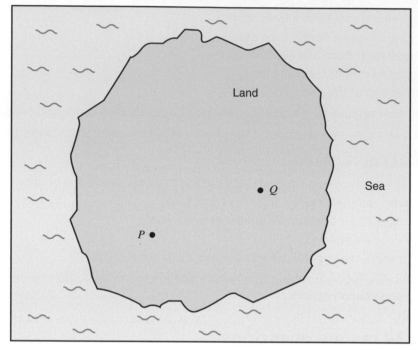

Scale: 1 cm represents 5 m

The treasure chest is buried the same distance from *P* as it is from *Q*.

**a** On the diagram, draw accurately the locus of points which are the same distance from *P* as they are from *Q*.

On the diagram, 1 centimetre represents 5 metres.
The treasure chest is buried 20 metres from *P*.

**b** On the diagram, draw accurately the locus which represents all the points which are 20 metres from *P*.

**c** Find the point where the chest is buried.

On the diagram, mark the point clearly with a *T*.

(1385 November 1998)

**7** The map shows part of a coastline and a coastguard station.
1 cm on the map represents 2 km.
A ship is 12 km from the coastguard station on a bearing of 160°.

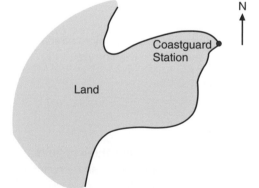

**a** Plot the position of the ship from the coastguard station, using a scale of 1 cm to represent 2 km.

It is not safe for ships to come within 6 km of the coastguard station.

**b** Shade the area on the map which is less than 6 km from the coastguard station.

The distance of a buoy from the coastguard station is 14 km to the nearest km.

**c** **i** Write down the maximum distance it could be.
**ii** Write down the minimum distance it could be.

(1384 June 1994)

**8** The scale drawing below shows the positions of an airport tower, $T$, and a radio mast, $M$.
1 cm on the diagram represents 20 km.

**a**   **i**  Measure, in centimetres, the distance $TM$.
     **ii**  Work out the distance in km of the airport tower from the radio mast.
**b**   **i**  Measure and write down the bearing of the airport tower from the radio mast.
     **ii**  Write down the bearing of the radio mast from the airport tower.

A plane is 80 km from the radio mast on a bearing of 220°.

**c** Plot the position of the plane, using a scale of 1 cm to 20 km.

Signals from the radio mast can be received up to a distance of 100 km.

**d** Shade the region on the scale diagram in which signals from the radio mast can be received.

The distance of a helicopter from the radio mast is 70 km correct to the nearest kilometre.

**e** Write down
    **i**  the maximum distance the helicopter could be from the mast,
    **ii**  the minimum distance the helicopter could be from the mast.     (1384 June 1997)

**9** $AED$ and $CEB$ are straight lines.
$AE = CE$ and $BE = DE$
Explain why triangles $ABE$ and
$CDE$ are congruent.

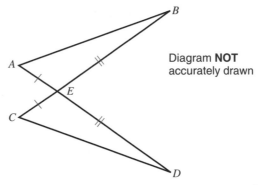

Diagram **NOT**
accurately drawn

(1384 November 1995)

**10** In the diagram, $PQ = PS$ and $PR = PT$.
Angle $RPT$ = angle $SPQ$
  **a** Prove that triangles $PRQ$ and $PTS$ are
    congruent.
  **b** Hence, prove that $PS$ bisects angle $QST$.

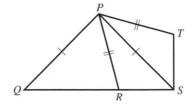

Diagram **NOT**
accurately drawn

(1384 June 1995)

# Bounds and surds

## 35.1 Lower bounds and upper bounds

When carrying out calculations with numbers that have been rounded it is possible to calculate the range of answers that can be produced depending on the accuracy to which the numbers have been given.

If a number has been written as 5.6 correct to one decimal place then the true value of the number lies between 5.55 and 5.65 The value at the lower boundary of the interval is called the **lower bound**. The value at the upper boundary is called the **upper bound**. In this case, 5.55 is the lower bound and 5.65 is the upper bound.

If a number has been written as 3.44 correct to 3 significant figures, then the true value of the number lies between 3.435 to 3.445 so the lower bound is 3.435 and the upper bound is 3.445

---

### Example 1

Write down **i** the lower bound **ii** the upper bound of

**a**  64 correct to 2 significant figures        **b**  32.8 correct to 1 decimal place.

### Solution 1

**a**  **i**  lower bound = 63.5
    **ii**  upper bound = 64.5

| | | | | |
|---|---|---|---|---|
| 63 | 63.5 | 64 | 64.5 | 65 |

Numbers in the red interval from 63.5 to 64.5 are closer to 64 than to 63 or to 65 The lower bound (the value at the lower boundary of this interval) is 63.5  The upper bound (the value at the upper boundary of the interval) is 64.5

**b**  **i**  lower bound = 32.75
    **ii**  upper bound = 32.85

| | | | | |
|---|---|---|---|---|
| 32.7 | 32.75 | 32.8 | 32.85 | 32.9 |

Numbers in the red interval from 32.75 to 32.85 are closer to 32.8 than to 32.7 or to 32.9

---

If numbers in calculations have been rounded, the final answer will not be exact. It is, however, possible to find the lower bound and the upper bound of the answer.

For example, in the product $4.5 \times 6.4$ where both numbers have been rounded to 1 decimal place

> to find the upper bound of the product, work out $4.55 \times 6.45 = 29.3475$
> to find the lower bound of the product, work out $4.45 \times 6.35 = 28.2575$

For any product, the lower bound of the product is worked out using the lower bound of each number in the product. Similarly, for the upper bound, the upper bound of the product is worked out using the upper bound of each number in the product.

However, the lower bound of the difference $6.0 - 3.8$, where both numbers are written correct to 1 decimal place, is **not** found by subtracting the lower bound of 3.8 from the lower bound of 6.0. The lower bound of the difference is the difference between the lower bound of 6.0 and the **upper** bound of 3.8

Lower bound $= 5.95 - 3.85 = 2.1$ (**not** $6.05 - 3.85 = 2.2$)

## Example 2

Correct to 1 decimal place, $x = 4.8$ and $y = 2.4$

Work out the lower bounds of

**a** $xy$      **b** $x - y$      **c** $x + y$      **d** $\dfrac{x}{y}$

### Solution 2

**a** $4.75 \times 2.35$

         $= 11.1625$      Lower bound × Lower bound

**b** $4.75 - 2.45$

         $= 2.3$      Lower bound − Upper bound

**c** $4.75 + 2.35$

         $= 7.1$      Lower bound + Lower bound

**d** $4.75 \div 2.45$

         $= 1.938\ 775\ 5$      Lower bound ÷ Upper bound

## Example 3

$H = \dfrac{v^2}{2g}$ is a formula used to find the height $H$, of a stone thrown upwards at a speed $v$.

$v = 10$ correct to the nearest integer, $g = 9.8$ correct to 2 significant figures.

**a** Write down the upper bound of $g$.

**b** Work out the lower bound of $H$.

Give your answer correct to 3 decimal places.

### Solution 3

**a** Upper bound of $g = 9.85$

**b** Lower bound of $H = \dfrac{9.5^2}{2 \times 9.85}$      Lower bound of $v^2 \div (2 \times$ Upper bound of $g)$

         $= 4.5812...$

         $= 4.581$      An answer correct to 3 decimal places is required.

## Exercise 35A

**1** $p = 18$ correct to the nearest integer, $q = 12$ correct to the nearest integer.

  **a** Write down the lower bound of    **i** $p$      **ii** $q$

  **b** Write down the upper bound of    **i** $p$      **ii** $q$

  **c** Work out the lower bound of    **i** $p + q$    **ii** $p - q$    **iii** $pq$    **iv** $\dfrac{p}{q}$

  **d** Work out the upper bound of    **i** $p + q$    **ii** $p - q$    **iii** $pq$    **iv** $\dfrac{p}{q}$

**2** $r = 16.4$ correct to 1 decimal place, $t = 4.7$ correct to 1 decimal place.

  **a** Work out the lower bound of    **i** $r + t$    **ii** $r - t$    **iii** $rt$    **iv** $\dfrac{r}{t}$

  **b** Work out the upper bound of    **i** $r + t$    **ii** $r - t$    **iii** $rt$    **iv** $\dfrac{r}{t}$

**3** $x = 6.4$ correct to 1 decimal place, $y = 8.3$ correct to 1 decimal place.

   **a** Work out the lower bound of     **i** $x + y$    **ii** $y - x$    **iii** $xy$    **iv** $\dfrac{y}{x}$

   **b** Work out the upper bound of     **i** $x + y$    **ii** $y - x$    **iii** $xy$    **iv** $\dfrac{y}{x}$

**4** $k = 2.45$ correct to 2 decimal places.

   **a** Work out the lower bound of     **i** $4k$    **ii** $\dfrac{1}{k}$    **iii** $k^2$    **iv** $\sqrt{k}$

   **b** Work out the upper bound of     **i** $4k$    **ii** $\dfrac{1}{k}$    **iii** $k^2$    **iv** $\sqrt{k}$

**5** $x = 4.62$ correct to 2 decimal places, $y = 2.5$ correct to 1 decimal place.

   **a** Work out the lower bound of     **i** $x + y$    **ii** $x - y$    **iii** $xy$    **iv** $\dfrac{y}{x}$

   **b** Work out the upper bound of     **i** $x + y$    **ii** $x - y$    **iii** $xy$    **iv** $\dfrac{y}{x}$

**6** $p = 3.8$ correct to 1 decimal place, $q = 4.60$ correct to 2 decimal places.

   **a** Work out the lower bound of     **i** $p^2$    **ii** $3p + 2q$    **iii** $\dfrac{p + q}{p}$

   **b** Work out the upper bound of     **i** $p^2$    **ii** $3p + 2q$    **iii** $\dfrac{p + q}{p}$

**7** Scott cycles at a steady speed of 10 metres per second correct to the nearest metre per second.

   **a** Work out the lower bound of the time Scott takes to cycle exactly 100 metres.

   **b** Work out the upper bound of the time Scott takes to cycle exactly 100 metres.

**8** The length of a rectangle is 16 cm. The width of the rectangle is 12 cm. Both measurements have been given correct to the nearest cm.

   **a** Find the lower bound of the perimeter.    **b** Find the upper bound of the perimeter.

   **c** Find the lower bound of the area.    **d** Find the upper bound of the area.

**9** $h = \dfrac{v^2}{2g}$, $v = 10.2$ correct to 3 significant figures. $g = 9.8$ correct to 2 significant figures.

Work out the difference between the lower bound of $h$ and the upper bound of $h$.

**10** The population of a country is 14.6 million correct to 3 significant figures.
8.5 million, correct to 2 significant figures, of the population are under 30 years of age.
Calculate the difference between the lower bound and the upper bound of the percentage of the population which are under 30 years of age.

## 35.2 Surds

A **surd** is a root of a number which does not have an exact value.

For example $\sqrt{2}$ is a surd but $\sqrt{4}$ ($=2$) is not.

Similarly $\sqrt[3]{7}$ is a surd but $\sqrt[3]{1000}$ ($=10$) is not.

These two laws are often used to simplify expressions involving surds.

$$\sqrt{m} \times \sqrt{n} = \sqrt{mn} \qquad \frac{\sqrt{m}}{\sqrt{n}} = \sqrt{\frac{m}{n}}$$

> **For non-zero values of $m$ and $n$**
> $$\sqrt{m} + \sqrt{n} \neq \sqrt{m + n}$$
> $$\sqrt{m} - \sqrt{n} \neq \sqrt{m - n}$$

For example

$$\sqrt{75} = \sqrt{25 \times 3} = \sqrt{25} \times \sqrt{3} = 5\sqrt{3} \text{ and } \sqrt{\frac{32}{49}} = \frac{\sqrt{16 \times 2}}{\sqrt{49}} = \frac{\sqrt{16} \times \sqrt{2}}{\sqrt{49}} = \frac{4\sqrt{2}}{7}$$

To simplify surds of the form $\sqrt{n}$ write $n$ as a product including a square number.

## Example 4

**a** Write $\sqrt{48}$ in the form $k\sqrt{3}$ where $k$ is an integer.

**b** Expand $(2 + \sqrt{3})(4 + 5\sqrt{3})$. Give your answer in the form $a + b\sqrt{3}$ where $a, b$ are integers.

*Solution 4*

**a** $\sqrt{48} = \sqrt{16 \times 3}$

| Write 48 as the product of 16 (a square number) and 3 |

$\qquad = \sqrt{16} \times \sqrt{3}$

| Use $\sqrt{mn} = \sqrt{m} \times \sqrt{n}$ |

$\qquad = 4\sqrt{3}$

**b** $(2 + \sqrt{3})(4 + \sqrt{3}) = 8 + 2\sqrt{3} + 4\sqrt{3} + \sqrt{3} \times \sqrt{3}$

| Expand the brackets. |

$\qquad\qquad\qquad = 8 + 2\sqrt{3} + 4\sqrt{3} + 3$

| $\sqrt{3} \times \sqrt{3} = 3$ |

$\qquad\qquad\qquad = 11 + 6\sqrt{3}$

| Simplify |

For a fraction with a surd as its denominator it is often useful to **rationalise the denominator**.

For a fraction of the form $\dfrac{a}{\sqrt{b}}$ where $a$ and $b$ are positive integers, rationalising the denominator

means multiplying the fraction by $\dfrac{\sqrt{b}}{\sqrt{b}}$, so that

$$\frac{a}{\sqrt{b}} = \frac{a}{\sqrt{b}} \times \frac{\sqrt{b}}{\sqrt{b}} = \frac{a \times \sqrt{b}}{\sqrt{b} \times \sqrt{b}} = \frac{a\sqrt{b}}{b}$$

The final expression, $\dfrac{a\sqrt{b}}{b}$, now has an integer as the denominator.

## Example 5

**a** Rationalise the denominator of $\dfrac{2}{\sqrt{3}}$

**b** Rationalise the denominator of $\dfrac{3}{\sqrt{6}}$ and simplify your answer.

*Solution 5*

**a** $\dfrac{2}{\sqrt{3}} = \dfrac{2}{\sqrt{3}} \times \dfrac{\sqrt{3}}{\sqrt{3}}$

| Multiply the fraction by $\dfrac{\sqrt{3}}{\sqrt{3}}$ |

$\qquad = \dfrac{2 \times \sqrt{3}}{\sqrt{3} \times \sqrt{3}}$

$\qquad = \dfrac{2\sqrt{3}}{3}$

| $\sqrt{3} \times \sqrt{3} = 3$ |

**b** $\dfrac{3}{\sqrt{6}} = \dfrac{3}{\sqrt{6}} \times \dfrac{\sqrt{6}}{\sqrt{6}}$

Multiply the fraction by $\dfrac{\sqrt{6}}{\sqrt{6}}$

$= \dfrac{3\sqrt{6}}{6}$

$\sqrt{6} \times \sqrt{6} = 6$

$= \dfrac{\sqrt{6}}{2}$

Simplify

### Exercise 35B

**1** Find the value of the integer $k$.

**a** $\sqrt{8} = k\sqrt{2}$    **b** $\sqrt{18} = k\sqrt{2}$    **c** $\sqrt{50} = k\sqrt{2}$    **d** $\sqrt{80} = k\sqrt{5}$    **e** $\sqrt{72} = k\sqrt{2}$

**2** Expand these expressions. Write your answers in the form $a + b\sqrt{c}$ where $a$, $b$ and $c$ are integers.

**a** $\sqrt{3}(2 + \sqrt{3})$      **b** $(\sqrt{3} + 1)(2 + \sqrt{3})$      **c** $(\sqrt{5} - 1)(2 + \sqrt{5})$

**d** $(\sqrt{7} + 1)(2 - 2\sqrt{7})$      **e** $(2 - \sqrt{3})^2$

**3** Rationalise the denominators.

**a** $\dfrac{1}{\sqrt{2}}$    **b** $\dfrac{1}{\sqrt{5}}$    **c** $\dfrac{2}{\sqrt{7}}$    **d** $\dfrac{3}{\sqrt{2}}$    **e** $\dfrac{5}{\sqrt{11}}$

**4** Rationalise the denominators and simplify your answers.

**a** $\dfrac{2}{\sqrt{6}}$    **b** $\dfrac{3}{\sqrt{12}}$    **c** $\dfrac{5}{\sqrt{10}}$    **d** $\dfrac{2}{\sqrt{2}}$    **e** $\dfrac{10}{\sqrt{5}}$

**5** Rationalise the denominators and give your answers in the form $a + b\sqrt{c}$ where $a$, $b$ and $c$ are integers.

**a** $\dfrac{2 + \sqrt{2}}{\sqrt{2}}$    **b** $\dfrac{2 - \sqrt{2}}{\sqrt{2}}$    **c** $\dfrac{10 + \sqrt{5}}{\sqrt{5}}$    **d** $\dfrac{5 - \sqrt{5}}{\sqrt{5}}$    **e** $\dfrac{14 + \sqrt{7}}{\sqrt{7}}$

**6** The lengths of the two shorter sides of a right-angled triangle are $\sqrt{7}$ cm and 3 cm. Find the length of the hypotenuse.

**7** The length of the side of a square is $(1 + \sqrt{2})$ cm. Work out the area of the square. Give your answer in the form $(a + b\sqrt{2})$ cm² where $a$ and $b$ are integers.

**8** The length of a rectangle is $(3 + \sqrt{5})$ cm. The width of the rectangle is $(4 - \sqrt{5})$ cm.

Work out   **a** the perimeter of the rectangle   **b** the area of the rectangle.

## Chapter summary

**You should now be able to:**

★ write down the **lower bound** and the **upper bound** of a value written to a given degree of accuracy

★ work out the lower bound and the upper bound of an expression

★ rationalise denominators

★ simplify expressions involving surds.

# Chapter 35 review questions

1 For each of these rounded numbers, write down    **i** the lower bound    **ii** the upper bound
    **a** 6.4 (1 decimal place.)           **b** 5.08 (2 decimal places.)
    **c** 6400 (3 significant figures)      **d** 0.0148 (3 significant figures)
    **e** 21 (nearest whole number)

2 A square has sides of length 8.5 cm correct to the nearest millimetre. Calculate
    **a** the lower bound of the perimeter       **b** the upper bound of the area

3 $y = \dfrac{a}{b}$, $a = 4.8$ (2 significant figures), $b = 6$ (1 significant figure).

    Calculate    **i** the lower bound of $y$    **ii** the upper bound of $y$.
    Give your answers correct to 4 significant figures.

4 The mass of a solid shape is 6460 kg correct to 3 significant figures. The volume of the solid
    shape is 2.8 m³ correct to 2 significant figures. Calculate **i** the lower bound of the density,
    **ii** the upper bound of the density. Give your answers correct to 4 significant figures.

5 The time taken for a car to cover a distance of 3100 m was 20.0 seconds. The distance was
    measured correct to 2 significant figures and the time correct to 3 significant figures.
    Work out the upper bound of the average speed of the car.

6 John weighs 88 kg and Sophie weighs 65 kg.
    Both weights have been rounded to the nearest kg.
    Explain why their minimum combined weight is 152 kg.

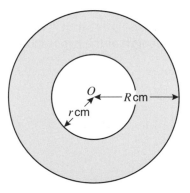

7 Correct to 2 significant figures, the area of a rectangle is 470 cm².
    Correct to 2 significant figures, the length of the rectangle is 23 cm.
    Calculate the upper bound for the width of the rectangle.        (1388 March 2004)

8 $O$ is the centre of both circles.
    The radius of the outer circle is $R$ cm.
    The radius of the inner circle is $r$ cm.
    $R = 15.8$ correct to 1 decimal place.
    $r = 14.2$ correct to 1 decimal place.

    **a** John says that the minimum possible diameter of
      the inner circle is 28.35 cm.
      Explain why John is wrong.

    The upper bound for the area, in cm², of the shaded region is $k\pi$.

    **b** Find the **exact** value of $k$.

9 The time period, $T$ seconds, of a pendulum is calculated using the formula

$$T = 6.283 \times \sqrt{\dfrac{L}{g}}$$

    where $L$ metres is the length of the pendulum and $g$ m/s² is the acceleration due to gravity.

    $L = 1.36$ correct to 2 decimal places
    $g = 9.8$ correct to 1 decimal place

    Find the difference between the lower bound of $T$ and the upper bound of $T$.     (1387 November 2004)

**10 a** Find the value of  **i** $64^0$    **ii** $64^{\frac{1}{2}}$    **iii** $64^{-\frac{2}{3}}$

**b** $3 \times \sqrt{27} = 3^n$. Find the value of $n$.                                           (1387 June 2005)

**11 a** Express $\dfrac{6}{\sqrt{2}}$ in the form $a\sqrt{b}$ where

$a$ and $b$ are positive integers.

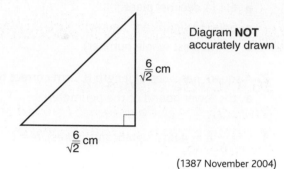

Diagram **NOT** accurately drawn

$\dfrac{6}{\sqrt{2}}$ cm

$\dfrac{6}{\sqrt{2}}$ cm

The diagram shows a right-angled isosceles triangle.

The length of each of its equal sides is $\dfrac{6}{\sqrt{2}}$ cm.

**b** Find the area of the triangle.
Give your answer as an integer.                                       (1387 November 2004)

**12 a** Evaluate  **i** $(\sqrt{5})^2$    **ii** $9^{\frac{1}{2}}$    **iii** $16^{-\frac{3}{4}}$

**b** Work out the value of $(5 - \sqrt{3})^2$  Give your answer in the form $a + b\sqrt{3}$ where $a$ and $b$ are integers.

**13** Work out

$$\dfrac{(5 + \sqrt{3})(5 - \sqrt{3})}{\sqrt{22}}$$

Give your answer in its simplest form.                                         (1387 June 2003)

# Circle geometry (2)

## 36.1 Circle theorems

### Theorem 1 – the angle subtended by an arc at the centre of a circle is twice the angle subtended at the circumference

Angle $AOB = 2 \times$ angle $ACB$

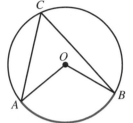

**Proof**

Draw the line $CO$ and produce it to $D$.
$OA = OB = OC$ (radii).

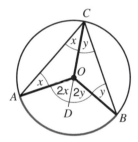

Triangle $OAC$ is isosceles so angle $OAC =$ angle $OCA = x$ (say).

Triangle $OBC$ is isosceles so angle $OBC =$ angle $OCB = y$ (say).

Angle $AOD =$ angle $OAC +$ angle $OCA$ (exterior angle of triangle), i.e. angle $AOD = 2x$.

Similarly, angle $BOD = 2y$.

Angle $AOB = 2x + 2y = 2(x + y) = 2 \times$ angle $ACB$,

Angle $AOB = 2 \times$ angle $ACB$.

---

### Example 1

$P$, $Q$ and $R$ are points on a circle, centre $O$.
Angle $PRQ = 41°$.

Work out the size of angle $POQ$.
Give a reason for your answer.

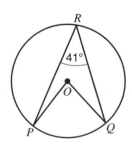

**Solution 1**

Angle $POQ = 2 \times 41°$

$\qquad = 82°$

> Double angle $PRQ$.

The angle at the centre of a circle is twice the angle at the circumference.

> The reason may be shortened to this.

---

## Theorem 2 – the angle in a semicircle is a right angle

Angle $ACB = 90°$

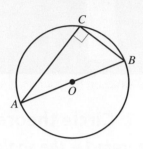

### Proof
The angle subtended at $O$, the centre of the circle, by the arc $AB$ is 180°, that is, angle $AOB = 180°$.

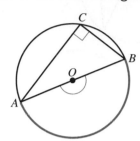

Angle $AOB = 2 \times$ angle $ACB$ (angle at the centre of a circle is twice the angle at the circumference).

Angle $ACB = \frac{1}{2}$ angle $AOB$

$\qquad = \frac{1}{2} \times 180°$

$\qquad = 90°$

## Example 2

$A$, $B$ and $C$ are points on a circle.
$AB$ is a diameter of the circle.
Angle $BAC = 58°$.

Work out the size of angle $ABC$.
Give a reason for each step in your working.

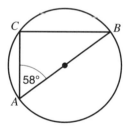

### Solution 2
Angle $ACB = 90°$

State the size of angle $ACB$.

The angle in a semicircle is a right angle.

Give the reason.

Angle $ABC = 180° - (90° + 58°)$

$\qquad = 180° - 148°$

$\qquad = 32°$

Add 90° and 58°.
Subtract the sum from 180°.

The angle sum of a triangle $= 180°$.

Give the reason.

## Theorem 3 – angles in the same segment are equal

Angle $APB =$ angle $AQB$

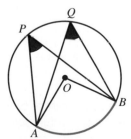

### Proof
Angle $APB = \frac{1}{2} \times$ angle $AOB$ (angle at the centre of a circle is twice the angle at the circumference).

Similarly, angle $AQB = \frac{1}{2}$ angle $AOB$.

So angle $APB =$ angle $AQB$.

## Example 3

$A$, $B$, $C$ and $D$ are points on a circle.
Angle $ADB = 63°$.

Find the size of angle $ACB$.
Give a reason for your answer.

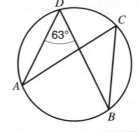

### Solution 3

Angle $ACB = 63°$

| State the size of angle $ACB$, which is equal in size to angle $ADB$. |

The angles in the same segment are equal.

| Give the reason. |

## Cyclic quadrilaterals

A quadrilateral whose vertices (corners) all lie on the circumference of a circle is called a **cyclic quadrilateral**.

The diagram below shows a cyclic quadrilateral $PQRS$.

### Theorem 4 – the sum of the opposite angles of a cyclic quadrilateral is 180°

Angle $SPQ +$ angle $SRQ = 180°$
and
angle $PSR +$ angle $PQR = 180°$.

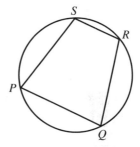

### Proof

$PQRS$ is a cyclic quadrilateral whose vertices lie on a circle, centre $O$.

Let angle $SPQ = a$, angle $SRQ = b$, angle $SOQ = x$ and reflex angle $SOQ = y$.

Then $x = 2a$ (angle at the centre of a circle is twice the angle at the circumference).
Similarly, $y = 2b$.

$x + y = 360°$ (sum of angles at a point $= 360°$) so $2a + 2b = 360°$.
Dividing both sides by 2, $a + b = 180°$.

That is, angle $SPQ +$ angle $SRQ = 180°$.

Also, angle $PSR +$ angle $PQR = 180°$ (the sum of the angles of a quadrilateral is 360°).

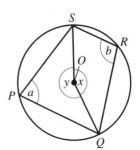

## Example 4

$ABC$ is a straight line.
$B$, $C$, $D$ and $E$ are points on a circle.
Angle $ABE = 81°$.

Work out the size of angle $CDE$.
Give a reason for each step in your working.

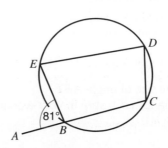

*Solution 4*
Angle $CBE = 180° - 81°$

         $= 99°$

> Subtract 81° from 180°.

The sum of angles on a straight line $= 180°$.

> Give the reason.

$BCDE$ is a cyclic quadrilateral

so angle $CDE = 180° - 99°$

             $= 81°$

> Subtract 99° from 180°.

The sum of opposite angles of a cyclic quadrilateral $= 180°$.

> Give the reason.

Notice that angle $ABE$ = angle $CDE$.
Angle $ABE$ is an exterior angle of the cyclic quadrilateral and it is the same size as the opposite interior angle.

## Theorem 5 – the angle between a chord and the tangent at the point of contact is equal to the angle in the alternate segment

Angle $PTB$ = angle $BAT$

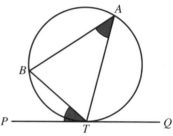

### Proof
$PTQ$ is a tangent to the circle at $T$.

$TB$ is a chord of the circle.

Angle $BAT$ is any angle in the alternate (opposite) segment to angle $PTB$.

Let angle $PTB = x$ and angle $BAT = y$.

Draw the diameter $TC$.

Angle $CTB = 90° - x$ (tangent is perpendicular to a radius).

Angle $CBT = 90°$ (angle in a semicircle is a right angle).

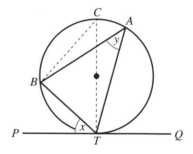

In triangle $CBT$,
$90° + 90° - x +$ angle $BCT = 180°$ (angle sum of triangle).
So angle $BCT = x$.

Angle $BCT$ = angle $BAT$ (angles in the same segment).
That is $x = y$ and angle $PTB$ = angle $BAT$.

This theorem is known as the **alternate segment theorem**.

## Example 5

$A$, $B$ and $T$ are points on a circle.
$PTQ$ is a tangent to the circle.
Angle $PTB = 37°$.
Angle $ATB = 68°$.

Work out the size of angle $ABT$.
Give a reason for each step in your working.

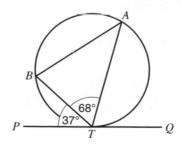

*Solution 5*
*Method 1*
Angle $PTB$ = angle $BAT$
            = 37°

Alternate segment theorem.

> The reason may be shortened to this.

Angle $ABT$ = 180° − (37° + 68°)
            = 75°

> Add 37° and 68°.
> Subtract the sum from 180°.

The angle sum of triangle = 180°.

> Give the reason.

*Method 2*
Angle $ATQ$ = 180° − (37° + 68°)
            = 75°

> Add 37° and 68°.
> Subtract the sum from 180°.

The sum of angles on a straight line = 180°.

> Give the reason.

Angle $ATQ$ = angle $ABT$
            = 75°

Alternate segment theorem.

> The reason may be shortened to this.

---

## Example 6

$ABCD$ is a cyclic quadrilateral.
Angle $ADB$ = 36°. Angle $BDC$ = 47°.
**a** Find the size of    **i** angle $BAC$    **ii** angle $ABC$.
     Give reasons for your answers.
**b** Is $AC$ a diameter? Explain your answer.

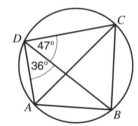

*Solution 6*
**a   i** Angle $BAC$ = 47°

     Angles in the same segment.

> The reason may be shortened to this.

   **ii** Angle $ADC$ = 36° + 47°
                     = 83°

> Add 36° and 47° to find the size of angle $ADC$.

       Angle $ABC$ = 180° − 83°
                   = 97°

> Subtract 83° (the size of angle $ADC$) from 180°.

       The sum of opposite angles of a
       cyclic quadrilateral = 180°.

> Give the reason.

**b** $AC$ is not a diameter.
     If it were, angle $ADC$ would be 90°
     (the angle in a semicircle) but it is 83°.

> The full answer consists of a statement
> and an explanation.

**Exercise 36A**

The diagrams are **NOT** accurately drawn.
Dots show the centres of some of the circles.

In Questions **1–9**, find the size of the angles marked with letters.
Give a reason for each answer.

**1**

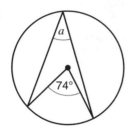

**2**

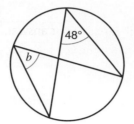

**3**

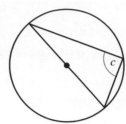

**4**

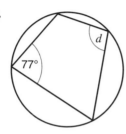

**5**

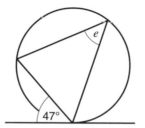

**6**

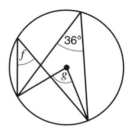

**7**

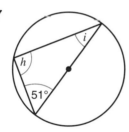

**8**

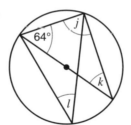

**9**

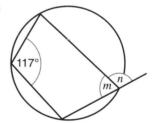

**10** $A$, $B$ and $T$ are points on the circle.
$PT$ is a tangent to the circle at $T$.
Angle $PTB = 38°$.
$AB = AT$.

Work out the size of angle $ABT$.
Give a reason for each step in your working.

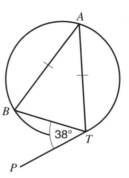

**11** $A$, $B$ and $C$ are points on a circle.
Angle $ABC = 28°$.
Angle $BAC = 62°$.

Is $AB$ a diameter? Explain your answer.

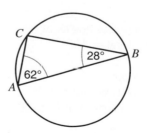

**12** $A$, $B$, $C$ and $D$ are points on a circle.
Angle $ABC = 76°$.
Angle $ADB = 31°$.

Work out the size of    **i** angle $BDC$    **ii** angle $CAB$.
Give a reason for each step in your working.

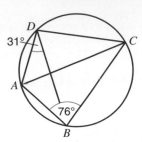

**13 a** Is a rectangle a cyclic quadrilateral? Explain your answer.
 **b** Is this quadrilateral cyclic? Explain your answer.

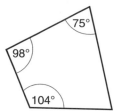

**14** $A$, $B$, $C$ and $T$ are points on the circle.
$PTQ$ is a tangent to the circle.
Angle $PTC = 51°$.
Angle $BAC = 23°$.

Work out the size of angle $BCT$.
Give a reason for each step in your working.

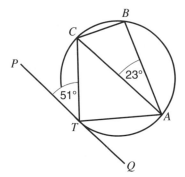

**15** $A$, $Q$ and $R$ are points on the circle.
$PQ$ and $PR$ are tangents to the circle.
Angle $QPR = 48°$.

Work out the size of angle $QAR$.
Give a reason for each step in your working.

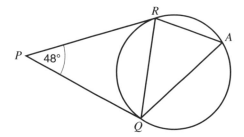

## Chapter summary

**You should now know these geometric facts and be able to prove them:**

★  the angle subtended by an arc at the centre of a circle is
   twice the angle subtended at the circumference

   $b = 2a$

★ the angle in a semicircle is a right angle.

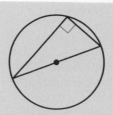

★ angles in the same segment are equal

★ a quadrilateral whose vertices (corners) all lie on the circumference of a circle is called a cyclic quadrilateral. The sum of the opposite angles of a cyclic quadrilateral is 180°.

$$a + c = 180° \text{ and } b + d = 180°$$

★ the angle between a chord and the tangent at the point of contact is equal to the angle in the alternate segment.

## Chapter 36 review questions

The diagrams are **NOT** accurately drawn.

**1** $A$, $B$, $C$ and $D$ are points on a circle centre $O$.
Angle $ADB = 38°$.

  **a** Give a reason why angle $ACB = 38°$.
  **b**  **i** Find the size of angle $AOB$.
    **ii** Give a reason for your answer.

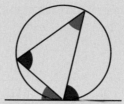

**2** The diagram shows a circle with its centre at $O$.
$A$, $B$, and $C$ are points on the circumference of the circle.
At $C$, a tangent to the circle has been drawn.
$D$ is a point on this tangent.
Angle $OCB = 24°$.

  **a** Find the size of angle $BCD$.
    Give a reason for your answer.
  **b** Find the size of angle $CAB$.
    Give a reason for your answer.

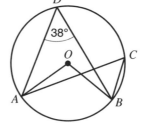
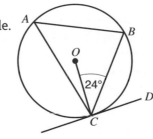

Diagram **NOT** accurately drawn

(1384 June 1995)

**3** $A$, $B$, $C$ and $D$ are four points on the circumference of a circle.
$TA$ is a tangent to the circle at $A$.
Angle $DAT = 30°$.
Angle $ADC = 132°$.

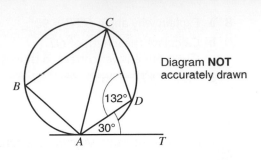

Diagram **NOT** accurately drawn

  **a**   **i** Calculate the size of angle $ABC$.

      **ii** Explain your method.

  **b**   **i** Calculate the size of angle $CBD$.

      **ii** Explain your method.

  **c** Explain why $AC$ cannot be a diameter of the circle.

(1385 June 2000)

**4** $A$, $B$, $C$ and $D$ are points on a circle.
$AP$ and $BP$ are tangents to the circle.
Angle $BAD = 80°$.
Angle $BAP = 70°$.

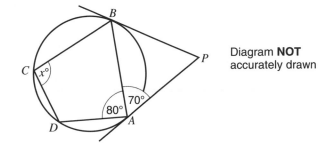

Diagram **NOT** accurately drawn

  **a** Find the size of angle $BCD$, marked $x°$ in the diagram.

  **b** Find the size of angle $APB$.
    Give reasons for your answer.

  **c** Find the size of angle $DCA$.
    Give reasons for your answer.

**5** $A$, $B$, $C$ and $T$ are points on the circumference of a circle.
Angle $BAC = 25°$.
The line $PTS$ is the tangent at $T$ to the circle.
$AT = AP$.
$AB$ is parallel to $TC$.

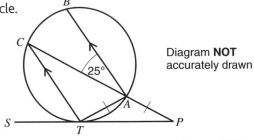

Diagram **NOT** accurately drawn

  **a** Calculate the size of angle $APT$.
    Give reasons for your answer.

  **b** Calculate the size of angle $BTS$.
    Give reasons for your answer.

(1384 June 1997)

**6** $A$, $B$, $C$ and $D$ are points on the circumference of a circle centre $O$.
$AC$ is a diameter of the circle.
Angle $BDO = x°$.
Angle $BCA = 2x°$.
Express, in terms of $x$, the size of
**i** angle $BDA$    **ii** angle $AOD$    **iii** angle $ABD$.

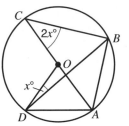

Diagram **NOT** accurately drawn

(1385 November 1998)

**7** The diagram shows a triangle $ABC$ and a circle, centre $O$.
$A$, $B$ and $C$ are points on the circumference of the circle.
$AB$ is a diameter of the circle.
$AC = 16$ cm and $BC = 12$ cm.

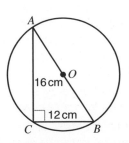

Diagram **NOT** accurately drawn

  **a** Angle $ACB = 90°$. Give a reason why.

  **b** Work out the diameter $AB$ of the circle.

  **c** Work out the area of the circle.
    Give your answer correct to three significant figures.

(1387 June 2005)

**8 a** Explain why angle $OTP = 90°$.

  **b** Calculate the length of $OT$.
Give your answer correct to three significant figures.

  **c** Angle $QOT = 36°$.
Calculate the length of $OQ$.
Give your answer correct to three
significant figures.

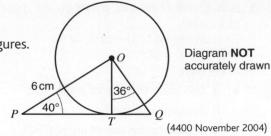

Diagram **NOT**
accurately drawn

(4400 November 2004)

**9** $A, B$ and $C$ are three points on the circumference of a circle.
Angle $ABC$ = Angle $ACB$.
$PB$ and $PC$ are tangents to the circle from the point $P$.

  **a** Prove that triangle $APB$ and triangle $APC$
are congruent.

Angle $BPA = 10°$.

  **b** Find the size of angle $ABC$.

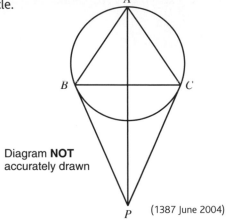

Diagram **NOT**
accurately drawn

(1387 June 2004)

**10** The diagram shows a circle with centre $O$ and a triangle $OPT$.
$P$ is a point on the circumference of the circle and $TP$ is a
tangent to the circle.

  **a** Angle $OPT = 90°$. Give a reason why.

The radius of the circle is 50 cm. $TP = 92$ cm.

  **b** Calculate the length of $OT$.
Give your answer correct to three significant figures.

  **c** Calculate the size of the angle marked $x°$.
Give your answer correct to three significant figures.

The region that is inside the triangle but outside
the circle is shown shaded in the diagram.

  **d** Calculate the area of the shaded region.
Give your answer correct to two significant figures.

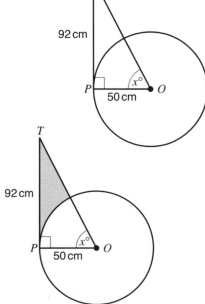

# Completing the square

## 37.1 Completing the square

The diagram shows $x^2 + 6x$.
The $+6x$ has been split in half.

$$
\begin{array}{c|c|c}
 & x & +3 \\
\hline
x & x^2 & +3x \\
\hline
+3 & +3x & \\
\end{array}
$$

To complete the square $+9\ (= +3 \times +3)$ needs to be added on.

$$
\begin{array}{c|c|c}
 & x & +3 \\
\hline
x & x^2 & +3x \\
\hline
+3 & +3x & +9 \\
\end{array}
$$

The diagram now shows $x^2 + 6x + 9 = (x + 3)^2$

Removing the $+9$ shows $x^2 + 6x = (x + 3)^2 - 9$

Writing $x^2 + 6x$ in the form $(x + 3)^2 - 9$ is called **completing the square** for $x^2 + 6x$.

In general writing the quadratic expression $ax^2 + bx + c$ in the form $a(x + p)^2 + q$ is called **completing the square.**

In Exercise 11C these expansions of perfect squares were found

$$(x + 7)^2 = x^2 + \mathbf{14}x + 49 \qquad (x - 5)^2 = x^2 - \mathbf{10}x + 25$$

In each case the **final number** in the bracket is half the **coefficient of $x$**.

In general $(x + A)^2 = x^2 + \mathbf{2A}x + A^2$ so $x^2 + \mathbf{2A}x = (x + A)^2 - A^2$

To complete the square for the expression $x^2 + \mathbf{2A}x$

Step 1: Write $(x \qquad )^2$
Leave enough space inside the bracket for writing in a constant term

Step 2: Write $(x + A)^2$
Since $+A$ is half the coefficient of $x$ ($\frac{1}{2}$ of $+ \mathbf{2A} = +A$)

Step 3: Write $(x + A)^2 - A^2$
Subtract the square of half the coefficient of $x$ (subtract $A^2$)

> To complete the square for $x^2 + \mathbf{2A}x$ find half the coefficient of $x$ then square it to give $A^2$
>
> Then write $x^2 + \mathbf{2A}x$ as $x^2 + \mathbf{2A}x + A^2 - A^2$
> $= (x + A)^2 - A^2$

---

### Example 1

The expression $x^2 - 8x$ can be written in the form $(x + p)^2 + q$ for all values of $x$.
Find the value of $p$ and the value of $q$.

## Solution 1

$x^2 - 8x$

> Complete the square for $x^2 - 8x$.

$\frac{1}{2}$ of $-8 = -4$

> Halve the coefficient of $x$.

$x^2 - 8x = (x - 4)^2 - (-4)^2$

> Write $(x\quad)^2$ with half the coefficient of $x$ before the end bracket then subtract the square of half the coefficient of $x$.

$x^2 - 8x = (x - 4)^2 - 16$

> $(-4)^2 = 16$

$(x + p)^2 + q$ is the same as
$(x - 4)^2 - 16$ when
$p = -4$ and $q = -16$

> Compare $(x - 4)^2 - 16$ with $(x + p)^2 + q$ to read off the value of $p$ and the value of $q$.

In Section 32.1 quadratic graphs were drawn and maximum and minimum points were often estimated by reading from the graphs.

The method of completing the square can be used to find the exact values for maximum and minimum points on quadratic graphs.

For example the graph of $y = x^2 - 8x$ has a minimum point. The coordinates of this minimum point can be found by using the result of example 11

Completing the square for $x^2 - 8x$ gives $x^2 - 8x = (x - 4)^2 - 16$
so the equation of the graph can be written as $y = (x - 4)^2 - 16$

Squaring any value always gives an answer which is positive or zero so for any value of $x$, the smallest value of $(x - 4)^2$ is 0

When $(x - 4)^2 = 0$, $x = 4$ and $y = 0 - 16 = -16$

The minimum point of the graph $y = x^2 - 8x$ has coordinates $(4, -16)$

The minimum value of $x^2 - 8x$ is $-16$

## Example 2

The graph of the curve with equation $y = x^2 + x - 6$ has a minimum point.
**a** Write the expression $x^2 + x - 6$ in the form $(x + p)^2 + q$.

> See Section 32.2

**b** Hence find the coordinates of the minimum point.

## Solution 2

**a** $(x^2 + x) - 6$

> Separate the constant term and complete the square for $x^2 + 1x$.

$\frac{1}{2}$ of $+1 = \frac{1}{2}$

> Halve the coefficient of $x$.

$(x^2 + x) - 6 = [(x + \frac{1}{2})^2 - (\frac{1}{2})^2] - 6$

> Write $(x\quad)^2$ with half the coefficient of $x$ before the end bracket then subtract the square of half the coefficient of $x$.

$x^2 + x - 6 = (x + \frac{1}{2})^2 - \frac{1}{4} - 6$

$x^2 + x - 6 = (x + \frac{1}{2})^2 - 6.25$

> Write the answer in the required form $(x + p)^2 + q$ so $p = \frac{1}{2}$ and $q = -6.25$

**b** $y = (x + \frac{1}{2})^2 - 6.25$

> Write $y = x^2 + x - 6$ in the completed square form using the answer to part **a**.

The minimum value of $y$ is $-6.25$ and occurs when $(x + \frac{1}{2})^2 = 0$ so $x = -\frac{1}{2}$

> The least value of $(x + \frac{1}{2})^2$ is 0

The minimum point is $(-0.5, -6.25)$

Note: The minimum value of $x^2 + x - 6$ is $-6.25$

## Example 3

The expression $17 + 20x - 2x^2$ can be written in the form $a(x + p)^2 + q$.

**a** Find the value of $a$, the value of $p$ and the value of $q$.

**b** Use the answers to part **a**

   **i** to find the maximum value of $17 + 20x - 2x^2$

   **ii** to find the value of $x$ for which $17 + 20x - 2x^2$ has its maximum value.

### Solution 3
### Method 1

**a**   $17 + 20x - 2x^2 = -2(x^2 - 10x) + 17$

> Take out the coefficient of $x^2$ and separate the constant then complete the square for $x^2 - \mathbf{10}x$.

$\frac{1}{2}$ of $\mathbf{-10} = -5$

> Halve the coefficient of $x$.

$$-2(x^2 - 10x) + 17 = -2[(x - 5)^2 - (-5)^2] + 17$$
$$= -2[(x - 5)^2 - 25] + 17$$
$$= -2(x - 5)^2 + 50 + 17$$

$$17 + 20x - 2x^2 = -2(x - 5)^2 + 67$$

$a(x + p)^2 + q$ is the same as
$-2(x - 5)^2 + 67$ when
$a = -2, p = -5$ and $q = 67$

> Compare $-2(x - 5)^2 + 67$ with $a(x + p)^2 + q$ to read off the value of $a$, the value of $p$ and the value of $q$.

### Method 2

**a**   $a(x + p)^2 + q = 17 + 20x - 2x^2$

> Equate the two expressions.

$$a(x^2 + 2px + p^2) + q = 17 + 20x - 2x^2$$
$$ax^2 + 2apx + ap^2 + q = 17 + 20x - 2x^2$$

> Expand brackets using
> $(x + A)^2 = x^2 + 2Ax + A^2$

$ax^2 = -2x^2$     so $a = -2$

$+2apx = +20x$ so $2ap = 20, -4p = 20, p = -5$

$+ap^2 + q = 17$ so $-2 \times (-5)^2 + q = 17, -50 + q = 17$

> For the two expressions to be the same for all values of $x$
> the $x^2$ terms must be identical,
> the $x$ terms must be identical and
> the constant terms must be identical.

$a = -2, p = -5$ and $q = 67$

**b**   $17 + 20x - 2x^2 = -2(x - 5)^2 + 67$

> Using part **a**.

   **i** The maximum value of $-2(x - 5)^2 + 67$ is
     $-2 \times 0 + 67$
     The maximum value of $17 + 20x - 2x^2$ is 67

> $(x - 5)^2 \geqslant 0$ so $67 - 2(x - 5)^2$ is maximum when least value is subtracted from 67
> Least value of $(x - 5)^2$ is 0

   **ii** The maximum value of $17 + 20x - 2x^2$ occurs
     when $x = 5$

> The maximum value occurs when $(x - 5)^2 = 0$

## Exercise 37A

**1** Write in the form $(x + p)^2 + q$.

  **a** $x^2 + 2x$      **b** $x^2 + 4x$      **c** $x^2 + 12x$      **d** $x^2 + 20x$      **e** $x^2 + 24x$

  **f** $x^2 - 6x$      **g** $x^2 - 8x$      **h** $x^2 - 14x$      **i** $x^2 - 18x$      **j** $x^2 - 22x$

**2** Write in the form $(x + p)^2 + q$.

  **a** $x^2 + 6x + 10$   **b** $x^2 + 8x + 20$   **c** $x^2 + 14x + 10$   **d** $x^2 + 16x - 1$   **e** $x^2 + 24x - 8$

  **f** $x^2 - 2x + 16$   **g** $x^2 - 4x + 18$   **h** $x^2 - 10x - 17$   **i** $x^2 - 40x + 20$   **j** $x^2 - 26x - 10$

**3** Write in the form $(x + p)^2 + q$.

  **a** $x^2 + x$        **b** $x^2 - 3x + 2$      **c** $x^2 + 5x - 1$     **d** $x^2 + 3x - 1$     **e** $x^2 - 9x + 20$

**4** Write in the form $a(x + p)^2 + q$.

  **a** $2x^2 + 4x$               **b** $2x^2 - 12x + 28$            **c** $3x^2 + 24x - 10$

  **d** $5x^2 + 20x - 19$       **e** $6x^2 - 60x + 149$

**5** Write in the form $p - (x + q)^2$.

  **a** $16 + 2x - x^2$    **b** $9 + 4x - x^2$     **c** $10 - 6x - x^2$    **d** $4 + x - x^2$     **e** $1 - 3x - x^2$

**6** For all values of $x$, $x^2 + 10x + 32 = (x + p)^2 + q$.

  **a** Find the value of $p$ and the value of $q$.

  **b** Write down the minimum value of $x^2 + 10x + 32$

**7** The curve with equation $y = x^2 + 2x - 3$ has a minimum point.

  **a** Write the expression $x^2 + 2x - 3$ in the form $(x + p)^2 + q$.

  **b** Hence find the coordinates of the minimum point.

**8** The curve with equation $y = 6 + 2x - x^2$ has a maximum point.

  **a** Write the expression $6 + 2x - x^2$ in the form $p - (x + q)^2$.

  **b** Hence find the coordinates of the maximum point.

**9** The curve with equation $y = 1 + 2x - 2x^2$ has a maximum point.

  **a** Write the expression $1 + 2x - 2x^2$ in the form $a(x + p)^2 + q$.

  **b** Hence find the coordinates of the maximum point.

**10** Show that the minimum value of $x^2 - 4x + 13$ is the same as the maximum value of $8 + 2x - x^2$

## Chapter summary

**You should now:**

★ how to **complete the square** by writing the quadratic expression $ax^2 + bx + c$ in the form $a(x + p)^2 + q$

★ that the minimum value of $(x + p)^2 + q$ is $q$ and occurs when $x + p = 0$

★ that the maximum value of $q - (x + p)^2$ is $q$ and occurs when $x + p = 0$

## Chapter 37 review questions

**1** The expression $8x - x^2$ can be written in the form $p - (x - q)^2$ for all values of $x$.

  **a** Find the value of $p$ and the value of $q$.

  **b** The expression $8x - x^2$ has a maximum value.

   **i** Find the maximum value of $8x - x^2$.

   **ii** State the value of $x$ for which this maximum value occurs.

**2** Given that $x^2 - 14x + a = (x + b)^2$ for all values of $x$, find the value of $a$ and the value of $b$.

(1388 November 2005)

**3 a** Simplify   **i** $(3x^2y)^3$    **ii** $(2t^{-3})^{-2}$

  **b** Show that $x^2 - 4x + 15$ can be written as $(x + p)^2 + q$ for all values of $x$. State the values of $p$ and $q$.                                                    (1387 November 2005)

**4** The expression $x^2 - 8x + 18$ can be written in the form $(x - p)^2 + q$ for all values of $x$.

  **a** Find the value of   **i** $p$         **ii** $q$.

Here is a sketch of the graph of $y = x^2 - 8x + 18$
The minimum point on the curve is $M$.

  **b** Write down the coordinates of $M$.

The line with equation $y = k$ has two points of intersection with the graph of $y = x^2 - 8x + 18$ when $k > a$.

  **c** Write down the least possible value of $a$.

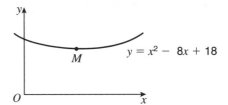

# Quadratic equations

## 38.1 Introduction to solving quadratic equations

In Chapter 32 we saw that $x^2 + x - 6 = 0$ is an example of a quadratic equation.

All quadratic equations can be written in the form $ax^2 + bx + c = 0$ where $a$ ($\neq 0$), $b$ and $c$ represent numbers.

A quadratic equation always contains an $x^2$ term and 2 is the highest power of $x$ in the equation. ($a$ cannot be zero but $b$ or $c$ may be zero.)

Here are five examples of quadratic equations.

$$2x^2 - 6x + 3 = 0 \qquad x^2 - 3 = 0 \qquad x(x + 4) = 7 \qquad (2x - 1)(3x - 2) = 0 \qquad 2x^2 = 50$$

$2x^2 = 50$ can be solved easily.

$$2x^2 = 50$$
$$x^2 = 25$$

so $x = 5$ or $x = -5$, which can be written as $x = \pm 5$

There are several other methods of solving quadratic equations.

## 38.2 Solving by factorisation

If the product of two numbers is 0 then at least one of the numbers must be 0

This means that if $pq = 0$ then $p = 0$ or $q = 0$

So if $(x - 1)(x + 2) = 0$ then $x - 1 = 0$ or $x + 2 = 0$

that is, $x = 1$ or $x = -2$

---

### Example 1

Solve    **a** $2x(x - 3) = 0$        **b** $(2z - 1)(z + 3) = 0$

*Solution 1*

**a** $2x(x - 3) = 0$ $\qquad$ The product of $2x$ and $x - 3$ is 0

$\quad$ $2x = 0$ or $x - 3 = 0$

$\quad$ $x = 0$ or $x = 3$ $\qquad$ This can be written as $x = 0, 3$

**b** $(2z - 1)(z + 3) = 0$ $\qquad$ The product of $2z - 1$ and $z + 3$ is 0

$\quad$ $2z - 1 = 0$ or $z + 3 = 0$

$\quad$ $z = \frac{1}{2}$ or $z = -3$

---

If the left-hand side of a quadratic equation of the form $ax^2 + bx + c = 0$ can be factorised, the equation can be solved.

## Example 2

Solve    **a** $x^2 - 9x + 20 = 0$    **b** $y^2 + 8y = 0$    **c** $z^2 - 64 = 0$

**Solution 2**

**a** $(x - 4)(x - 5) = 0$      | Factorise $x^2 - 9x + 20$ |

  $x - 4 = 0$ or $x - 5 = 0$

  $x = 4$ or $x = 5$

**b** $y(y + 8) = 0$      | Factorise $y^2 + 8y$ |

  $y = 0$ or $y + 8 = 0$

  $y = 0$ or $y = -8$

**c** **Method 1**

  $(z + 8)(z - 8) = 0$      | $z^2 - 64$ is the difference of two squares (See Section 11.6) |

  $z + 8 = 0$ or $z - 8 = 0$

  $z = -8$ or $z = 8$

  **Method 2**

  $z^2 = 64$

  $z = \pm\sqrt{64}$      | Take the square root of both sides |

  $z = 8$ or $z = -8$

---

Solving quadratic equations by factorisation depends on having 0 on one side of the equation and so sometimes it is necessary to rearrange the equation into the form $ax^2 + bx + c = 0$ first.

## Example 3

Solve $(2x - 1)(x - 2) = 5$

**Solution 3**

$2x^2 - 5x + 2 = 5$      | Expand the left-hand side |

$2x^2 - 5x - 3 = 0$      | Write the equation in the form $ax^2 + bx + c = 0$ |

$(2x + 1)(x - 3) = 0$      | Factorise $2x^2 - 5x - 3$ |

$2x + 1 = 0$ or $x - 3 = 0$

$x = -\frac{1}{2}$ or $x = 3$

---

## Exercise 38A

Solve these equations.

**1** $x^2 = 9$                  **2** $4x^2 = 64$            **3** $2x^2 - 8 = 0$

**4** $x^2 - 100 = 0$        **5** $x^2 - 81 = 0$        **5** $4x^2 - 9 = 0$

**7** $(x - 3)(x - 5) = 0$    **8** $x(x - 4) = 0$       **9** $(2x - 3)(x + 2) = 0$

**10** $(2x - 1)(3x + 4) = 0$   **11** $(4x - 1)(2x - 5) = 0$   **12** $x^2 - 7x = 0$

**13** $x^2 + 5x = 0$        **14** $x^2 - x = 0$         **15** $2x^2 - x = 0$

**16** $3x^2 - 5x = 0$      **17** $x^2 - 5x + 4 = 0$    **18** $x^2 - 6x + 8 = 0$

**19** $x^2 - 7x - 8 = 0$    **20** $x^2 - 7x + 12 = 0$   **21** $x^2 + 6x + 8 = 0$

**22** $x^2 - 3x - 10 = 0$   **23** $x^2 + 7x + 12 = 0$   **24** $x^2 - 11x - 12 = 0$

**25** $x^2 + 8x + 7 = 0$    **26** $2x^2 - 5x + 2 = 0$   **27** $2x^2 + 7x + 3 = 0$

**28** $2x^2 - 9x + 4 = 0$   **29** $2x^2 - 3x - 5 = 0$   **30** $2x^2 + 11x + 5 = 0$

**31** $3x^2 - 5x - 2 = 0$   **32** $2x^2 + 3x + 1 = 0$   **33** $3x^2 - 4x + 1 = 0$

**34** $5x^2 - 6x + 1 = 0$            **35** $x^2 + 8 = 6x$            **36** $x^2 - 4x = 5$

**37** $x^2 + 3x = 4$            **38** $x^2 - 6x = 7$            **39** $8 - x^2 = 2x$

**40** $7x - x^2 = 6$            **41** $2x^2 - 6 = 4x$            **42** $2x^2 - 5x = 3$

**43** $3x^2 + 5 = 8x$            **44** $3x^2 - 4x = 15$            **45** $11 - 2x^2 = 9x$

**46** $21 - 2x^2 = x$            **47** $(x + 2)(x + 3) = 2$            **48** $(x + 2)(x - 3) = 6$

**49** $(x - 2)(x - 4) = 8$            **50** $(x - 5)(x + 3) + 15 = 0$            **51** $(x - 1)(x + 4) = 6$

**52** $(x - 1)(x - 5) = 12$

## 38.3 Solving by completing the square

When the left-hand side of the quadratic equation cannot be factorised, the method of **completing the square** can be used (See Section 37.1).

In general,

$$x^2 + px + q = \left(x + \frac{p}{2}\right)^2 - \left(\frac{p}{2}\right)^2 + q$$

### Example 4

Solve $x^2 - 6x + 7 = 0$ by completing the square.
Give your solutions **a** in surd form, **b** correct to 3 significant figures.

*Solution 4*

$(x - 3)^2 - 3^2 + 7 = 0$        | Complete the square for $x^2 - 6x$ |

$\qquad (x - 3)^2 = 9 - 7$
$\qquad (x - 3)^2 = 2$
$\qquad\quad x - 3 = \pm\sqrt{2}$        | Take the square root of both sides. |

**a**  $x = 3 \pm\sqrt{2}$
    so $x = 3 +\sqrt{2}$ or $x = 3 -\sqrt{2}$        | Leave your solutions in surd form. (See Section 26.4) |

**b**    $x = 3 + 1.4142...$
    or $x = 3 - 1.4142...$        | Use your calculator. |
    so $x = 4.4142...$  or $x = 1.5857...$
        $x = 4.41$ or $x = 1.59$ (to 3 s.f.)        | Give your solutions correct to 3 significant figures. |

### Exercise 38B

**1** Solve these equations by completing the square.
Give your solutions **i** in surd form, **ii** correct to 3 significant figures.

   **a** $x^2 + 6x - 10 = 0$        **b** $x^2 - 4x - 7 = 0$        **c** $x^2 - 6x + 1 = 0$

   **d** $x^2 - 10x + 5 = 0$        **e** $x^2 + 2x - 5 = 0$        **f** $x^2 - 10x + 4 = 0$

**2** Solve these equations by completing the square. Give your solutions correct to 3 significant figures.

   **a** $x^2 + 6x = 12$        **b** $x^2 - 4x = 10$        **c** $x^2 - 12x = 6$

   **d** $8x - x^2 = 11$        **e** $x^2 - 9x = 5$        **f** $x(x - 5) = 7$

## 38.4 Solving using the quadratic formula

The general quadratic equation $ax^2 + bx + c = 0$ can be solved using the method of completing the square. This gives a formula which can be used to solve any quadratic equation.

$$ax^2 + bx + c = 0$$

Divide both sides by $a$
$$x^2 + \frac{b}{a}x + \frac{c}{a} = 0$$

Complete the square
$$\left(x + \frac{b}{2a}\right)^2 - \left(\frac{b}{2a}\right)^2 + \frac{c}{a} = 0$$

Rearrange
$$\left(x + \frac{b}{2a}\right)^2 = \left(\frac{b}{2a}\right)^2 - \frac{c}{a}$$

Write the right-hand side as a single term
$$\left(x + \frac{b}{2a}\right)^2 = \frac{b^2}{4a^2} - \frac{4ac}{4a^2} = \frac{b^2 - 4ac}{4a^2}$$

Take the square root of both sides
$$x + \frac{b}{2a} = \pm\sqrt{\frac{b^2 - 4ac}{4a^2}}$$

$$x + \frac{b}{2a} = \pm\frac{\sqrt{b^2 - 4ac}}{\sqrt{4a^2}}$$

$$x = -\frac{b}{2a} \pm \frac{\sqrt{b^2 - 4ac}}{2a}$$

This formula, called the **quadratic formula**, is usually written as

$$x = \frac{-b \pm \sqrt{b^2 - 4ac}}{2a}$$

---

### Example 5

Use the quadratic formula to solve $x^2 + 6x + 2 = 0$
Give your solutions correct to 3 significant figures.

*Solution 5*

$$x = \frac{-6 \pm \sqrt{6^2 - 4 \times 1 \times 2}}{2 \times 1}$$

$$x = \frac{-6 \pm \sqrt{28}}{2}$$

$$x = \frac{-6 + \sqrt{28}}{2} \text{ or } x = \frac{-6 - \sqrt{28}}{2}$$

$$x = \frac{-0.70849...}{2} \text{ or } x = \frac{-11.291...}{2}$$

$x = -0.354$ or $x = -5.65$ (to 3 s.f.)

Compare $x^2 + 6x + 2 = 0$
with $ax^2 + bx + c = 0$
which gives $a = 1, b = 6, c = 2$
Substitute $a = 1, b = 6, c = 2$ into

$$x = \frac{-b \pm \sqrt{b^2 - 4ac}}{2a}$$

---

### Example 6

Use the quadratic formula to solve $2x^2 - 7x - 6 = 0$
Give your solutions correct to 3 significant figures.

**Solution 6**

$$x = \frac{-(-7) \pm \sqrt{(-7)^2 - 4 \times 2 \times (-6)}}{2 \times 2}$$

Substitute $a = 2$, $b = (-7)$, $c = (-6)$ into the quadratic formula

$$x = \frac{7 \pm \sqrt{97}}{4}$$

$(-7)^2 - 4 \times 2 \times (-6) = 49 - (-48) = 97$

$$x = \frac{7 + 9.848...}{4} = \frac{16.848...}{4} \quad \text{or} \quad x = \frac{7 - 9.848...}{4} = \frac{-2.848...}{4}$$

$x = 4.21$ or $x = -0.712$ (to 3 s.f.)

It is sometimes necessary to rewrite an equation in the form $ax^2 + bx + c = 0$ before the quadratic formula can be used.

## Example 7

Solve $5x(x - 1) = 18$
Give your answers correct to 3 significant figures.

**Solution 7**

$5x^2 - 5x = 18$

Rewrite $5x(x - 1) = 18$ in the form $ax^2 + bx + c = 0$

$5x^2 - 5x - 18 = 0$

Compare $5x^2 - 5x - 18 = 0$ with $ax^2 + bx + c = 0$
$a = 5$, $b = (-5)$, $c = (-18)$

$$x = \frac{-(-5) \pm \sqrt{(-5)^2 - 4 \times 5 \times (-18)}}{2 \times 5}$$

Substitute into the quadratic formula

$$x = \frac{5 \pm \sqrt{385}}{10}$$

$(-5)^2 - 4 \times 5 \times (-18) = 25 - -360 = 385$

$$x = \frac{5 + \sqrt{385}}{10} = 2.462... \text{ or}$$

$$x = \frac{5 - \sqrt{385}}{10} = -1.462...$$

$x = 2.46$ or $x = -1.46$ (to 3 s.f.)

Make sure you know how to find square roots on your calculator.

## Exercise 38C

Solve these equations. Give your solutions correct to 3 significant figures.

| | | |
|---|---|---|
| **1** $x^2 + 6x + 4 = 0$ | **2** $x^2 + 8x + 5 = 0$ | **3** $x^2 + 10x + 6 = 0$ |
| **4** $x^2 + 6x - 14 = 0$ | **5** $x^2 + 8x - 4 = 0$ | **6** $x^2 - 4x - 6 = 0$ |
| **7** $x^2 - 5x - 9 = 0$ | **8** $x^2 + 4x - 7 = 0$ | **9** $x^2 - 3x - 5 = 0$ |
| **10** $2x^2 + 5x + 1 = 0$ | **11** $2x^2 + 6x + 3 = 0$ | **12** $2x^2 + 9x + 8 = 0$ |
| **13** $2x^2 + 7x - 5 = 0$ | **14** $5x^2 - 10x + 3 = 0$ | **15** $2x^2 - 6x - 3 = 0$ |
| **16** $3x^2 + 9x - 5 = 0$ | **17** $4x^2 + 7x + 1 = 0$ | **18** $3x^2 + 7x - 1 = 0$ |
| **19** $x(x + 1) - 5 = 0$ | **20** $x(x + 3) - 20 = 0$ | **21** $x(x - 4) = 16$ |
| **22** $x^2 + 5x = 8$ | **23** $x^2 - 5x = 18$ | **24** $2x^2 + 6x = 7$ |
| **25** $x^2 + 8 = 10x$ | **26** $x + 1 = x^2$ | **27** $8x - 2x^2 = 5$ |

## 38.5 Solving equations with algebraic fractions

Equations with algebraic fractions sometimes lead to quadratic equations, which can then be solved using one of the methods from the earlier sections in this chapter.

### Example 8

Solve $\dfrac{4}{3x-1} + \dfrac{2}{x+1} = 3$

### Solution 8

$$(3x-1)(x+1)\frac{4}{(3x-1)} + (3x-1)(x+1)\frac{2}{(x+1)} = 3(3x-1)(x+1)$$

Multiply both sides by $(3x-1)(x+1)$

$$\cancel{(3x-1)}(x+1)\frac{4}{\cancel{(3x-1)}} + (3x-1)\cancel{(x+1)}\frac{2}{\cancel{(x+1)}} = 3(3x-1)(x+1)$$

Cancel on the left-hand side.

$$4(x+1) + 2(3x-1) = 3(3x-1)(x+1)$$

$$4x + 4 + 6x - 2 = 3(3x^2 + 2x - 1)$$

Expand the brackets.

$$10x + 2 = 9x^2 + 6x - 3$$

Simplify both sides.

$$9x^2 - 4x - 5 = 0$$

$$(9x+5)(x-1) = 0$$

Rearrange and rewrite the equation in the form $ax^2 + bx + c = 0$
Solve the quadratic equation by factorisation.

$$x = -\frac{5}{9} \text{ or } x = 1$$

In Example 8, each side was multiplied by the product of the denominators. It is not necessary to do this when the denominators have a common factor. In general, multiply both sides by the simplest common denominator.

### Example 9

Solve $\dfrac{4}{x-1} - \dfrac{2}{x+1} = \dfrac{30}{(x+2)(x+1)}$

The simplest common denominator is $(x-1)(x+1)(x+2)$
Multiply both sides by $(x-1)(x+1)(x+2)$

### Solution 9

$$(x-1)(x+1)(x+2)\frac{4}{(x-1)} - (x-1)(x+1)(x+2)\frac{2}{(x+1)} = (x-1)(x+1)(x+2)\frac{30}{(x+2)(x+1)}$$

Cancel on both sides.

$$\cancel{(x-1)}(x+1)(x+2)\frac{4}{\cancel{(x-1)}} - (x-1)\cancel{(x+1)}(x+2)\frac{2}{\cancel{(x+1)}} = (x-1)\cancel{(x+1)}\cancel{(x+2)}\frac{30}{\cancel{(x+2)}\cancel{(x+1)}}$$

$$4(x+1)(x+2) - 2(x-1)(x+2) = 30(x-1)$$

$$4(x^2 + 3x + 2) - 2(x^2 + x - 2) = 30(x-1)$$

$$4x^2 + 12x + 8 - 2x^2 - 2x + 4 = 30x - 30$$

Expand the brackets.

$$2x^2 - 20x + 42 = 0$$

Rewrite the equation in the form $ax^2 + bx + c = 0$

$$x^2 - 10x + 21 = 0$$

Divide both sides by 2

$$(x-7)(x-3) = 0$$

Solve the quadratic equation by factorisation.

$$x = 7 \text{ or } x = 3$$

**Exercise 38D**

**1** Solve these equations.

**a** $\dfrac{3}{x} - \dfrac{4}{x+1} = 1$     **b** $\dfrac{3}{x} + \dfrac{4}{x+2} = 1$     **c** $\dfrac{3}{x+1} + \dfrac{4}{x+2} = 2$

**d** $\dfrac{6}{x} - \dfrac{5}{x+2} = 1$     **e** $\dfrac{3}{2x+1} - \dfrac{2}{x+1} = 1$     **f** $\dfrac{4}{x+2} + \dfrac{3}{2x-1} = 2$

**g** $\dfrac{5}{x} + \dfrac{6}{x+1} = 2$     **h** $\dfrac{6}{x} - \dfrac{8}{x+2} = 1$     **i** $\dfrac{1}{x} + \dfrac{4}{x+3} = 1$

**2** Solve these equations. Give your solutions correct to 3 significant figures.

**a** $\dfrac{5}{x+2} + \dfrac{2}{x-1} = 1$     **b** $\dfrac{3}{x} + \dfrac{4}{x+1} = 1$     **c** $\dfrac{1}{x} + \dfrac{2}{2x+1} = 1$

**d** $3x - \dfrac{4}{x+2} = 1$     **e** $\dfrac{3}{2x} - \dfrac{4}{x+1} = 4$     **f** $\dfrac{3}{x+2} - \dfrac{4}{x-2} = 1$

**3** Solve these equations.

**a** $\dfrac{1}{x} + \dfrac{3}{x-1} = \dfrac{7}{2}$     **b** $\dfrac{6}{x+1} + \dfrac{6}{x-1} = 2\dfrac{1}{2}$     **c** $\dfrac{9}{x+3} + \dfrac{4}{x} = \dfrac{1}{6}$

**d** $\dfrac{4}{x+3} - \dfrac{1}{x+2} = \dfrac{1}{2x+3}$     **e** $\dfrac{3}{x+2} + \dfrac{1}{x-1} = \dfrac{1}{(x+1)(x+2)}$

## 38.6 Problems involving quadratic equations

### Example 10

The sum of the squares of two consecutive integers is 145 Find the two integers.

*Solution 10*

Let $x$ be the smaller of the two integers. | Define any letters you introduce.

Then the other integer is $x + 1$ | $x$ and $x + 1$ are consecutive integers.

$x^2 + (x+1)^2 = 145$
$x^2 + x^2 + 2x + 1 = 145$
$2x^2 + 2x - 144 = 0$

Write down an expression for the sum of the squares of the two integers. Write this equal to 145 to form a quadratic equation.

Rewrite the equation in the form $ax^2 + bx + c = 0$

$x^2 + x - 72 = 0$ | Divide both sides by 2

$(x+9)(x-8) = 0$ | Solve the quadratic equation by factorisation.
$x = -9$ or $x = 8$

when $x = -9$, $x + 1 = -8$ and when $x = 8$, $x + 1 = 9$
The two integers are $-9$ and $-8$ or 8 and 9

### Example 11

The diagram shows a rectangle.
All the measurements are in centimetres.
The area of the rectangle is 60 cm$^2$.
**a** Show that $2x^2 - 5x - 63 = 0$
**b** Solve the equation to find the shortest side
of the rectangle.

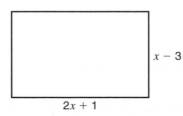

$x - 3$

$2x + 1$

## Solution 11

**a**    $(2x + 1)(x - 3) = 60$

length × width = area of a rectangle

$2x^2 - 6x + x - 3 = 60$

Expand the brackets.

$2x^2 - 5x - 3 = 60$

$2x^2 - 5x - 63 = 0$

Collect terms and obtain the given equation.

**b**   $2x^2 - 5x - 63 = 0$

$(2x + 9)(x - 7) = 0$

Solve the quadratic equation by factorisation.

$x = -4.5$ or $x = 7$

Reject $x = -4.5$, as this gives a negative value for both $x - 3$ and $2x + 1$. (Lengths of the sides of a rectangle cannot be negative.)

When $x = 7$

$x - 3 = 4$

and   $2x + 1 = 15$

Length of the shortest side $= 4$ cm

Some problems involve equations with algebraic fractions which lead to quadratic equations.

## Example 12

Trevor walks a distance of 20 km at an average speed of $x$ km/h.
Tony walks the same distance at an average speed 1 km/h faster than Trevor's average speed.
Tony takes 1 hour less than Trevor.

**a**  Show that $\dfrac{20}{x} - \dfrac{20}{x + 1} = 1$

**b**  Solve the equation to find Trevor's average speed.

## Solution 12

**a**  Tony's average speed is $(x + 1)$ km/h

Trevor's time $= \dfrac{20}{x}$ hours

Tony's time $= \dfrac{20}{x + 1}$ hours

time $= \dfrac{\text{distance}}{\text{average speed}}$

$\dfrac{20}{x} - \dfrac{20}{x + 1} = 1$

Tony's time is 1 hour less than Trevor's. This gives the required equation.

**b**  $20(x + 1) - 20x = x(x + 1)$

Multiply both sides by $x(x + 1)$ and cancel.

$20x + 20 - 20x = x^2 + x$

Expand the brackets and simplify.

$20 = x^2 + x$

$x^2 + x - 20 = 0$

$(x - 4)(x + 5) = 0$

Rearrange and solve by factorising.

$x = 4$ or $x = -5$

Trevor's average speed $= 4$ km/h

Reject the negative value of $x$, as speed cannot be negative.

### Exercise 38E

In **Questions 1–6**, form a suitable quadratic equation and then solve it.

1  The sum of the squares of two consecutive integers is 265. Find the two integers.

2  The product of two numbers is 117   One number is 4 more than the other. Find the two numbers.

3  In a rectangle the length is 3 cm longer than the width. The area of the rectangle is 208 cm². Find the length.

4  The length of the hypotenuse of a right-angled triangle is 17 cm. The base of the triangle is 7 cm more than the height. Find the height of the triangle.

5  The area of a right-angled triangle is 48 cm². The base of the triangle is 4 cm more than the height. Find the base and height of the triangle.

6  The product of the three numbers 4, $x$ and $(x + 4)$ is 33   Find the possible values of $x$.

7  The diagram shows a trapezium. All the measurements are in centimetres.

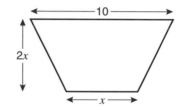

   **a**  Write down an expression, in terms of $x$, for the area of the trapezium.

  The area of the trapezium is 75 cm².

   **b**  Show that $x^2 + 10x - 75 = 0$

   **c**  Solve the equation $x^2 + 10x - 75 = 0$

   **d**  Write down the height of the trapezium.

8  The diagram shows a shape. All the corners are right angles. All the measurements are in cm.

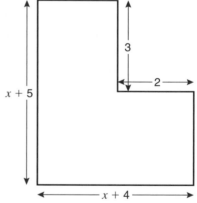

   **a**  Explain why $(x + 4)(x + 5) - 6$ is an expression for the area, in cm², of the shape.

  The area of the shape is 104 cm².

   **b**  Show that $x^2 + 9x - 90 = 0$

   **c**  Solve the equation $x^2 + 9x - 90 = 0$

   **d**  Write down the length of the longest side of the shape.

9  The diagram shows a path around a rectangular lawn. All the measurements shown are in metres.
The width of the path is 1 m.
The area of the lawn is 220 m².

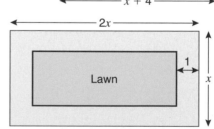

   **a**  Show that $x^2 - 3x - 108 = 0$

   **b**  Solve this equation.

   **c**  Find the length of the longest side of the path.

10  The diagram shows a trapezium.
All the measurements of the trapezium are in centimetres. The height is $x$ cm.
The area of the trapezium is 27 cm².

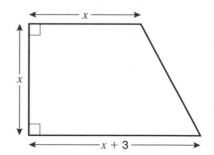

   **a**  Show that $2x^2 + 3x - 54 = 0$

   **b**  Solve this equation.

   **c**  Write down the height of the trapezium.

**11** The sum of a number and its reciprocal is 2.9 Find all possible values of the number.

**12** John walks a distance of 15 km from his home to the town at an average speed of $x$ km/h. He walks back from town to his home at an average speed 2 km/h faster than before. The total time John takes to walk to town and back is 8 hours.

**a** Show that $\dfrac{15}{x} + \dfrac{15}{x+2} = 8$

**b** Show that $\dfrac{15}{x} + \dfrac{15}{x+2} = 8$ can be written as $4x^2 - 7x - 15 = 0$

**c** Solve $4x^2 - 7x - 15 = 0$

**d** Find John's average speed when he walks back from town to his home.

**13** Mijan has £30 to spend on pens. The normal price of a pen is £$x$. In a sale, the price of the pen is reduced by £1 If he spends the £30 on pens at the sale price, he gets 8 more pens than if he spends £30 on pens at the normal price.

**a** Explain why $\dfrac{30}{x-1} - \dfrac{30}{x} = 8$

**b** Show that $x$ satisfies $4x^2 - 4x - 15 = 0$

**c** Solve the equation.

**d** Find the sale price of a pen.

## Chapter summary

> **You should now be able to:**
> ★ solve quadratic equations using where appropriate
>
> - **factorisation**
> - **completing the square**
> - **the quadratic formula** $x = \dfrac{-b \pm \sqrt{b^2 - 4ac}}{2a}$
>   which gives the solutions of $ax^2 + bx + c = 0$ where $a \neq 0$
>
> ★ solve equations with algebraic fractions which lead to quadratic equations
> ★ solve problems involving quadratic equations.

## Chapter 38 review questions

**1** Solve $x^2 - 6 = 30$

**2 a** Factorise $x^2 - 6x - 7$ **b** Solve $x^2 - 6x - 7 = 0$

**3 a** Factorise $x^2 - 7x + 12$ **b** Solve the equation $x^2 - 7x + 12 = 0$ (1387 November 2003)

**4 a** Factorise **i** $2x^2 - 7x + 3$ **ii** $4x^2 - 9$
 **b** Solve **i** $(3x+2)(2x+1) = 0$ **ii** $(3x+2)(2x+1) = 1$ (1385 November 2001)

**5** Solve the equation $(2x-3)^2 = 100$ (1385 November 2000)

**6 a** Factorise $3x^2 - 4x - 4$ **b** Solve $3x^2 - 4x - 4 = 0$

**7 a** Factorise $2x^2 - 35x + 98$ **b** Solve the equation $2x^2 - 35x + 98 = 0$ (1387 June 2004)

**8** Solve $6x^2 - 5x - 6 = 0$

**9 a** Write $x^2 + 6x + 3$ in the form $(x + p)^2 + q$.
   **b** Use your answer to part **a** to solve $x^2 + 6x + 3 = 0$
   Give your answers in the form $c \pm \sqrt{6}$ where $c$ is an integer.

**10** Solve $x^2 + 6x = 4$
   Give your answers in the form $p \pm \sqrt{q}$, where $p$ and $q$ are integers. (1388 January 2005)

**11** Find the solutions of the equation $x^2 - 4x - 1 = 0$
   Give your solutions correct to 3 decimal places. (1385 November 2000)

**12** Solve $x^2 + x + 11 = 14$
   Give your solutions correct to 3 significant figures. (1387 November 2004)

**13** Solve the equation $2x^2 - 7x + 4 = 0$
   Give your answers correct to 2 decimal places. (1385 November 2002)

**14 a** Show that the equation $\dfrac{2}{(x + 1)} - \dfrac{1}{(x + 2)} - \dfrac{1}{2}$ can be written in the form $x^2 + x - 4 = 0$

   **b** Hence, or otherwise, find the values of $x$ correct to 2 decimal places, that satisfy the equation
   $$\frac{2}{(x + 1)} - \frac{1}{(x + 2)} = \frac{1}{2}$$
   (1385 June 1999)

**15** Solve the equation $\dfrac{7}{x + 2} + \dfrac{1}{x - 1} = 4$

**16** The diagram shows a 6-sided shape. All the corners are right angles.
   All measurements are given in centimetres.
   The area of the shape is 25 cm².
   **a** Show that $6x^2 + 17x - 39 = 0$
   **b i** Solve the equation $6x^2 + 17x - 39 = 0$
   **ii** Hence work out the length of the longest side of the shape.

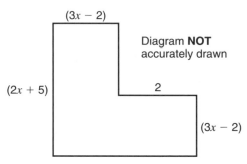

Diagram **NOT** accurately drawn

(1387 June 2005)

**17** The diagram shows a trapezium.
   The measurements on the diagram are in centimetres.
   The lengths of the parallel sides are $x$ cm and 20 cm.
   The height of the trapezium is $2x$ cm.
   The area of the trapezium is 400 cm².
   **a** Show that $x^2 + 20x = 400$
   **b** Find the value of $x$. Give your answer correct to 3 decimal places.

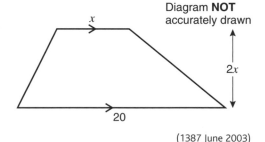

Diagram **NOT** accurately drawn

(1387 June 2003)

**18** $AT$ is a tangent at $T$ to a circle, centre $O$.
   $OT = x$ cm, $AT = (x + 5)$ cm, $OA = (x + 8)$ cm.
   **a** Show that $x^2 - 6x - 39 = 0$
   **b** Solve the equation $x^2 - 6x - 39 = 0$ to find the radius of the circle.
   Give your answer correct to 3 significant figures.

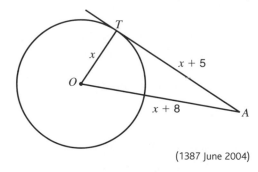

(1387 June 2004)

# Pythagoras' theorem and trigonometry (2)

In Chapter 29, Pythagoras' theorem and trigonometry were used to find the lengths of sides and the sizes of angles in right-angled triangles. These methods will now be used with three-dimensional shapes.

## 39.1 Problems in three dimensions

In a cuboid all the edges are perpendicular to each other.
Problems with cuboids and other 3-D shapes involve identifying suitable right-angled triangles and applying Pythagoras' theorem and trigonometry to them.

### Example 1

*ABCDEFGH* is a cuboid with length 8 cm, breadth 6 cm and height 9 cm.

**a**  **i** Calculate the length of *AC*.

   **ii** Calculate the length of *AG*.
       Give your answer correct to 3 significant figures.

**b** Calculate the size of angle *GAC*.
   Give your answer correct to the nearest degree.

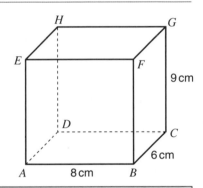

### Solution 1

**a**  **i**

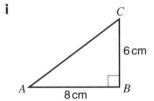

| | |
|---|---|
| Look for a right-angled triangle where *AC* is one side and the lengths of the other two sides are known. | |

*ABC* is a suitable triangle.

So draw triangle *ABC* marking the known lengths.

$AC^2 = AB^2 + BC^2$

     $= 8^2 + 6^2$

     $= 64 + 36$

     $= 100$

$AC\ = 10$ cm

> Use Pythagoras' theorem for this triangle.

> Look for a right-angled triangle where *AG* is one side and the lengths of the other two sides are known.
>
> *ACG* is a suitable triangle.
>
> So draw triangle *ACG* marking the known lengths.

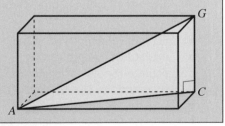

**ii**

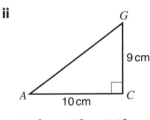

$AG^2 = AC^2 + CG^2$

$AG^2 = 10^2 + 9^2$

     $= 100 + 81$

     $= 181$

$AG\ = \sqrt{181} = 13.4536\ \ldots$

$AG\ = 13.5$ cm (to 3 s.f.)

> Use Pythagoras' theorem for this triangle.

**b**

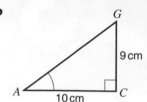

For angle *GAC*.
9 cm is the opposite side.
10 cm is the adjacent side.

tan (angle *GAC*) = $\frac{9}{10}$ = 0.9
angle *GAC* = 41.987 ...°
Angle *GAC* = 42°
(to the nearest degree)

$\tan = \dfrac{\text{opp}}{\text{adj}}$

## Exercise 39A

Where necessary give lengths correct to 3 significant figures and angles correct to one decimal place.

**1** *ABCDEFGH* is a cuboid of length 8 cm, breadth 4 cm and height 13 cm.

   **a** Calculate the length of
     **i** *AC*         **ii** *GB*
     **iii** *FA*      **iv** *GA*.

   **b** Calculate the size of
     **i** angle *FAB*     **ii** angle *GBC*     **iii** angle *GAC*.

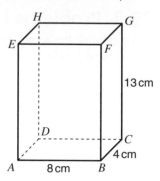

**2** *ABCDEF* is a triangular prism.
In triangle *ABC* angle *CAB* = 90°,
*AB* = 5 cm and *AC* = 12 cm.
In rectangle *ABED* the length
of *BE* = 15 cm.

   **a** Calculate the length of *CB*.

   **b** Calculate the length of
     **i** *CE*        **ii** *AF*.

   **c** Calculate the size of
     **i** angle *FED*     **ii** angle *FAD*.

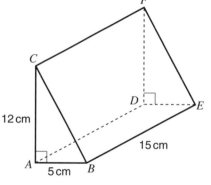

**3** The diagram shows a square-based pyramid.
The lengths of sides of the square base, *ABCD*, are
10 cm and the base is on a horizontal plane.
The centre of the base is the point *M* and the vertex of
the pyramid is *O*, so that *OM* is vertical.
The point *E* is the midpoint of the side *AB*.
*OA* = *OB* = *OC* = *OD* = 15 cm.

   **a** Calculate the length of   **i** *AC*   **ii** *AM*.
   **b** Calculate the length of *OM*.
   **c** Calculate the size of angle *OAM*.
   **d** Hence find the size of angle *AOC*.
   **e** Calculate the length of *OE*.
   **f** Calculate the size of angle *OAB*.

## Angle between a line and a plane

Imagine a light shining directly above *AB* onto the plane.
*AN* is the shadow of *AB* on the plane.
A line drawn from point *B* perpendicular to the plane will meet
the line *AN* and form a right angle with this line.
Angle *BAN* is the angle between the line *AB* and the plane.

## Example 2

The diagram shows a pyramid.
The base, $ABCD$, is a horizontal rectangle in which $AB = 12$ cm and $AD = 9$ cm.
The vertex, $O$, is vertically above the midpoint of the base and $OB = 18$ cm.
Calculate the size of the angle that $OB$ makes with the horizontal plane.
Give your answer correct to one decimal place.

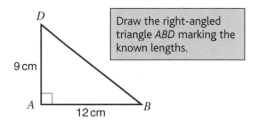

## Solution 2

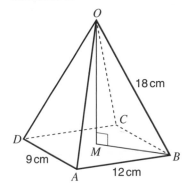

The base, $ABCD$, of the pyramid is horizontal so the angle that $OB$ makes with the horizontal plane is the angle that $OB$ makes with the base $ABCD$.

Let $M$ be the midpoint of the base which is directly below $O$.

Join $O$ to $M$ and $M$ to $B$.

As $OM$ is perpendicular to the base of the pyramid the angle $OBM$ is the angle between $OB$ and the base and so is the required angle.

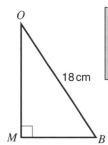

Draw triangle $OBM$ marking $OB = 18$ cm.

To find the size of angle $OBM$ find the length of either $MB$ or $OM$.

Calculate the length of $MB$ which is $\frac{1}{2}DB$.

Draw the right-angled triangle $ABD$ marking the known lengths.

$$DB^2 = 9^2 + 12^2 = 81 + 144$$

$$DB^2 = 225$$

$$DB = \sqrt{225} = 15$$

$$MB = \tfrac{1}{2}DB = 7.5$$

Use Pythagoras' theorem to calculate the length of $DB$.

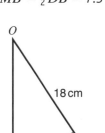

For angle $OBM$, 18 cm is the hypotenuse, 7.5 cm is the adjacent side.

$$\cos(\text{angle } OBM) = \frac{7.5}{18}$$

$$\cos = \frac{\text{adj}}{\text{hyp}}$$

$$\text{angle } OBM = 65.37 \ldots°$$

The angle between $OB$ and the
horizontal plane is 65.4° (to one d.p.)

## Exercise 39B

Where necessary give lengths correct to 3 significant figures and angles correct to one decimal place.

**1** The diagram shows a pyramid.
The base, $ABCD$, is a horizontal rectangle in which
$AB = 15$ cm and $AD = 8$ cm.
The vertex, $O$, is vertically above the centre of the
base and $OA = 24$ cm.
Calculate the size of the angle that $OA$ makes with
the horizontal plane.

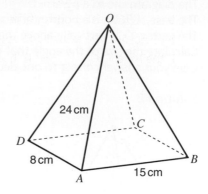

**2** $ABCDEFGH$ is a cuboid with a rectangular base in which
$AB = 12$ cm and $BC = 5$ cm.
The height, $AE$, of the cuboid is 15 cm.
Calculate the size of the angle

**a** between $FA$ and $ABCD$

**b** between $GA$ and $ABCD$

**c** between $BE$ and $ADHE$

**d** Write down the size of the angle between $HE$ and $ABFE$.

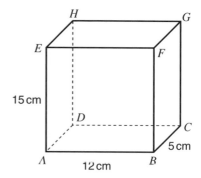

**3** $ABCDEF$ is a triangular prism.
In triangle $ABC$, angle $CAB = 90°$, $AB = 8$ cm and $AC = 10$ cm.
In rectangle $ABED$, the length of $BE = 5$ cm.
Calculate the size of the angle between

**a** $CB$ and $ABED$

**b** $CD$ and $ABED$

**c** $CE$ and $ABED$

**d** $BC$ and $ADFC$.

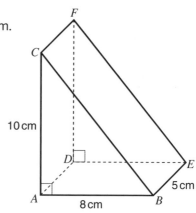

**4** The diagram shows a square-based pyramid.
The lengths of sides of the square base, $ABCD$,
are 8 cm and the base is on a horizontal plane.
The centre of the base is the point $M$ and the vertex
of the pyramid is $O$ so that $OM$ is vertical.
The point $E$ is the midpoint of the side $AB$.
$OA = OB = OC = OD = 20$ cm
Calculate the size of the angle between $OE$
and the base $ABCD$.

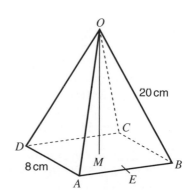

**5** $ABCD$ is a horizontal rectangular lawn in a garden and $TC$ is a vertical pole. Ropes run from the top of the pole, $T$, to the corners, $A$, $B$ and $D$, of the lawn.

  **a** Calculate the length of the rope $TA$.

  **b** Calculate the size of the angle made with the lawn by

  **i** the rope $TB$  **ii** the rope $TD$  **iii** the rope $TA$.

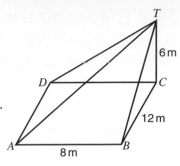

**6** The diagram shows a learner's ski slope, $ABCD$, of length, $AB$, 500 m. Triangles $BAF$ and $CDE$ are congruent right-angled triangles and $ABCD$, $AFED$ and $BCEF$ are rectangles. The rectangle $BCEF$ is horizontal and the rectangle $AFED$ is vertical. The angle between $AB$ and $BCEF$ is 20° and the angle between $AC$ and $BCEF$ is 10°.

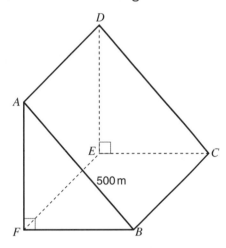

Calculate

  **a** the length of $FB$  **b** the height of $A$ above $F$

  **c** the distance $AC$  **d** the width, $BC$, of the ski slope.

**7** Diagram 1 shows a square-based pyramid $OABCD$. Each side of the square is of length 60 cm and $OA = OB = OC = OD = 50$ cm.

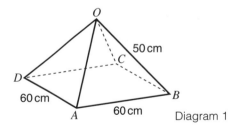

Diagram 1

Diagram 2 shows a cube, $ABCDEFGH$, in which each edge is of length 60 cm.

A solid is made by placing the pyramid on top of the cube so that the base, $ABCD$, of the pyramid is on the top, $ABCD$, of the cube. The solid is placed on a horizontal table with the face, $EFGH$, on the table.

  **a** Calculate the height of the vertex $O$ above the table.

  **b** Calculate the size of the angle between $OE$ and the horizontal.

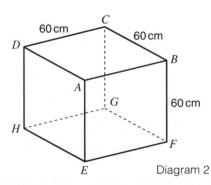

Diagram 2

## 39.2 Trigonometric ratios for any angle

The diagram shows a circle, centre the origin $O$ and radius 1 unit. Imagine a line, $OP$, of length 1 unit fixed at $O$, rotating in an **anticlockwise** direction about $O$, starting from the $x$-axis.

The diagram shows $OP$ when it has rotated through $40°$.

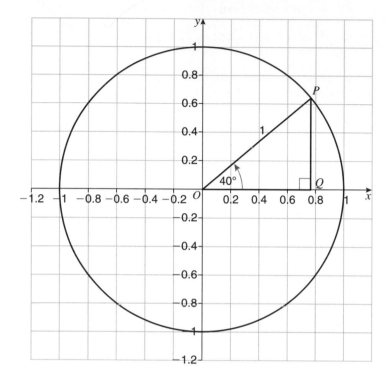

The right-angled triangle $OPQ$ has hypotenuse $OP = 1$

Relative to angle $POQ$, side $PQ$ is the opposite side and side $OQ$ is the adjacent side. This means that

$$OQ = \cos 40° \text{ and } PQ = \sin 40°$$

For $P$, $x = \cos 40°$ and $y = \sin 40°$ so the coordinates of $P$ are $(\cos 40°, \sin 40°)$.

In general when $OP$ rotates through any angle $\theta°$, the position of $P$ on the circle, radius $= 1$ is given by $x = \cos \theta°$, $y = \sin \theta°$.

The coordinates of $P$ are $(\cos \theta°, \sin \theta°)$.

So when $OP$ rotates through $400°$ the coordinates of $P$ are $(\cos 400°, \sin 400°)$.
A rotation of $400°$ is 1 complete revolution of $360°$ plus a further rotation of $40°$.

The position of $P$ is the same as in the previous diagram so $(\cos 400°, \sin 400°)$ is the same point as $(\cos 40°, \sin 40°)$, therefore $\cos 400° = \cos 40°$ and $\sin 400° = \sin 40°$.

If $OP$ rotates through $-40°$ this means $OP$ rotates through $40°$ in a **clockwise** direction.

For $\theta = 136$, $\theta = 225$, $\theta = 304$ and $\theta = -40$ the position of $P$ is shown on the diagram.

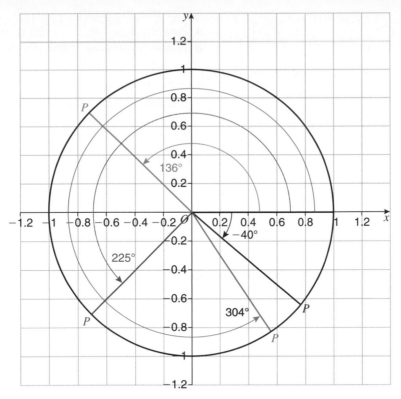

For $P$ when $\theta = 136$, $x = \cos 136°$ and $y = \sin 136°$.
From the diagram, $\cos 136° < 0$ and $\sin 136° > 0$

For $P$ when $\theta = 225$, $x = \cos 225°$ and $y = \sin 225°$.
From the diagram, $\cos 225° < 0$ and $\sin 225° < 0$

For $P$ when $\theta = 304$, $x = \cos 304°$ and $y = \sin 304°$.
From the diagram, $\cos 304° > 0$ and $\sin 304° < 0$

For $P$ when $\theta = -40$, $x = \cos -40°$ and $y = \sin -40°$.
From the diagram, $\cos -40° > 0$ and $\sin -40° < 0$

The diagram shows for each quadrant whether the sine and cosine of angles in that quadrant are positive or negative.

|  | 2nd | 1st |  |
|---|---|---|---|
| sin +<br>cos − |  |  | sin +<br>cos + |
|  | 3rd | 4th |  |
| sin −<br>cos − |  |  | sin −<br>cos + |

The sine and cosine of any angle can be found using your calculator. The following table shows some of these values corrected where necessary to 3 decimal places.

| $\theta$ | 0 | 30 | 40 | 45 | 60 | 90 | 136 | 180 | 225 | 270 | 304 | 360 |
|---|---|---|---|---|---|---|---|---|---|---|---|---|
| $\sin \theta°$ | 0 | 0.5 | 0.643 | 0.707 | 0.866 | 1 | 0.695 | 0 | −0.707 | −1 | −0.829 | 0 |
| $\cos \theta°$ | 1 | 0.866 | 0.766 | 0.707 | 0.5 | 0 | −0.719 | −1 | −0.707 | 0 | 0.559 | 1 |

Using these values and others from a calculator the graphs of $y = \sin \theta°$ and $y = \cos \theta°$ can be drawn. A graphical calculator would be useful here.

## Graph of $y = \sin \theta°$

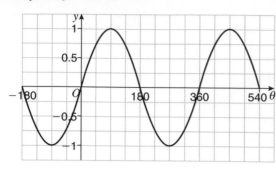

Notice that the graph:
- cuts the $\theta$-axis at ... , $-180, 0, 180, 360, 540, ...$
- repeats itself every 360°, that is, it has a **period** of 360°
- has a maximum value of 1 at $\theta = ...$ , 90, 450, ...
- has a minimum value of $-1$ at $\theta = ...$ , $-90, 270, ...$

## Graph of $y = \cos \theta°$

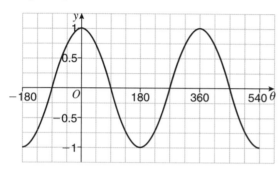

Notice that the graph:
- cuts the $\theta$-axis at ... $-90, 90, 270, 450, ...$
- repeats itself every 360°, that is it has a **period** of 360°
- has a maximum value of 1 at $\theta = ...$ , 0, 360, ...
- has a minimum value of $-1$ at $\theta = ...$ , $-180, 180, 540, ...$

Notice also that the graph of $y = \sin \theta°$ and the graph of $y = \cos \theta°$ are horizontal translations of each other.

To find the value of the tangent of any angle, use $\tan \theta° = \dfrac{\sin \theta°}{\cos \theta°}$

From the graph of $y = \cos \theta°$ it can be seen that $\cos \theta° = 0$ at $\theta = 90, 270, 450, ...$ for example. As it is not possible to divide by 0 there are no values of $\tan \theta°$ at $\theta = 90, 270, 450, ...$ that is, the graph is discontinuous at these values of $\theta$.

## Graph of $y = \tan \theta°$

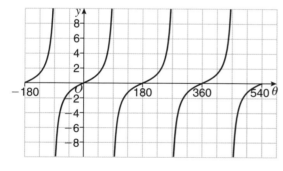

Notice that the graph:
- cuts the $\theta$-axis where $\tan \theta° = 0$, that is, at ... $-180, 0, 180, 360, 540 ...$
- repeats itself every 180°, that is it has a **period** of 180°
- does not have values at $\theta = \pm 90, \pm 270, \pm 450, ...$
- does not have any maximum or minimum points.

Notice also that $\tan \theta°$ can take any value.

## Example 3

For values of $\theta$ in the interval $-180$ to $360$ solve the equation

**i** $\sin \theta° = 0.7$

**ii** $5 \cos \theta° = 2$

Give each answer correct to one decimal place.

### Solution 3

**i** $\sin \theta° = 0.7$                    Use a calculator to find one value of $\theta$.

$\quad \theta = 44.4$

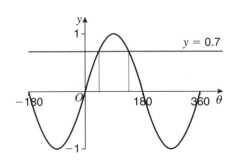

To find the other solutions draw a sketch of $y = \sin \theta°$ for $\theta$ from $-180$ to $360$

The sketch shows that there are two values of $\theta$ in the interval $-180$ to $360$ for which $\sin \theta° = 0.7$

One solution is $\theta = 44.4$ and by symmetry the other solution is $\theta = 180 - 44.4$

$\quad \theta = 44.4, 180 - 44.4$

$\quad \theta = 44.4, 135.6$

**ii** $5 \cos \theta° = 2$                    Divide each side of the equation by 5

$\quad \cos \theta° = \frac{2}{5} = 0.4$

$\quad \theta = 66.4$                         Use a calculator to find one value of $\theta$.

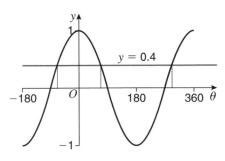

To find the other solutions draw a sketch of $y = \cos \theta°$ for $\theta$ from $-180$ to $360$

The sketch shows that there are three values of $\theta$ in the interval $-180$ to $360$ for which $\cos \theta° = 0.4$

One solution is $\theta = 66.4$ and by symmetry another solution is $\theta = -66.4$

Using the period of the graph the other solution is $\theta = 360 + -66.4$

$\quad \theta = 66.4, -66.4, 360 + -66.4$

$\quad \theta = 66.4, -66.4, 293.6$

## Exercise 39C

**1** For $-360 \leqslant \theta \leqslant 360$ sketch the graph of

    **a** $y = \sin \theta°$                 **b** $y = \cos \theta°$               **c** $y = \tan \theta°$.

**2** Find all values of $\theta$ in the interval 0 to 360 for which

    **a** $\sin \theta° = 0.5$           **b** $\cos \theta° = 0.1$          **c** $\tan \theta° = 1$

**3 a** Show that one solution of the equation $3 \sin \theta° = 1$ is 19.5, correct to 1 decimal place.

    **b** Hence solve the equation $3 \sin \theta° = 1$ for values of $\theta$ in the interval 0 to 720

**4 a** Show that one solution of the equation $10 \cos \theta° = -3$ is 107.5 correct to 1 decimal place.

  **b** Hence find all values of $\theta$ in the interval $-360$ to $360$ for which $10 \cos \theta° = -3$

**5** Solve $4 \tan \theta° = 3$ for values of $\theta$ in the interval $-180$ to $360$

## 39.3 Area of a triangle

### Labelling sides and angles

The vertices of a triangle are labelled with capital letters.
The triangle shown is triangle $ABC$.

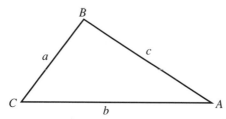

The sides opposite the angles are labelled so that $a$ is the length of the side opposite angle $A$, $b$ is the length of the side opposite angle $B$ and $c$ is the length of the side opposite angle $C$.

Area of a triangle $= \frac{1}{2}$ base $\times$ height

Area of triangle $ABC = \frac{1}{2} bh$

In the right-angled triangle $BCN$    $h = a \sin C$

So area of triangle $ABC = \frac{1}{2} b \times a \sin C$ that is

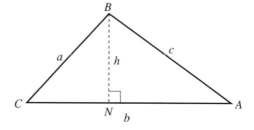

> **area of triangle $ABC = \frac{1}{2} ab \sin C$**

The angle $C$ is the angle between the sides of length $a$ and $b$ and is called the **included angle**. The formula for the area of a triangle means that

Area of a triangle $= \frac{1}{2}$ product of two sides $\times$ sine of the included angle.

For triangle $ABC$ there are other formulae for the area.

Area of triangle $ABC = \frac{1}{2} ab \sin C = \frac{1}{2} bc \sin A = \frac{1}{2} ac \sin B$.

These formulae give the area of a triangle whether the included angle is acute or obtuse.

---

### Example 4

Find the area of each of the triangles correct to 3 significant figures.

**a**

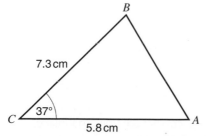

**b**

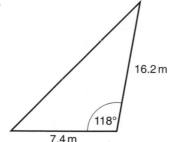

## Solution 4

**a** Area $= \frac{1}{2} \times 7.3 \times 5.8 \times \sin 37°$    Substitute $a = 7.3$ cm, $b = 5.8$ cm, $C = 37°$ into area $= \frac{1}{2} ab \sin C$

Area $= 12.74 ...$

Area $= 12.7$ cm$^2$    Give the area correct to 3 significant figures and state the units.

**b** Area $= \frac{1}{2} \times 7.4 \times 16.2 \times \sin 118°$    Substitute into area of a triangle $= \frac{1}{2}$ product of two sides $\times$ sine of the included angle.

Area $= 52.92 ...$

Area $= 52.9$ m$^2$

---

## Example 5

The area of this triangle is 20 cm$^2$.
Find the size of the acute angle $x°$.
Give your angle correct to one decimal place.

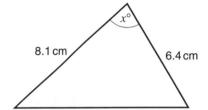

## Solution 5

$\frac{1}{2} \times 8.1 \times 6.4 \times \sin x° = 20$    Use area of a triangle $= \frac{1}{2}$ product of two sides $\times$ sine of the included angle.

$\sin x° = \dfrac{2 \times 20}{8.1 \times 6.4} = 0.7716$    Find the value of $\sin x°$.

$x° = 50.49 ...°$

$x° = 50.5°$    Give the angle correct to one decimal place.

---

## Exercise 39D

Give lengths and areas correct to 3 significant figures and angles correct to one decimal place.

**1** Work out the area of each of these triangles.

**i**

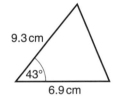

**ii**

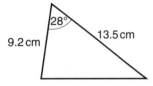

**iii**

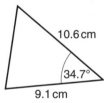

**iv**

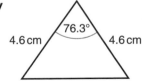

**v**

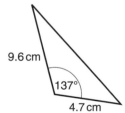

**vi**

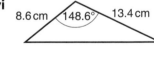

**2** *ABCD* is a quadrilateral.
Work out the area of the quadrilateral.

**3** The area of triangle $ABC$ is 15 cm²
Angle $A$ is acute.
Work out the size of angle $A$.

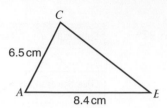

**4** The area of triangle $ABC$ is 60.7 m²
Work out the length of $BC$.

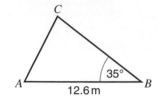

**5 a** Triangle $ABC$ is such that $a = 6$ cm, $b = 9$ cm and angle $C = 25°$.
Work out the area of triangle $ABC$.

  **b** Triangle $PQR$ is such that $p = 6$ cm, $q = 9$ cm and angle $R = 155°$.
Work out the area of triangle $PQR$.

  **c** What do you notice about your answers? Why do you think this is true?

**6** The diagram shows a regular octagon with centre $O$.

  **a** Work out the size of angle $AOB$.
$OA = OB = 6$ cm.

  **b** Work out the area of triangle $AOB$.

  **c** Hence work out the area of the octagon.

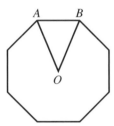

**7** Work out the area of the parallelogram.

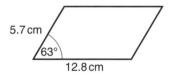

**8 a** An equilateral triangle has sides of length 12 cm.
Calculate the area of the equilateral triangle.

  **b** A regular hexagon has sides of length 12 cm.
Calculate the area of the regular hexagon.

**9** The diagram shows a sector, $AOB$, of a circle, centre $O$.
The radius of the circle is 8 cm and the size of angle $AOB$ is 50°.

  **a** Work out the area of triangle $AOB$.

  **b** Work out the area of the sector $AOB$.

  **c** Hence work out the area of the segment shown shaded in
the diagram.

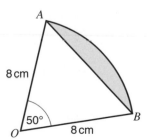

## 39.4 The sine rule

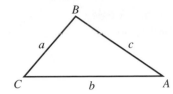

The last section showed that

Area of triangle $ABC = \frac{1}{2}ab\sin C = \frac{1}{2}bc\sin A = \frac{1}{2}ca\sin B$

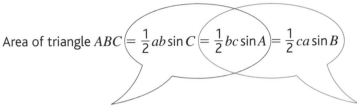

$\frac{1}{2}ab\sin C = \frac{1}{2}bc\sin A$     and    $\frac{1}{2}bc\sin A = \frac{1}{2}ca\sin B$

cancelling $\frac{1}{2}$ and $b$ from both sides     cancelling $\frac{1}{2}$ and $c$ from both sides

$a\sin C = c\sin A$     and    $b\sin A = a\sin B$

or                             or

$\dfrac{a}{\sin A} = \dfrac{c}{\sin C}$     and    $\dfrac{b}{\sin B} = \dfrac{a}{\sin A}$

Combining these results

$$\frac{a}{\sin A} = \frac{b}{\sin B} = \frac{c}{\sin C}$$

This result is known as the **sine rule** and can be used in *any* triangle.

### *Using the sine rule to calculate a length*

**Example 6**

Find the length of the side marked $a$ in the triangle.
Give your answer correct to 3 significant figures.

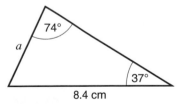

**Solution 6**

$\dfrac{a}{\sin 37°} = \dfrac{8.4}{\sin 74°}$

> Substitute $A = 37°$, $b = 8.4$, $B = 74°$ into $\dfrac{a}{\sin A} = \dfrac{b}{\sin B}$.

$a = \dfrac{8.4 \times \sin 37°}{\sin 74°}$

> Multiply both sides by $\sin 37°$.

$a = 5.258\ ...$

$a = 5.26$ cm

## Example 7

Find the length of the side marked $x$ in the triangle.
Give your answer correct to 3 significant figures.

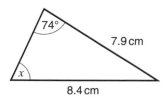

### Solution 7

Missing angle $= 180 - (18 + 124)$

$\qquad = 38°$

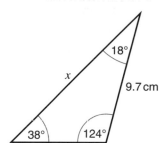

| The angle opposite 9.7 cm must be found before the sine rule can be used. Use the angle sum of a triangle. |

$$\frac{x}{\sin 124°} = \frac{9.7}{\sin 38°}$$

| Write down the sine rule with $x$ opposite 124° and 9.7 opposite 38°. |

$$x = \frac{9.7 \times \sin 124°}{\sin 38°}$$

| Multiply both sides by $\sin 124°$. |

$$x = 13.06 \ldots$$

$$x = 13.1 \text{ cm}$$

## Using the sine rule to calculate an angle

When the sine rule is used to calculate an angle it is a good idea to turn each fraction upside down (the reciprocal). This gives

$$\frac{\sin A}{a} = \frac{\sin B}{b} = \frac{\sin C}{c}$$

## Example 8

Find the size of the acute angle $x$ in the triangle.
Give your answer correct to one decimal place.

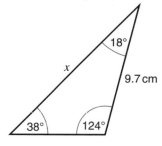

### Solution 8

$$\frac{\sin x}{7.9} = \frac{\sin 74°}{8.4}$$

| Write down the sine rule with $x$ opposite 7.9 and 74° opposite 8.4 |

$$\sin x = \frac{7.9 \times \sin 74°}{8.4}$$

| Multiply both sides by 7.9 |

$$\sin x = 0.904 \ldots$$

| Find the value of $\sin x$. |

$$x = 64.69 \ldots°$$

$$x = 64.7°$$

**Exercise 39E**

Give lengths and areas correct to 3 significant figures and angles correct to 1 decimal place.

**1** Find the lengths of the sides marked with letters in these triangles.

**a**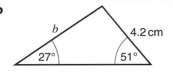

**b**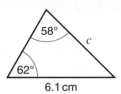

**c**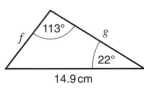

**d**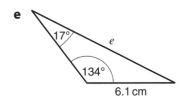

**e**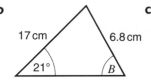

**f**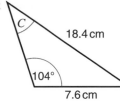

**2** Calculate the size of each of the *acute* angles marked with a letter.

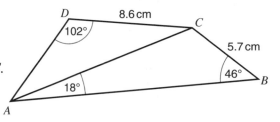  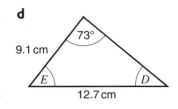

**a**  **b**  **c**  **d**

**3** The diagram shows quadrilateral *ABCD* and its diagonal *AC*.

  **a** In triangle *ABC*, work out the length of *AC*.

  **b** In triangle *ACD*, work out the size of angle *DAC*.

  **c** Work out the size of angle *BCD*.

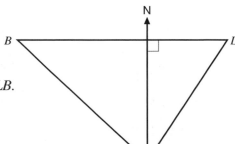

**4** In triangle *ABC*, *BC* = 8.6 cm, angle *BAC* = 52° and angle *ABC* = 63°.

  **a** Calculate the length of *AC*.

  **b** Calculate the length of *AB*.

  **c** Calculate the area of triangle *ABC*.

**5** In triangle *PQR* all the angles are acute. *PR* = 7.8 cm and *PQ* = 8.4 cm. Angle *PQR* = 58°.

  **a** Work out the size of angle *PRQ*.

  **b** Work out the length of *QR*.

**6** The diagram shows the position of a port (*P*), a lighthouse (*L*) and a buoy (*B*). The lighthouse is due east of the buoy. The lighthouse is on a bearing of 035° from the port and the buoy is on a bearing of 312° from the port.

  **a** Work out the size of   **i** angle *PBL*   **ii** angle *PLB*.

The lighthouse is 8 km from the port.

  **b** Work out the distance *PB*.

  **c** Work out the distance *BL*.

  **d** Work out the shortest distance from the port (*P*) to the line *BL*.

## 39.5 The cosine rule

The diagram shows triangle $ABC$.
The line $BN$ is perpendicular to $AC$ and meets the line $AC$ at $N$ so that $AN = x$ and $NC = (b - x)$.
The length of $BN$ is $h$.

In triangle $ANB$
Pythagoras' theorem
gives
$$c^2 = x^2 + h^2 \quad \text{(1)}$$

In triangle $BNC$
Pythagoras' theorem
gives
$$a^2 = (b - x)^2 + h^2$$
$$a^2 = b^2 - 2bx + x^2 + h^2$$
Using ① substitute $c^2$ for $x^2 + h^2$
$$a^2 = b^2 - 2bx + c^2 \quad \text{(2)}$$

In the right-angled triangle $ANB$, $x = c \cos A$

Substituting this into ②      $a^2 = b^2 + c^2 - 2bc \cos A$

This result is known as the **cosine rule** and can be used in *any* triangle.

Similarly                      $b^2 = a^2 + c^2 - 2ac \cos B$
and                          $c^2 = a^2 + b^2 - 2ab \cos C$

## Using the cosine rule to calculate a length

### Example 9

Find the length of the side marked with a letter in each triangle.
Give your answers correct to 3 significant figures.

**a**
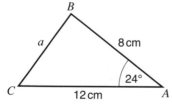

**b**
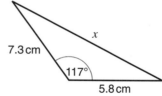

*Solution 9*

**a** $a^2 = 12^2 + 8^2 - 2 \times 12 \times 8 \times \cos 24°$

> Substitute $b = 12$, $c = 8$, $A = 24°$ into $a^2 = b^2 + c^2 - 2bc \cos A$.
> Evaluate each term separately.

$a^2 = 144 + 64 - 175.4007 \ldots$

$a^2 = 32.599\,27 \ldots$

$a = \sqrt{32.599\,27 \ldots}$

> Take the square root.

$a = 5.709\,577 \ldots$

$a = 5.71$ cm

**b** $x^2 = 7.3^2 + 5.8^2 - 2 \times 7.3 \times 5.8 \times \cos 117°$

> Substitute the two given lengths and the included angle into the cosine rule.

$x^2 = 53.29 + 33.64 - 84.68 \times (-0.4539 \ldots)$

> $\cos 117° < 0$

$x^2 = 86.93 + 38.44 \ldots$

$x^2 = 125.37 \ldots$

$x = \sqrt{125.37 \ldots}$

> Take the square root.

$x = 11.19 \ldots$

$x = 11.2$ cm

## Using the cosine rule to calculate an angle

To find an angle using the cosine rule, when the lengths of all three sides of a triangle are known, rearrange $a^2 = b^2 + c^2 - 2bc \cos A$.

$$2bc \cos A = b^2 + c^2 - a^2$$

$$\cos A = \frac{b^2 + c^2 - a^2}{2bc}$$

Similarly $\quad \cos B = \dfrac{a^2 + c^2 - b^2}{2ac}$

and $\quad \cos C = \dfrac{a^2 + b^2 - c^2}{2ab}$

---

### Example 10

Find the size of    **a** angle $BAC$    **b** angle $X$.
Give your answers correct to one decimal place.

**a**

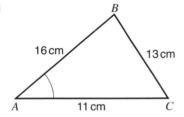

**b**

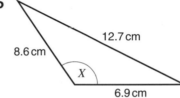

### Solution 10

**a** $\cos A = \dfrac{11^2 + 16^2 - 13^2}{2 \times 11 \times 16}$

    Substitute $b = 11$, $c = 16$, $a = 13$ into $\cos A = \dfrac{b^2 + c^2 - a^2}{2bc}$.

$\cos A = \dfrac{208}{352}$

$\cos A = 0.590\,909\ldots$

$\quad A = 53.77\ldots^\circ$

$\quad A = 53.8^\circ$

**b** $\cos X = \dfrac{8.6^2 + 6.9^2 - 12.7^2}{2 \times 8.6 \times 6.9}$

    Substitute the three lengths into the cosine rule noting that 12.7 cm is opposite the angle to be found.

$\cos X = \dfrac{-39.72}{118.68}$

$\cos X = -0.334\,68\ldots$

    The value of $\cos X$ is negative so $X$ is an obtuse angle.

$\quad X = 109.55\ldots^\circ$

$\quad X = 109.6^\circ$

---

## Exercise 39F

Where necessary give lengths and areas correct to 3 significant figures and angles correct to 1 decimal place.

**1** Calculate the length of the sides marked with letters in these triangles.

**a**
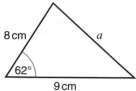
8 cm, a, 62°, 9 cm

**b**
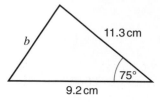
11.3 cm, b, 75°, 9.2 cm

**c**
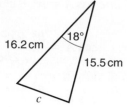
16.2 cm, 18°, 15.5 cm, c

**d**
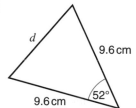
d, 9.6 cm, 52°, 9.6 cm

**e**
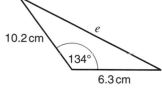
10.2 cm, e, 134°, 6.3 cm

**f**
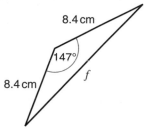
8.4 cm, 147°, 8.4 cm, f

**2** Calculate the size of each of the angles marked with a letter in these triangles.

**a**
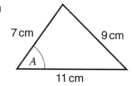
7 cm, 9 cm, A, 11 cm

**b**
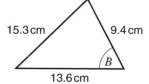
15.3 cm, 9.4 cm, B, 13.6 cm

**c**
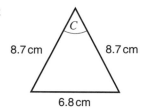
C, 8.7 cm, 8.7 cm, 6.8 cm

**d**
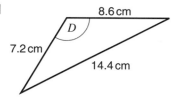
8.6 cm, D, 7.2 cm, 14.4 cm

**3** The diagram shows the quadrilateral $ABCD$.
  **a** Work out the length of $DB$.
  **b** Work out the size of angle $DAB$.
  **c** Work out the area of quadrilateral $ABCD$.

26.4 cm, C, 56°, D, 8.4 cm, 9.8 cm, A, 16.3 cm, B

**4** Work out the perimeter of triangle $PQR$.

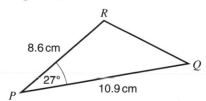
R, 8.6 cm, Q, 27°, 10.9 cm, P

**5** In triangle $ABC$, $AB = 10.1$ cm, $AC = 9.4$ cm and $BC = 8.7$ cm.
   Calculate the size of angle $BAC$.

**6** In triangle $XYZ$, $XY = 20.3$ cm, $XZ = 14.5$ cm and angle $YXZ = 38°$.
   Calculate the length of $YZ$.

**7** $AB$ is a chord of a circle with centre $O$.
The radius of the circle is 7 cm and the length of the chord is 11 cm.
Calculate the size of angle $AOB$.

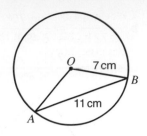

**8** The region $ABC$ is marked on a school field.
The point $B$ is 70 m from $A$ on a bearing of 064°.
The point $C$ is 90 m from $A$ on a bearing of 132°.

**a** Work out the size of angle $BAC$.

**b** Work out the length of $BC$.

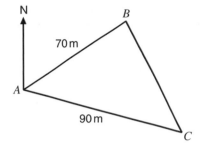

**9** Chris ran 4 km on a bearing of 036° from $P$ to $Q$. He then ran in a straight line from $Q$ to $R$
where $R$ is 7 km due East of $P$. Chris then ran in a straight line from $R$ to $P$.
Calculate the total distance run by Chris.

**10** The diagram shows a parallelogram.
Work out the length of each diagonal of the parallelogram.

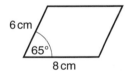

## 39.6  Solving problems using the sine rule, the cosine rule and $\frac{1}{2}ab \sin C$

The area of triangle $ABC$ is 12 cm²
$AB = 3.8$ cm and angle $ABC = 70°$.

**a** Find the length of    **i** $BC$    **ii** $AC$.
Give your answers correct to 3 significant figures.

**b** Find the size of angle $BAC$.
Give your answer correct to 1 decimal place.

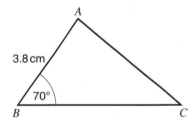

**Solution 11**

**a**  **i** $\frac{1}{2} \times 3.8 \times BC \sin 70° = 12$          Substitute $c = 3.8$, $B = 70°$ into area $= \frac{1}{2}ac \sin B$.

$$BC = \frac{2 \times 12}{3.8 \sin 70°}$$

$BC = 6.721 \dots$

$BC = 6.72$ cm

**ii** $b^2 = 6.721 \dots^2 + 3.8^2 - 2 \times 6.721\dots \times 3.8 \cos 70°$     Substitute $a = 6.721 \dots$, $c = 3.8$ and $B = 70°$
into $b^2 = a^2 + c^2 - 2ac \cos B$.

$b^2 = 59.613 \dots - 17.470 \dots$

$b^2 = 42.142 \dots$

$b = 6.491 \dots$

$AC = 6.49$ cm

**b** $\dfrac{\sin A}{6.721\ldots} = \dfrac{\sin 70°}{6.491\ldots}$

$\sin A = \dfrac{6.721\ldots \times \sin 70°}{6.491\ldots}$

$\sin A = 0.9728\ldots$

$A = 76.62\ldots°$

Angle $BAC = 76.6°$

> Substitute $a = 6.721\ldots$, $b = 6.491\ldots$ and $B = 70°$ into $\dfrac{\sin A}{a} = \dfrac{\sin B}{b}$.

### Exercise 39G

Where necessary give lengths and areas correct to 3 significant figures and angles correct to 1 decimal place, unless the question states otherwise.

**1** A triangle has sides of lengths 9 cm, 10 cm and 11 cm.
  **a** Calculate the size of each angle of the triangle.
  **b** Calculate the area of the triangle.

**2** In the diagram $ABC$ is a straight line.
  **a** Calculate the length of $BD$.
  **b** Calculate the size of angle $DAB$.
  **c** Calculate the length of $AC$.

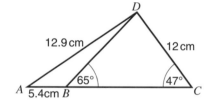

**3** The area of triangle $ABC$ is 15 cm². $AB = 4.6$ cm and angle $BAC = 63°$.
  **a** Work out the length of $AC$.
  **b** Work out the length of $BC$.
  **c** Work out the size of angle $ABC$.

**4** $ABCD$ is a kite with diagonal $DB$.
  **a** Calculate the length of $DB$.
  **b** Calculate the size of angle $BDC$.
  **c** Calculate the value of $x$.
  **d** Calculate the length of $AC$.

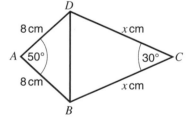

**5** Kultar walked 9 km due South from point $A$ to point $B$.
He then changed direction and walked 5 km to point $C$.
Kultar was then 6 km from his starting point $A$.
  **a** Work out the bearing of point $C$ from point $B$. Give your answer correct to the nearest degree.
  **b** Work out the bearing of point $C$ from point $A$. Give your answer correct to the nearest degree.

**6** The diagram shows a pyramid. The base of the pyramid, $ABCD$, is a rectangle in which $AB = 15$ cm and $AD = 8$ cm.
The vertex of the pyramid is $O$ where $OA = OB = OC = OD = 20$ cm.
Work out the size of angle $DOB$ correct to the nearest degree.

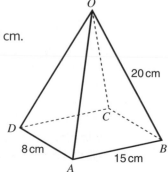

**7** The diagram shows a vertical pole, $PQ$, standing on a hill.
The hill is at an angle of 8° to the horizontal.
The point $R$ is 20 m downhill from $Q$ and the line $PR$ is at 12° to the hill.

  **a** Calculate the size of angle $RPQ$.

  **b** Calculate the length, $PQ$, of the pole.

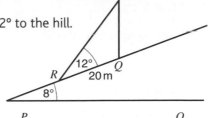

**8** $A$, $B$ and $C$ are points on horizontal ground so that
$AB = 30$ m, $BC = 24$ m and angle $CAB = 50°$.
$AP$ and $BQ$ are vertical posts, where $AP = BQ = 10$ m.

  **a** Work out the size of angle $ACB$.

  **b** Work out the length of $AC$.

  **c** Work out the size of angle $PCQ$.

  **d** Work out the size of the angle between
    $QC$ and the ground.

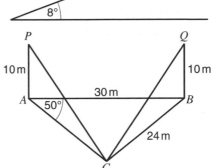

## Chapter summary

**You should now be able to:**

★ use Pythagoras' theorem to solve problems in 3 dimensions

★ use trigonometry to solve problems in 3 dimensions

★ work out the size of the angle between a line and a plane

★ draw sketches of the graphs of $y = \sin x°$, $y = \cos x°$, $y = \tan x°$ and use these graphs to solve simple trigonometric equations

★ use the formula area $= \frac{1}{2} ab \sin C$ to calculate the area of any triangle

★ use the sine rule $\dfrac{a}{\sin A} = \dfrac{b}{\sin B} = \dfrac{c}{\sin C}$ and the cosine rule $a^2 = b^2 + c^2 - 2bc \cos A$ in triangles and in solving problems.

## Chapter 39 review questions

**1** In the diagram, $XY$ represents a vertical tower
on level ground.
$A$ and $B$ are points due West of $Y$.
The distance $AB$ is 30 metres.
The angle of elevation of $X$ from $A$ is 30°.
The angle of elevation of $X$ from $B$ is 50°.
Calculate the height, in metres, of the tower $XY$.
Give your answer correct to 2 decimal places.

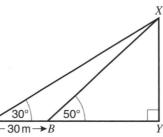

Diagram **NOT**
accurately drawn

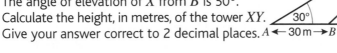

(1384 June 1996)

**2** The diagram shows triangle $ABC$.
$AC = 7.2$ cm    $BC = 8.35$ cm
Angle $ACB = 74°$.

  **a** Calculate the area of triangle $ABC$.
    Give your answer correct to 3 significant figures.

  **b** Calculate the length of $AB$.
    Give your answer correct to 3 significant figures.

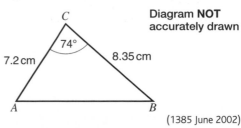

Diagram **NOT**
accurately drawn

(1385 June 2002)

**3** In triangle $ABC$
$AC = 8$ cm
$CB = 15$ cm
Angle $ACB = 70°$.

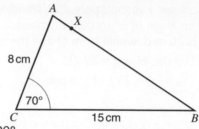

Diagram **NOT**
accurately drawn

**a** Calculate the area of triangle $ABC$.
Give your answer correct to
3 significant figures.

$X$ is the point on $AB$ such that angle $CXB = 90°$.

**b** Calculate the length of $CX$.
Give your answer correct to 3 significant figures.

(1387 June 2003)

**4** The diagram shows a cuboid.
$A, B, C, D$ and $E$ are five vertices of the cuboid.
$AB = 5$ cm
$BC = 8$ cm
$CE = 3$ cm.

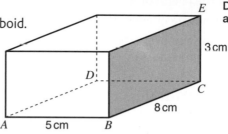

Diagram **NOT**
accurately drawn

Calculate the size of the angle the
diagonal $AE$ makes with the plane
$ABCD$.
Give your answer correct to 1 decimal place.

**5** In triangle $ABC$
$AC = 8$ cm      $BC = 15$ cm
Angle $ACB = 70°$.

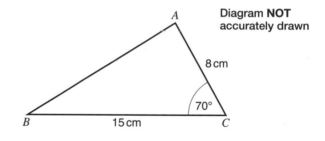

Diagram **NOT**
accurately drawn

**a** Calculate the length of $AB$.
Give your answer correct to
3 significant figures.

**b** Calculate the size of angle $BAC$.
Give your answer correct to 1 decimal
place.

(1387 June 2003)

**6** This is a sketch of the graph of $y = \cos x°$
for values of $x$ between 0 and 360.
Write down the coordinates of the point

**i** $A$

**ii** $B$.

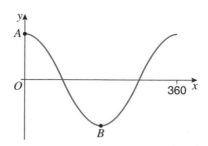

**7** Angle $ACB = 150°$
$BC = 60$ m.
The area of triangle $ABC$ is 450 m²
Calculate the perimeter of triangle $ABC$.
Give your answer correct to 3 significant figures.

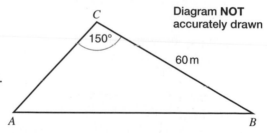

Diagram **NOT**
accurately drawn

(1385 November 2000)

**8** The diagram shows a quadrilateral $ABCD$.
$AB = 4.1$ cm
$BC = 7.6$ cm
$AD = 5.4$ cm
Angle $ABC = 117°$
Angle $ADC = 62°$.

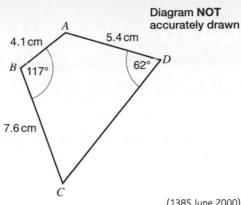

Diagram **NOT** accurately drawn

**a** Calculate the length of $AC$.
Give your answer correct to 3 significant figures.

**b** Calculate the area of triangle $ABC$.
Give your answer correct to 3 significant figures.

**c** Calculate the area of the quadrilateral $ABCD$.
Give your answer correct to 3 significant figures.

(1385 June 2000)

**9** This is a graph of the curve $y = \sin x°$ for $0 \leqslant x \leqslant 180$

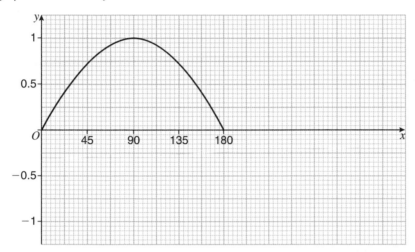

**a** Using the graph or otherwise, find estimates of the solutions in the interval $0 \leqslant x \leqslant 360$ of the equation
**i** $\sin x° = 0.2$     **ii** $\sin x° = -0.6$.

$\cos x° = \sin (x + 90)°$ for all values of $x$.

**b** Write down two solutions of the equation $\cos x° = 0.2$

(1385 November 2002)

**10** In the diagram, $ABCD$, $ABFE$ and $EFCD$ are rectangles.
The plane $EFCD$ is horizontal and the plane $ABFE$ is vertical.
$EA = 10$ cm
$DC = 20$ cm
$ED = 20$ cm.

Calculate the size of the angle that the line $AC$ makes with the plane $EFCD$.

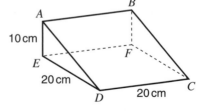

**11** In triangle $ABC$
$AB = 10$ cm
$AC = 14$ cm
$BC = 16$ cm.

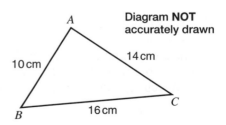

Diagram **NOT** accurately drawn

**a** Calculate the size of the smallest angle in the triangle.
Give your answer correct to the nearest $0.1°$.

**b** Calculate the area of triangle $ABC$.
Give your answer correct to 3 significant figures.

## 40.1 Solving simultaneous equations

Section 23.4 covered solving simultaneous equations where both equations are linear, that is, finding a pair of values of $x$ and $y$ which are solutions of both equations.

This section covers solving a pair of simultaneous equations where one is linear and one is quadratic.

Consider the equations $y = 2x - 1$ and $y = x^2 - 4$

> $y = 2x - 1$ is the linear equation, because it has the form $y = mx + c$ and
> $y = x^2 - 4$ is the quadratic equation, because it has the form $y = ax^2 + bx + c$

To solve these simultaneous equations substitute $2x - 1$ for $y$ in $y = x^2 - 4$

This gives $2x - 1 = x^2 - 4$

This quadratic equation in $x$ can be solved using one of the methods covered in Chapter 38.

Firstly, write $2x - 1 = x^2 - 4$ in the form $ax^2 + bx + c = 0$
$$x^2 - 2x - 3 = 0$$
$$(x - 3)(x + 1) = 0$$

so $x = 3$ or $x = -1$

To find the values of $y$, substitute into the linear equation $y = 2x - 1$
When $x = 3$, $y = 5$ and when $x = -1$, $y = -3$

Sometimes the linear equation must be rearranged to make $y$ the subject.

---

### Example 1

Solve the simultaneous equations $y = 2x^2$ and $y + 2x = 4$

**Solution 1**

| | |
|---|---|
| $y = 4 - 2x$ | Make $y$ the subject of $y + 2x = 4$ |
| $4 - 2x = 2x^2$ | Substitute $4 - 2x$ for $y$ in $y = 2x^2$ |
| $2x^2 + 2x - 4 = 0$ | Rearrange into the form $ax^2 + bx + c = 0$ |
| $x^2 + x - 2 = 0$ | Divide both sides by 2 |
| $(x + 2)(x - 1) = 0$ | Factorise. |
| $x = -2$ or $x = 1$ | |
| When $x = -2$, $y = 8$ | To find the values of $y$, substitute into $y = 4 - 2x$ |
| When $x = 1$, $y = 2$ | |

---

### Example 2

Solve $y = 4x^2$ and $3y + 2x = 4$

**Solution 2**

$y = \dfrac{4 - 2x}{3}$ | Make $y$ the subject of $3y + 2x = 4$

$\dfrac{4 - 2x}{3} = 4x^2$ | Substitute $\dfrac{4 - 2x}{3}$ for $y$ in $y = 4x^2$

$4 - 2x = 12x^2$ | Multiply both sides by 3

$12x^2 + 2x - 4 = 0$ | Rearrange.

$6x^2 + x - 2 = 0$ | Divide both sides by 2

$(3x + 2)(2x - 1) = 0$ | Factorise.

$x = -\frac{2}{3}$ or $x = \frac{1}{2}$

When $x = -\frac{2}{3}$, $y = 4 \times \left(-\frac{2}{3}\right)^2 = \frac{16}{9}$ | To find the values of $y$, substitute into $y = 4x^2$ as the working

When $x = \frac{1}{2}$, $y = 4 \times \left(\frac{1}{2}\right)^2 = 1$ | required is easier than substituting into $y = \dfrac{4 - 2x}{3}$

## Using simultaneous equations to find the intersection of a line and a quadratic curve

Just as the solution of two simultaneous linear equations represents the point of intersection of two straight lines, so the solution of a pair of simultaneous equations where one is linear and one is quadratic represents the points of intersection of a straight line and a quadratic curve.

The sketch shows the straight line $y = x + 2$ and the quadratic curve $y = x^2$. The line and the curve intersect at two points.
The coordinates of these points of intersection satisfy both equations $y = x + 2$ and $y = x^2$ simultaneously (at the same time) and can be found by solving the simultaneous equations $y = x + 2$ and $y = x^2$.

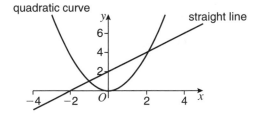

$$x^2 = x + 2$$
$$x^2 - x - 2 = 0$$
$$(x - 2)(x + 1) = 0$$
so $$x = 2 \text{ or } x = -1$$

When $x = 2$, $y = 4$ and when $x = -1$, $y = 1$
So the coordinates of the points of intersection of the line and the curve are $(2, 4)$ and $(-1, 1)$.

### Example 3

Find the coordinates of the points of intersection of the straight line $y = x$ and the curve $y = 3x^2 - 2$

**Solution 3**

$x = 3x^2 - 2$ | Solve the simultaneous equations $y = x$ and $y = 3x^2 - 2$

$3x^2 - x - 2 = 0$

$(3x + 2)(x - 1) = 0$

$x = -\frac{2}{3}$ or $x = 1$

When $x = -\frac{2}{3}, y = -\frac{2}{3}$    | To find the values of $y$, substitute into $y = x$ |

When $x = 1, y = 1$
Points of intersection are $\left(-\frac{2}{3}, -\frac{2}{3}\right)$ and $(1, 1)$

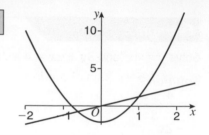

## Exercise 40A

**1** Solve these simultaneous equations.

    **a** $y = 2x$ and $y = 2x^2$

    **c** $y = 3x - 1$ and $y = x^2 - 5$

    **b** $y = x + 3$ and $y = x^2 + 3x$

    **d** $y = 6 - x^2$ and $y = 4x + 1$

**2** Solve

    **a** $y - 3x = 4$ and $y = 2x^2 - 5$

    **c** $y + 2x = 5$ and $y = 2x^2 + x$

    **e** $2x - y = -8$ and $y = x^2$

    **b** $x + y = 2$ and $y = 3x^2 - 2$

    **d** $x - y = 3$ and $y = x^2 - 2x - 1$

    **f** $2x + 3y = 13$ and $y = x^2 - 1$

**3** Find the coordinates of the points of intersection of these lines and quadratic curves.

    **a** $y = 3$ and $y = x^2 + 2x$

    **c** $y = -4$ and $y = x^2 - 5x$

    **b** $y = 5$ and $y = x^2 - 4x$

    **d** $y = -1$ and $y = 2x^2 + 5x + 1$

**4** Find the coordinates of the points of intersection of these lines and quadratic curves.

    **a** $y = x + 6$ and $y = x^2$

    **c** $y = x + 1$ and $y = 2x^2$

    **e** $y = x$ and $y = x^2 + 7x + 5$

    **b** $y = x$ and $y = x^2 - 2$

    **d** $y = 4 - x$ and $y = 2x^2 + 3$

    **f** $x + 2y = 0$ and $y = 2x^2 - 4x - 1$

# 40.2 Loci and equations

**Locus** (plural 'loci') was considered in Section 34.2 A locus is the path of a point which moves according to some mathematical rule. Alternatively, it is the set of points which obey a mathematical rule.

The following examples show how to find the equations of some loci.

## Example 4

Find the equation of the locus of a point which moves so that it is equidistant from the points $A(2, 2)$ and $B(8, 2)$.

*Solution 4*
The equation of the locus is $x = 5$

| The locus of a point which moves so that it is equidistant from two fixed points $A$ and $B$ is the perpendicular bisector of $AB$. Draw this perpendicular bisector and write down its equation. |

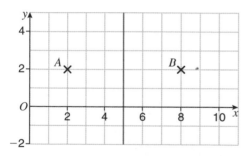

## Example 5

**a** Sketch the locus of a point which is equidistant from the points $P(4, 0)$ and $Q(0, 4)$.

**b** Find the equation of this locus.

### Solution 5

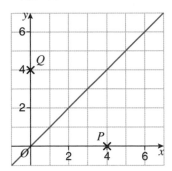

The equation of the locus is $y = x$

The locus is the line which passes through $(0, 0)$, $(1, 1)$, etc.

---

Sometimes the equation of a locus is obtained by considering a point with coordinates $(x, y)$ representing any point on the locus.

## Example 6

$P$ is the point $(x, y)$.

**a** Write down, in terms of $y$, an expression for the distance of $P$ from the line $y = 2$

**b** Write down, in terms of $x$ and $y$, an expression for distance of $P$ from the point $(0, 4)$

The distance of the point $P$ from the line $y = 2$ is the same as its distance from the point $(0, 4)$.

**c** Write down an equation which describes this locus.

**d** Show that the equation of the locus can be written as

$$y = \frac{x^2 + 12}{4}$$

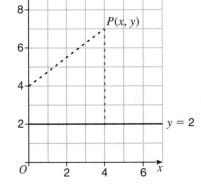

### Solution 6

**a**

Distance of $P$ from the line $y = 2$ is $y - 2$

The perpendicular distance is the difference between the $y$ values.

**b** Distance of $P$ from the point $(0, 4) = \sqrt{x^2 + (y-4)^2}$

> Use Pythagoras' theorem.

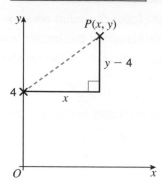

**c** $\sqrt{x^2 + (y-4)^2} = y - 2$

> The distances in **a** and **b** are the same.

**d** $\quad x^2 + (y-4)^2 = (y-2)^2$

> Square both sides of the equation in part **c**.

$x^2 + y^2 - 8y + 16 = y^2 - 4y + 4$

> Expand the brackets.

$x^2 + 12 = 4y$

> Simplify.

$$y = \frac{x^2 + 12}{4}$$

> Divide both sides by 4.

> $y = \dfrac{x^2 + 12}{4}$ may be written as $y = \dfrac{x^2}{4} + 3$

---

The locus of a point $P$ which moves so that its distance from a fixed point $O$ is always 4 cm is a circle, centre $O$ and radius 4 cm.

To find the equation of a circle, centre $O(0, 0)$ and radius 4, let $P(x, y)$ be any point on the circle. Then $OP = 4$

Using Pythagoras' theorem $x^2 + y^2 = 4^2$

So the equation of the circle is $x^2 + y^2 = 16$

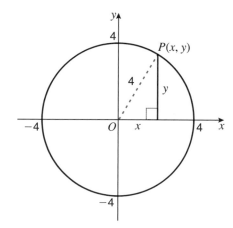

In general,

> **the equation of a circle, centre $O$ and radius $r$, is $x^2 + y^2 = r^2$**

In Chapter 32, the $x$-coordinates of the points of intersection of a line and a quadratic curve were used to estimate the solutions of the related quadratic equation.
In a similar way, the coordinates of the points of intersection of a straight line and a circle, centre the origin, can be used to find estimates of the solutions to the pair of simultaneous equations representing the line and the circle.

## Example 7

**a** Draw the circle with equation $x^2 + y^2 = 25$

**b** On the same axes, draw the line with equation $y = 2x + 1$

**c** Hence find estimates of the solutions to the simultaneous equations

$$x^2 + y^2 = 25$$
$$y = 2x + 1$$

Give your answers correct to 1 decimal place.

### Solution 7

**a**

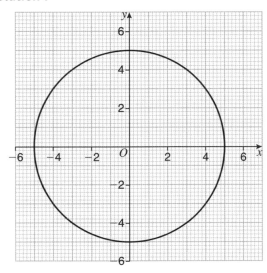

> The circle has centre $O$ and radius 5

**b**

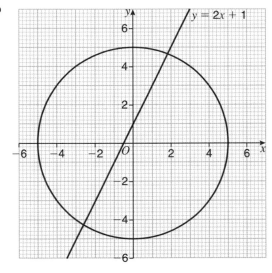

> Draw the straight line with equation $y = 2x + 1$

**c** Estimates of the solutions to the simultaneous equations are $x = 1.8$, $y = 4.6$ or $x = -2.6$, $y = -4.2$

> The estimates are obtained from the coordinates of the points of intersection of the straight line and the circle.

---

## Exercise 40B

**1** In each case sketch the locus of the point which moves so that it is equidistant from the given points. Give the equation of each locus.

    **a** $(0, 0)$ and $(6, 0)$             **b** $(-2, 0)$ and $(6, 0)$

    **c** $(0, 0)$ and $(0, 4)$             **d** $(0, -4)$ and $(0, 2)$

**2** In each case sketch the locus of the point which moves so that it is equidistant from the given points. Give the equation of each locus.

   **a** $(3, 0)$ and $(0, 3)$                **b** $(-4, 0)$ and $(0, 4)$

   **c** $(2, 3)$ and $(3, 2)$                **d** $(-1, 4)$ and $(-4, 1)$

**3** Write down the equation of the circle, centre $(0, 0)$, radius 6

**4** $P$ is the point $(x, y)$

   **a** Write down the distance of $P$ from the $y$-axis.

   **b** Write down the distance of $P$ from the $x$-axis.

   **c** Find the equation of the locus of $P$ for each of these rules.

      **i** The point $P$ moves so that its distance from the $y$-axis is the same as its distance from the $x$-axis.

      **ii** The point $P$ moves so that its distance from the $y$-axis is twice its distance from the $x$-axis.

      **iii** The point $P$ moves so that its distance from the $y$-axis is 1 more than its distance from the $x$-axis.

**5** $P$ is the point $(x, y)$.

   **a** Write down an expression for the distance of $P$ from the line $y = 1$

   **b** Write down an expression for the distance of $P$ from the point $(0, 3)$.

   $P$ moves so that its distance from the line $y = 1$ is the same as its distance from the point $(0, 3)$.

   **c** Find the equation of the locus of $P$. Give your answer in its simplest form.

**6** The point $P(x, y)$ moves so that its distance from the line $x = 1$ is the same as its distance from the line $y = 2$. Find the equation of the locus of $P$.

**7**  **a** Draw on graph paper the circle with equation $x^2 + y^2 = 16$

   **b** Using the same axes, draw the straight line with equation $y = x + 1$

   **c** Hence find estimates of the solutions to the simultaneous equations $x^2 + y^2 = 16$ and $y = x = 1$

**8** Draw suitable graphs to find estimates of the solutions to these pairs of simultaneous equations.

   **a** $x^2 + y^2 = 16$ and $y = 2x$          **b** $x^2 + y^2 = 64$ and $3x + 2y = 12$

**9** By drawing suitable graphs, solve these simultaneous equations.

   **a** $x^2 + y^2 = 25$ and $y = x + 1$        **b** $x^2 + y^2 = 100$ and $x + y = 2$

## 40.3 Intersection of lines and circles – algebraic solutions

The coordinates of the points of intersection of a straight line and a circle can be found by solving the simultaneous equations representing the line and the circle.

For example, the solutions of the simultaneous equations $x^2 + y^2 = 17$ and $y = x + 3$ represent the coordinates of the points of intersection of the circle and the line.

Substitute $x + 3$ for $y$ in $x^2 + y^2 = 17$

$$x^2 + (x + 3)^2 = 17$$
$$x^2 + x^2 + 6x + 9 = 17$$
$$2x^2 + 6x - 8 = 0$$
$$x^2 + 3x - 4 = 0$$
$$(x + 4)(x - 1) = 0$$

so $x = -4$ or $x = 1$

Substitute into the linear equation $y = x + 3$
to obtain the corresponding values of $y$.

When $x = -4$, $y = -1$
When $x - 1$, $y = 4$

So the line and circle intersect at $(-4, -1)$ and $(1, 4)$.

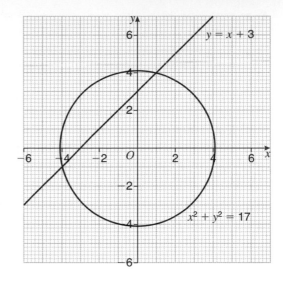

## Example 8

Solve $x^2 + y^2 = 20$ and $x + 2y = 5$
Give your answers correct to 3 significant figures.

### Solution 8

$x = 5 - 2y$  | Make $x$ the subject of $x + 2y = 5$

$(5 - 2y)^2 + y^2 = 20$  | Substitute $5 - 2y$ for $x$ in $x^2 + y^2 = 20$

$25 - 20y + 4y^2 + y^2 = 20$
$5y^2 - 20y + 5 = 0$
$y^2 - 4y + 1 = 0$  | Simplify.

$y = \dfrac{-(-4) \pm \sqrt{(-4)^2 - 4 \times 1 \times 1}}{2 \times 1}$  | Substitute $a = 1$, $b = (-4)$, $c = 1$ into the quadratic formula $x = \dfrac{-b \pm \sqrt{b^2 - 4ac}}{2a}$

$y = \dfrac{4 \pm \sqrt{12}}{2}$

$y = 3.7320...$ or $y = 0.2679...$  | Find $y$ with a calculator.

When $y = 3.73$, $x = -2.46$  | Substitute into $x = 5 - 2y$

When $y = 0.268$, $x = 4.46$

> It is equally correct to make $y$ the subject of $x + 2y = 5$ and then substitute $\dfrac{5 - x}{2}$ for $y$ in $x^2 + y^2 = 20$ but the algebra is more complicated.

## Exercise 40C

**1** Solve these simultaneous equations.

**a** $x^2 + y^2 = 10$    **b** $x^2 + y^2 = 20$    **c** $x^2 + y^2 = 26$    **d** $x^2 + y^2 = 40$
   $y = x - 4$          $y = x - 6$          $y = 2x + 3$        $y = x + 8$

**2** Find algebraically the coordinates of the points of intersection of each circle and straight line.

**a** $x^2 + y^2 = 45$    **b** $x^2 + y^2 = 25$    **c** $x^2 + y^2 = 34$    **d** $x^2 + y^2 = 13$
   $y = x + 3$          $x + y = 7$          $x + 2y = 1$        $2x + y = 7$

**3** Solve these simultaneous equations. Give your answers correct to 3 significant figures.

**a** $x^2 + y^2 = 25$    **b** $x^2 + y^2 = 30$    **c** $x^2 + y^2 = 40$    **d** $x^2 + y^2 = 50$
   $x + y = 6$          $x + 2y = 7$        $y = x + 2$        $2x + 4y = 9$

**4 a** $x^2 + y^2 = 36$
$x = 7$

   **i** By eliminating $x$, show that these simultaneous equations have no solution.

   **ii** On the same axes, sketch the graphs of $x^2 + y^2 = 36$ and $x = 7$

**b** $x^2 + y^2 = 36$
$x = 6$

   **i** By eliminating $x$, show that these simultaneous equations have only one solution.

   **ii** On the same axes, sketch the graphs of $x^2 + y^2 = 36$ and $x = 6$

## Chapter summary

**You should know and be able to use these facts:**

★   that the solutions of a pair of simultaneous equations correspond to the coordinates of the points of intersection of the related graphs

★   that the equation of a circle, centre the origin, radius $r$ is $x^2 + y^2 = r^2$

★   that a circle is the locus of the set of points which are equidistant from a fixed point (the centre).

**You should also be able to:**

★   solve algebraically or graphically a pair of simultaneous equations
   ● when one is linear and one is quadratic
   ● when one is linear and one is of the form $x^2 + y^2 = r^2$ (the equation of a circle, centre the origin)

★   find the equation of a locus from its rule.

## Chapter 40 review questions

**1** Solve

  **a** $y = 2x$ and $y = x^2$                 **b** $y = 2x$ and $y = x^2 - 3$

  **c** $y = 2x$ and $y = x^2 + x$            **d** $y = x$ and $y = 2x^2 - 3$

**2** Solve these simultaneous equations. Give your answers correct to 3 significant figures.

  **a** $y = 2x$ and $y = x^2 - 2$            **b** $y = 2x$ and $y = 2x^2 - 1$

  **c** $y = 2x$ and $y = x^2 - x - 1$       **d** $y = 2x$ and $y = 4 - x^2$

**3** Find the coordinates of the points of intersection of each straight line and curve.

  **a** $y = 2x$ and $y = 2x(x - 1)$        **b** $y = 3x - 2$ and $y = x^2$

  **c** $y = 3x + 3$ and $y = 4x^2 - x$      **d** $x + y = 2$ and $y = 4x^2 - 3$

**4** Solve $y = 12x^2 + 7$
        $5x + y = 10$

**5** Write down the equation of the circle, centre the origin and diameter 12

**6** Find the equation of the locus of the point $P$ which moves so that its distance from the point $(5, 0)$ is the same as its distance from the point $(0, 5)$.

**7** Find the equation of the locus of the point $P$ which moves such that its distance from the $x$-axis is the same as its distance from the line $y = 3$

**8** Find the equation of the locus of the point $P$ which moves such that its distance from the $x$-axis is the same as its distance from the line $x = 3$

**9** Find the equation of the locus of the point $P$ which moves such that its distance from the $x$-axis is three times its distance from the $y$-axis.

**10** $P$ is the point $(x, y)$.

  **a** Write down an expression for the distance of $P$ from the line $y = 6$

  **b** Write down an expression for the distance of $P$ from the origin.

  The point $P$ moves such that its distance from the line $y = 6$ is the same as its distance from the origin.

  **c** Find the equation of the locus of $P$. Give your answer in its simplest form.

**11** **a** Draw the circle with radius 6 and centre the origin.

  **b** On the same axes, draw the straight line with equation $y = 2x - 1$

  **c** Use the graphs to find estimates of the solutions to the simultaneous equations
    $x^2 + y^2 = 36$ and $y = 2x - 1$

**12** Draw suitable graphs to find estimates of the solutions to these simultaneous equations.

  **a** $x^2 + y^2 = 25$     **b** $x^2 + y^2 = 49$     **c** $x^2 + y^2 = 64$
      $y = x + 2$           $y = 2x + 2$           $x + y = 10$

**13** Solve these simultaneous equations.

  **a** $x^2 + y^2 = 45$     **b** $x^2 + y^2 = 80$     **c** $x^2 + y^2 = 58$
      $y = x - 3$           $y = x + 4$             $y = 2x + 1$

**14** The diagram shows a sketch of a curve.
The point $P(x, y)$ lies on the curve.
The locus of $P$ has the following property:
the distance of the point $P$ from the point $(0, 2)$ is the same as the distance of the point $P$ from the $x$-axis.
Show that $y = \frac{1}{4}x^2 + 1$

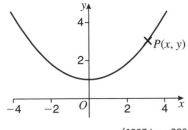

(1387 June 2004)

**15** The line $y = 4 - 4x$ intersects the curve $y = 3(x^2 - x)$ at the points $A$ and $B$.
Use an algebraic method to find the coordinates of $A$ and $B$.     (1387 November 2005)

**16** Use the resource sheet which shows a grid where $-4 < x < 4$ and $-4 < y < 4$

  **a** On the grid, draw the curve with equation $x^2 + y^2 = 9$

  **b** By drawing a suitable straight line find estimates for the solutions of the equations
    $x^2 + y^2 = 9$
    $y = x - 1$

**17** Bill said that the line $y = 6$ cuts the curve $x^2 + y^2 = 25$ at two points.

  **a** By eliminating $y$ show that Bill is incorrect.

  **b** By eliminating $y$ find the solutions of the simultaneous equations
    $x^2 + y^2 = 25$
    $y = 2x - 2$     (1387 June 2004)

**18** Solve the simultaneous equations

    $x^2 + y^2 = 29$
    $y - x = 3$     (1387 November 2003)

# Similar shapes

## 41.1 Similar triangles

Triangle $ABC$ and triangle $A'B'C'$ have the same shape but not the same size. They are called **similar** triangles. The angles in triangle $ABC$ are the same as the angles in triangle $A'B'C'$, and so two similar triangles have equal angles.

$AB$ and $A'B'$ are a pair of **corresponding** sides. $AC$ and $A'C'$ and $BC$ and $B'C'$ are also pairs of corresponding sides.

In general, if two shapes are similar, the lengths of pairs of corresponding sides are in the same proportion.

In this case, that means $\dfrac{A'B'}{AB} = \dfrac{A'C'}{AC} = \dfrac{B'C'}{BC}$

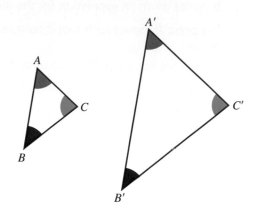

---

### Example 1

Triangles $ABC$ and $DEF$ are similar.

**a** Find the value of $x$.

**b** Work out the length of

  **i** $DE$

  **ii** $BC$.

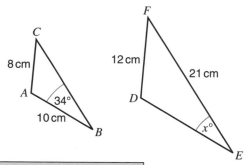

*Solution 1*

**a** $x = 34$     Similar triangles have equal angles.

**b** **i** $\dfrac{DE}{AB} = \dfrac{DF}{AC}$     The lengths of pairs of corresponding sides are in the same proportion.

$\dfrac{DE}{10} = \dfrac{12}{8}$     Substitute the known lengths.

$DE = \dfrac{10 \times 12}{8} = 15$ cm     Rearrange the equation to work out the length of $DE$.

**ii** $\dfrac{EF}{BC} = \dfrac{DF}{AC}$     The lengths of pairs of corresponding sides are in the same proportion.

$\dfrac{21}{BC} = \dfrac{12}{8}$     Substitute the known lengths.

$BC = \dfrac{8 \times 21}{12} = 14$ cm     Rearrange the equation to work out the length of $BC$.

In the diagram, $ABC$ and $AED$ are straight lines.
The line $BE$ is parallel to the line $CD$.

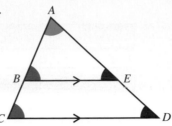

Angle $ABE$ = angle $ACD$.

Angle $AEB$ = angle $ADC$.

They are both pairs of corresponding angles.

Also, angle $A$ is common to both triangle $ABE$ and triangle $ACD$.
So triangle $ABE$ and triangle $ACD$ have equal angles and are similar triangles.

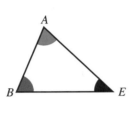 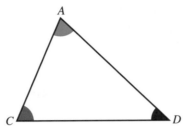

The lengths of their corresponding sides are in the same proportion, that is

$$\frac{AC}{AB} = \frac{AD}{AE} = \frac{CD}{BE}$$

## Example 2

$RS$ is parallel to $QT$.
$PQR$ and $PTS$ are straight lines.
$PQ$ = 5 cm, $QR$ = 3 cm,
$QT$ = 6 cm, $PT$ = 4.5 cm.

**a** Calculate the length of $RS$.

**b** Calculate the length of $ST$.

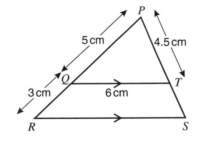

### Solution 2

**a** $\dfrac{PR}{PQ} = \dfrac{RS}{QT}$

The lengths of pairs of corresponding sides are in the same proportion.

$\dfrac{8}{5} = \dfrac{RS}{6}$

Substitute the known lengths.

$RS = \dfrac{6 \times 8}{5} = 9.6$ cm

Rearrange the equation to work out the length of $RS$.

**b** *Method 1*

$\dfrac{PR}{PQ} = \dfrac{PS}{PT}$

The lengths of pairs of corresponding sides are in the same proportion.

$\dfrac{8}{5} = \dfrac{PS}{4.5}$

Substitute the known lengths.

$PS = \dfrac{8 \times 4.5}{5} = 7.2$ cm

Rearrange the equation to work out the length of $PS$.

$ST = 7.2$ cm $- 4.5$ cm

$ST = 2.7$ cm

To find the length of $ST$, subtract the length of $PT$ from the length of $PS$.

*Method 2*

$$\frac{PR}{PQ} = \frac{PS}{PT}$$

> Let $x$ cm represent the length of $ST$ so $(x + 4.5)$ cm will represent the length of $PS$.

$$\frac{8}{5} = \frac{x + 4.5}{4.5}$$

> Substitute the known lengths and $x + 4.5$ for $PS$.

$$8 \times 4.5 = 5(x + 4.5)$$

> Solve the equation for $x$.

$$36 = 5x + 22.5$$
$$5x = 13.5$$
$$x = 2.7$$

$$ST = 2.7 \text{ cm}$$

> State the length of $ST$.

In the diagram, $ACE$ and $BCD$ are straight lines. The line $AB$ is parallel to the line $DE$.

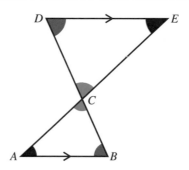

Now,     angle $BAC$ = angle $CED$

angle $ABC$ = angle $CDE$

as they are both pairs of alternate angles.

Also,     angle $ACB$ = angle $DCE$

because, where two straight lines cross, the opposite angles are equal.

So triangle $ABC$ and triangle $EDC$ have equal angles and are similar triangles.

The lengths of their corresponding sides are in the same proportion, that is

$$\frac{ED}{AB} = \frac{EC}{AC} = \frac{CD}{CB}$$

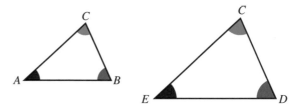

---

## Example 3

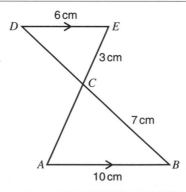

$AB$ is parallel to $DE$.
$ACE$ and $BCD$ are straight lines.
$AB = 10$ cm, $BC = 7$ cm,
$DE = 6$ cm, $CE = 3$ cm.

**a** Calculate the length of $AC$.

**b** Calculate the length of $CD$.

*Solution 3*

**a**  $$\frac{AB}{DE} = \frac{AC}{CE}$$

> The lengths of pairs of corresponding sides are in the same proportion.

$$\frac{10}{6} = \frac{AC}{3}$$

> Substitute the known lengths.

$$AC = \frac{3 \times 10}{6} = 5 \text{ cm}$$

> Rearrange the equation to work out the length of $AC$.
> Notice that in triangle $CDE$, $CE = \frac{1}{2}DE$ and so in triangle $ABC$, $AC = \frac{1}{2}AB$.

$$\mathbf{b} \quad \frac{AB}{DE} = \frac{BC}{CD}$$

| The lengths of pairs of corresponding sides are in the same proportion. |

$$\frac{10}{6} = \frac{7}{CD}$$

| Substitute the known lengths. |

$$CD = \frac{6 \times 7}{10} = 4.2 \text{ cm}$$

| Rearrange the equation to work out the length of $CD$. |

## Exercise 41A

**1** Triangles $ABC$ and $DEF$ are similar.

  **a** Find the size of angle $DFE$.

  **b** Work out the length of

    **i** $DF$         **ii** $BC$.

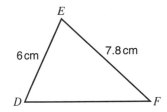

**2** Triangles $PQR$ and $STU$ are similar. Calculate the length of

  **a** $SU$         **b** $QR$.

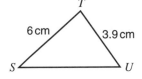

**3** Triangles $ABC$ and $DEF$ are similar. Calculate the length of

  **a** $DF$         **b** $AB$.

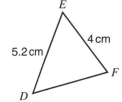

In Questions **4–7**, $BE$ is parallel to $CD$. $ABC$ and $AED$ are straight lines.

**4**

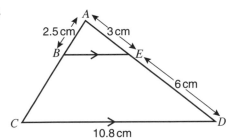

Calculate the length of   **a** $BC$   **b** $BE$.

**5**

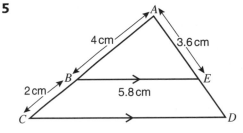

Calculate the length of   **a** $DE$   **b** $CD$.

**6**

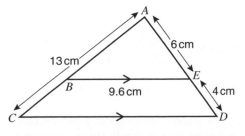

Calculate the length of   **a** $CD$   **b** $AB$.

**7**

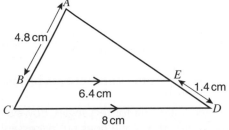

Calculate the length of   **a** $BC$   **b** $AE$.

543

In Questions **8–10**, *AB* is parallel to *DE*.   *ACE* and *BCD* are straight lines.

**8**

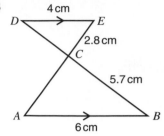

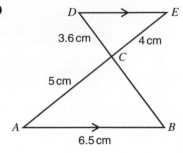

Calculate the length of   **a** *AC*   **b** *CD*.          Calculate the length of   **a** *BC*   **b** *DE*.

**10**

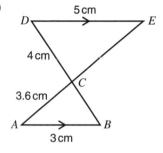

Calculate the length of   **a** *CE*   **b** *BC*.

## 41.2 Similar polygons

Two polygons are similar if the lengths of pairs of corresponding sides are in the same proportion.

For example, a square of side 2 cm and a square of side 6 cm are similar because they have equal angles **and** corresponding sides are in the proportion $\dfrac{6}{2} = 3$

The same argument applies to any pair of squares and so all squares are similar. It also apples to any pair of **regular polygons** with the same number of sides. So, for example, all regular hexagons are similar. (All circles are also similar.)

Two rectangles may be similar.
For example, rectangle **R** and rectangle **S**
are similar because

$$\frac{10}{4} = \frac{5}{2} = 2\tfrac{1}{2} \text{ and } \frac{15}{6} = \frac{5}{2} = 2\tfrac{1}{2}.$$

So, corresponding sides are in the same proportion.

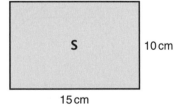

Another way of finding whether two rectangles are similar is to work out the value of $\dfrac{\text{length}}{\text{width}}$ for each of them.

For rectangle **R**,   $\dfrac{\text{length}}{\text{width}} = \dfrac{6}{4} = \dfrac{3}{2} = 1\tfrac{1}{2}$   and for rectangle **S**,   $\dfrac{\text{length}}{\text{width}} = \dfrac{15}{10} = \dfrac{3}{2} = 1\tfrac{1}{2}$

The value of $\dfrac{\text{length}}{\text{width}}$ is the same for both rectangles and so rectangles **R** and **S** are similar.

This is not generally true for rectangles, however, as Example 4 shows. Although any two rectangles have equal angles, this alone does not necessarily mean that they are similar. In this respect, shapes with more than three sides differ from triangles.

## Example 4

Show that rectangle $ABCD$ and rectangle $EFGH$ are not similar.

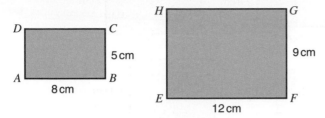

### Solution 4

**Method 1**

$$\frac{EF}{AB} = \frac{12}{8} = 1.5$$

> Divide the length of rectangle EFGH by the length of rectangle $ABCD$.

$$\frac{FG}{BC} = \frac{9}{5} = 1.8$$

> Divide the width of rectangle $EFGH$ by the width of rectangle $ABCD$.

The lengths of pairs of corresponding sides are not in the same proportion and so the rectangles are not similar.

**Method 2**

$$\frac{AB}{BC} = \frac{8}{5} = 1.6$$

> Work out the value of $\frac{\text{length}}{\text{width}}$ for rectangle $ABCD$.

$$\frac{EF}{FG} = \frac{12}{9} = 1.3...$$

> Work out the value of $\frac{\text{length}}{\text{width}}$ for rectangle $EFGH$.

The value of $\frac{\text{length}}{\text{width}}$ is different for each rectangle and so they are not similar.

---

The methods used to find the lengths of sides in pairs of similar triangles can be used to find the lengths of sides in other pairs of similar shapes.

## Example 5

Pentagons **P** and **Q** are similar.
Calculate the value of
**a** $x$ **b** $y$.

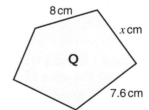

### Solution 5

**a** $$\frac{8}{6} = \frac{x}{4.8}$$

> The lengths of pairs of corresponding sides are in the same proportion.

$$x = \frac{8 \times 4.8}{6} = 6.4$$

> Rearrange the equation to work out the value of $x$.

**b** $$\frac{8}{6} = \frac{7.6}{y}$$

> The lengths of pairs of corresponding sides are in the same proportion.

$$y = \frac{6 \times 7.6}{8} = 5.7$$

> Rearrange the equation to work out the value of $y$.

**Exercise 41B**

**1** Rectangles **P** and **Q** are similar.
   Work out the value of $x$.

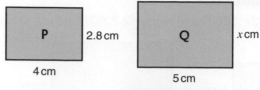

**2** Rectangles **R** and **S** are similar.
   Work out the value of $y$.

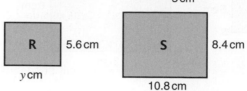

**3** Show that rectangles **T** and **U** are
   not similar.

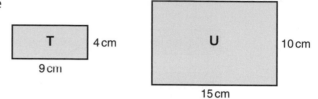

**4** Are rectangles **V** and **W** similar?
   You must show working to
   explain your answer.

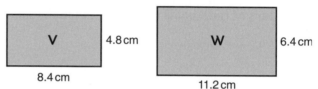

**5** A photograph is 15 cm long and 10 cm wide.
   It is mounted on a rectangular piece of card so that there
   is a border 5 cm wide all around the photograph.
   Are the photograph and the card similar shapes?
   Show working to explain your answer.

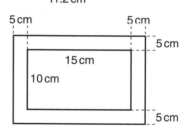

**6** Quadrilaterals **P** and **Q** are similar.
   Calculate the value of
   **a** $x$    **b** $y$.

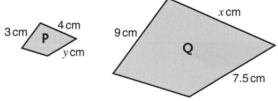

**7** Pentagons **R** and **S** are similar.
   Calculate the value of
   **a** $x$    **b** $y$.

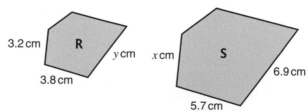

**8** Hexagons **T** and **U** are similar.
   Calculate the value of
   **a** $x$    **b** $y$.

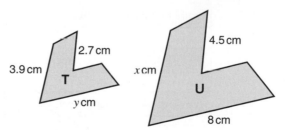

# 41.3 Areas of similar shapes

The diagram shows a square of side 1 cm and a square of side 2 cm.
Area of the square of side 1 cm = 1 cm².
Area of the square of side 2 cm = 4 cm².

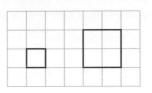

When lengths are multiplied by 2, area is multiplied by 4

The diagram shows a cube of side 1 cm and a cube of side 2 cm.
Surface area of the cube of side 1 cm = 6 × 1 cm² = 6 cm².
Surface area of the cube of side 2 cm = 6 × 4 cm² = 24 cm².

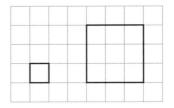

Again, when lengths are multiplied by 2, area is multiplied by 4

The diagram shows a square of side 1 cm and a square of side 3 cm.
Area of the square of side 1 cm = 1 cm².
Area of the square of side 3 cm = 9 cm².

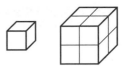

When lengths are multiplied by 3, area is multiplied by 9

In general, for similar shapes,

> **when lengths are multiplied by $k$, area is multiplied by $k^2$.**

For example, if the lengths of a shape are multiplied by 5, its area is multiplied by $5^2$, that is 25

---

## Example 6

Quadrilaterals **P** and **Q** are similar.
The area of quadrilateral **P** is 10 cm².

Calculate the area of quadrilateral **Q**.

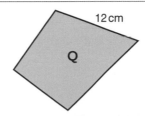

### Solution 6

| | |
|---|---|
| $\dfrac{12}{3} = 4$ | Work out $\dfrac{\text{length of side in } \mathbf{Q}}{\text{length of corresponding side in } \mathbf{P}}$ to find the number by which lengths have been multiplied, that is, find the scale factor. |
| $4^2 = 16$ | Square the scale factor to find the number by which the area has to be multiplied. |
| 10 cm² × 16 = 160 cm² | Multiply the area of quadrilateral **P** by 16 to find the area of quadrilateral **Q**. |

---

## Example 7

Cylinders **R** and **S** are similar.
The surface area of cylinder **R** is 40 cm².

Calculate the surface area of cylinder **S**.

### Solution 7

| | |
|---|---|
| $\dfrac{35}{14} = 2.5$ | Work out $\dfrac{\text{height of cylinder } \mathbf{S}}{\text{height of cylinder } \mathbf{R}}$ to find the number by which lengths have been multiplied, that is, find the scale factor. |
| $2.5^2 = 6.25$ | Square the scale factor to find the number by which the area has to be multiplied. |
| 40 cm² × 6.25 = 250 cm² | Multiply the surface area of cylinder **R** by 6.25 to find the surface area of cylinder **S**. |

If the areas of two similar shapes are known, the scale factor can be found.

For example, if two similar shapes **T** and **U** have areas 5 cm² and 320 cm², the area of shape **T** has been multiplied by $\frac{320}{5} = 64$

So, if the scale factor is $k$, then $k^2 = 64$ and $k = \sqrt{64} = 8$

---

### Example 8

Pentagons **V** and **W** are similar.
The area of pentagon **V** is 40 cm² and the area of pentagon **W** is 90 cm².

Calculate the value of    **a** $x$    **b** $y$.

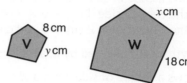

*Solution 8*

$\frac{90}{40} = 2.25$ | Work out $\frac{\text{area of } \mathbf{W}}{\text{area of } \mathbf{V}}$ to find the number by which the area has been multiplied.

$k^2 = 2.25$ | This number is (scale factor)².

$k = \sqrt{2.25} = 1.5$ | The scale factor is the square root of this number.

**a** $x = 8 \times 1.5 = 12$ | Multiply the 8 cm length on **V** by the scale factor to find the corresponding length $x$ cm on **W**.

**b** $y \times 1.5 = 18$ | The length $y$ cm on **V** multiplied by the scale factor gives the corresponding length 18 cm on **W**.

$y = \frac{18}{1.5} = 12$ | Divide 18 by the scale factor to find the value of $y$.

---

### Exercise 41C

**1** Quadrilaterals **P** and **Q** are similar.
The area of quadrilateral **P** is 20 cm².

Calculate the area of quadrilateral **Q**.

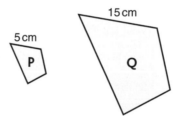

**2** Triangles **P** and **Q** are similar.
The area of triangle **P** is 5 cm².

Calculate the area of triangle **Q**.

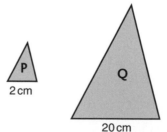

**3** Pentagons **P** and **Q** are similar.
The area of pentagon **Q** is 250 cm².

Calculate the area of pentagon **P**.

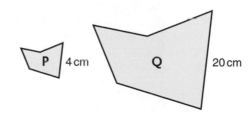

**4** Cylinders **P** and **Q** are similar.
The surface area of cylinder **P** is 60 cm².

Calculate the surface area of cylinder **Q**.

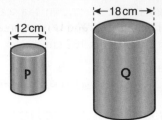

**5** Cuboids **P** and **Q** are similar.
The surface area of cuboid **P** is 72 cm².

Calculate the surface area of cuboid **Q**.

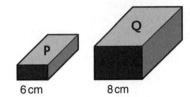

**6** Cones **P** and **Q** are similar.
The surface area of cone **Q** is 64 cm².

Calculate the surface area of cone **P**.

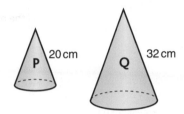

**7** Parallelograms **P** and **Q** are similar.
The area of parallelogram **Q** is 36 times the
area of parallelogram **P**.

Calculate the value of $x$.

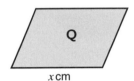

**8** Pyramids **P** and **Q** are similar.
The surface area of pyramid **Q** is 64 times the
surface area of pyramid **P**.

Calculate the value of   **a** $x$   **b** $y$.

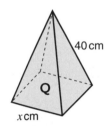

**9** Quadrilaterals **P** and **Q** are similar.
The area of quadrilateral **P** is 10 cm².
The area of quadrilateral **Q** is 360 cm².

Calculate the value of   **a** $x$   **b** $y$.

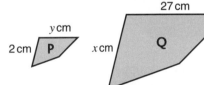

**10** Cylinders **P** and **Q** are similar.
The surface area of cylinder **P** is 50 cm².
The surface area of cylinder **Q** is 72 cm².

Calculate the value of $h$.

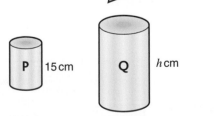

**11** Trapeziums **P** and **Q** are similar.
The area of trapezium **P** is 36 cm².
The area of trapezium **Q** is 100 cm².

Calculate the value of   **a** $x$   **b** $y$.

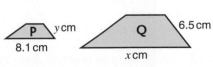

**12** Cuboids **P** and **Q** are similar.
The surface area of cuboid **P** is 50 cm².
The surface area of cuboid **Q** is 162 cm².

Calculate the value of      **a** $x$      **b** $y$.

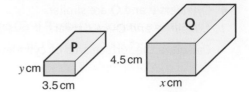

## 41.4 Volumes of similar solids

The diagram shows a cube of side 1 cm and a cube of side 2 cm.
Volume of the cube of side 1 cm = $(1 \times 1 \times 1)$ cm³ = 1 cm³.
Volume of the cube of side 2 cm = $(2 \times 2 \times 2)$ cm³ = 8 cm³.

When lengths are multiplied by 2, volume is multiplied by 8

The diagram shows a cube of side 1 cm and a cube of side 3 cm.
Volume of the cube of side 1 cm = $(1 \times 1 \times 1)$ cm³ = 1 cm³.
Volume of the cube of side 2 cm = $(3 \times 3 \times 3)$ cm³ = 27 cm³.

When lengths are multiplied by 3, volume is multiplied by 27

In general, for similar solids,

> **when lengths are multiplied by $k$, volume is multiplied by $k^3$.**

For example, if the lengths of a shape are multiplied by 5, its volume is multiplied by $5^3$, that is 125.

### Example 9

Cuboids **R** and **S** are similar.
The volume of cuboid **R** is 50 cm³.

Calculate the volume of cuboid **S**.

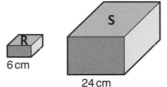

**Solution 9**

$\dfrac{24}{6} = 4$

| Work out $\dfrac{\text{length of side in S}}{\text{length of corresponding side in R}}$ to find the number by which lengths have been multiplied, that is, find the scale factor. |
|---|

$4^3 = 64$

| Cube the scale factor to find the number by which the volume has to be multiplied. |
|---|

50 cm³ × 64 = 3200 cm³

| Multiply the volume of cuboid **R** by 64 to find the volume of cuboid **S**. |
|---|

### Example 10

Cylinders **T** and **U** are similar.
The volume of cylinder **T** is 250 cm³.
The volume of cylinder **U** is 432 cm³.

Calculate the value of $h$.

**Solution 10**

$\dfrac{432}{250} = 1.728$

| Work out $\dfrac{\text{volume of U}}{\text{volume of T}}$ to find the number by which the volume has been multiplied. |
|---|

$k^3 = 1.728$

| This number is (scale factor)³. |
|---|

$k = \sqrt[3]{1.728} = 1.2$

| The scale factor is the cube root of this number. |
|---|

$h = 35 \times 1.2 = 42$

| Multiply the height of cylinder **T** by the scale factor to find the height $h$ cm of cylinder **U**. |
|---|

## Exercise 41D

**1** Cuboids **P** and **Q** are similar.
The volume of cuboid **P** is 20 cm³.

Calculate the volume of cuboid **Q**.

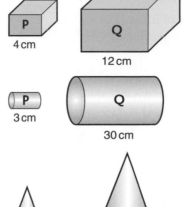

4 cm

12 cm

**2** Cylinders **P** and **Q** are similar.
The volume of cylinder **P** is 7 cm³.

Calculate the volume of cylinder **Q**.

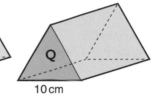

3 cm

30 cm

**3** Cones **P** and **Q** are similar.
The volume of cone **Q** is 40 cm³.

Calculate the volume of cone **P**.

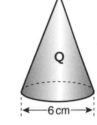

3 cm

6 cm

**4** Prisms **P** and **Q** are similar.
The volume of prism **P** is 80 cm³.

Calculate the volume of prism **Q**.

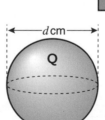

4 cm

10 cm

**5** Pyramids **P** and **Q** are similar.
The volume of pyramid **P** is 320 cm³.

Calculate the volume of pyramid **Q**.

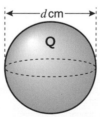

12 cm

15 cm

**6** Cuboids **P** and **Q** are similar.
The volume of cuboid **P** is 250 cm³.

Calculate the volume of cuboid **Q**.

27 cm

15 cm

**7** Spheres **P** and **Q** are similar.
The volume of sphere **Q** is
216 times the volume
of sphere **P**.

Calculate the value of $d$.

5 cm

$d$ cm

All spheres are similar
(so are all cuboids).

**8** Prisms **P** and **Q** are similar.
The volume of prism **Q** is 1000 times the volume
of prism **P**.

Calculate the value of  **a** $x$   **b** $y$.

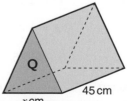

3 cm

$y$ cm

45 cm

$x$ cm

**9** Cylinders **P** and **Q** are similar.
The volume of prism **P** is 30 cm³.
The volume of prism **Q** is 810 cm³.

Calculate the value of $h$.

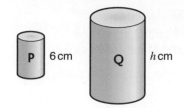

**10** Pyramids **P** and **Q** are similar.
The volume of pyramid **P** is 40 cm³.
The volume of pyramid **Q** is 135 cm³.

Calculate the value of　　**a** $x$　　**b** $y$.

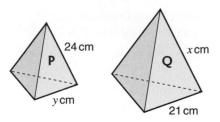

**11** Cuboids **P** and **Q** are similar.
The volume of cuboid **P** is 125 cm³.
The volume of cuboid **Q** is 729 cm³.

Calculate the value of　　**a** $x$　　**b** $y$.

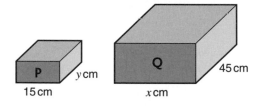

**12** Cones **P** and **Q** are similar.
The volume of cone **P** is 128$\pi$ cm³.
The volume of cone **Q** is 250$\pi$ cm³.

Calculate the value of $x$.

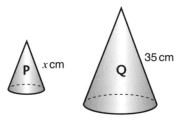

## 41.5　Lengths, areas and volumes of similar solids

Sometimes both area and volume are involved in questions on similar solids.

---

### Example 11

Cuboids **R** and **S** are similar.
The surface area of cuboid **R** is 60 cm².
The surface area of cuboid **S** is 1500 cm².
The volume of cuboid **R** is 40 cm³.

Calculate the volume of cuboid **S**.

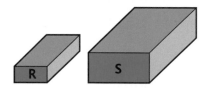

*Solution 11*

$$\frac{1500}{60} = 25$$

Work out $\dfrac{\text{area of } \mathbf{S}}{\text{area of } \mathbf{R}}$ to find the number by which the area has been multiplied.

$$k^2 = 25$$

This number is (scale factor)².

$$k = \sqrt{25} = 5$$

The scale factor is the square root of this number.

$$5^3 = 125$$

Cube the scale factor to find the number by which the volume has to be multiplied.

$$40 \text{ cm}^3 \times 125 = 5000 \text{ cm}^3$$

Multiply the volume of cuboid **R** by 125 to find the volume of cuboid **S**.

---

## Example 12

Cylinders **T** and **U** are similar.
The volume of cylinder **U** is 512 times the volume of cylinder **T**.
The surface area of cylinder **U** is 1600 cm².

Calculate the surface area of cylinder **T**.

### Solution 12

$k^3 = 512$ | The number by which the volume has been multiplied is (scale factor)³.

$k = \sqrt[3]{512} = 8$ | The scale factor is the cube root of this number.

$8^2 = 64$ | Square the scale factor.
So the surface area of cylinder **U** is 64 times the surface area of cylinder **T**.

$1600 \text{ cm}^2 \div 64 = 25 \text{ cm}^2$ | Divide the surface area of cylinder **U** by 64 to find the surface area of cylinder **T**.

## Exercise 41E

**1** Cuboids **P** and **Q** are similar.
The surface area of cuboid **Q** is 25 times the surface area of cuboid **P**.
The volume of cuboid **P** is 40 cm³.

Calculate the volume of cuboid **Q**.

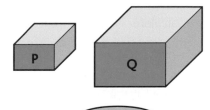

**2** Cylinders **P** and **Q** are similar.
The volume of cylinder **Q** is 27 times the volume of cylinder **P**.
The surface area of cylinder **P** is 20 cm².

Calculate the surface area of cylinder **Q**.

**3** Cones **P** and **Q** are similar.
The surface area of cone **P** is 24 cm².
The surface area of cone **Q** is 54 cm².
The volume of cone **P** is 16 cm³.

Calculate the volume of cone **Q**.

**4** Prisms **P** and **Q** are similar.
The volume of prism **P** is 250 cm³.
The volume of prism **Q** is 686 cm³.
The surface area of prism **P** is 300 cm².

Calculate the surface area of prism **Q**.

**5** Pyramids **P** and **Q** are similar.
The surface area of pyramid **P** is 108 cm².
The surface area of pyramid **Q** is 300 cm².
The volume of pyramid **Q** is 375 cm³.

Calculate     **a** the value of $x$

               **b** the volume of pyramid **P**.

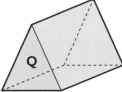

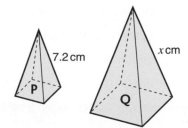

**6** Cuboids **P** and **Q** are similar.
The volume of cuboid **P** is 24 cm³.
The volume of cuboid **Q** is 375 cm³.
The surface area of cuboid **P** is 56 cm².

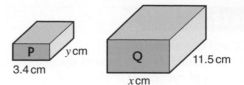

Calculate    **a**   the value of $x$

             **b**   the value of $y$

             **c**   the surface area of cuboid **Q**.

**7** Cylinders **P** and **Q** are similar.
The surface area of cylinder **P** is 90 cm².
The surface area of cylinder **Q** is 250 cm².
The volume of cylinder **Q** is 375 cm³.

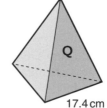

Calculate    **a**   the value of $d$

             **b**   the volume of cylinder **P**.

**8** Pyramids **P** and **Q** are similar.
The volume of pyramid **P** is 1000 cm³.
The volume of pyramid **Q** is 1728 cm³.
The surface area of pyramid **P** is 450 cm².

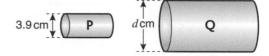

Calculate    **a**   the value of $x$

             **b**   the surface area of pyramid **Q**.

**9** Cones **P** and **Q** are similar.
The surface area of cone **P** is $10\pi$ cm².
The surface area of cone **Q** is $160\pi$ cm².
The volume of cone **P** is $6\pi$ cm³.

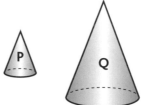

Calculate the volume of cone **Q**.
Give you answer as a multiple of $\pi$.

**10** Frustums **P** and **Q** are similar.
The volume of frustum **P** is $256\pi$ cm³.
The volume of frustum **Q** is $500\pi$ cm³.
The surface area of frustum **Q** is $225\pi$ cm².

Calculate the surface area of frustum **P**.
Give your answer as a multiple of $\pi$.

## Chapter summary

**You should now know that:**

★   triangles which have the same shape but not the same size are called **similar** triangles

★   similar triangles have equal angles

★   all squares are similar and all circles are similar, as are all cubes and all spheres

★   all regular polygons with the same number of sides are similar

★   for similar shapes, the lengths of pairs of corresponding sides are in the same proportion

★   for similar shapes, when lengths are multiplied by $k$, area is multiplied by $k^2$

★   for similar shapes, when lengths are multiplied by $k$, volume is multiplied by $k^3$.

# Chapter 41 review questions

**1** Triangles *ABC* and *DEF* are similar.

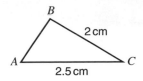

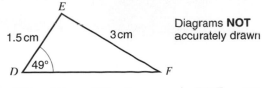

Diagrams **NOT** accurately drawn

$AC = 2.5$ cm $BC = 2$ cm

$DE = 1.5$ cm $EF = 3$ cm, angle $EDF = 49°$

**a** Find the size of angle *BAC*.

**b** Work out the length of **i** *DF* **ii** *AB*.

(4400 May 2005)

**2** In the triangle *ADE*:
*BC* is parallel to *DE*
$AB = 8$ cm, $AC = 5$ cm,
$BD = 4$ cm, $BC = 9$ cm.

**a** Work out the length of *DE*.

**b** Work out the length of *CE*.

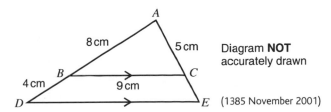

Diagram **NOT** accurately drawn

(1385 November 2001)

**3** *ADB* and *AEC* are straight lines.
*DE* is parallel to *BC*.
$AD = 4$ cm, $DE = 3$ cm, $BC = 4.5$ cm.

Work out the length of *DB*.

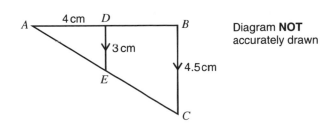

Diagram **NOT** accurately drawn

**4** *AB* is parallel to *CD*.
The lines *AD* and *BC* intersect at point *O*.
$AB = 11$ cm, $AO = 8$ cm, $OD = 6$ cm.

Calculate the length of *CD*.

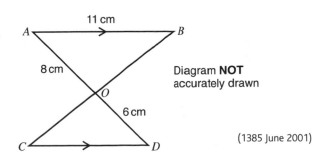

Diagram **NOT** accurately drawn

(1385 June 2001)

**5** $AB : AC = 1 : 3$

**a** Work out the length of *CD*.

**b** Work out the length of *BC*.

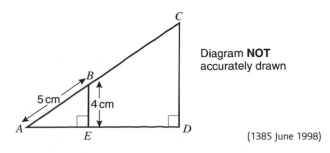

Diagram **NOT** accurately drawn

(1385 June 1998)

**6**

Pictures **NOT**
accurately drawn

A 20 Euro note is a rectangle 133 mm long and 72 mm wide.
A 500 Euro note is a rectangle 160 mm long and 82 mm wide.

Show that the two rectangles are not mathematically similar.     (1387 June 2004)

**7** Shapes *ABCD* and *EFGH* are mathematically similar.

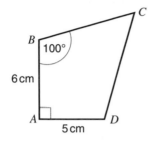
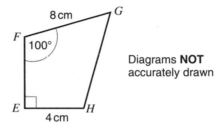

Diagrams **NOT**
accurately drawn

**a** Calculate the length of *BC*.

**b** Calculate the length of *EF*.     (1387 November 2003)

**8** Quadrilateral **P** is mathematically similar to quadrilateral **Q**.

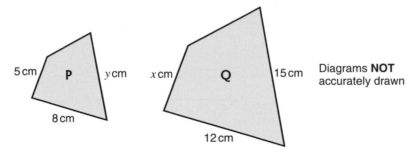

Diagrams **NOT**
accurately drawn

**a** Calculate the value of $x$.

**b** Calculate the value of $y$.

The area of quadrilateral **P** is 60 cm².

**c** Calculate the area of quadrilateral **Q**.     (4400 May 2004)

**9** Cylinder **A** and cylinder **B** are mathematically similar.
The length of cylinder **A** is 4 cm and the length of cylinder **B** is 6 cm.
The volume of cylinder **A** is 80 cm³.

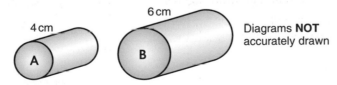

Diagrams **NOT**
accurately drawn

Calculate the volume of cylinder **B**.     (1387 November 2003)

**10** Two cuboids, **S** and **T**, are mathematically similar.
The total surface area of cuboid **S** is 157 cm² and the total surface area of cuboid **T** is 2512 cm².

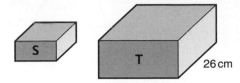

Diagrams **NOT**
accurately drawn

26 cm

   **a** The length of cuboid **T** is 26 cm.
   Calculate the length of cuboid **S**.

   **b** The volume of cuboid **S** is 130 cm³.
   Calculate the volume of cuboid **T**.

(4400 May 2005)

**11** A solid plastic toy is made in the shape
of a cylinder which is joined to a
hemisphere at both ends.
The diameter of the toy at the joins is 5 cm.
The length of the toy is 15 cm.

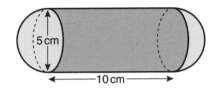

5 cm

Diagram **NOT**
accurately drawn

10 cm

   **a** Calculate the volume of plastic needed to make the toy.
   Give your answer correct to three significant figures.

A similar toy has a volume of 5500 cm³.

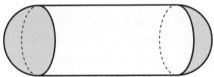

   **b** Calculate the diameter of this toy.
   Give your answer correct to three
   significant figures.

(1385 November 1998)

**12** $ABC$ is a right-angled triangle.
$ED$ is parallel to $BC$.
$BC = 7.5$ cm, $ED = 2.5$ cm, $EB = 2.6$ cm.

   **a** Calculate the area of trapezium $BCDE$.

$AE = x$ cm.

   **b** Calculate the value of $x$.

The smallest angle of the trapezium is $\theta$.
The lengths shown in the diagram are correct to
the nearest millimetre.

   **c** Calculate the least possible value of $\tan \theta$.

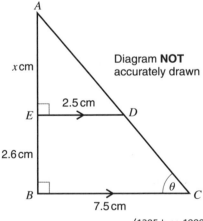

$A$

$x$ cm

Diagram **NOT**
accurately drawn

2.5 cm

$E$          $D$

2.6 cm

$B$          $\theta$   $C$

7.5 cm

(1385 June 1999)

**13** A solid statue is contained within a hemisphere of diameter 500 cm.

   **a** Calculate the total surface area of the hemisphere,
   including the base.
   Give your answer in m², correct to three significant figures.

The solid statue has a height of 80 cm and a mass of 1.5 kg.
A larger solid statue is geometrically similar and is made of
the same material.
It has a height of 128 cm.

   **b** Calculate the mass of the larger solid statue.
   Give your answer in kg, correct to three significant figures.

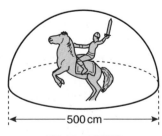

500 cm

Diagram **NOT**
accurately drawn

(1385 June 1999)

**14** The diagram represents a prism. It has a cross-
section in the shape of a sector $AOB$ of a circle,
centre $O$. The radius of the sector is 8 cm and the
length of the prism is 12 cm. Angle $AOB = 54°$.

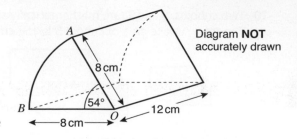

Diagram **NOT**
accurately drawn

  **a** Calculate **i** the area of sector $AOB$
            **ii** the total surface area of the prism.

A geometrically similar prism is made of the same
material as the original prism and has a mass 64
times the mass of the original prism.

  **b** Calculate the length of the heavier prism.

(1384 November 1995)

**15** The diagram shows a frustum.
The diameter of the base is $3d$ cm and the
diameter of the top is $d$ cm.
The height of the frustum is $h$ cm.
The formula for the curved surface area, $S$ cm$^2$,
of the frustum is

$$S = 2\pi d\sqrt{h^2 + d^2}$$

Diagram **NOT**
accurately drawn

  **a** Rearrange the formula to make $h$ the subject.

Two mathematically similar frustums have heights of 20 cm and 30 cm.
The surface area of the smaller frustum is 450 cm$^2$.

  **b** Calculate the surface area of the larger frustum.

(1387 June 2003)

# Direct and inverse proportion

## 42.1 Direct proportion

When one quantity increases in the same proportion as another quantity, the quantities are said to be directly proportional to each other (see Section 30.4).

For example, the cost of a bag of potatoes is directly proportional to the weight of the potatoes.

The symbol $\propto$ is used to denote direct proportion.

If the cost of the bag of potatoes is $C$ pence and the weight of the potatoes is $W$ kg then $C \propto W$.

If the potatoes cost $k$ pence per kilogram then $C = kW$.

In general if $y$ is directly proportional to $x$

$y \propto x$ and $y = kx$

where $k$ is a number known as the **constant of proportionality**.

Since $y = kx$ the graph of $y$ against $x$ is a straight line passing through the origin.

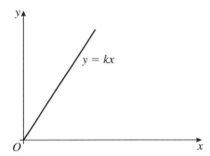

The constant of proportionality, $k$, is the gradient of this straight line.

---

### Example 1

$y$ is directly proportional to $x$.
When $x = 30$, $y = 45$
Find $y$ when $x = 40$

*Solution 1*
$y \propto x$ so $y = kx$

$45 = k \times 30$      Substitute $x = 30$ and $y = 45$

$k = \dfrac{45}{30} = 1.5$      Find $k$

$y = 1.5x$      This is the formula for $y$ in terms of $x$

$y = 1.5 \times 40$      Substitute $x = 40$ into $y = 1.5x$

$y = 60$

---

### Example 2

The voltage, $V$ volts, across an electrical circuit is directly proportional to the current, $I$ amps, flowing through the circuit.
When $I = 1.2$, $V = 78$

a Express $V$ in terms of $I$.

b Find $V$ when $I = 2$

c Find $I$ when $V = 162.5$

### Solution 2

a $V \propto I$ so $V = kI$

$78 = k \times 1.2$      | Substitute $V = 78$ and $I = 1.2$ |

$k = \dfrac{78}{1.2} = 65$      | Find $k$ |

$V = 65I$

b $V = 65 \times 2 = 130$      | Substitute $I = 2$ into $V = 65I$ |

c $162.5 = 65 \times I$      | Substitute $V = 162.5$ into $V = 65I$ |

$I = \dfrac{162.5}{65} = 2.5$

---

### Exercise 42A

**1** $y$ is directly proportional to $x$.

  a $y = 10$ when $x = 2$ Find $y$ when $x = 3$

  b $y = 5$ when $x = 3$ Find $y$ when $x = 4.5$

  c $y = 6$ when $x = 2$ Find $y$ when $x = 3.3$

  d $y = 3$ when $x = 8$ Find $y$ when $x = 6$

**2** $y$ is directly proportional to $x$.

  a $y = 8$ when $x = 2$ Find $x$ when $y = 10$

  b $y = 6$ when $x = 4$ Find $x$ when $y = 7.5$

  c $y = 7$ when $x = 2$ Find $x$ when $y = 3$

  d $y = 8$ when $x = 5$ Find $x$ when $y = 13$

**3** The height, $T$ mm, of a pile of paper is directly proportional to the number of sheets, $N$, in the pile.
When $N = 250$, $T = 28$

  a Find a formula for $T$ in terms of $N$.

  b Find the value of $T$ when $N = 300$

  c Find the value of $N$ when $T = 98$

**4** The distance, $D$ km, travelled by a car is directly proportional to the amount, $A$ litres, of petrol used.
When $A = 5$, $D = 270$

  a Express $D$ in terms of $A$.

  b Find the value of $D$ when $A = 4.5$

  c How many litres of petrol are needed for a journey of 324 km?

**5** The time, $T$ seconds, taken for a pan of water to boil is directly proportional to the amount, $A$ litres, of water in the pan.
When $A = 2.4$, $T = 150$

  a Find a formula for $T$ in terms of $A$.

  b Find the value of $T$ when $A = 1.8$

  c If the pan takes 3 *minutes* to boil, how much water is in it?

**6** The perimeter, $P$ cm, of a regular hexagon is proportional to the length, $l$ cm, of its longest diagonal. When $l = 3.6$, $P = 10.8$

  a Find a formula for $P$ in terms of $l$

  b The value of $l$ increases from 4.8 to 5.4 Find the increase in the value of $P$.

  c The value of $l$ increases by 20%. Find the percentage increase in the value of $P$.

## 42.2 Further direct proportion

Sometimes one quantity is directly proportional to the *square* or the *cube* of another quantity. For example, the area, $A$ cm², of a circle is proportional to the square of its radius, $r$ cm.
That is, $A \propto r^2$ or $A = kr^2$

In general if $y$ is proportional to the square of $x$

$$y \propto x^2 \text{ and } y = kx^2$$

where $k$ is the constant of proportionality.

Since $y = kx^2$, the graph is a quadratic curve passing through the origin. $k$ is generally positive.

Similarly if $y$ is proportional to the cube of $x$

$$y \propto x^3 \text{ and } y = kx^3$$

where $k$ is the constant of proportionality.

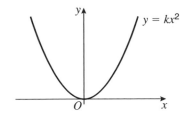

### Example 3

The area, $A$ cm², of a square is proportional to the square of its perimeter, $P$ cm.
When $P = 8$, $A = 4$ Find a formula for $A$ in terms of $P$.

*Solution 3*
$A \propto P^2$ so $A = kP^2$

$4 = k \times 8^2$      | Substitute $A = 4$, $P = 8$ into $A = kP^2$. |

$k = \dfrac{4}{64} = \dfrac{1}{16}$

$A = \dfrac{1}{16}P^2$      | This is the formula for $A$ in terms of $P$. |

### Example 4

$y$ is proportional to the square of $x$. $y = 60$ when $x = 6$
  **a** Find a formula for $y$ in terms of $x$.        **b** Find $y$ when $x = 4.5$
  **c** Find a value of $x$ for which $y = 135$

*Solution 4*
**a** $y \propto x^2$ so $y = kx^2$

    $60 = k \times 6^2$      | Substitute $y = 60$, $x = 6$ into $y = kx^2$ |

    $k = \dfrac{60}{36} = \dfrac{5}{3}$

    $y = \dfrac{5}{3}x^2$

**b** $y = \dfrac{5}{3} \times 4.5^2 = \dfrac{5}{3} \times 20.25$      | Substitute $x = 4.5$ into $y = \dfrac{5}{3}x^2$ |

    $y = 33.75$

**c** $135 = \dfrac{5}{3}x^2$      | Substitute $y = 135$ into $y = \dfrac{5}{3}x^2$ |

    $\dfrac{135 \times 3}{5} = x^2$

    $x^2 = 81$

    $x = 9$ (or $x = -9$)

## Example 5

The mass, $M$ kg, of a solid cube made from lead is proportional to the cube of the length, $L$ cm, of an edge.
When $L = 0.2$, $M = 90$

a Find a formula for $M$ in terms of $L$.
b Find the value of $M$ when $L = 0.3$
c Find the value of $L$ when $M = 2000$ Give your answer correct to 3 significant figures.

### Solution 5

a $M \propto L^3$ so $M = kL^3$

$$90 = k \times 0.2^3$$

> Substitute $M = 90$, $L = 0.2$ into $M = kL^3$

$$k = \frac{90}{0.008} = 11\,250$$

$$M = 11\,250L^3$$

b $M = 11\,250 \times 0.3^3$

> Substitute $L = 0.3$ into $M = 11\,250L^3$

$$M = 303.75$$

c $2000 = 11\,250 \times L^3$

> Substitute $M = 2000$ into $M = 11\,250L^3$

$$L^3 = \frac{2000}{11\,250}$$

$$L = \sqrt[3]{\frac{2000}{11\,250}} = 0.5622\ldots$$

$$L = 0.562 \text{ (to 3 s.f.)}$$

### Exercise 42B

1  $y$ is proportional to the square of $x$.
   a  When $x = 2$, $y = 8$ Find $y$ when $x = 3$
   b  When $x = 2$, $y = 10$ Find $y$ when $x = 8$
   c  When $x = 2$, $y = 7$ Find $y$ when $x = 6$
   d  When $x = 3$, $y = 12$ Find $y$ when $x = 15$

2  $y$ is proportional to the square of $x$.
   a  When $x = 2$, $y = 12$ Find $x$ when $y = 108$
   b  When $x = 3$, $y = 18$ Find $x$ when $y = 162$
   c  When $x = 4$, $y = 40$ Find $x$ when $y = 160$
   d  When $x = 4$, $y = 200$ Find $x$ when $y = 32$

3  The area, $A$ cm$^2$, of a regular hexagon is proportional to the square of the length, $l$ cm, of the longest diagonal.
   When $A = 65$, $l = 10$
   a  Find a formula for $A$ in terms of $l$.
   b  Find the value of $A$ when $l = 4$
   c  Find the value of $l$ when $A = 200$
      Give your answer correct to 3 significant figures.

4  The rate of heat loss, $H$ calories per second, from a sphere is proportional to the square of the radius, $r$ cm, of the sphere.
   When $H = 2.5$, $r = 5$
   a  Find a formula for $H$ in terms of $r$.
   b  Find the value of $H$ when $r = 4$
   c  Find the value of $r$ when $H = 90$

5  The quantity of light, $Q$, given out by a lamp is proportional to the square of the current, $I$, passing through the lamp.
   When $Q = 1000$, $I = 2$
   a  Find a formula for $Q$ in terms of $I$.
   b  Find the value of $Q$ when $I = 3$
   c  Find the value of $I$ when $Q = 2000$
      Give your answer correct to 3 significant figures.

6  The power, $P$ watts, of an engine is proportional to the square of the speed, $s$ m/s, of the engine.
   When $s = 30$, $P = 1260$
   a  Find a formula for $P$ in terms of $s$.
   b  Find the value of $P$ when $s = 25$
   c  Find the value of $s$ when $P = 1715$

**7** $y$ is proportional to the cube of $x$.

   **a** When $x = 2$, $y = 16$ Find $y$ when $x = 3$

   **b** When $x = 2$, $y = 10$ Find $y$ when $x = 3$

   **c** When $x = 4$, $y = 20$ Find $y$ when $x = 6$

   **d** When $x = 5$, $y = 800$ Find $y$ when $x = 8$

**8** $y$ is proportional to the cube of $x$.
When $x = 8$, $y = 1000$

   **a** Find a formula for $y$ in terms of $x$.

   **b** Find the value of $x$ when $y = 1728$

## 42.3 Inverse proportion

When one quantity increases at the same rate as another quantity decreases, the quantities are said to be **inversely proportional** to each other (see Section 30.5).

In general if $y$ is inversely proportional to $x$

$$y \propto \frac{1}{x} \text{ and } y = \frac{k}{x}$$

where $k$ is the constant of proportionality.

The graph of $y = \dfrac{k}{x}$ when $k$ is positive has a similar

shape to the graph of $y = \dfrac{1}{x}$ (see Section 25.1).

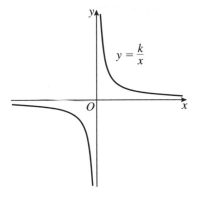

Similarly if $y$ is inversely proportional to the *square* of $x$

$$y \propto \frac{1}{x^2} \text{ and } y = \frac{k}{x^2}$$

where $k$ is the constant of proportionality.

Here is the graph of $y = \dfrac{k}{x^2}$ when $k$ is positive.

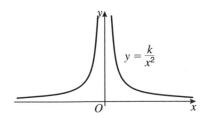

## Example 6

When a fixed volume of water is poured into a cylindrical jar, the depth, $D$ cm, of the water is inversely proportional to the cross-sectional area, $A$ cm$^2$, of the cylindrical jar.
When $A = 40$, $D = 120$

**a** Find a formula for $A$ in terms of $D$.

**b** Find $A$ when $D = 150$

**c** Find $D$ when $A = 60$

*Solution 6*

**a** $A \propto \dfrac{1}{D}$ so $A = \dfrac{k}{D}$

$40 = \dfrac{k}{120}$

$k = 40 \times 120 = 4800$

$A = \dfrac{4800}{D}$

> Substitute $A = 40$, $D = 120$ into $A = \dfrac{k}{D}$

**b** $A = \dfrac{4800}{150} = 32$

> Substitute $D = 150$ into $A = \dfrac{4800}{D}$

**c** $60 = \dfrac{4800}{D}$

$D = \dfrac{4800}{60} = 80$

> Substitute $A = 60$ into $A = \dfrac{4800}{D}$

---

## Example 7

The force of attraction, $F$ newtons, between two spheres is inversely proportional to the square of the distance, $d$ m, between the centres of the spheres.
When $d = 2$, $F = 0.006$

**a** Express $F$ in terms of $d$.

**b** Find $F$ when $d = 2.5$

**c** Find $d$ when $F = 0.001$ Give your answer correct to 3 significant figures.

*Solution 7*

**a** $F \propto \dfrac{1}{d^2}$ so $F = \dfrac{k}{d^2}$

$0.006 = \dfrac{k}{4}$

$k = 0.006 \times 4 = 0.024$

$F = \dfrac{0.024}{d^2}$

> Substitute $F = 0.006$, $d = 2$ into $F = \dfrac{k}{d^2}$

**b** $F = \dfrac{0.024}{2.5^2} = 0.003\,84$

> Substitute $d = 2.5$ into $F = \dfrac{0.024}{d^2}$

**c** $0.001 = \dfrac{0.024}{d^2}$

$d^2 = \dfrac{0.024}{0.001} = 24$

$d = \sqrt{24} = 4.898...$

$d = 4.90$ (to 3 s.f.)

> Substitute $F = 0.001$ into $F = \dfrac{0.024}{d^2}$

---

## Exercise 42C

**1** $y$ is inversely proportional to $x$.

    **a** $y = 8$ when $x = 2$ Find $y$ when $x = 4$    **b** $y = 10$ when $x = 4$ Find $y$ when $x = 16$

    **c** $y = 16$ when $x = 10$ Find $y$ when $x = 8$    **d** $y = 21$ when $x = 10$ Find $y$ when $x = 15$

**2** $y$ is inversely proportional to $x$.

   **a** $y = 20$ when $x = 4$ Find $x$ when $y = 5$      **b** $y = 25$ when $x = 8$ Find $x$ when $y = 10$

   **c** $y = 30$ when $x = 9$ Find $x$ when $y = 20$    **d** $y = 45$ when $x = 4$ Find $x$ when $y = 54$

**3** The time, $T$ seconds, taken for a pan of water to boil on a gas ring is inversely proportional to the setting, $N$, of the gas ring.
When $N = 4.5$, $T = 108$

   **a** Find a formula for $T$ in terms of $N$.

   **b** Find the value of $T$ when $N = 6$

   **c** If the pan takes 3 minutes to boil, what was the setting?

**4** For rectangles with the same area, the length, $l$ metres, of the rectangle is inversely proportional to the width, $w$ metres, of the rectangle.
When $l = 2.5$, $w = 2.4$

   **a** Express $l$ in terms of $w$.

   **b** Find the value of $w$ when $l = 3.2$

   **c** Given that the values of $l$ and $w$ are the same, find the value of $l$.

**5** The frequency, $f$ cycles per second, of a sound wave is inversely proportional to the wavelength, $l$ cm, of the sound wave.
When $f = 256$, $l = 133$

   **a** Find a formula for $f$ in terms of $l$.

   **b** Find $f$ when $l = 250$ Give your answer correct to 3 significant figures.

   **c** Find $l$ when $f = 300$ Give your answer correct to 3 significant figures.

**6** $y$ is inversely proportional to the square of $x$.

   **a** $y = 4$ when $x = 2$ Find $y$ when $x = 4$      **b** $y = 10$ when $x = 2$ Find $y$ when $x = 4$

   **c** $y = 20$ when $x = 3$ Find $y$ when $x = 2$    **d** $y = 45$ when $x = 4$ Find $y$ when $x = 5$

**7** $y$ is inversely proportional to the square of $x$.

   **a** $y = 12$ when $x = 2$ Find $x$ when $y = 0.75$    **b** $y = 10$ when $x = 4$ Find $x$ when $y = 6.4$

   **c** $y = 0.5$ when $x = 6$ Find $x$ when $y = 1800$   **d** $y = 12.5$ when $x = 2$ Find $x$ when $y = 2$

**8** When a fixed volume of liquid is poured into any cylinder, the depth, $D$ cm, of the liquid is inversely proportional to the square of the radius, $r$ cm, of the cylinder.
When $r = 5$, $D = 40$

   **a** Find a formula for $D$ in terms of $r$.

   **b** Find the value of $D$ when $r = 4$

   **c** Find the value of $r$ when $D = 15$ Give your answer correct to 3 significant figures.

   **d** For what value of $r$ is the depth equal to the diameter of the cylinder? Give your answer correct to 3 significant figures.

**9** The intensity, $I$, of the light at a distance, $d$, from a lamp is inversely proportional to the square of the distance.
When $I = 4.5$, $d = 2.4$

   **a** Find a formula for $I$ in terms of $d$.

   **b** Find $I$ when $d = 1.8$

   **c** Find $d$ when $I = 6$ Give your answer correct to 3 significant figures.

**10** The pressure, $P$ pascals, that a constant force exerts on a square with an edge of length, $x$ m, is inversely proportional to $x$.
When $x = 0.4$, $P = 50$

  **a** Find a formula for $P$ in terms of $x$.

  **b** Find $P$ when $x = 0.5$

  **c** Find $x$ when $P = 600$ Give your answer correct to 3 significant figures.

## 42.4 Proportion and square roots

Sometimes one quantity is directly proportional to the *square root* of another quantity.
In general if $y$ is proportional to the square root of $x$

$$y \propto \sqrt{x} \text{ and } y = k\sqrt{x}$$

where $k$ is the constant of proportionality.

Here is the graph of $y = k\sqrt{x}$ when $k$ is positive.

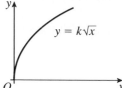

### Example 8

The speed, $s$, of a particle is directly proportional to the square root of its kinetic energy, $E$.
When $E = 225$, $s = 40$

**a** Find a formula for $s$ in terms of $E$.

**b** Find $s$ when $E = 900$

**c** Rearrange the formula to find $E$ in terms of $s$.

### Solution 8

**a** $s \propto \sqrt{E}$ so $s = k\sqrt{E}$

$$40 = k\sqrt{225} = k \times 15$$

$$k = \frac{40}{15} = \frac{8}{3}$$

$$s = \frac{8}{3}\sqrt{E}$$

> Substitute $s = 40$, $E = 225$ into $s = k\sqrt{E}$

**b** $s = \frac{8}{3} \times \sqrt{900} = \frac{8}{3} \times 30$

$$s = 80$$

> Substitute $E = 900$ into $s = \frac{8}{3}\sqrt{E}$

**c** $s = \frac{8}{3}\sqrt{E}$

$$\sqrt{E} = \frac{3s}{8}$$

$$E = \left(\frac{3s}{8}\right)^2 \text{ or } E = \frac{9s^2}{64}$$

> Multiply both sides by 3 and then divide both sides by 8

> Square both sides.

> Either formula is acceptable.

Sometimes one quantity is inversely proportional to the *square root* of another quantity.
In general if $y$ is inversely proportional to the square root of $x$

$$y \propto \frac{1}{\sqrt{x}} \text{ and } y = \frac{k}{\sqrt{x}}$$      where $k$ is the constant of proportionality.

Here is the graph of $y = \frac{k}{\sqrt{x}}$ when $k$ is positive.

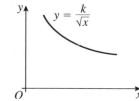

## Example 9

*y* is inversely proportional to the square root of *x*.
When $x = 64$, $y = 20$

**a** Find a formula for *y* in terms of *x*.

**b** Find *y* when $x = 100$

**c** Find *x* when $y = 5$

### Solution 9

**a** $y \propto \dfrac{1}{\sqrt{x}}$ so $y = \dfrac{k}{\sqrt{x}}$

$20 = \dfrac{k}{\sqrt{64}} = \dfrac{k}{8}$     $\boxed{\text{Substitute } y = 20, x = 64 \text{ into } y = \dfrac{k}{\sqrt{x}}}$

$k = 20 \times 8 = 160$

$y = \dfrac{160}{\sqrt{x}}$

**b** $y = \dfrac{160}{\sqrt{100}} = 16$     $\boxed{\text{Substitute } x = 100 \text{ into } y = \dfrac{160}{\sqrt{x}}}$

**c** $5 = \dfrac{160}{\sqrt{x}}$     $\boxed{\text{Substitute } y = 5 \text{ into } y = \dfrac{160}{\sqrt{x}}}$

$5\sqrt{x} = 160$     $\boxed{\text{Multiply both sides by } \sqrt{x}}$

$\sqrt{x} = \dfrac{160}{5}, \ \sqrt{x} = 32$

$x = 32^2$     $\boxed{\text{Square both sides.}}$

$x = 1024$

---

### Exercise 42D

**1** *y* is directly proportional to the square root of *x*.

   **a** When $x = 4$, $y = 6$ Find *y* when $x = 25$     **b** When $x = 16$, $y = 20$ Find *y* when $x = 49$

   **c** When $x = 9$, $y = 4$ Find *y* when $x = 81$     **d** When $x = 100$, $y = 40$ Find *y* when $x = \frac{1}{4}$

**2** *y* is directly proportional to the square root of *x*.

   **a** When $x = 1$, $y = 4$ Find *x* when $y = 8$     **b** When $x = 4$, $y = 10$ Find *x* when $y = 25$

   **c** When $x = 16$, $y = 10$ Find *x* when $y = 25$     **d** When $x = 49$, $y = 21$ Find *x* when $y = 27$

**3** *y* is inversely proportional to the square root of *x*.

   **a** When $x = 4$, $y = 2$ Find *y* when $x = 25$     **b** When $x = 1$, $y = 5$ Find *y* when $x = 16$

   **c** When $x = 4$, $y = 4$ Find *y* when $x = 1$     **d** When $x = 100$, $y = 0.3$ Find *y* when $x = 900$

**4** *y* is inversely proportional to the square root of *x*.

   **a** When $x = 4$, $y = 2.5$ Find *x* when $y = 2$     **b** When $x = 4$, $y = \frac{1}{2}$ Find *x* when $y = 2$

   **c** When $x = 9$, $y = 2$ Find *x* when $y = 6$     **d** When $x = 25$, $y = 0.8$ Find *x* when $y = \frac{1}{2}$

**5** When a ball is thrown upwards, the time, $T$ seconds, the ball remains in the air is directly proportional to the square root of the height, $h$ metres, reached.
When $h = 25$, $T = 4.47$

   **a** Find a formula for $T$ in terms of $h$.

   **b** Find the value of $T$ when $h = 50$ Give your answer correct to 3 significant figures.

   The ball is thrown upwards and remains in the air for 5 seconds.

   **c** Find the height reached. Give your answer correct to 3 significant figures.

## Chapter summary

### You should now know:

★  how to set up and use equations to solve problems involving direct proportion, for example
  • if $y$ is directly proportional to $x$, $y \propto x$ and $y = kx$
  • if $y$ is directly proportional to the square of $x$, $y \propto x^2$ and $y = kx^2$

★  how to set up and use equations to solve problems involving inverse proportion, for example
  • if $y$ is inversely proportional to $x$, $y \propto \dfrac{1}{x}$ and $y = \dfrac{k}{x}$
  • if $y$ is inversely proportional to the square of $x$, $y \propto \dfrac{1}{x^2}$ and $y = \dfrac{k}{x^2}$

★  that $k$ is a number known as the **constant of proportionality**

★  the shapes of the graphs that represent the different types of proportionality.

## Chapter 42 review questions

**1** Here are three examples of proportionality.
   **i** $y$ is directly proportional to $x$.
   **ii** $V$ is directly proportional to the cube of $r$.
   **iii** $T$ is inversely proportional to the square root of $s$.

   **a** Express each of **i** to **iii** as a formula. Include a constant of proportionality.

   **b** Draw a sketch of the graph that represents the type of proportionality described in each of **i** to **iii**.

**2** $V$ is directly proportional to $r$.
   When $r = 2$, $V = 8$

   **a** Find $V$ when $r = 6$          **b** Find $r$ when $V = 2$

**3** $y$ is inversely proportional to $x$.
   When $x = 10$, $y = 12$

   **a** Find a formula for $y$ in terms of $x$.

   **b** Find the value of $y$ when $x = 20$

   **c** Find the value of $x$ when $y = 25$

**4** The time, $T$ seconds, it takes a pendulum to swing once is proportional to the square root of the length, $l$ metres, of the pendulum.
   When $l = 0.16$, $T = 0.8$

   **a** Find a formula for $T$ in terms of $l$.

   **b** Find the value of $T$ when $l = 1.44$

   **c** What length of pendulum will give a swing of 1 second? Give your answer correct to 3 significant figures.

**5** The drag force, $F$ newtons, on an object moving with a speed, $s$ metres per second, is proportional to the square of the speed.
When $s = 20$, $F = 80$

   **a** Express $F$ in terms of $s$.

   **b** Find $F$ when $s = 30$

   **c** Find the speed when the drag force is 300 newtons.

**6** $d$ is directly proportional to the square of $t$.
$d = 80$ when $t = 4$

   **a** Express $d$ in terms of $t$.

   **b** Work out the value of $d$ when $t = 7$

   **c** Work out the positive value of $t$ when $d = 45$                  (1387 June 2005)

**7** $P$ is inversely proportional to $V$.
When $V = 2$, $P = 7.5$

   **a** Find a formula for $P$ in terms of $V$.

   The value of $V$ is increased by 25%.

   **b** Work out the percentage change in the value of $P$.

**8** The temperature, $T°$, at a distance, $d$ metres, from a heat source is inversely proportional to the square of the distance.
When $d = 4$, $T = 275$

   **a** Find $T$ when $d = 6$

   **b** Find $d$ when $T = 1000$ Give your answer correct to 3 significant figures.

**9** The oscillation frequency, $f$ cycles per second, of a spring is inversely proportional to the square root of the mass, $m$ kg, of the spring.
When $m = 2.56$, $f = 2$
Find $f$ when $m = 4$

**10** The time taken, $T$ seconds, for a particle to slide down a smooth slope of length, $l$ m, is directly proportional to the square root of the length.
When $T = 1.5$, $l = 6.25$

   **a** Find a formula for $T$ in terms of $l$.

   **b** Rearrange the formula to find $l$ in terms of $T$.

**11** The rate of melting, $M$ grams per second, of a sphere of ice is proportional to the square of the radius, $r$ cm.
When $r = 20$, $M = 0.6$

   **a** Show that $M = 0.0015 \times r^2$

   **b** Find the rate of melting when the radius is 40 cm.

   **c** Find the radius when the rate of melting is 1 gram per second. Give your answer correct to 3 significant figures.

   **d** Hannah claims that the rate of melting is directly proportional to the surface area, $A$ cm², of the sphere. Is Hannah correct? You must justify your answer.

**12** The force, $F$, between two magnets is inversely proportional to the square of the distance, $x$, between them.
When $x = 3$, $F = 4$

   **a** Find an expression for $F$ in terms of $x$.

   **b** Calculate $F$ when $x = 2$

   **c** Calculate $x$ when $F = 64$                        (1387 June 2003)

**13** In a factory, chemical reactions are carried out in spherical containers.
The time, $T$ minutes, the chemical reaction takes is directly proportional to the square of the radius, $R$ cm, of the spherical container.
When $R = 120$, $T = 32$
Find the value of $T$ when $R = 150$                                    (1387 November 2004)

**14** The shutter speed, $S$, of a camera varies inversely as the square of the aperture setting, $f$.
When $f = 8$, $S = 125$

**a** Find a formula for $S$ in terms of $f$.

**b** Hence, or otherwise, calculate the value of $S$ when $f = 4$      (1387 June 2004)

**15**

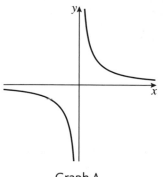

Graph A                                   Graph B

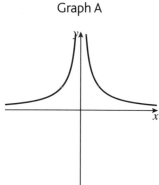

Graph C                                   Graph D

The graphs of $y$ against $x$ represent four different types of proportionality.
Copy the table and write down the letter of the graph which represents the type of proportionality.

| Type of proportionality | Graph letter |
|---|---|
| $y$ is directly proportional to $x$ | .............................. |
| $y$ is inversely proportional to $x$ | .............................. |
| $y$ is proportional to the square of $x$ | .............................. |
| $y$ is inversely proportional to the square of $x$ | .............................. |

(1387 November 2004)

# Vectors

## 43.1 Vectors and vector notation

When a ball is thrown, the direction of the throw is as important as the strength of the throw. So if a netball player wants to pass the ball to another player to her right it would be no use throwing the ball to her left.

Similarly it would not be sensible to travel 60 km due north in order to go from London to Brighton when Brighton is 60 km due south of London.

In mathematics there are many quantities that need a **direction** as well as a size in order to describe them completely.

The netball example illustrates the fact that to describe a force the direction is important. How a body moves when it is pushed or pulled will depend on the direction of the push or pull as well as the **size** or **magnitude** of the push or pull.

The second example illustrates the fact that to describe a **change in position** or a **displacement**, it is necessary to give the direction of the movement as well as the distance moved. To tell someone how to get from London to Brighton it is not enough to tell them that Brighton is 60 kilometres from London. It would be necessary to tell them that Brighton is also due south of London.

Forces and displacements that need a size and a direction to describe them are examples of **vectors**.

A **vector** needs a **magnitude** and a **direction** to describe it completely.

Velocity is another example of a vector. To describe a velocity it is necessary to give its magnitude (speed) and a direction, for example 50 km/h north.

In this chapter, only displacement vectors will be considered but the results apply to other vectors.

The displacement from $A$ to $B$ is 4 cm on a bearing of 030°.

This displacement is written $\overrightarrow{AB}$ to show that it is a vector and it has a direction from $A$ to $B$.

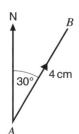

In the diagram the line from $A$ to $B$ is drawn 4 cm long in a direction of 030° and it is marked with an arrow to show that the direction is from $A$ to $B$.

Vectors can also be labelled with single bold letters such as **a**, **b** and **c**.

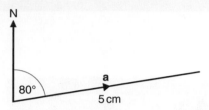

## Example 1

A vector **a** has magnitude 5 cm and direction 080°. Draw the vector **a**.

### Solution 1

> Draw a line 5 cm long (magnitude) on a bearing of 080° (direction marked with an arrow).

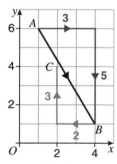

The vector **b** has been drawn on a centimetre grid. The displacement represented by **b** can be described as 4 to the right and 2 up. As with translations this can be written as the **column vector** $\begin{pmatrix} 4 \\ 2 \end{pmatrix}$.

So we can write $\mathbf{b} = \begin{pmatrix} 4 \\ 2 \end{pmatrix}$.

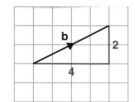

## Example 2

Point $A$ has coordinates $(1, 6)$ and point $B$ has coordinates $(4, 1)$.

**a** Write $\overrightarrow{AB}$ as a column vector.

The point $C$ is such that $\overrightarrow{BC} = \begin{pmatrix} -2 \\ 3 \end{pmatrix}$.

**b** Find the coordinates of $C$.

### Solution 2

**a**

> Mark the points $A$ and $B$ on a grid.

$$\overrightarrow{AB} = \begin{pmatrix} 3 \\ -5 \end{pmatrix}$$

> To move from $A$ to $B$ go 3 to the right and 5 down.

**b** The coordinates of $C$ are $(2, 4)$

> For $\overrightarrow{BC} = \begin{pmatrix} -2 \\ 3 \end{pmatrix}$, from $B$ go 2 to the left and 3 up to find $C$.

### Exercise 43A

**1** Draw accurately and label the following vectors.

  **i** Vector **a** with magnitude 3 cm and direction east.

  **ii** Vector **b** with magnitude 5 cm and direction with bearing 030°

  **iii** Vector **c** with magnitude 4.5 cm and direction with bearing 240°

  **iv** Vector $\overrightarrow{AB}$ with magnitude 6 cm and direction 335°

  **v** Vector $\overrightarrow{PQ}$ with magnitude 4 cm and direction 140°.

**2** On squared paper draw and label the following vectors.

**i** $a = \begin{pmatrix} 1 \\ 2 \end{pmatrix}$    **ii** $b = \begin{pmatrix} 4 \\ -2 \end{pmatrix}$    **iii** $c = \begin{pmatrix} -5 \\ -3 \end{pmatrix}$    **iv** $\overrightarrow{AB} = \begin{pmatrix} -4 \\ 3 \end{pmatrix}$    **v** $\overrightarrow{CD} = \begin{pmatrix} 0 \\ 5 \end{pmatrix}$.

**3** The point $A$ is $(1, 3)$, the point $B$ is $(6, 9)$ and the point $C$ is $(5, -3)$.

   **a** Write as column vectors

     **i** $\overrightarrow{AB}$      **ii** $\overrightarrow{BC}$      **iii** $\overrightarrow{AC}$.

   **b** What do you notice about your answers in **a**?

**4** The points $A$, $B$, $C$ and $D$ are the vertices of a quadrilateral where

   $A$ has coordinates $(2,1)$, $\overrightarrow{AB} = \begin{pmatrix} 2 \\ 4 \end{pmatrix}$, $\overrightarrow{BC} = \begin{pmatrix} 3 \\ 1 \end{pmatrix}$ and $\overrightarrow{CD} = \begin{pmatrix} 4 \\ -2 \end{pmatrix}$.

   **a** On squared paper draw quadrilateral $ABCD$.

   **b** Write as a column vector $\overrightarrow{AD}$.

   **c** What type of quadrilateral is $ABCD$?

   **d** What do you notice about $\overrightarrow{BC}$ and $\overrightarrow{AD}$?

**5** The points $A$, $B$, $C$ and $D$ are the vertices of a parallelogram.

   $A$ has coordinates $(0, 1)$, $\overrightarrow{AB} = \begin{pmatrix} 4 \\ 0 \end{pmatrix}$ and $\overrightarrow{AD} = \begin{pmatrix} 2 \\ 3 \end{pmatrix}$.

   **a** On squared paper draw the parallelogram $ABCD$.

   **b** Write as a column vector  **i** $\overrightarrow{DC}$      **ii** $\overrightarrow{CB}$.

   **c** What do you notice about  **i** $\overrightarrow{AB}$ and $\overrightarrow{DC}$      **ii** $\overrightarrow{AD}$ and $\overrightarrow{CB}$?

## 43.2 Equal vectors

As vectors need magnitude (size) and direction to describe them, vectors are equal only when they have equal magnitudes **and** the same direction.

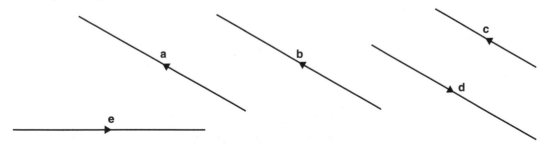

The vectors **a** and **b** are equal, that is **a** = **b**. They have the same magnitude and direction.

The vectors **a** and **c** are not equal. They do not have the same magnitude although they have the same direction.

The vectors **a** and **d** are not equal. They have the same magnitude and are parallel but they are in opposite directions and so do not have the same direction.

The vectors **a** and **e** are not equal. They have the same magnitude but they do not have the same direction.

## 43.3 The magnitude of a vector

The magnitude of the vector **a** is written $a$ or $|a|$.

The magnitude of the vector $\overrightarrow{AB}$ is $AB$, that is, the length of the line segment $AB$.

---

### Example 3

Find the magnitude of the vector $\mathbf{a} = \begin{pmatrix} 4 \\ -6 \end{pmatrix}$.

Give your answer **i** as a surd **ii** correct to 3 significant figures.

**Solution 3**

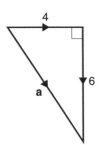

> $\begin{pmatrix} 4 \\ -6 \end{pmatrix}$ means 4 to the right and 6 down.
>
> Draw a right-angled triangle to show this.

$a^2 = 4^2 + 6^2 = 16 + 36$

$a^2 = 52$

> Use Pythagoras' theorem to find the length, $a$, of the hypotenuse.

**i** $a = \sqrt{52} = 2\sqrt{13}$

**ii** $a = 7.21$ (to 3 s.f.)

---

Notice that in this example

the magnitude of the vector $\begin{pmatrix} 4 \\ -6 \end{pmatrix}$ is $\sqrt{(4)^2 + (-6)^2} = \sqrt{16 + 36} = \sqrt{52}$

In general

> **the magnitude of the vector $\begin{pmatrix} x \\ y \end{pmatrix}$ is $\sqrt{x^2 + y^2}$**

---

### Example 4

Find the magnitude of the vector $\overrightarrow{AB} = \begin{pmatrix} -3 \\ -4 \end{pmatrix}$.

**Solution 4**

$AB = \sqrt{(-3)^2 + (-4)^2}$

$\qquad = \sqrt{9 + 16}$

$AB = 5$

> Substitute $x = -3$ and $y = -4$ into $\sqrt{x^2 + y^2}$

---

## Exercise 43B

**1** Here are 8 vectors.

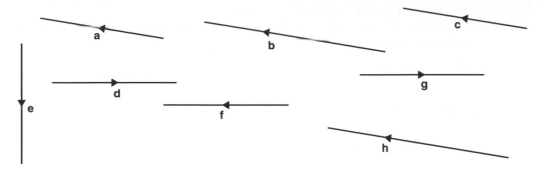

There are 3 pairs of equal vectors.
Name the equal vectors.

**2** Work out the magnitude of each of these vectors. Where necessary, answers may be left as surds.

**i** $a = \begin{pmatrix} 5 \\ 12 \end{pmatrix}$       **ii** $b = \begin{pmatrix} 12 \\ -5 \end{pmatrix}$       **iii** $c = \begin{pmatrix} 1 \\ 3 \end{pmatrix}$

**iv** $d = \begin{pmatrix} -5 \\ -7 \end{pmatrix}$       **v** $\overrightarrow{AB} = \begin{pmatrix} 8 \\ -15 \end{pmatrix}$       **vi** $\overrightarrow{PQ} = \begin{pmatrix} -8 \\ 4 \end{pmatrix}$

**3** In triangle $ABC$, $\overrightarrow{AB} = \begin{pmatrix} -20 \\ -15 \end{pmatrix}$ and $\overrightarrow{AC} = \begin{pmatrix} 24 \\ -7 \end{pmatrix}$.

     **a** Work out the length of the side $AB$ of the triangle.

     **b** Show that the triangle is an isosceles triangle.

**4** In quadrilateral $ABCD$, $\overrightarrow{AB} = \begin{pmatrix} 3 \\ 4 \end{pmatrix}$, $\overrightarrow{BC} = \begin{pmatrix} 5 \\ 0 \end{pmatrix}$, $\overrightarrow{CD} = \begin{pmatrix} -3 \\ -4 \end{pmatrix}$, $\overrightarrow{DA} = \begin{pmatrix} -5 \\ 0 \end{pmatrix}$.

     What type of quadrilateral is $ABCD$?

## 43.4 Addition of vectors

The two-stage journey from $A$ to $B$ and then from $B$ to $C$ has the same starting point and the same finishing point as the single journey from $A$ to $C$.

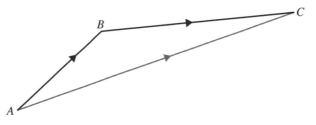

That is,             $A$ to $B$ followed by $B$ to $C$ is equivalent to $A$ to $C$

or                 $\overrightarrow{AB}$ followed by $\overrightarrow{BC}$ is equivalent to $\overrightarrow{AC}$.

This is written as       $\overrightarrow{AB} + \overrightarrow{BC} = \overrightarrow{AC}$.

Notice the pattern here $\overrightarrow{AB}\ \overrightarrow{BC}$ gives $AC$.

This leads to the triangle law of vector addition.

> This does not mean that $AB + BC = AC$. The sum of the lengths of $AB$ and $BC$ is not equal to the length of $AC$.

## Triangle law of vector addition

Let $\overrightarrow{AB}$ represent the vector **a** and $\overrightarrow{BC}$ represent the vector **b**.

Then if $\overrightarrow{AC}$ represents the vector **c**

$$\mathbf{a} + \mathbf{b} = \mathbf{c}$$

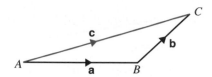

## Parallelogram law of vector addition

$PQRS$ is a parallelogram.

In a parallelogram, opposite sides are equal in length and are parallel.

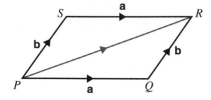

So since $\overrightarrow{PQ}$ and $\overrightarrow{SR}$ are also in the same direction $\overrightarrow{PQ} = \overrightarrow{SR}$ (= **a**).

Similarly $\overrightarrow{PS} = \overrightarrow{QR}$ (= **b**).

From the triangle law $\overrightarrow{PQ} + \overrightarrow{QR} = \overrightarrow{PR}$ so that $\overrightarrow{PR} = \mathbf{a} + \mathbf{b}$.

Hence $\overrightarrow{PR} = \overrightarrow{PQ} + \overrightarrow{PS}$ as $\overrightarrow{PQ} = \mathbf{a}$ and $\overrightarrow{PS} = \mathbf{b}$.

So if in parallelogram $PQRS$, $\overrightarrow{PQ}$ represents the vector **a** and $\overrightarrow{PS}$ represents the vector **b**, the diagonal $\overrightarrow{PR}$ of the parallelogram represents the vector **a** + **b**.

Both of these laws allow vectors to be added but the triangle law is the easier to use.

When **c** = **a** + **b** the vector **c** is said to be the **resultant** of the two vectors **a** and **b**.

### Example 5

Find, by drawing, the sum of the vectors **a** and **b**.

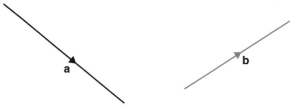

### Solution 5

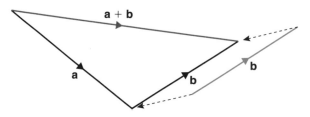

Use the triangle law of vector addition.

Move vector **b** to the end of vector **a** so that the arrows follow on.

Draw and label the vector **a** + **b** to complete the triangle.

**a** + **b** could also have been found by moving the vector **a** to the beginning of vector **b**. The answer is the same as the two triangles are congruent.

## Example 6

In the quadrilateral $ABCD$, $\overrightarrow{AB} = \mathbf{a}$, $\overrightarrow{BC} = \mathbf{b}$ and $\overrightarrow{CD} = \mathbf{c}$.

Find the vectors  **i** $\overrightarrow{AC}$    **ii** $\overrightarrow{AD}$.

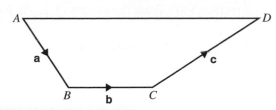

### Solution 6

> Use the triangle law of vector addition.

**i** $\overrightarrow{AC} = \overrightarrow{AB} + \overrightarrow{BC}$

> Make sure that the $B$s follow each other.

so $\overrightarrow{AC} = \mathbf{a} + \mathbf{b}$

**ii** $\overrightarrow{AD} = \overrightarrow{AC} + \overrightarrow{CD}$

> Use $\overrightarrow{AC} = \mathbf{a} + \mathbf{b}$.

so $\overrightarrow{AD} = (\mathbf{a} + \mathbf{b}) + \mathbf{c}$

$\overrightarrow{AD} = \mathbf{a} + \mathbf{b} + \mathbf{c}$

> Vector expressions like this can be treated as in ordinary algebra. The brackets can be removed.

## Example 7

$\overrightarrow{AB} = \begin{pmatrix} 3 \\ 5 \end{pmatrix}$ and $\overrightarrow{BC} = \begin{pmatrix} 8 \\ -4 \end{pmatrix}$      Find $\overrightarrow{AC}$.

### Solution 7

$\overrightarrow{AC} = \overrightarrow{AB} + \overrightarrow{BC}$

> Use the triangle law of vector addition.

$\overrightarrow{AC} = \begin{pmatrix} 3 \\ 5 \end{pmatrix} + \begin{pmatrix} 8 \\ -4 \end{pmatrix}$

> Draw a sketch.

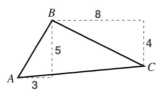

> From $A$ to $B$ is 3 to the right.
> From $B$ to $C$ is 8 to the right.
> So from $A$ to $C$ is $3 + 8 = 11$ to the right.

$\overrightarrow{AC} = \begin{pmatrix} 11 \\ 1 \end{pmatrix}$

> From $A$ to $B$ is 5 up.
> From $B$ to $C$ is 4 down.
> So from $A$ to $C$ is $5 + -4 = 1$ up.

Example 7 shows that $\begin{pmatrix} a \\ b \end{pmatrix} + \begin{pmatrix} c \\ d \end{pmatrix} = \begin{pmatrix} a + c \\ b + d \end{pmatrix}$.

## Example 8

$\mathbf{a} = \begin{pmatrix} 5 \\ -6 \end{pmatrix}$ and $\mathbf{b} = \begin{pmatrix} -3 \\ 4 \end{pmatrix}$      Find $\mathbf{a} + \mathbf{b}$.

### Solution 8

$\mathbf{a} + \mathbf{b} = \begin{pmatrix} 5 \\ -6 \end{pmatrix} + \begin{pmatrix} -3 \\ 4 \end{pmatrix} = \begin{pmatrix} 5 + -3 \\ -6 + 4 \end{pmatrix}$

> Add across.

$\mathbf{a} + \mathbf{b} = \begin{pmatrix} 2 \\ -2 \end{pmatrix}$

**Exercise 43C**

1 A vector **a** has magnitude 5 cm and direction 030°. A vector **b** has magnitude 7 cm and direction 140°. Draw the vector **i a ii b iii a + b**.

2 Work out

**a** $\begin{pmatrix} 2 \\ 6 \end{pmatrix} + \begin{pmatrix} 4 \\ 2 \end{pmatrix}$ **b** $\begin{pmatrix} 6 \\ 3 \end{pmatrix} + \begin{pmatrix} -2 \\ 5 \end{pmatrix}$ **c** $\begin{pmatrix} -5 \\ 8 \end{pmatrix} + \begin{pmatrix} 3 \\ -4 \end{pmatrix}$ **d** $\begin{pmatrix} 6 \\ 0 \end{pmatrix} + \begin{pmatrix} 3 \\ -5 \end{pmatrix}$ **e** $\begin{pmatrix} -5 \\ 3 \end{pmatrix} + \begin{pmatrix} -3 \\ -6 \end{pmatrix}$.

3 $\overrightarrow{PQ} = \begin{pmatrix} 3 \\ 1 \end{pmatrix}$  $\overrightarrow{QR} = \begin{pmatrix} 7 \\ -6 \end{pmatrix}$

Work out $\overrightarrow{PR}$.

4 $\mathbf{p} = \begin{pmatrix} 3 \\ 6 \end{pmatrix}$  $\mathbf{q} = \begin{pmatrix} 1 \\ -3 \end{pmatrix}$  $\mathbf{r} = \begin{pmatrix} 4 \\ 7 \end{pmatrix}$

**a** Work out **i p + q ii q + p**
**b** What do you notice?
**c** Work out **i (p + q) + r ii p + (q + r)**
**d** What do you notice?

5 *ABCDEF* is a regular hexagon.
$\overrightarrow{AB} = \mathbf{n}$

**a** Explain why $\overrightarrow{ED} = \mathbf{n}$.
$\overrightarrow{BC} = \mathbf{m}$  $\overrightarrow{CD} = \mathbf{p}$

**b** Find **i** $\overrightarrow{AC}$  **ii** $\overrightarrow{AD}$.

**c** What is $\overrightarrow{FD}$?

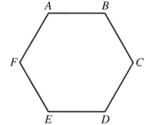

# 43.5 Parallel vectors

Using the ordinary rules of algebra **a + a = 2a** but what does this mean?

Here is the vector **a**.

Here are **a + a** and 2**a**.

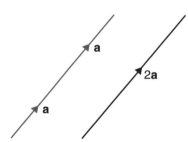

2**a** is a vector in the same direction as **a** and with twice the magnitude.

For $\mathbf{a} = \begin{pmatrix} 2 \\ 5 \end{pmatrix}$, $\mathbf{a} + \mathbf{a} = \begin{pmatrix} 2 \\ 5 \end{pmatrix} + \begin{pmatrix} 2 \\ 5 \end{pmatrix} = \begin{pmatrix} 2+2 \\ 5+5 \end{pmatrix} = \begin{pmatrix} 2 \times 2 \\ 2 \times 5 \end{pmatrix}$

that is, $2\mathbf{a} = 2\begin{pmatrix} 2 \\ 5 \end{pmatrix} = \begin{pmatrix} 2 \times 2 \\ 2 \times 5 \end{pmatrix} = \begin{pmatrix} 4 \\ 10 \end{pmatrix}$.

Similarly 3**a** is a vector in the same direction as **a** and with magnitude 3 times the magnitude of **a**.

And $3\mathbf{a} = 3\begin{pmatrix} 2 \\ 5 \end{pmatrix} = \begin{pmatrix} 3 \times 2 \\ 3 \times 5 \end{pmatrix} = \begin{pmatrix} 6 \\ 15 \end{pmatrix}$.

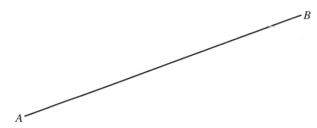

The vector $\overrightarrow{AB}$ is the displacement from $A$ to $B$ and $\overrightarrow{BA}$ is the displacement from $B$ to $A$.

These displacements have the same magnitudes but are in opposite directions so $\overrightarrow{AB}$ followed by $\overrightarrow{BA}$ is the zero displacement (**0**) as there is no overall change in position.

This is written $\overrightarrow{AB} + \overrightarrow{BA} = \mathbf{0}$

Using the usual rules of algebra it follows that $\overrightarrow{BA} = -\overrightarrow{AB}$.

$\overrightarrow{AB}$ and $\overrightarrow{BA}$ have the same magnitude but opposite directions.

A negative sign in front of a vector, reverses the direction of the vector.

$$\overrightarrow{AB} = \begin{pmatrix} 3 \\ -5 \end{pmatrix} \text{ so } \overrightarrow{BA} = -\begin{pmatrix} 3 \\ -5 \end{pmatrix} = -1\begin{pmatrix} 3 \\ -5 \end{pmatrix} = \begin{pmatrix} -1 \times 3 \\ -1 \times -5 \end{pmatrix} = \begin{pmatrix} -3 \\ 5 \end{pmatrix}$$

showing that the reverse of 3 to the right and 5 down is 3 to the left and 5 up.

The vector $-\mathbf{a}$ has the same magnitude as **a** but is in the opposite direction.
The vector $-3\mathbf{a}$ has the same magnitude as 3**a** but is in the opposite direction.
So the vector $-3\mathbf{a}$ has 3 times the magnitude as **a** but is in the opposite direction.

Vectors that are parallel either have the same direction or have opposite directions.

**For any non-zero value of $k$, the vectors a and $k$a are parallel.**

The number $k$ is called a **scalar**; it has magnitude only.

**If $a = \begin{pmatrix} p \\ q \end{pmatrix}$ then $ka = k\begin{pmatrix} p \\ q \end{pmatrix} = \begin{pmatrix} kp \\ kq \end{pmatrix}$**

### Example 9

Here is the vector **a**.

Draw the vectors   **i**   3**a**     **ii**   $-2$**a**.

## Solution 9

**i**

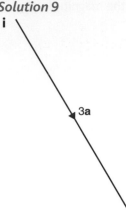

3**a** is in the same direction as **a** but with 3 times the magnitude.

Draw a line in the same direction as **a** but 3 times longer.

**ii**

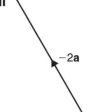

−2**a** is in the opposite direction to **a** but with 2 times the magnitude.

Draw a line in the opposite direction to **a** and twice as long.

## Example 10

Draw the vector **a** − **b**.

## Solution 10

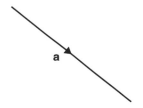

**a** − **b** means **a** + −**b**

−**b** is the vector **b** reversed in direction.

Use the triangle law of vector addition to add **a** and −**b**

Move vector −**b** to the end of vector **a** so that the arrows follow on.

Draw and label the vector **a** − **b** to complete the triangle.

## Example 11

With origin $O$, the points $A$, $B$, $C$ and $D$ have coordinates $(1, 3)$, $(2, 7)$, $(-6, -10)$ and $(-1, 10)$ respectively.

**a** Write down as a column vector   **i** $\overrightarrow{OA}$     **ii** $\overrightarrow{OB}$.

**b** Work out   **i** $\overrightarrow{AB}$ as a column vector     **ii** $\overrightarrow{CD}$ as a column vector.

**c** What do these results show about $AB$ and $CD$?

### Solution 11

**a**  **i** $\overrightarrow{OA} = \begin{pmatrix} 1 \\ 3 \end{pmatrix}$

> From $O$ to $A$ is 1 across and 3 up.

**ii** $\overrightarrow{OB} = \begin{pmatrix} 2 \\ 7 \end{pmatrix}$

> From $O$ to $B$ is 2 across and 7 up.

**b**  **i**

**Method 1**

$\overrightarrow{AB} = \begin{pmatrix} 1 \\ 4 \end{pmatrix}$

> $A$ to $B$, that is, $(1, 3)$ to $(2, 7)$ is 1 across and 4 up.

**Method 2**

$\overrightarrow{AB} = \overrightarrow{AO} + \overrightarrow{OB}$

> Another way to obtain $\overrightarrow{AB}$ is to use the triangle law of vector addition.

$\overrightarrow{AB} = -\overrightarrow{OA} + \overrightarrow{OB}$

> $\overrightarrow{AO} = -\overrightarrow{OA}$.

$\overrightarrow{AB} = -\begin{pmatrix} 1 \\ 3 \end{pmatrix} + \begin{pmatrix} 2 \\ 7 \end{pmatrix}$

$= \begin{pmatrix} -1 \\ -3 \end{pmatrix} + \begin{pmatrix} 2 \\ 7 \end{pmatrix}$

$= \begin{pmatrix} -1 + 2 \\ -3 + 7 \end{pmatrix}$

$\overrightarrow{AB} = \begin{pmatrix} 1 \\ 4 \end{pmatrix}$

**ii** $\overrightarrow{CD} = \begin{pmatrix} 5 \\ 20 \end{pmatrix}$

> $C$ to $D$ is 5 to the right and 20 up.

**c** $\overrightarrow{CD} = \begin{pmatrix} 5 \\ 20 \end{pmatrix} = 5\begin{pmatrix} 1 \\ 4 \end{pmatrix}$

$\overrightarrow{CD} = 5\overrightarrow{AB}$

The lines $CD$ and $AB$ are parallel and the length of the line $CD$ is 5 times the length of the line $AB$.

> **a** and $k$**a** are parallel vectors.

With the origin $O$, the vectors $\overrightarrow{OA}$ and $\overrightarrow{OB}$ are called the **position vectors** of the points $A$ and $B$.

In general the point $(p, q)$ has position vector $\begin{pmatrix} p \\ q \end{pmatrix}$.

## Example 12

Simplify  **i** $3\mathbf{a} + 5\mathbf{b} + 2\mathbf{a} - 3\mathbf{b}$   **ii** $2\mathbf{a} + \frac{1}{2}(4\mathbf{a} - 2\mathbf{b})$.

### Solution 12

> The ordinary rules of algebra can be applied to vector expressions like this.

**i** $3\mathbf{a} + 5\mathbf{b} + 2\mathbf{a} - 3\mathbf{b}$

$\quad = 5\mathbf{a} + 2\mathbf{b}$

> $3\mathbf{a} + 2\mathbf{a} = 5\mathbf{a}$

> $5\mathbf{b} - 3\mathbf{b} = 2\mathbf{b}$

**ii** $2\mathbf{a} + \frac{1}{2}(4\mathbf{a} - 2\mathbf{b})$

$\quad = 2\mathbf{a} + 2\mathbf{a} - \mathbf{b} = 4\mathbf{a} - \mathbf{b}$

> $\frac{1}{2}(4\mathbf{a} - 2\mathbf{b}) = 2\mathbf{a} - \mathbf{b}$

## Example 13

$ABC$ is a straight line where $BC = 3AB$.

$\overrightarrow{OA} = \mathbf{a}$    $\overrightarrow{AB} = \mathbf{b}$

Express $\overrightarrow{OC}$ in terms of $\mathbf{a}$ and $\mathbf{b}$.

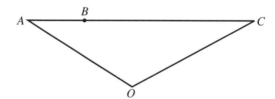

### Solution 13

$\overrightarrow{OC} = \overrightarrow{OA} + \overrightarrow{AC}$

$\overrightarrow{OC} = \overrightarrow{OA} + 4\overrightarrow{AB}$

$\overrightarrow{OC} = \mathbf{a} + 4\mathbf{b}$

> Use the triangle law of vector addition.

> As $BC = 3AB, \overrightarrow{AC} = 4\overrightarrow{AB}$.

## Exercise 43D

**1** The vector **a** has magnitude 4 cm and direction 130°. The vector **b** has magnitude 5 cm and direction 220°.
Draw the vector  **i a**   **ii b**   **iii** −**b**   **iv a** − **b**.

**2** Here is the vector **p**.
Draw the vector  **i** $2\mathbf{p}$   **ii** $-\frac{1}{2}\mathbf{p}$.

**3** $\mathbf{m} = \begin{pmatrix} 4 \\ 3 \end{pmatrix}$   $\mathbf{n} = \begin{pmatrix} 6 \\ -3 \end{pmatrix}$   $\mathbf{p} = \begin{pmatrix} -2 \\ 6 \end{pmatrix}$

  **a** Find as a column vector  **i** $5\mathbf{m}$   **ii** $-2\mathbf{n}$   **iii** $4\mathbf{m} + 3\mathbf{p}$   **iv** $2\mathbf{m} - 4\mathbf{n} + 5\mathbf{p}$.
  **b** Find  **i** the magnitude of the vector **m**   **ii** the magnitude of the vector $2\mathbf{m} - \mathbf{p}$.

**4** The points $P, Q, R$ and $S$ have coordinates $(-2, 5), (3, 1), (-6, -9)$ and $(14, -25)$ respectively.
  **a** Write down the position vector, $\overrightarrow{OP}$, of the point $P$.
  **b** Write down as a column vector  **i** $\overrightarrow{PQ}$   **ii** $\overrightarrow{RS}$.
  **c** What do these results show about the lines $PQ$ and $RS$?

**5** The point $A$ has coordinates $(1, 3)$, the point $B$ has coordinates $(4, 5)$, the point $C$ has coordinates $(-2, -4)$. Find the coordinates of the point $D$ where $\overrightarrow{CD} = 6\overrightarrow{AB}$.

**6**

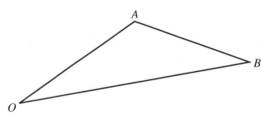

$\overrightarrow{OA} = \mathbf{a}$   $\overrightarrow{OB} = \mathbf{b}$

**i** Express $\overrightarrow{AB}$ in terms of $\mathbf{a}$ and $\mathbf{b}$.

**ii** Where is the point $C$ such that $\overrightarrow{OC} = \frac{1}{2}\mathbf{b}$?

**7** Here are 5 vectors.

$$\overrightarrow{AB} = 2\mathbf{m} + 4\mathbf{n}, \ \overrightarrow{CD} = 6\mathbf{m} - 12\mathbf{n}, \ \overrightarrow{EF} = 4\mathbf{m} + 8\mathbf{n}, \ \overrightarrow{GH} = -\mathbf{m} - 2\mathbf{n}, \ \overrightarrow{IJ} = 6\mathbf{m} + 16\mathbf{n}$$

**a** Three of these vectors are parallel. Which are the parallel vectors?

**b** Simplify   **i** $8\mathbf{p} + 5\mathbf{q} - 3\mathbf{p} - 8\mathbf{q}$     **ii** $2(2\mathbf{m} - 5\mathbf{n}) + \frac{2}{3}(3\mathbf{m} - 6\mathbf{n})$.

**8** Here is a regular hexagon $ABCDEF$. In the hexagon $FC$ is parallel to $AB$ and twice as long.

$\overrightarrow{AB} = \mathbf{m}$

**a** Express $\overrightarrow{FC}$ in terms of $\mathbf{m}$.

$\overrightarrow{CD} = \mathbf{n}$

**b** Express $\overrightarrow{FD}$ in terms of $\mathbf{m}$ and $\mathbf{n}$.

$\overrightarrow{BC} = \mathbf{x}$

**c** Express $\overrightarrow{AC}$ in terms of $\mathbf{m}$ and $\mathbf{x}$.

The lines $AC$ and $FD$ are parallel and equal in length.

**d** Find an expression for $\mathbf{x}$ in terms of $\mathbf{m}$ and $\mathbf{n}$.

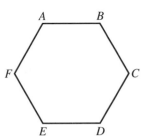

## 43.6 Solving geometric problems in two dimensions

To solve geometric problems the following results are useful.

**i** Triangle law of vector addition so that $\overrightarrow{PQ} + \overrightarrow{QR} = \overrightarrow{PR}$.

**ii** When $\overrightarrow{PQ} = \mathbf{a}$, $\overrightarrow{QP} = -\mathbf{a}$.

**iii** When $\overrightarrow{PQ} = k\overrightarrow{RS}$, $k$ a scalar (number), the lines $PQ$ and $RS$ are parallel and the length of $PQ$ is $k$ times the length of $RS$.

**iv** When $\overrightarrow{PQ} = k\overrightarrow{PR}$ then the lines $PQ$ and $PR$ are parallel. But these lines have the point $P$ in common so that $PQ$ and $PR$ are part of the same straight line. That is, the points $P$, $Q$ and $R$ lie on the same straight line.

### Example 14

In triangle $OAB$ the point $M$ is the midpoint of $OA$ and the point $N$ is the midpoint of $OB$.

$\overrightarrow{OA} = 2\mathbf{a}$   $\overrightarrow{OB} = 2\mathbf{b}$

**i** Express $\overrightarrow{AB}$ in terms of $\mathbf{a}$ and $\mathbf{b}$

**ii** Express $\overrightarrow{MN}$ in terms of $\mathbf{a}$ and $\mathbf{b}$.

**iii** Explain what the answers in **i** and **ii** show about $AB$ and $MN$.

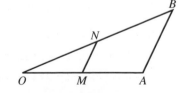

*Solution 14*

**i** $\overrightarrow{AB} = \overrightarrow{AO} + \overrightarrow{OB}$

| Use the triangle law of vector addition. |
| --- |

$\overrightarrow{AB} = -2\mathbf{a} + 2\mathbf{b}$

| $\overrightarrow{OA} = 2\mathbf{a}$ so $\overrightarrow{AO} = -2\mathbf{a}$. |
| --- |

**ii** $\overrightarrow{OM} = \frac{1}{2}2\mathbf{a} = \mathbf{a}$

| $M$ is the midpoint of $OA$ so $\overrightarrow{OM} = \frac{1}{2}\overrightarrow{OA}$. |
| --- |

Similarly

$\overrightarrow{ON} = \mathbf{b}$

$\overrightarrow{MN} = \overrightarrow{MO} + \overrightarrow{ON}$

| Use the triangle law of vector addition. |
| --- |

$\overrightarrow{MN} = -\mathbf{a} + \mathbf{b}$

| $\overrightarrow{OM} = \mathbf{a}$ so that $\overrightarrow{MO} = -\mathbf{a}$. |
| --- |

**iii** $\overrightarrow{AB} = 2\overrightarrow{MN}$

| $\overrightarrow{AB} = -2\mathbf{a} + 2\mathbf{b}$ and $\overrightarrow{MN} = -\mathbf{a} + \mathbf{b}$. |
| --- |

This means that $AB$ and $MN$ are parallel and that the length of $AB$ is twice the length of $MN$.

---

## Example 15

$OABC$ is a quadrilateral in which $\overrightarrow{OA} = \mathbf{a}$, $\overrightarrow{OB} = \mathbf{a} + 2\mathbf{b}$ and $\overrightarrow{OC} = 4\mathbf{b}$.

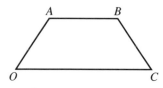

**i** Find $\overrightarrow{AB}$ in terms of $\mathbf{a}$ and $\mathbf{b}$ and explain what this answer means.

**ii** Find $\overrightarrow{CB}$ in terms of $\mathbf{a}$ and $\mathbf{b}$.

$D$ is the point such that $\overrightarrow{BD} = \overrightarrow{OC}$ and $X$ is the midpoint of $BC$.

Find in terms of $\mathbf{a}$ and $\mathbf{b}$  **iii** $\overrightarrow{OD}$  **iv** $\overrightarrow{OX}$  and  **v** explain what these results mean.

*Solution 15*

**i** $\overrightarrow{AB} = \overrightarrow{AO} + \overrightarrow{OB}$

| Express $\overrightarrow{AB}$ in terms of known vectors using the triangle law of vector addition. |
| --- |

$\overrightarrow{AB} = -\mathbf{a} + \mathbf{a} + 2\mathbf{b} = 2\mathbf{b}$

| $\overrightarrow{OA} = \mathbf{a}$ so $\overrightarrow{AO} = -\mathbf{a}$ |
| --- |

$\overrightarrow{OC} = 2\overrightarrow{AB}$

| $\overrightarrow{OC} = 4\mathbf{b}$ |
| --- |

$OC$ and $AB$ are parallel and the length of $OC$ is twice the length of $AB$.

**ii** $\overrightarrow{CB} = \overrightarrow{CO} + \overrightarrow{OB}$

| Express $\overrightarrow{CB}$ in terms of known vectors using the triangle law of vector addition. |
| --- |

$\overrightarrow{CB} = -4\mathbf{b} + \mathbf{a} + 2\mathbf{b} = \mathbf{a} - 2\mathbf{b}$

| $\overrightarrow{OC} = 4\mathbf{b}$ so $\overrightarrow{CO} = -4\mathbf{b}$ |
| --- |

| $\overrightarrow{CB} = \overrightarrow{CO} + \overrightarrow{OA} + \overrightarrow{AB}$ could also have been used. |
| --- |

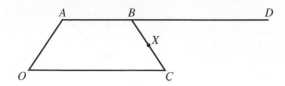

**iii** $\overrightarrow{OD} = \overrightarrow{OA} + \overrightarrow{AD}$

> $\overrightarrow{BD} = \overrightarrow{OC}$ means that the point $D$ is on $AB$ extended so that $BD$ and $OC$ have the same length. Redraw the diagram with $BD$ in and $X$ the midpoint of $BC$.

> Use the triangle law of vector addition for $\overrightarrow{OD}$.

> $\overrightarrow{OA} = $ **a**, $\overrightarrow{BD} = \overrightarrow{OC} = $ 4**b**.

$\overrightarrow{OD} = $ **a** + 6**b**

> $\overrightarrow{AD} = \overrightarrow{AB} + \overrightarrow{BD} = $ 2**b** + 4**b** = 6**b**.

**iv** As $X$ is the midpoint of $BC$

$\overrightarrow{CX} = \dfrac{1}{2}\overrightarrow{CB} = \dfrac{1}{2}(\mathbf{a} - 2\mathbf{b})$

> $\overrightarrow{CB} = $ **a** − 2**b**.

$\overrightarrow{CX} = \dfrac{1}{2}\mathbf{a} - \mathbf{b}$

$\overrightarrow{OX} = \overrightarrow{OC} + \overrightarrow{CX} = 4\mathbf{b} + \dfrac{1}{2}\mathbf{a} - \mathbf{b}$

> Use the triangle law of vector addition for $\overrightarrow{OX}$.

$\overrightarrow{OX} = \dfrac{1}{2}\mathbf{a} + 3\mathbf{b}$

> $\overrightarrow{OD} = $ **a** + 6**b**.

**v** $\overrightarrow{OD} = 2\overrightarrow{OX}$

> $\overrightarrow{OX} = \dfrac{1}{2}\mathbf{a} + 3\mathbf{b}$.

So the lines $OD$ and $OX$ are parallel with the point $O$ in common.
This means that $OX$ and $OD$ are part of the same straight line.

That is, $OXD$ is a straight line such that the length of $OD$ is 2 times the length of $OX$.
In other words $X$ is the midpoint of $OD$.

---

## Exercise 43E

**1** The points $A$, $B$ and $C$ have coordinates $(2, 13)$, $(5, 22)$ and $(11, 40)$ respectively.

  **a** Find as column vectors  **i** $\overrightarrow{AB}$   **ii** $\overrightarrow{AC}$.

  **b** What do these results show about the points $A$, $B$ and $C$?

**2** $ABCD$ is a quadrilateral.
  $A$ is the point $(1, 2)$, $B$ is the point $(3, 6)$ and $P$ is the midpoint of $AB$.

  **a** Write down the coordinates of $P$.

  $C$ is the point $(7, 4)$ and $Q$ is the midpoint of $BC$.

  **b** Write down the coordinates of $Q$.

  $D$ is the point $(7, 2)$ and $R$ is the midpoint of $CD$.

  **c** Write down the coordinates of $R$.

  $S$ is the midpoint of $AD$.

  **d** Write down the coordinates of $S$.

  **e** Find as column vectors  **i** $\overrightarrow{PQ}$   **ii** $\overrightarrow{SR}$.

  **f** Explain with reasons what the answers to **e** show about the quadrilateral $PQRS$.

**3** In triangle $OAB$, $\overrightarrow{OA} = \mathbf{a}$ and $\overrightarrow{OB} = \mathbf{b}$.

   **i** Find in terms of $\mathbf{a}$ and $\mathbf{b}$, the vector $\overrightarrow{AB}$.

   $P$ is the midpoint of $AB$.

   **ii** Find in terms of $\mathbf{a}$ and $\mathbf{b}$, the vector $\overrightarrow{AP}$

   **iii** Find in terms of $\mathbf{a}$ and $\mathbf{b}$, the vector $\overrightarrow{OP}$.

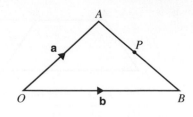

**4** $OACB$ is a parallelogram with $\overrightarrow{OA} = \mathbf{a}$ and $\overrightarrow{OB} = \mathbf{b}$.
   $P$ is the midpoint of $AB$.

   **i** Use the result of question **3** to write down $\overrightarrow{OP}$ in terms of $\mathbf{a}$ and $\mathbf{b}$.

   **ii** Express $\overrightarrow{OC}$ in terms of $\mathbf{a}$ and $\mathbf{b}$.

   $Q$ is the midpoint of $OC$.

   **iii** Express $\overrightarrow{OQ}$ in terms of $\mathbf{a}$ and $\mathbf{b}$.

   **iv** What do your answers to **i** and **iii** show about the points $P$ and $Q$?

   **v** What property of a parallelogram has been proved in this question?

**5** $KLMN$ is a quadrilateral where $\overrightarrow{KL} = \mathbf{k}$, $\overrightarrow{LM} = \mathbf{m}$, $\overrightarrow{MN} = \mathbf{n}$ and $\overrightarrow{KN} = 3\mathbf{m}$.

   **a** What type of quadrilateral is $KLMN$?

   **b** Express $\mathbf{n}$ in terms of $\mathbf{k}$ and $\mathbf{m}$.

**6** $OACB$ is a parallelogram with $\overrightarrow{OA} = \mathbf{a}$ and $\overrightarrow{OB} = \mathbf{b}$.

   $E$ is the point on $AC$ such that $AE = \frac{1}{4}AC$.

   $F$ is the point on $BC$ such that $BF = \frac{1}{4}BC$.

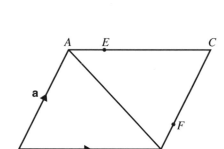

   **a** Find in terms of $\mathbf{a}$ and $\mathbf{b}$

     **i** $\overrightarrow{AB}$    **ii** $\overrightarrow{AE}$    **iii** $\overrightarrow{OE}$    **iv** $\overrightarrow{OF}$    **v** $\overrightarrow{EF}$.

   **b** Write down two geometric properties connecting $EF$ and $AB$.

**7**

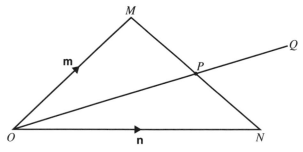

In triangle $OMN$, $\overrightarrow{OM} = \mathbf{m}$ and $\overrightarrow{ON} = \mathbf{n}$. The point $P$ is the midpoint of $MN$ and $Q$ is the point such that $\overrightarrow{OQ} = \frac{3}{2}\overrightarrow{OP}$.

   **a** Find in terms of $\mathbf{m}$ and $\mathbf{n}$   **i** $\overrightarrow{OP}$    **ii** $\overrightarrow{OQ}$    **iii** $\overrightarrow{MQ}$.

The point $R$ is such that $\overrightarrow{OR} = 3\overrightarrow{ON}$.

   **b** Find in terms of $\mathbf{m}$ and $\mathbf{n}$, the vector $\overrightarrow{MR}$.

   **c** Explain why $MQR$ is a straight line and give the value of $\dfrac{MR}{MQ}$.

**8** In the diagram $\overrightarrow{OR} = 6\mathbf{a}$, $\overrightarrow{OP} = 2\mathbf{b}$ and $\overrightarrow{PQ} = 3\mathbf{a}$.

The point $M$ is on $PQ$ such that $\overrightarrow{PM} = 2\mathbf{a}$.

The point $N$ is on $OR$ such that $\overrightarrow{ON} = \frac{1}{3}\overrightarrow{OR}$.

The midpoint of $MN$ is the point $S$.

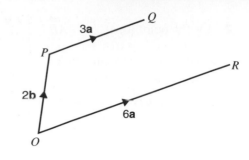

**i** Find in terms of $\mathbf{a}$ and/or $\mathbf{b}$ the vector $\overrightarrow{NM}$.

**ii** Find in terms of $\mathbf{a}$ and/or $\mathbf{b}$ the vector $\overrightarrow{OS}$.

$T$ is the point such that $\overrightarrow{QT} = \mathbf{a}$.

**iii** Find in terms of $\mathbf{a}$ and $\mathbf{b}$, the vector $\overrightarrow{OT}$.

**iv** Give a geometric fact about the point $S$ and the line $OT$.

**v** When $\mathbf{a} = \begin{pmatrix} 8 \\ 2 \end{pmatrix}$ and $\mathbf{b} = \begin{pmatrix} 3 \\ 15 \end{pmatrix}$ find the length of $QR$.

## Chapter summary

> **You should now:**
>
> ★ understand and be able to use the vector notation, $\overrightarrow{AB}$, $\mathbf{a}$ and $\begin{pmatrix} p \\ q \end{pmatrix}$
>
> ★ know that a vector has magnitude and direction
>
> ★ know that $\mathbf{a} - \mathbf{b} = \mathbf{a} + (-\mathbf{b})$.
>
> **You should also be able to:**
>
> ★ recognise equal vectors
>
> ★ find the magnitude of a vector $\begin{pmatrix} p \\ q \end{pmatrix}$ by using $\sqrt{p^2 + q^2}$
>
> ★ add vectors using the triangle and parallelogram laws of vector addition
>
> ★ simplify vector expressions including those involving scalar multiples
>
> ★ recognise parallel vectors
>
> ★ solve geometrical problems using vector methods including recognising when three points lie on a straight line.

## Chapter 43 review questions

**1** The diagram shows two vectors $\mathbf{a}$ and $\mathbf{b}$.

$$\overrightarrow{PQ} = \mathbf{a} + 2\mathbf{b}$$

Use the resource sheet to draw the vector $\overrightarrow{PQ}$ on the grid.

(1388 March 2005)

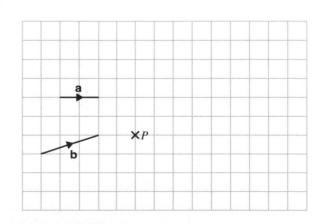

**2** $A$ is the point $(0, 4)$.     $\overrightarrow{AB} = \begin{pmatrix} 3 \\ 2 \end{pmatrix}$

    **a** Find the coordinates of $B$.

    $C$ is the point $(3, 4)$. $BD$ is a diagonal of the parallelogram $ABCD$.

    **b** Express $\overrightarrow{BD}$ as a column vector.

$\overrightarrow{CE} = \begin{pmatrix} 1 \\ -3 \end{pmatrix}$

    **c** Calculate the length of $AE$.                                        (1385 June 1999)

**3** $OPQR$ is a trapezium. $PQ$ is parallel to $OR$.

    $\overrightarrow{OP} = \mathbf{b}$     $\overrightarrow{PQ} = 2\mathbf{a}$     $\overrightarrow{OR} = 6\mathbf{a}$

    $M$ is the midpoint of $PQ$.
    $N$ is the midpoint of $OR$.

    **i** Find in terms of $\mathbf{a}$ and $\mathbf{b}$ the vector $\overrightarrow{OM}$.

    **ii** Find in terms of $\mathbf{a}$ and $\mathbf{b}$ the vector $\overrightarrow{MN}$.

    $X$ is the midpoint of $MN$.

    **iii** Find in terms of $\mathbf{a}$ and $\mathbf{b}$, the vector $\overrightarrow{OX}$.

    The lines $OX$ and $PQ$ are extended to meet at the point $Y$.

    **iv** Find in terms of $\mathbf{a}$ and $\mathbf{b}$, the vector $\overrightarrow{NY}$.         (1385 June 2000)

Diagram **NOT** accurately drawn

**4** $PQRS$ is a parallelogram.
    $T$ is the midpoint of $QR$.
    $U$ is the point on $SR$ for
    which $SU : UR = 1 : 2$

      $\overrightarrow{PQ} = \mathbf{a}$     $\overrightarrow{PS} = \mathbf{b}$

    Write down in terms of $\mathbf{a}$ and $\mathbf{b}$,
    expressions for

      **i** $\overrightarrow{PT}$     **ii** $\overrightarrow{TU}$.                      (1385 November 2000)

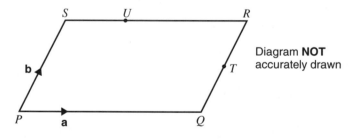

Diagram **NOT** accurately drawn

**5** $A$ is the point $(2, 3)$ and $B$ is the point $(-2, 0)$.

    **a** **i** Write $\overrightarrow{AB}$ as a column vector.

       **ii** Find the length of the vector $\overrightarrow{AB}$.

    $D$ is the point such that $\overrightarrow{BD}$ is parallel to $\begin{pmatrix} 0 \\ 1 \end{pmatrix}$ and the length of $\overrightarrow{AD}$ = the length of $\overrightarrow{AB}$.

    $O$ is the point $(0, 0)$.

    **b** Find $\overrightarrow{OD}$ as a column vector.

    $C$ is the point such that $ABCD$ is a rhombus.

    $AC$ is a diagonal of the rhombus.

    **c** Find the coordinates of $C$.                                        (1385 June 2001)

**6** The diagram shows a regular hexagon, $ABCDEF$, with centre $O$.

$$\overrightarrow{OA} = 6a \qquad \overrightarrow{OB} = 6b$$

**a** Express in terms of **a** and/or **b**

   **i** $\overrightarrow{AB}$

   **ii** $\overrightarrow{EF}$.

$X$ is the midpoint of $BC$.

**b** Express $\overrightarrow{EX}$ in terms of **a** and/or **b**.

$Y$ is the point on $AB$ extended such that $AB : BY = 3 : 2$

**c** Prove that $E, X$ and $Y$ lie on the same straight line.

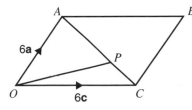

Diagram **NOT** accurately drawn

(1387 June 2003)

**7** $OABC$ is a parallelogram.

$P$ is the point on $AC$ such that $AP = \dfrac{2}{3}AC$.

$$\overrightarrow{OA} = 6a \qquad \overrightarrow{OC} = 6c$$

**i** Find the vector $\overrightarrow{OP}$.

   Give your answer in terms of **a** and **c**.

The midpoint of $CB$ is $M$.

**ii** Prove that $OPM$ is a straight line.

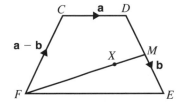

Diagram **NOT** accurately drawn

(1387 June 2004)

**8** $CDEF$ is a quadrilateral with $\overrightarrow{CD} = \mathbf{a}$, $\overrightarrow{DE} = \mathbf{b}$ and $\overrightarrow{FC} = \mathbf{a} - \mathbf{b}$.

**i** Express $\overrightarrow{CE}$ in terms of **a** and **b**.

**ii** Prove that $FE$ is parallel to $CD$.

$M$ is the midpoint of $DE$.

**iii** Express $\overrightarrow{FM}$ in terms of **a** and **b**.

$X$ is the point on $FM$ such that $FX : XM = 4 : 1$

**iv** Prove that $C, X$ and $E$ lie on the same straight line.

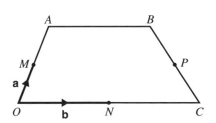

**9** $OABC$ is a trapezium.

$$\overrightarrow{OC} = 2\overrightarrow{AB}$$

$M$ is the midpoint of $OA$.

$N$ is the midpoint of $OC$.

$$\overrightarrow{OM} = \mathbf{a} \qquad \overrightarrow{ON} = \mathbf{b}$$

**i** Find $\overrightarrow{OC}$ in terms of **b**.

**ii** Find $\overrightarrow{OB}$ in terms of **a** and **b**.

**iii** Find $\overrightarrow{BC}$ in terms of **a** and **b**.

$P$ is the midpoint of $BC$.

**iv** Prove that $MP$ is parallel to $OC$.

Diagram **NOT** accurately drawn

**10** $OPQ$ is a triangle.
$R$ is the midpoint of $OP$.
$S$ is the midpoint of $PQ$.

$$\overrightarrow{OP} = \mathbf{p} \text{ and } \overrightarrow{OQ} = \mathbf{q}$$

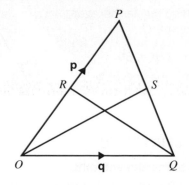

Diagram **NOT**
accurately drawn

  **i** Find $\overrightarrow{OS}$ in terms of $\mathbf{p}$ and $\mathbf{q}$.

  **ii** Show that $RS$ is parallel to $OQ$.

(1387 November 2004)

**11** $OPQR$ is a trapezium with $PQ$ parallel to $OR$.

$$\overrightarrow{OP} = 2\mathbf{b} \qquad \overrightarrow{PQ} = 2\mathbf{a} \qquad \overrightarrow{OR} = 6\mathbf{a}$$

$M$ is the midpoint of $PQ$ and $N$ is the
midpoint of $OR$.

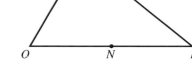

Diagram **NOT**
accurately drawn

  **i** Find the vector $\overrightarrow{MN}$ in terms of $\mathbf{a}$ and $\mathbf{b}$.

$X$ is the midpoint of $MN$ and $Y$ is the midpoint of $QR$.

  **ii** Prove that $XY$ is parallel to $OR$.

(1387 June 2005)

# Transformations of functions

## 44.1 Function notation

A function is a rule that changes a number.

A function can be thought of as a machine.
A number called the input is put into the machine.
The machine then changes the number and gives the result (the output).

Input $\longrightarrow$ function $\longrightarrow$ Output

Let the letter f stand for the rule 'double the input and then add 3'

If $x$ is the input, the output will be $2x + 3$

$x \longrightarrow$ f $\longrightarrow 2x + 3$

This rule or function can be written as $f(x) = 2x + 3$

f(4), read as 'f of 4', means work out the output when the input is 4

$4 \longrightarrow$ f $\longrightarrow 2 \times 4 + 3 = 11$

This is written as $f(4) = 2 \times 4 + 3$
$$= 11$$

In general terms, if $x$ is the input then $f(x)$ is the output

$x \longrightarrow$ f $\longrightarrow f(x)$

$f(x)$ is an example of function notation and is read as 'f of $x$'.
Any letter can be used for a function although f, g and h are the letters most commonly used.

---

### Example 1

$f(x) = 3x - 5$ and $g(x) = x^2 + 7$
Find the value of     **a** f(2)    **b** g($-3$)    **c** f(0) + g(0)

**Solution 1**
**a**   $f(x) = 3x - 5$

    $f(2) = 3 \times 2 - 5$        | Substitute **2** for $x$ |

      $= 1$

**b**    $g(x) = x^2 + 7$

    $g(-3) = (-3)^2 + 7$      | Substitute **$-3$** for $x$ |

       $= 16$

**c**   $f(x) + g(x) = (3x - 5) + (x^2 + 7)$

    $f(0) + g(0) = (3 \times 0 - 5) + (0^2 + 7)$    | Substitute **0** for $x$ |

        $= 2$

---

## Example 2

$f(x) = 2x + 1$
Find an expression for     **a** $f(x) + 4$     **b** $3f(x)$     **c** $f(-x)$

**Solution 2**
**a**   $f(x) + 4 = 2x + 1 + 4$       | Replace $f(x)$ with $2x + 1$ |

           $= 2x + 5$            | Simplify. |

**b**   $3f(x) = 3(2x + 1)$       | Replace $f(x)$ with $2x + 1$ |

**c**     $f(x) = 2x + 1$

    $f(-x) = 2 \times (-x) + 1$       | Replace $x$ by $-x$ |

        $= 1 - 2x$

## Exercise 44A

**1**   $f(x) = 5x - 2$ Find the value of
    **a** $f(2)$        **b** $f(6)$        **c** $f(-3)$        **d** $f(0)$        **e** $f(\tfrac{1}{2})$

**2**   $g(x) = x^2 + 3$ Find the value of
    **a** $g(4)$        **b** $g(0)$        **c** $g(-1)$        **d** $g(\tfrac{1}{4})$        **e** $g(-\tfrac{1}{2})$

**3**   $f(x) = x^2$ Find an expression for
    **a** $f(x) + 1$        **b** $f(x) - 3$        **c** $2f(x)$        **d** $\tfrac{1}{4}f(x)$        **e** $3f(x) - 1$

**4**   $f(x) = x^2$ Find an expression for
    **a** $f(x + 1)$        **b** $f(x - 3)$        **c** $f(2x)$        **d** $f(\tfrac{1}{4}x)$        **e** $f(3x - 1)$

**5**   $h(x) = 3x - 2$ Find an expression for
    **a** $h(x) + 3$        **b** $h(x + 3)$        **c** $h(2x)$        **d** $h(-x)$        **e** $-h(x)$

# 44.2 Applying vertical translations

The graphs of $y = x$, $y = x + 1$ and
$y = x - 2$ are shown.

The three equations are all in the form $y = x + c$
so all the lines have a gradient of 1

Therefore the lines are parallel.

The line $y = x$ can be moved vertically
onto each of the other lines.

If the line $y = x$ is moved **up** by **1** unit
then it becomes the line $y = x + 1$

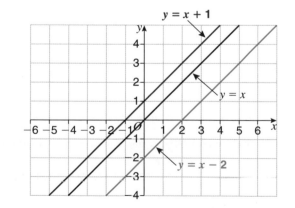

You can confirm this using tracing paper:
- trace the graph of $y = x$
- move the tracing paper **one unit vertically upwards**
- the line drawn on the tracing paper will be directly over the graph of $y = x + 1$

So the **transformation** that **maps** the graph of $y = x$ onto the graph of $y = x + 1$ is a **translation** of **1** unit vertically in the **positive** $y$ direction.

This can be written as a **translation** with vector $\begin{pmatrix} 0 \\ 1 \end{pmatrix}$.

Tracing paper can be used to confirm that if the graph of $y = x$ is translated 2 units vertically in the **negative** $y$ direction or by $\begin{pmatrix} 0 \\ -2 \end{pmatrix}$, then its equation becomes $y = x - 2$

## Example 3

**a** The graph of $y = x$ is translated by 5 units vertically in the positive $y$ direction.
Write down the equation of the new graph.

**b** The graph of $y = x$ is translated by 3 units vertically in the negative $y$ direction.
Write down the equation of the new graph.

**c** Describe the transformation that will map the graph of $y = x$ onto the graph of $y = x + 8$

### Solution 3

**a** $y = x + 5$

> The graph is translated **5** units vertically in the **positive** $y$ direction.

**b** $y = x - 3$

> The graph is translated **3** units vertically in the **negative** $y$ direction.

**c** Translation of 8 units vertically in the positive $y$ direction

or

Translation of $\begin{pmatrix} 0 \\ 8 \end{pmatrix}$

---

The graphs of $y = x^2$, $y = x^2 + 6$ and $y = x^2 - 2$ are shown opposite.

Function notation can be used when drawing graphs.

If $f(x) = x^2$ then    $y = x^2$
can be written as    $y = f(x)$

                 $y = x^2 + 6$
can be written as    $y = f(x) + 6$

                 $y = x^2 - 2$
can be written as    $y = f(x) - 2$

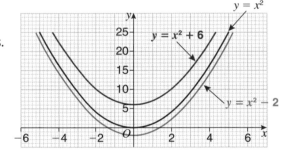

You can use tracing paper to confirm that the transformation that maps the graph of $y = f(x)$ onto the graph of $y = f(x) + 6$ is a translation of $\begin{pmatrix} 0 \\ 6 \end{pmatrix}$ or **6** units vertically in the **positive** $y$ direction

and

the transformation that maps the graph of $y = f(x)$ onto the graph of $y = f(x) - 2$ is a translation of $\begin{pmatrix} 0 \\ -2 \end{pmatrix}$ or **2** units vertically in the **negative** $y$ direction.

For any function, f, the transformation which maps the graph of $y = f(x)$ onto the graph of $y = f(x) + a$ is a translation of $\begin{pmatrix} 0 \\ a \end{pmatrix}$.

If $a > 0$ this is a translation in the positive $y$ direction.
If $a < 0$ this is a translation in the negative $y$ direction.

## Example 4

The graph of $y = x^2 - 2x$ is shown in black.

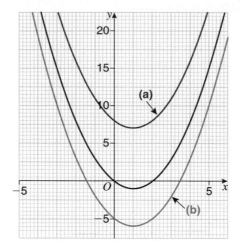

Trace the graph $y = x^2 - 2x$ and use the tracing to help you write down the equation for each of the other two graphs.

*Solution 4*

**a** $y = x^2 - 2x + 8$

> Move the tracing paper up **8** units $y = x^2 - 2x$ has been translated **8** units vertically in the **positive** $y$ direction.

**b** $y = x^2 - 2x - 5$

> Move the tracing paper down **5** units $y = x^2 - 2x$ has been translated **5** units vertically in the **negative** $y$ direction.

It is possible to transform a graph without knowing its equation.

## Example 5

The graph of $y = f(x)$ is shown in black. Sketch on the same axes the graph of
**a** $y = f(x) + 3$
**b** $y = f(x) - 2$

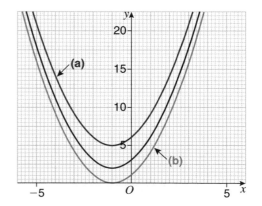

*Solution 5*
**a** The equation of $y = f(x) + 3$ is of the form $y = f(x) + a$ with $a = 3$ so to draw the graph of $y = f(x) + 3$, translate the graph of $y = f(x)$ by **3** units vertically in the **positive** $y$ direction

(or by $\begin{pmatrix} 0 \\ 3 \end{pmatrix}$). The graph of $y = f(x) + 3$ is drawn in red.

**b** Similarly the graph of $y = f(x) - 2$ is a translation of $y = f(x)$ by **2** units in the **negative**

$y$ direction (or by $\begin{pmatrix} 0 \\ -2 \end{pmatrix}$).

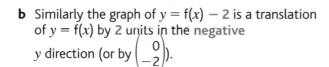

The graph of $y = f(x) - 2$ is drawn in blue.

## Exercise 44B

**1** The equation of the graph drawn in black is given.
Write down the equation of each of the other two graphs. You may find it useful to use tracing paper and trace the original graph each time.

**i**

**ii**

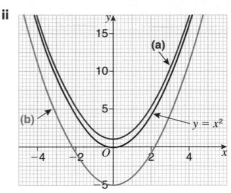

**2 a** The graph of $y = x^2$ is translated by 12 units vertically in the positive $y$ direction.
Write down the equation of the new graph.

**b** The graph of $y = x^2$ is translated by 8 units vertically in the negative $y$ direction.
Write down the equation of the new graph.

**c** Describe the transformation that will map the graph of $y = x^2$ onto the graph of $y = x^2 + 7$

**3** The graph of $y = f(x)$ where $f(x) = x^2 + 2x$ is shown.
Copy and sketch on the same axes the graph of

**a** $y = f(x) + 1$    **b** $y = f(x) - 2$

**4** The graph of $y = g(x)$ is shown.
Copy and sketch on the same axes the graph of

**a** $y = g(x) - 7$    **b** $y = g(x) + 4$

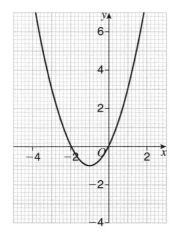

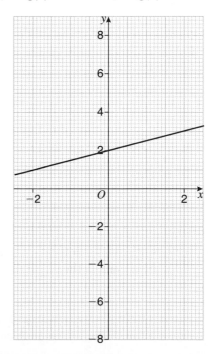

**5** The graph of $y = f(x)$ is shown.
On the resoure sheet sketch on the same axes
the graph of
  **a** $y = f(x) + 2$
  **b** $y = f(x) - 1$

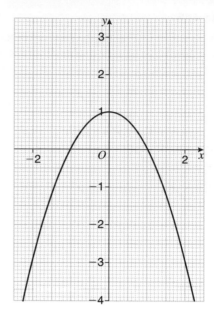

## 44.3 Applying horizontal translations

Graphs can be translated horizontally.

The graphs of $y = f(x)$, $y = f(x + 3)$ and $y = f(x - 2)$ where $f(x) = x^2$ are shown below:

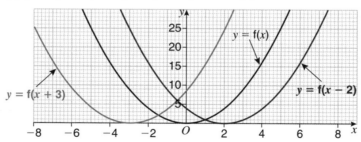

- trace the graph of $y = f(x)$
- move the tracing paper **3 units to the left** which is the **negative** $x$ direction
- the curve drawn on the tracing paper is directly over the graph of $y = f(x + 3)$

The transformation that maps the graph of $y = f(x)$ onto the graph of $y = f(x + 3)$ is a translation of
**3** units horizontally in the **negative** $x$ direction.

This can be written as a translation of $\begin{pmatrix} -3 \\ 0 \end{pmatrix}$.

Tracing paper can be used to confirm that if the graph of $y = f(x)$ is translated **2** units horizontally in
the **positive** $x$ direction or by $\begin{pmatrix} 2 \\ 0 \end{pmatrix}$, then it will coincide with the graph of $y = f(x - 2)$.

For any function, f, the transformation which maps the graph of $y = f(x)$ onto the graph of
$y = f(x + a)$ is a translation of $\begin{pmatrix} -a \\ 0 \end{pmatrix}$.

If $a > 0$ this is a translation in the negative $x$ direction.
If $a < 0$ this is a translation in the positive $x$ direction.

## Example 6

Here is a sketch of the curve with equation $y = f(x)$.

The vertex, $A$, of the curve is $(1, 2)$.

Write down the coordinates of the vertex of each of the curves with these equations.

**a** $y = f(x + 3)$

**b** $y = f(x - 1)$

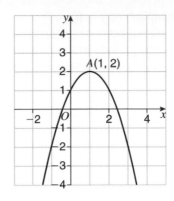

## Solution 6

**a** The vertex of the curve $y = f(x + 3)$ will have coordinates $(1 - 3, 2) = (-2, 2)$

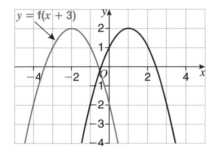

The graph of $y = f(x + 3)$ is a translation of $y = f(x)$ by **3** units horizontally in the **negative** $x$ direction. The vertex will move **3** units horizontally in the negative $x$ direction. So its $y$-coordinate will remain the same but its $x$-coordinate will **decrease** by **3**

**b** The vertex of the curve $y = f(x - 1)$ will have coordinates $(1 + 1, 2) = (2, 2)$

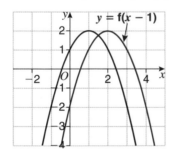

The graph of $y = f(x - 1)$ is a translation of $y = f(x)$ by **1** unit horizontally in the **positive** $x$ direction. The vertex will move **1** unit horizontally in the positive $x$ direction. So its $y$-coordinate will remain the same but its $x$-coordinate will **increase** by **1**

Transformations can be combined.

## Example 7

The graph of $y = f(x)$ is shown.
Sketch the graph of $y = f(x + 3) + 2$

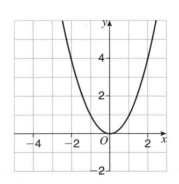

*Solution 7*

**Step 1**

Sketch the graph of $y = f(x + 3)$.

$y = f(x + 3)$ is a horizontal translation of $y = f(x)$ by **3** units horizontally in the **negative** $x$ direction.

$y = f(x + 3)$ is drawn in blue.

**Step 2**

Sketch the graph of $y = f(x + 3) + 2$

$y = f(x + 3) + 2$ is a vertical translation of $y = f(x + 3)$ by **2** units vertically in the **positive** $y$ direction

$y = f(x + 3) + 2$ is drawn in red.

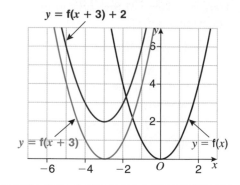

## Exercise 44C

**1** The equation of the graph drawn in black is given each time. Write down the equation of each of the other two graphs.

**i**

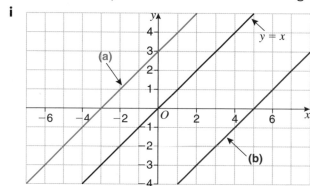

**ii**

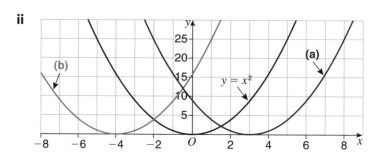

**2** The graph of $y = f(x)$ where $f(x) = x^2 + 2x$ is shown.
On the resource sheet sketch on the same axes the graph of

**a** $y = f(x + 4)$

**b** $y = f(x - 3)$

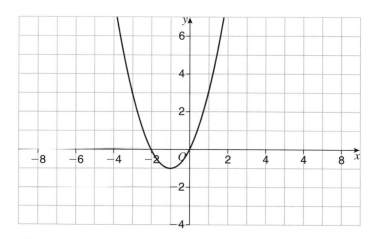

**3** The graph of $y = f(x)$ is shown.
On the resource sheet sketch on the same axes the graph of

**a** $y = f(x + 2)$

**b** $y = f(x - 4)$

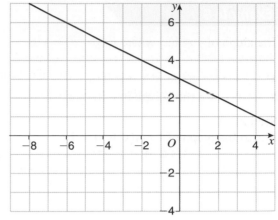

**4** This is a sketch of the curve with equation $y = f(x)$.
The vertex of the curve is $(0, 2)$. Write down the coordinates of the vertex for each of the curves with these equations.

**a** $y = f(x + 3)$

**b** $y = f(x - 1)$

**c** $y = f(x) + 2$

**d** $y = f(x) - 3$

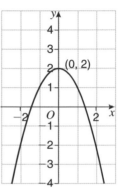

**5** The graph of $y = f(x)$ is shown.
On the resource sheet sketch the graph of

**a** $y = f(x + 2) - 1$

**b** $y = f(x - 2) + 1$

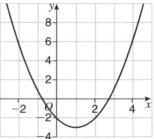

# 44.4 Applying reflections

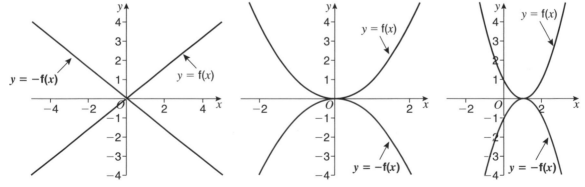

The diagrams show the graphs of $y = f(x)$ in black and $y = -f(x)$ in red for $f(x) = x$, $f(x) = x^2$ and $f(x) = (x - 1)^3$

In each case the graph of $y = -f(x)$ is the reflection in the $x$-axis of the graph of $y = f(x)$.

When a graph is reflected in the $x$-axis the $x$-coordinate of every point remains the same whilst the sign of the $y$-coordinate changes.

For any function, f, the transformation which maps the graph of $y = f(x)$ onto the graph of $y = -f(x)$ is a reflection in the $x$-axis.

## Example 8

The graph of $y = f(x)$ where $f(x) = x^2 - 4$ is shown in black.

**a** Sketch the graph of $y = -f(x)$.

**b** Write down the equation of the graph of $y = -f(x)$.

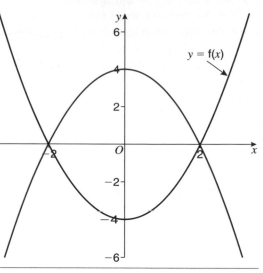

## Solution 8

**a** The graph of $y = -f(x)$ is a reflection in the $x$-axis of the graph of $y = f(x)$.
The graph of $y = -f(x)$ is drawn in red.

**b**     $f(x) = (x^2 - 4)$

so   $-f(x) = -(x^2 - 4)$

$= 4 - x^2$

The equation of graph of $y = -f(x)$ is $y = 4 - x^2$

---

The diagrams below show the graphs of $y = f(x)$ and $y = f(-x)$ for $f(x) = x$, $f(x) = (x - 1)^2$ and $f(x) = (x - 1)^3$

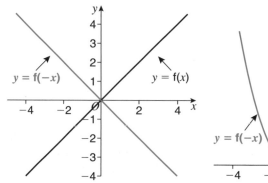
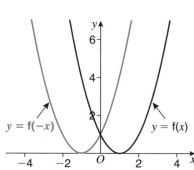
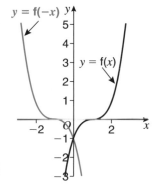

In each case the graph of $y = f(-x)$ is the reflection in the $y$-axis of the graph of $y = f(x)$.

When a graph is reflected in the $y$-axis the $y$-coordinate of every point remains the same whilst the sign of the $x$-coordinate changes.

For any function, f, the transformation which maps the graph of $y = f(x)$ onto the graph of $y = f(-x)$ is a reflection in the $y$-axis.

## Example 9

The diagram shows the graph of $y = f(x)$ where $f(x) = x^2 - 4x + 3$ drawn in black.

**a** Sketch the graph of $y = f(-x)$.

**b** Write down the equation of the graph of $y = f(-x)$.

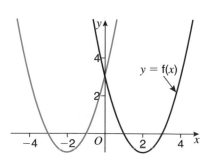

## Solution 9

**a** The graph of $y = f(-x)$ is a reflection in the $y$-axis of the graph of $y = f(x)$. The graph of $y = f(-x)$ is drawn in blue.

**b**     $f(x) = x^2 - 4x + 3$

so   $f(-x) = (-x)^2 - 4(-x) + 3$

$= x^2 + 4x + 3$

The equation of the graph of $y = f(-x)$ is $y = x^2 + 4x + 3$

## Example 10

Here is a sketch of the curve with equation $y = f(x)$.
The vertex of the curve is (2, 1).
Write down the coordinates of the vertex for each of
the curves with these equations.

**a** $y = -f(x)$
**b** $y = f(-x)$

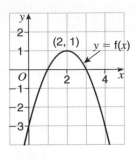

## Solution 10

**a** The vertex of the curve $y = -f(x)$
will have coordinates $(2, -1)$

> The graph of $y = -f(x)$ is a reflection in the $x$-axis of the graph of $y = f(x)$. So the $x$-coordinate of the vertex will remain the same but the $y$-coordinate will change sign.

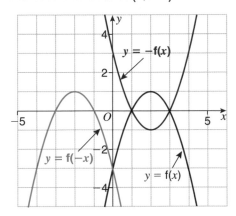

**b** The vertex of the curve $y = f(-x)$
will have coordinates $(-2, 1)$

> The graph of $y = f(-x)$ is a reflection in the $y$-axis of the graph of $y = f(x)$. So the $y$-coordinate of the vertex will remain the same but the $x$-coordinate will change sign.

## Exercise 44D

**1** Use the resource sheet and for each graph draw on the same axes the graph of
**i** $y = -f(x)$  **ii** $y = f(-x)$

**a**   **b** 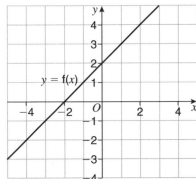 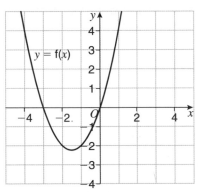  **c**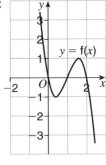

**2** The graph of $y = 2x + 3$ is reflected in the $x$-axis. Write down the equation of the new graph.

**3** The graph of $y = x(x + 2)$ is reflected in the $y$-axis. Write down the equation of the new graph.

**4** The graph of $y = f(x)$ has a vertex at $(3, -1)$. Write down the coordinates of the vertex for each of the curves with these equations  **a** $y = -f(x)$  **b** $y = f(-x)$

## 44.5 Applying stretches

The two diagrams show the graphs of $y = f(x)$, $y = 2f(x)$ and $y = \frac{1}{2}f(x)$ for the functions $f(x) = x$ and $f(x) = x^2$

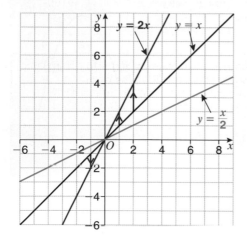

Points on the graph of $y = x$ are mapped onto corresponding points on the graph of $y = 2x$. For example $(1, 1) \rightarrow \mathbf{(1, 2)}$, $(2, 2) \rightarrow \mathbf{(2, 4)}$ and $(-1, -1) \rightarrow \mathbf{(-1, -2)}$.

Similarly points on the graph of $y = x^2$ are mapped onto corresponding points on the graph of $y = 2x^2$ For example $(1, 1) \rightarrow \mathbf{(1, 2)}$, $(2, 4) \rightarrow \mathbf{(2, 8)}$ and $(-1, 1) \rightarrow \mathbf{(-1, 2)}$.

In all cases the $x$-coordinate has remained unchanged and the $y$-coordinate has been multiplied by **2**

This transformation is called a **stretch**. The scale factor of the stretch is **2** and the direction of the stretch is parallel to the $y$-axis.

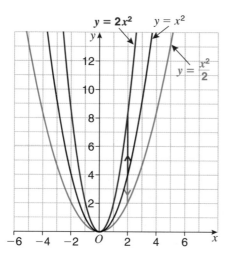

Points on the graph of $y = x$ are mapped onto corresponding points on the graph of $y = \frac{1}{2}x$. For example $(1, 1) \rightarrow (1, \frac{1}{2})$, $(2, 2) \rightarrow \mathbf{(2, 1)}$ and $(-1, -1) \rightarrow (-1, -\frac{1}{2})$.

Points on the graph of $y = x^2$ are mapped onto corresponding points on the graph of $y = \frac{1}{2}x^2$ For example $(1, 1) \rightarrow (1, \frac{1}{2})$, $(2, 4) \rightarrow \mathbf{(2, 2)}$ and $(-1, 1) \rightarrow (-1, \frac{1}{2})$.

In all cases the $x$-coordinate has again remained unchanged and the $y$-coordinate has been multiplied by $\frac{1}{2}$

The transformation is again a stretch. The scale factor of the stretch is $\frac{1}{2}$ and the direction of the stretch is parallel to the $y$-axis.

For any function, f, the transformation which maps the graph of $y = f(x)$ onto the graph of $y = af(x)$ for $a > 0$, is a stretch of scale factor $a$ parallel to the $y$-axis.

### Example 11

The graph $y = f(x)$ where $f(x) = x^2 - 4x + 1$ is drawn in black.

**a** Sketch the graph of $y = 2f(x)$.
**b** Sketch the graph of $y = \frac{1}{3}f(x)$.

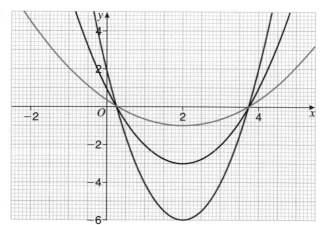

## Solution 11

**a** The graph of $y = 2f(x)$ is drawn in red.

> The transformation that maps $y = f(x)$ onto $y = 2f(x)$ is a stretch of scale factor 2 parallel to the $y$-axis.
> So the $x$-coordinate of each point on the graph will remain the same but each $y$-coordinate will be multiplied by 2 In particular
> $(0, 1) \rightarrow (0, 2)\ (2, -3) \rightarrow (2, -6)\ (4, 1) \rightarrow (4, 2)$

**b** The graph of $y = \frac{1}{3}f(x)$ is drawn in blue.

> The transformation that maps $y = f(x)$ onto $y = \frac{1}{3}f(x)$ is a stretch of scale factor $\frac{1}{3}$ parallel to the $y$-axis.
> So the $x$-coordinate of each point on the graph will remain the same but each $y$-coordinate will be multiplied by $\frac{1}{3}$ In particular
> $(0, 1) \rightarrow (0, \frac{1}{3})\ (2, -3) \rightarrow (2, -1)\ (4, 1) \rightarrow (4, \frac{1}{3})$

---

The two diagrams show the graphs of $y = f(x), y = f(\frac{1}{2}x)$ and $y = f(2x)$ for the functions $f(x) = x$ and $f(x) = x^2$

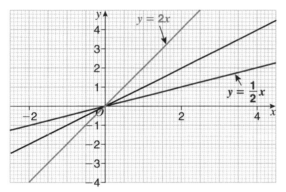

These diagrams show stretches which are parallel to the $x$-axis.

Points on the graph of $y = x$ are mapped onto corresponding points on the graph of $y = \frac{1}{2}x$.
For example $(1, 1) \rightarrow$ **(2, 1)**, $(2, 2) \rightarrow$ **(4, 2)** and $(-1, -1) \rightarrow$ **(-2, -1)**.

Points on the graph of $y = x^2$ are mapped onto corresponding points on the graph of $y = \frac{1}{4}x^2$
For example, $(1, 1) \rightarrow$ **(2, 1)**, $(2, 4) \rightarrow$ **(4, 4)** and $(-1, 1) \rightarrow$ **(-2, 1)**.

In all cases the $y$-coordinate has remained unchanged and the $x$-coordinate has been multiplied by **2**
This transformation is a stretch of scale factor **2** parallel to the $x$-axis.

Points on the graph of $y = x$ are mapped onto corresponding points on the graph of $y = 2x$
For example $(1, 1) \rightarrow (\frac{1}{2}, 1)$, $(2, 2) \rightarrow (1, 2)$ and $(-1, -1) \rightarrow (-\frac{1}{2}, -1)$.

Points on the graph of $y = x^2$ are mapped onto corresponding points on the graph of $y = 4x^2$
For example $(1, 1) \rightarrow (\frac{1}{2}, 1)$, $(2, 4) \rightarrow (1, 4)$ and $(-1, 1) \rightarrow (-\frac{1}{2}, 1)$.

In all cases the $y$-coordinate has again remained unchanged and the $x$-coordinate has been multiplied by $\frac{1}{2}$

The transformation is again a stretch. The scale factor of the stretch is $\frac{1}{2}$ and the direction of the stretch is parallel to the $x$-axis.

For any function, f, the transformation which maps the graph of $y = f(x)$ onto the graph of $y = f(ax)$, for $a > 0$, is a stretch of scale factor $\dfrac{1}{a}$ parallel to the $x$-axis.

## Example 12

The graph drawn in black has the equation $y = f(x)$.
Describe the stretch that will map

**a** $y = f(x)$ onto graph **(a)**

**b** $y = f(x)$ onto graph **(b)**

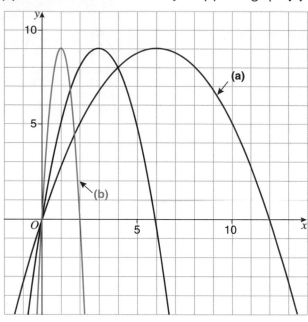

### Solution 12

**a** $(6, 0) \rightarrow \mathbf{(12, 0)}$  $(3, 9) \rightarrow \mathbf{(6, 9)}$  $(5, 5) \rightarrow \mathbf{(10, 5)}$
The $y$-coordinates remain the same each time
while the $x$-coordinates are multiplied by 2
The transformation is a stretch parallel to the
$x$-axis of scale factor 2

A stretch of scale factor $\dfrac{1}{a}$ parallel to the
$x$-axis maps the graph of $y = f(x)$ onto the
graph of $y = f(ax)$

**b** $(6, 0) \rightarrow \mathbf{(2, 0)}$  $(3, 9) \rightarrow \mathbf{(1, 9)}$
The transformation is a stretch parallel to the $x$-axis of scale factor $\frac{1}{3}$

## Exercise 44E

**1** On the resource sheet draw on the same
axes of the graph below the graphs of

**a** $y = 3f(x)$      **b** $y = \frac{1}{4}f(x)$

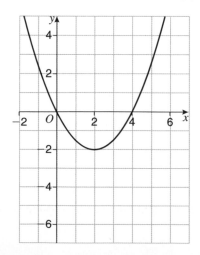

**2** On the resource sheet draw on the same
axes of the graph below the graphs of

**a** $y = f(\frac{1}{2}x)$      **b** $y = f(4x)$

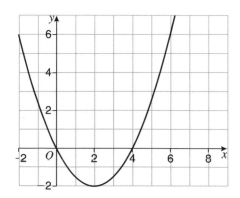

**3** The graph of $y = f(x)$ is shown in black.

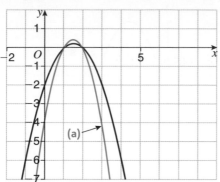  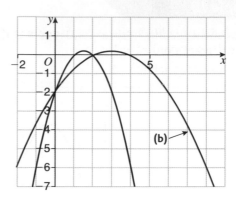

Describe the stretch that will map

**a** $y = f(x)$ onto the blue graph labelled **(a)**          **b** $y = f(x)$ onto the red graph labelled **(b)**

**4** The graph of $y = x(x + 2)$ is stretched by a scale factor of 3 parallel to the $y$-axis.
Write down the equation of the new graph.

**5** The graph of $y = x^2$ is stretched by a scale factor of 3 parallel to the $x$-axis.
Write down the equation of the new graph.

**6** The graph of $y = f(x)$ where $f(x) = x^2$ is stretched. Write down the equation of the new graph when the stretch is

**a** parallel to the $y$-axis with a scale factor of $\frac{1}{2}$

**b** parallel to the $x$-axis with a scale factor of $\frac{1}{2}$

**7** The graph of $y = f(x)$ has a vertex at $(5, -3)$.
Write down the coordinates of the vertex of

**a** $y = 2f(x)$                                              **b** $y = \frac{1}{2}f(x)$

**c** $y = f(3x)$                                              **d** $y = f(\frac{1}{3}x)$

## 44.6 Transformations applied to the graphs of sin $x$ and cos $x$

The graphs of $y = \sin x°$ and $y = \cos x°$ for $-360 \leqslant x \leqslant 360$ were introduced in Section 39.2 and are shown below.

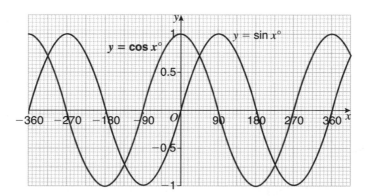

All the transformations described in the previous sections in this chapter can also be applied to the graphs of $y = \sin x°$ and $y = \cos x°$.

## Example 13

A sketch of the graph $y = \sin x°$ for $0 \leqslant x \leqslant 360$ is shown.
Sketch the graphs of

**a** $y = \sin x° + 2$       **b** $y = \sin 2x°$

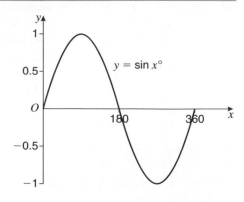

## Solution 13

**a**

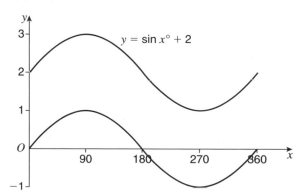

The transformation that will map the graph of $y = \sin x°$ onto the graph of $y = \sin x° + 2$ is a translation of **2** units **vertically** in the positive $y$ direction.

So the $y$-coordinate of every point on the graph of $y = \sin x°$ will increase by 2, in particular

$(0, 0) \rightarrow$ **(0, 2)** $(90, 1) \rightarrow$ **(90, 3)**
$(180, 0) \rightarrow$ **(180, 2)** $(270, -1) \rightarrow$ **(270, 1)**
$(360, 0) \rightarrow$ **(360, 2)**

**b**

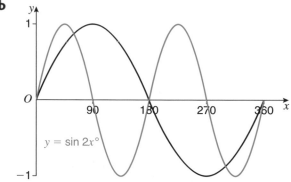

The transformation that will map the graph of $y = \sin x°$ onto the graph of $y = \sin 2x°$ is a stretch, scale factor $\frac{1}{2}$ parallel to the $x$-axis.

So, the $x$-coordinate of every point on the graph $y = \sin x°$ will be halved, in particular

$(0, 0) \rightarrow (0, 0)$ $(90, 1) \rightarrow (45, 1)$
$(180, 0) \rightarrow (90, 0)$
$(270, -1) \rightarrow (135, -1)$
$(360, 0) \rightarrow (180, 0)$

## Example 14

A sketch of the curve
$y = \cos x°$ for $0 \leqslant x \leqslant 360$
is drawn in black.

Using the sketch or otherwise
find the equations of the
graphs labelled **(a)** and **(b)**.

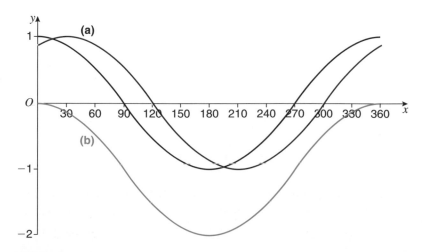

**Solution 14**

**a** $y = \cos(x - 30)°$

The transformation that would map the graph of $y = \cos x°$ onto the red graph is a translation of $\begin{pmatrix} 30 \\ 0 \end{pmatrix}$ (or **30** units horizontally in the **positive** $x$ direction).

**b** $y = \cos x° - 1$

The transformation that would map the graph of $y = \cos x°$ onto the blue graph is a translation of **1** unit vertically in the **negative** $y$ direction.

---

## Example 15

Describe fully the sequence of transformations that maps the graph of $y = \sin x°$ onto the graph of $y = \frac{1}{2}\sin(x + 60)°$.

**Solution 15**

Translation of $\begin{pmatrix} -60 \\ 0 \end{pmatrix}$

(or 60 units horizontally in the negative $x$ direction).

First consider the mapping of $y = \sin x°$ onto $y = \sin(x + 60)°$
This is a transformation of the type that maps $y = f(x)$ onto $y = f(x + a)$ so is a translation of $a$ units horizontally in the negative $x$ direction.

A stretch of scale factor $\frac{1}{2}$ parallel to the $y$-axis.

Next consider the mapping of $y = \sin(x + 60)°$ onto $y = \frac{1}{2}\sin(x + 60)°$.
This is a transformation of the type that maps $y = f(x)$ onto $y = af(x)$ so is a stretch of scale factor $a$ parallel to the $y$-axis.

---

## Exercise 44F

**1 a** Sketch the graph $y = \sin x°$ for $0 \leqslant x \leqslant 360$
   **b** On the same set of axes sketch the graphs of
      **i** $y = 2\sin x°$     **ii** $y = -\sin x°$

**2 a** Sketch the graph $y = \cos x°$ for $0 \leqslant x \leqslant 360$
   **b** On the same set of axes sketch the graphs of
      **i** $y = \cos\frac{1}{2}x°$     **ii** $y = 2 + \cos x°$

**3** Describe fully a sequence of transformations that maps the graph of $y = \sin x$ onto the graph of
   **a** $y = 5\sin(x - 30)$     **b** $y = -\sin 2x$     **c** $y = \frac{1}{3}\sin(-x)$

**4** A sketch of the curve $y = \cos x°$ for $0 \leqslant x \leqslant 360$ is drawn in black.
Using the sketch or otherwise find the equations of the graphs labelled **i** and **ii**.

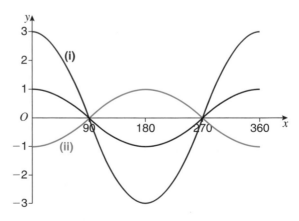

**5** The maximum possible value of $\sin x°$ is 1 Write down the maximum possible value of
   **a** $4\sin x°$     **b** $\sin x° + 2$     **c** $\sin x° - 1$     **d** $\frac{1}{4}\sin x°$
   **e** $\sin(x - 30)°$     **f** $5\sin x° + 2$     **g** $\sin 3x°$     **h** $\frac{1}{2}\sin x° - 3$

## Chapter summary

**You should now know that:**

★   a function is a rule that can be used to change numbers

★   $f(x)$ is read as 'f of $x$' and means that f is a function of $x$

★   $f(3)$ means work out the value of $f(x)$ when $x = 3$

★   function notation can be used when drawing graphs

★   for any function, f, the transformation which maps the graph of

    $y = f(x)$ onto the graph of $y = f(x) + a$ is a translation of $\begin{pmatrix} 0 \\ a \end{pmatrix}$.

    If $a > 0$ this is a translation in the positive $y$ direction
    If $a < 0$ this is a translation in the negative $y$ direction

★   for any function, f, the transformation which maps the graph of

    $y = f(x)$ onto the graph of $y = f(x + a)$ is a translation of $\begin{pmatrix} -a \\ 0 \end{pmatrix}$.

    If $a > 0$ this is a translation in the negative $x$ direction.
    If $a < 0$ this is a translation in the positive $x$ direction.

★   for any function, f, the transformation which maps the graph of
    $y = f(x)$ onto the graph of $y = -f(x)$ is a reflection in the $x$-axis
    $y = f(x)$ onto the graph of $y = f(-x)$ is a reflection in the $y$-axis

★   for any function, f, the transformation which maps the graph of
    $y = f(x)$ onto the graph of $y = af(x)$ is a stretch of scale factor $a$ parallel to the $y$-axis
    $y = f(x)$ onto the graph of $y = f(ax)$ is a stretch of scale factor $\dfrac{1}{a}$ parallel to the $x$-axis

★   All the above transformations can be applied to the graphs of $y = \sin x°$ and $y = \cos x°$

## Chapter 44 review questions

**1** Given that $f(x) = (x + 4)^2$ work out the value of
    **a** $f(0)$       **b** $f(5)$       **c** $f(-6)$

**2** Given that $g(x) = 5 - x$ write down an expression for
    **a** $g(2x)$     **b** $g(-x)$     **c** $g(x + 2)$     **d** $g(a - x)$

**3** This is a sketch of the curve with equation $y = f(x)$.

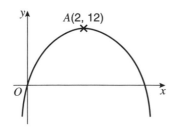

The vertex of the curve is $A(2, 12)$.
Write down the coordinates of the vertex for each of the curves having the following equations.
    **a** $y = f(x) + 6$     **b** $y = f(x + 3)$     **c** $y = f(-x)$     **d** $y = f(4x)$

(1385 June 1999)

**4** This is a sketch of the curve with equation $y = f(x)$.
It passes through the origin $O$.
The only vertex of the curve is at $A$ $(2, -4)$.

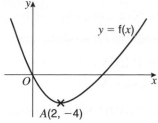

**a** Write down the coordinates of the vertex of the curve with equation
   **i** $y = f(x - 3)$   **ii** $y = f(x) - 5$   **iii** $y = -f(x)$   **iv** $y = f(2x)$

The curve with equation $y = x^2$ has been translated to give the curve $y = f(x)$.
**b** Find $f(x)$ in terms of $x$.     (1387 June 2003)

**5** A transformation has been applied to the graph of $y = x^2$ to give the graph of $y = -x^2$
**a** Describe fully the transformation.

For all values of $x$
$$x^2 + 4x = (x + p)^2 + q$$

**b** Find the values of $p$ and $q$.

A transformation has been applied to the graph of $y = x^2$ to give the graph of $y = x^2 + 4x$.
**c** Using your answer to part **b**, or otherwise, describe fully the transformation.
    (1385 June 2001)

**6** The expression $x^2 - 6x + 14$ can be written in the form $(x - p)^2 + q$ for all values of $x$.

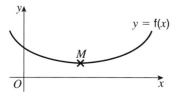

**a** Find the value of  **i** $p$  **ii** $q$
The equation of a curve is $y = f(x)$ where $f(x) = x^2 - 6x + 14$
Here is a sketch of the graph of $y = f(x)$.

**b** Write down the coordinates of the minimum point, $M$, of
the curve.

Here is a sketch of the graph of $y = f(x) - k$ where $k$ is a positive constant. The graph touches
the $x$-axis.

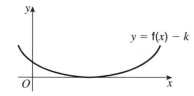

**c** Find the value of $k$.     **d** For the graph of $y = f(x - 1)$
  **i** write down the coordinates of the minimum point
  **ii** find the coordinates of the point where the curve crosses the $y$-axis.   (1387 November 2003)

**7** A sketch of the curve $y = \sin x°$ for $0 \leqslant x \leqslant 360$ is shown on the right.

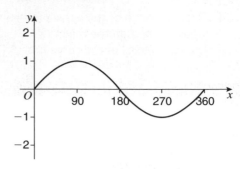

**a** Using the sketch above, or otherwise, find the equation of each of the following two curves.

**i**

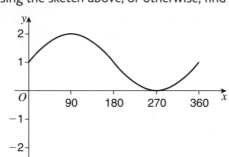

**ii**

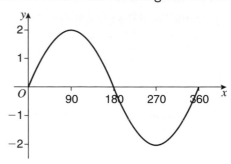

**b** Describe fully the sequence of two transformations that maps the graph of $y = \sin x°$ onto the graph of $y = 3 \sin 2x°$.

(1387 June 2004)

**8** The graph of $y = a - b \cos (kt)$ for values of $t$ between $0°$ and $120°$, is drawn on the grid.

Use the graph to find an estimate for the value of

    **i** $a$
    **ii** $b$
    **iii** $k$.

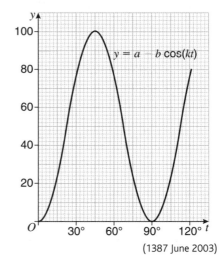

(1387 June 2003)

**9 a** On the grid on the resource sheet sketch the graphs of
    **i** $y = \sin x°$
    **ii** $y = \sin 2x°$
    for values of $x$ between 0 and 360
    Label each graph clearly.

**b** Calculate all the solutions to the equation

$$2 \sin 2x° = -1$$

between $x = 0$ and $x = 360$

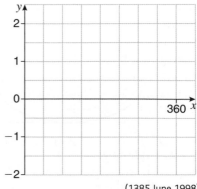

(1385 June 1998)

# Index

Published by: Edexcel Limited, One90 High Holborn, London WC1V 7BH

Distributed by: Pearson Education Limited, Edinburgh Gate, Harlow, Essex CM20 2JE, England
www.longman.co.uk

© Edexcel Limited 2006

First published 2006
ISBN-10:    1 846 901022
ISBN-13:    978 1 846 901027

Concept design by Mick Harris and Juice Creative Ltd. Cover design by Juice Creative Ltd. Index by John Holmes.

Typeset by Tech-Set, Gateshead

Printed in the U.K. by CPI

The publisher's policy is to use paper manufactured from sustainable forests.

Live Learning, Live Authoring and Live Player are all trademarks of Live Learning Ltd.

The Publisher wishes to draw attention to the Single-User Licence Agreement below.
Please read this agreement carefully before installing and using the CD-ROM.
We are grateful to the following for permission to reproduce photographs:

Every effort has been made to trace the copyright holders and we apologise in advance for any unintentional omissions. We would be pleased to insert the appropriate acknowledgement in any subsequent edition of this publication.

**Action Plus Sports Images:** pg571(©Glyn Kirk); **Alamy Images:** pg3 (©Aflo Foto Agency), pg23 (©PHOTOTAKE Inc.), pg31 (©Bob Thomas (Royalty-Free), pg241 (m) (©CuboImages srl), pg260 (m) (©numb (Royalty-Free), pg269 (©Hideo Kurihara), pg418 (Skyscraper) (©ImageState) (Royalty-Free), (Pool Balls) (©joeysworld.com), (Dice) (©Leander), (Louvre) (©isifa Image Service s.r.o.); **Corbis:** pg87 (©Royalty-Free), pg271 (©Danny Lehman), pg343 (©Royalty-Free), pg418 (©Klaus Hackenberg/zefa); **DK Images:** pg260 (r) (©Matthew Ward), pg405 (©Matthew Ward); **Pearson:** pg243 (©Pearson Education), pg418 (Toblerone) (©Pearson Education), (Smarties) (©Pearson Education); **Punchstock Royalty-Free Images:** pg4 (digitalvision), pg241 (l) (digitalvision), pg418 (Pyramids) (digitalvision); **Science Photo Library Ltd:** pg260 (l) (Tek Image)

Picture Research by Karen Jones

---